Studies in Systems, Decision and Control

Volume 329

The series "Studies in Systems, Decision and Control" (SSDC) covers both new developments and advances, as well as the state of the art, in the various areas of broadly perceived systems, decision making and control–quickly, up to date and with a high quality. The intent is to cover the theory, applications, and perspectives on the state of the art and future developments relevant to systems, decision making, control, complex processes and related areas, as embedded in the fields of engineering, computer science, physics, economics, social and life sciences, as well as the paradigms and methodologies behind them. The series contains monographs, textbooks, lecture notes and edited volumes in systems, decision making and control spanning the areas of Cyber-Physical Systems, Autonomous Systems, Sensor Networks, Control Systems, Energy Systems, Automotive Systems, Biological Systems, Vehicular Networking and Connected Vehicles, Aerospace Systems, Automation, Manufacturing, Smart Grids, Nonlinear Systems, Power Systems, Robotics, Social Systems, Economic Systems and other. Of particular value to both the contributors and the readership are the short publication timeframe and the world-wide distribution and exposure which enable both a wide and rapid dissemination of research output.

Indexed by SCOPUS, DBLP, WTI Frankfurt eG, zbMATH, SCImago.

All books published in the series are submitted for consideration in Web of Science.

More information about this series at http://www.springer.com/series/13304

Min Wu · Witold Pedrycz · Luefeng Chen
Editors

Developments in Advanced Control and Intelligent Automation for Complex Systems

Springer

Editors
Min Wu
School of Automation
China University of Geosciences
Wuhan, China

Luefeng Chen
School of Automation
China University of Geosciences
Wuhan, China

Witold Pedrycz
Department of Electrical
and Computer Engineering
University of Alberta
Edmonton, AB, Canada

ISSN 2198-4182 ISSN 2198-4190 (electronic)
Studies in Systems, Decision and Control
ISBN 978-3-030-62146-9 ISBN 978-3-030-62147-6 (eBook)
https://doi.org/10.1007/978-3-030-62147-6

This Springer imprint is published by the registered company Springer Nature Switzerland AG
The registered company address is: Gewerbestrasse 11, 6330 Cham, Switzerland

Preface

Advanced control and intelligent automation technology is indispensable in manufacturing and industry. It is of great significance to improve productivity and standard of living. However, due to limitations of developments in fundamental technology and emergence of new requirements in material and spiritual spheres, development level of these industries cannot meet people's needs. Advanced control and intelligent automation technology has emerged as a sound solution.

There are many complex systems in the actual industrial processes and living environment, such as metallurgical process, industrial robot arms, human-robot interaction, medical rehabilitation, and energy and resource exploration. They exhibit many common characteristics like time delay, non-linearity, underactuated behavior, uncertainty, etc. With social development, we are challenged by new demands for higher efficiency, comfort, reliability, lower cost, and sustainability encountered in those areas. To achieve these objectives, there are many problems to be solved which involve modeling, optimization, control, and decision. However, employing a conventional method to these complex systems usually may not produce satisfactory results. Advanced control and intelligent automation technology has been applied to these complex systems in industry and daily life to meet new demands.

In this volume, we summarized our work and experience in the advanced control and intelligent automation for complex systems completed over the last two decades. With the depth and breadth of coverage of this book, we hope that it would serve as a useful reference for engineers in the field of automation and complex process control, and graduate students interested in advanced control theory and computational intelligence and their applications to complex industrial process.

The monograph consists of 5 parts arranged into 15 chapters. Part I (i.e., Chapters "Stability Analysis and H_∞ Control of Time-Delay Systems", "Active Disturbance Rejection in Repetitive Control Systems", "Intelligent Control of Underactuated Mechanical System" and "Finite-Time Fault Detection and H_∞ State Estimation for Markov Jump Systems Under Dynamic Event-Triggered Mechanism ") covers the progress in advanced control and control method, achieved for some fundamental theoretical problems in complex system. Chapter "Stability Analysis and H_∞ Control of Time-Delay Systems" describes robust control of time-delay systems. Chapter "Active Disturbance Rejection in Repetitive Control Systems" describes repetitive control and active disturbance rejection. Chapter "Intelligent Control of Underactuated Mechanical System" focuses on underactuated mechanical systems control. Chapter "Finite-Time Fault Detection and H_∞ State Estimation for Markov Jump Systems Under Dynamic Event-Triggered Mechanism" introduces control and fault detection of networked systems. Part II (i.e., Chapters "Intelligent Control of Sintering Process" and "Decision-Making of Burden Distribution for Blast Furnace") introduces complex industrial process modeling and simulation, full-process optimization and control, and information processing. Chapter "Intelligent Control of Sintering Process" describes intelligent control of sintering process. Chapter "Decision-Making of Burden Distribution for Blast Furnace" describes decision-making of burden distribution for blast furnace. Part III (i.e., Chapters "Granular Computing: Fundamentals and System Modeling", "Distributed Consensus Control for Nonlinear Multi-agent Systems", "Stochastic Consensus Control of Multi-agent Systems under General Noises and Delays", "Multimodal Emotion Recognition and Intention Understanding in Human-Robot Interaction" and "Dynamic Multi-objective Optimization for Multi-objective Vehicle Routing Problem with Real-time Traffic Conditions") is devoted to intelligent systems with self-organization and self-adaptation, distributed computing technology of multi-agent system and its control, and affective computing for human-robot interaction. Chapter "Granular Computing: Fundamentals and System Modeling" focuses on intelligent systems. Chapter "Distributed Consensus Control for Nonlinear Multi-agent Systems" describes multi-agent systems and learning control. Chapter "Stochastic Consensus Control of Multi-agent Systems under General Noises and Delays" describes stochastic control and multi-agent systems. Chapter "Multimodal Emotion Recognition and Intention Understanding in Human-Robot Interaction" focuses on affective computing. Chapter "Dynamic Multi-objective Optimization for Multi-objective Vehicle Routing Problem with Real-time Traffic Conditions" focuses on evolutionary computation. Part IV (i.e., Chapters "Dielectric Elastomer Intelligent Devices for Soft Robots", "Design of a 2-DOF Compliant Micropositioning Stage with Large Workspace" and "Assistive Robots") describes soft robot adapting to various unstructured environments, design of multi-joint motion mechanism based on bionic, and foundation of rehabilitation robots. Chapter "Dielectric Elastomer Intelligent Devices for Soft Robots" focuses on soft robotics. Chapter "Design of a 2-DOF Compliant Micropositioning Stage with Large Workspace" describes design of multiple joint kinematic mechanisms. Chapter "Assistive Robots" focuses on rehabilitation robots. Part V (Chapter "A Short-Term Wind Power Forecasting Method Based on Hybrid-Kernel Least-Squares Support Vector Machine") introduces prediction and control technology in power systems. Chapter "A Short-Term Wind Power Forecasting Method Based on Hybrid-Kernel Least-Squares Support Vector Machine" describes advancement in prediction and control technology for renewable energy.

This work is supported by the National Natural Science Foundation of China under Grants 61733016, 61210011, 61333002, 61773353, 61773354, 61873248, 61973286, and 61603356; the National Key R&D Program of China under Grant

2018YFC0603405; the Hubei Provincial Technical Innovation Major Project under Grant 2018AAA035; the Hubei Provincial Natural Science Foundation of China under Grant 2015CFA010; and the 111 project under Grant B17040.

We are also grateful for the support of scholars, both domestic and foreign. We would like to thank Prof. Kouhei Ohnishi of Keio University; Profs. Joseph William Spencer and Lin Jiang of University of Liverpool; Profs. Ryuichi Yokoyama, Yosuke Nakanishi, and Yicheng Zhou of Waseda University; Prof. Kaoru Hirota of Tokyo Institute of Technology; Prof. Andres Kecskemethy of University of Duisburg-Essen; Profs. Krzysztof Galkowski and Wojciech Paszke of University of Zielona Gora; Prof. Seiichi Kawata of Advanced Institute of Industrial Technology; Prof. Takao Terano of Chiba University of Commerce; Profs. Yasuhiro Ohyama and Edwardo F. Fukushima of Tokyo University of Technology; Prof. Kangzhi Liu of Chiba University; Prof. Chunyi Su of Concordia University; Prof. Shengxiang Yang of De Montfort University; Prof. Zidong Wang of Brunel University London; Prof. Gang George Yin of Wayne State University; Prof. Ji-Feng Zhang of Chinese Academy of Sciences; Prof. Sho Yokota of Setsunan University; Profs. Jinhua She, Weihua Cao, Huafeng Ding, Yong He, Xuzhi Lai, Xin Chen, Chuanke Zhang, Changhe Li, Xiaofeng Zong, and Sanyou Zeng; Assoc. Profs. Jianqi An, Xiongbo Wan, and Yawu Wang; and Dr. Min Ding and Dr. Jinqiang Gan of China University of Geosciences.

We would also like to express our appreciations for the great efforts of Mrs. Sheng Du, Jie Hu, Xingchen Shangguan, Kuanlin Wang, Hao Fu, Zixin Huang, Peng Huang, Qingshan Tan, and Juncang Zhang; Ms. Wenjuan Lin, Yali Jin, Yu Feng, Xiaoling Shen, and Lulu Wu of China University of Geosciences; Mrs. Jundong Wu and Wenjun Ye of Concordia University.

Wuhan, China
July 2020

Min Wu
Witold Pedrycz
Luefeng Chen

Contents

Advanced Control Theory and Method

Stability Analysis and H_∞ Control of Time-Delay Systems

Chuan-Ke Zhang, Yong He, Joseph William Spencer, Lin Jiang, and Min Wu

Abstract With the development of networked control systems, time delays are frequently introduced into the control loops. The delay may cause performance degradation and even lead to instability and thus it should be taken into account during the analysis and design of control systems. This chapter introduces delay-dependent methods of stability analysis and H_∞ control design for time-delay systems with a time-varying delay based on a relaxed integral inequality and applies them to load frequency control problem of a single area power system.

Keywords Time-varying delay · Stability analysis · H_∞ control · Load frequency control

Abbreviations

LKF Lyapunov–Krasovskii functional
LMI Linear matrix inequality
FWM Free-weighting-matrix
LFC Load frequency control

C.-K. Zhang (✉) · Y. He · M. Wu
School of Automation, China University of Geosciences, Wuhan 430074, China
e-mail: ckzhang@cug.edu.cn

Hubei Key Laboratory of Advanced Control and Intelligent Automation for Complex Systems, Wuhan 430074, China

J. W. Spencer · L. Jiang
Department of Electrical Engineering and Electronics, University of Liverpool, Liverpool L69 3GJ, UK

M. Wu et al. (eds.), *Developments in Advanced Control and Intelligent Automation for Complex Systems*, Studies in Systems, Decision and Control 329,
https://doi.org/10.1007/978-3-030-62147-6_1

1 Introduction

With the great development of communication networks, control systems can be realized in more flexible forms. For example, the controllers can be equipped at the places fay away from the plants such that one can realize remote control. However, due to the finite communication speed of the related signals, time delays inevitably appear in the measure and control loops of closed-loop systems [1]. The existence of those delays may lead that the control performance of the systems is degraded and even the stability feature of the system is destroyed, both of which are undesirable situations. Therefore, for the control systems with long communication channels in loops, the influence of the delay on the system performance should be analyzed and such influence should be overcome during the control design of systems. In fact, the analysis and design of time-delay systems have become an ever-increasing topic in recent decades [2, 3].

The delay encountered in communication channels is commonly a time-varying function with the change of network state. Thus, the most common and powerful method for the time-delay systems is the Lyapunov–Krasovskii functional (LKF) method, which can lead to delay-independent results and delay-dependent results. In fact, the time delays of practical control systems are always bounded, which means that one just needs to analyze performance and design controllers by taking into account those bounded delays, instead of infinite delays. In other words, the delay-dependent method is required for those tasks. No matter analysis or design, to obtain delay-dependent results, a double integral term is usually needed during the construction of the LKF candidate. Then, when taking its time derivative, the following integral terms is obtained

$$\mathscr{J}(t) := -\int_{t-\rho(t)}^{t} \dot{x}^T(v)R\dot{x}(v)dv - \int_{t-h}^{t-\rho(t)} \dot{x}^T(v)R\dot{x}(v)dv \tag{1}$$

where $x(t)$ is the system state, $R > 0$ is the Lyapunov matrix, and h is the delay upper bound.

In order to develop a tractable criterion, which is usually expressed in the form of linear matrix inequality (LMI), the $\mathscr{J}(t)$ must be replaced by a new quadratic term, more specifically, its upper bound. Thus, how to obtain this upper bound with a small error as possible becomes an important and challenging problem [4].

Before 2004, many different types of model transformation were developed to handle the $\mathscr{J}(t)$ [5], whereas the transformed models may have different dynamics, and the bounding of cross-term is required after transformation, both of which lead to obviously conservative results [6]. In 2004, the free-weighting-matrix (FWM) approach was developed based on zero-value equation to overcome those drawbacks [7, 8], and its was further improved in [9–11]. The problem of such type of method is how to reasonably introduce the free matrices [12]. Moreover, another method was also used to estimates $\mathscr{J}(t)$. It applies bounding inequalities (such as Jensen-based, Wirtinger-based, auxiliary function-based, and Bessel-Legendre inequalities

[4, 13–15], to estimate two integral terms of $\mathscr{J}(t)$ at first, and then the resulted terms are further handled through reciprocally convex combination method. Since such a method does not require lots of decision variables, it has become the most popular method. However, most researches following this idea focus on developing improved bounding techniques [16–19], but the link between two parts of $\mathscr{J}(t)$ is not well taken into account.

Based on the above discussion, a relaxed integral inequality is introduced to estimate $\mathscr{J}(t)$ by taking into account two terms together, instead of estimating one by one, and then it is used to investigate the stability analysis of a linear system with a time-varying delay. Then, H_∞ control of such type of time-delay systems is conducted by applying the proposed inequality. Finally, the above-derived stability analysis and robust H_∞ control methods are applied to the problem of load frequency control (LFC) for the power system considering communication delays.

2 Stability Analysis Based on A Relaxed Integral Inequality

This section proposes a novel integral inequality and studies the stability problem of delayed linear systems via the proposed inequality.

2.1 System Description

Consider the following linear system with a time-varying delay:

$$\begin{cases} \dot{x}(t) = Ax(t) + A_d x(t - \rho(t)), \ t \geq 0 \\ x(t) = \phi(t), \quad t \in [-h, 0] \end{cases} \tag{2}$$

where $x(t)$ is the system state, $\phi(t)$ is the initial condition, A and A_d are all the system matrices, and time delay, $\rho(t)$, satisfies the following conditions

$$0 \leq \rho(t) \leq h \tag{3}$$

and

$$\tau_1 \leq \dot{\rho}(t) \leq \tau_2 \tag{4}$$

with h, τ_1, and τ_2 being known scalars.

For the time-delay systems, the first problem to be considered is how the stability feature of the systems is affected by the existing time delay. That is, the important issue is to find the maximal allowable delay interval under which the system to be stable. For this objective, this section derives a delay- and delay-rate dependent stability criterion. As discussed in Sect. 1, the important issue to be considered is to handle the following term

$$\mathscr{J}(t) = -\int_{t-\rho(t)}^{t} \dot{x}^T(v)R\dot{x}(v)dv - \int_{t-h}^{t-\rho(t)} \dot{x}^T(v)R\dot{x}(v)dv \tag{5}$$

The following inequality is introduced to estimate the above integral terms.

Lemma 1 *For a symmetric matrix* $\breve{R} = \text{diag}\{\text{R}, 3\text{R}\}$ *with* $R > 0$ *and any matrix* S, *the* $\mathscr{J}(t)$ *defined in (5) can be estimated as:*

$$\begin{aligned}\mathscr{J}(t) \leq & -\frac{1}{h}\zeta_1^T(t)\begin{bmatrix} E_1 \\ E_2 \end{bmatrix}^T \left(\begin{bmatrix} \breve{R} & S \\ * & \breve{R} \end{bmatrix}\right)\begin{bmatrix} E_1 \\ E_2 \end{bmatrix}\zeta_1(t) \\ & -\frac{1}{h}\zeta_1^T(t)\begin{bmatrix} E_1 \\ E_2 \end{bmatrix}^T \left(\begin{bmatrix} \frac{h-\rho(t)}{h}(\breve{R} - S\breve{R}^{-1}S^T) & 0 \\ 0 & \frac{\rho(t)}{h}(\breve{R} - S^T\breve{R}^{-1}S) \end{bmatrix}\right)\begin{bmatrix} E_1 \\ E_2 \end{bmatrix}\zeta_1(t)\end{aligned} \tag{6}$$

where

$$\begin{aligned}\zeta_1(t) &= \left[x^T(t), x^T(t-\rho(t)), x^T(t-h), \int_{t-\rho(t)}^{t} \frac{x^T(v)}{\rho(t)}ds, \int_{t-h}^{t-\rho(t)} \frac{x^T(v)}{h-\rho(t)}ds\right]^T \\ E_1 &= \begin{bmatrix} \varepsilon_1 - \varepsilon_2 \\ \varepsilon_1 + \varepsilon_2 - 2\varepsilon_4 \end{bmatrix} \\ E_2 &= \begin{bmatrix} \varepsilon_2 - \varepsilon_3 \\ \varepsilon_2 + \varepsilon_3 - 2\varepsilon_5 \end{bmatrix} \\ \varepsilon_i &= [0_{n\times(i-1)n}, I, 0_{n\times(5-i)n}],\ i = 1, 2, \ldots, 5\end{aligned}$$

2.2 A Stability Criterion

The stability criterion obtained via inequality (6) is summarized as follows.

Theorem 1 *For given scalars* h *and* $\tau_1 \leq 0 \leq \tau_2$, *system (2) is asymptotically stable if there exist a* $3n \times 3n$*-dimensional matrix* $P > 0$, $n \times n$*-dimensional matrices* $Q > 0$, $R > 0$, $Z > 0$, N_1, N_2, N_3, *and a* $2n \times 2n$*-dimensional matrix* S, *such that, for all* $\dot{\rho}(t) \in \{\tau_1, \tau_2\}$, *the following hold:*

$$\Phi_1 = \begin{bmatrix} \Psi|_{\rho(t)=0} & E_1^T S \\ * & -\breve{R} \end{bmatrix} < 0 \tag{7}$$

$$\Phi_2 = \begin{bmatrix} \Psi|_{\rho(t)=h} & E_2^T S^T \\ * & -\breve{R} \end{bmatrix} < 0 \tag{8}$$

where

$$
\begin{aligned}
\Psi &= \Xi_1 + \Xi_2 - \Xi_3 + \Xi_4 + \Xi_4^T \\
\Xi_1 &= \mathfrak{E}_1^T P \mathfrak{E}_2 + \mathfrak{E}_2^T P \mathfrak{E}_1 \\
\Xi_2 &= \bar{\varepsilon}_1^T (Q + Z)\bar{\varepsilon}_1 - (1 - \dot{d}(t))\bar{\varepsilon}_2^T Q \bar{\varepsilon}_2 - \bar{\varepsilon}_3^T Z \bar{\varepsilon}_3 + h^2 \bar{\varepsilon}_6^T R \bar{\varepsilon}_6 \\
\Xi_3 &= \begin{bmatrix} E_1 \\ E_2 \end{bmatrix}^T \begin{bmatrix} \frac{2h-\rho(t)}{h}\breve{R} & S \\ * & \frac{h+\rho(t)}{h}\breve{R} \end{bmatrix} \begin{bmatrix} E_1 \\ E_2 \end{bmatrix}, \quad \breve{R} = \begin{bmatrix} R & 0 \\ 0 & 3R \end{bmatrix} \\
\Xi_4 &= [N_1\bar{\varepsilon}_1 + N_2\bar{\varepsilon}_2 + N_3\bar{\varepsilon}_6]^T\,[A\bar{\varepsilon}_1 + A_d\bar{\varepsilon}_2 - \bar{\varepsilon}_6] \\
\bar{\varepsilon}_i &= [0_{n\times(i-1)n}, I, 0_{n\times(6-i)n}],\ i = 1, 2, \ldots, 6 \\
E_i &= \begin{bmatrix} \bar{\varepsilon}_i - \bar{\varepsilon}_{i+1} \\ \bar{\varepsilon}_i + \bar{\varepsilon}_{i+1} - 2\bar{\varepsilon}_{i+3} \end{bmatrix}, i = 1, 2 \\
\mathfrak{E}_1 &= \left[\bar{\varepsilon}_1^T,\ \rho(t)\bar{\varepsilon}_4^T,\ (h - \rho(t))\bar{\varepsilon}_5^T\right]^T \\
\mathfrak{E}_2 &= \left[\bar{\varepsilon}_6^T,\ \bar{\varepsilon}_1^T - (1 - \dot{\rho}(t))\bar{\varepsilon}_2^T,\ (1 - \dot{\rho}(t))\bar{\varepsilon}_2^T - \bar{\varepsilon}_3^T\right]^T
\end{aligned} \tag{9}
$$

Proof Construct the following LKF candidate:

$$
\begin{aligned}
V(t) = \eta^T(t)P\eta(t) + \int_{t-\rho(t)}^{t} x^T(v)Qx(v)dv + \int_{t-h}^{t} x^T(v)Zx(v)dv \\
+h\int_{-h}^{0}\int_{t+\theta}^{t} \dot{x}^T(v)R\dot{x}(v)dvd\theta
\end{aligned} \tag{10}
$$

where

$$
\eta(t) = \left[x^T(t), \int_{t-\rho(t)}^{t} x^T(v)dv, \int_{t-h}^{t-\rho(t)} x^T(v)dv\right]^T
$$

and $P > 0$, $Q > 0$, $Z > 0$, and $R > 0$, which means that

$$
V(t) \geq \rho_1||x(t)||^2 \tag{11}
$$

for a sufficiently small $\rho_1 > 0$.

Computing the time derivative of the above $V(t)$ yields

$$
\begin{aligned}
\dot{V}(t) &= 2\eta^T(t)P\dot{\eta}(t) + x^T(t)(Q + Z)x(t) - (1 - \dot{d}(t))x^T(t - \rho(t))Qx(t - \rho(t)) \\
&\quad -x^T(t - h)Zx(t - h) + h^2\dot{x}^T(t)R\dot{x}(t) - h\int_{t-\rho(t)}^{t} \dot{x}^T(v)R\dot{x}(v)dv \\
&\quad -h\int_{t-h}^{t-\rho(t)} \dot{x}^T(v)R\dot{x}(v)dv \\
&= \zeta_2^T(t)(\Xi_1 + \Xi_2)\zeta_2(t) - h\mathscr{J}(t)
\end{aligned} \tag{12}
$$

where Ξ_1 and Ξ_2 are defined in (9) and

$$\zeta_2(t) = \left[\zeta_1^T(t), \dot{x}^T(t)\right]^T \tag{13}$$

By applying inequality (6) to estimate $\mathscr{J}(t)$ in (12), one has:

$$-h\mathscr{J}(t) \le -\zeta_2^T(t)\begin{bmatrix} E_1 \\ E_2 \end{bmatrix}^T \left(\begin{bmatrix} \breve{R} + \frac{h-\rho(t)}{h}\mathscr{T}_1 & S \\ * & \breve{R} + \frac{\rho(t)}{h}\mathscr{T}_2 \end{bmatrix}\right)\begin{bmatrix} E_1 \\ E_2 \end{bmatrix}\zeta_2(t) \tag{14}$$

where $\mathscr{T}_1 = \breve{R} - S\breve{R}^{-1}S^T$ and $\mathscr{T}_2 = \breve{R} - S^T\breve{R}^{-1}S$.

For any matrices N_1, N_2 and N_3 of appropriately dimensional, it is found that

$$0 = 2\left[x^T(t)N_1 + x^T(t-\rho(t))N_2 + \dot{x}^T(t)N_3\right]\left[Ax(t) + A_d x(t-\rho(t)) - \dot{x}(t)\right] \tag{15}$$

Summing up (12)–(15), the $\dot{V}(t)$ is given as

$$\dot{V}(t) \le \zeta_2^T(t)(\Psi + \Xi_a)\zeta_2(t) \tag{16}$$

where Ψ is defined in (9), and

$$\Xi_a = \frac{h-\rho(t)}{h}E_1^T S\breve{R}^{-1}S^T E_1 + \frac{\rho(t)}{h}E_2^T S^T\breve{R}^{-1}SE_2 \tag{17}$$

It follows from Schur complement and convex combination method that $\Phi_i < 0, i = 1, 2$ is equivalent to $\Psi + \Xi_a < 0$, which further leads to $\dot{V}(t) \le -\rho_2||x(t)||^2$ for a sufficient small scalar $\rho_2 > 0$.

Therefore, if (7) and (8) holds, then the asymptotical stability of system (2) is guaranteed. The proof is finished. □

2.3 A Numerical Example

A pure numerical example is introduced to illustrate the advantages of the stability criterion proposed from the view of the point of conservatism and complexity.

Example 1 Consider the following time-delay system,

$$\dot{x}(t) = \begin{bmatrix} -2 & 0 \\ 0 & -0.9 \end{bmatrix}x(t) + \begin{bmatrix} -1 & 0 \\ -1 & -1 \end{bmatrix}x(t-\rho(t)) \tag{18}$$

The maximal allowable delay upper bounds (MADUPs) with respect to different preset μ provided by the proposed Theorem 1 are listed in Table 1, where the ones reported in some existing literature are also given. Note that, the conservatism is

Table 1 MADUPs for various $\mu = -\tau_1 = \tau_2$ (Example 1)

Methods	$\mu = -\tau_1 = \tau_2$						NoVs
	0	0.1	0.5	0.8	1.0	1000	
Corollary 3 [20]	4.472	3.669	2.337	1.934	1.868	1.868	$31.5n^2 + 7.5n$
Theorem 1 [21]	4.975	3.869	2.337	1.934	1.868	1.868	$49n^2 + 5n$
Theorem 2 [22]	5.120	4.081	2.528	2.152	1.991	–	$35.5n^2 + 3.5n$
Corollary 1 [11]	6.059	4.710	2.459	2.212	2.186	2.180	$54n^2 + 9n$
Theorem 7 [4]	6.059	4.703	2.420	2.137	2.128	2.113	$10n^2 + 3n$
Theorem 1	6.059	4.707	2.428	2.205	2.204	2.205	$10n^2 + 3n$

assessed through the calculated MADUPs, and the complexity is judged based on the number of decision variables (NoVs). It is found from the results that Theorem 1 provides bigger MADUPs than others and requires fewer decision variables. It well verifies the merits of the stability criterion developed.

3 H_∞ Control Design Based on A Parameter Tuning Method

This section discusses the H_∞ control problem of linear systems with a time-varying delay and presents a parameter tuning based control design method.

3.1 Problem Formulation

Consider the following linear system with a time-varying delay:

$$\begin{cases} \dot{x}(t) = Ax(t) + A_d x(t - \rho(t)) + Bu(t) + B_w \varpi(t) \\ z(t) = Cx(t) + Du(t) + D_w \varpi(t) \end{cases} \tag{19}$$

where $x(t) \in \mathbb{R}^n$ is the system state, $u(t) \in \mathbb{R}^p$ is the control input vector, $z(t) \in \mathbb{R}^m$ is the controlled output, and $\varpi(t) \in \mathbb{R}^r$ is the disturbance input and belongs to $\mathscr{L}_2[0, \infty)$. A, A_d, B, B_w, C, D and D_w are the system matrices, and $\rho(t)$ is the time-varying delay satisfying (3) and (4).

The purpose of this section is to design a H_∞ controller

$$u(t) = Kx(t) \tag{20}$$

such that

(1) the following disturbance-free closed-loop system

$$\begin{cases} \dot{x}(t) = (A + BK)x(t) + A_d x(t - \rho(t)) \\ z(t) = (C + DK)x(t) + C_d x(t - \rho(t)) \end{cases} \tag{21}$$

is asymptotically stable;

(2) under zero initial conditions, the following condition is satisfied

$$\|z(t)\|_2 \leq \gamma \|\varpi(t)\|_2 \tag{22}$$

for all non-zero $\varpi(t) \in \mathscr{L}_2[0, \infty)$ and a prescribed $\gamma > 0$.

3.2 H_∞ Performance Based Control Design

Based on the relaxed integral inequality (6), sufficient conditions for the H_∞ performance analysis of closed-loop system (21) are summarized.

Theorem 2 *For two scalars h and $\tau_1 \leq 0 \leq \tau_2$, system (21) is asymptotically stable with $\varpi(t) \equiv 0$ and satisfies a prescribed H_∞ performance index γ under zero initial conditions, if there exist a $3n \times 3n$-dimensional matrix $P > 0$, $n \times n$-dimensional matrices $Q > 0$, $R > 0$, $Z > 0, N_1$, N_2, N_3, and a $2n \times 2n$-dimensional matrix S, such that the following conditions hold for both $\dot{\rho}(t) = \tau_1$ and $\dot{\rho}(t) = \tau_2$:*

$$\begin{bmatrix} \bar{\Psi}|_{\rho(t)=0} & \Pi_1^T B_w & \Pi_2^T & \bar{E}_1^T S \\ * & -\gamma^2 I_r & D_w^T & 0 \\ * & * & -I_m & 0 \\ * & * & * & -\check{R} \end{bmatrix} < 0 \tag{23}$$

$$\begin{bmatrix} \bar{\Psi}|_{\rho(t)=h} & \Pi_1^T B_w & \Pi_2^T & \bar{E}_2^T S^T \\ * & -\gamma^2 I_r & D_w^T & 0 \\ * & * & -I_m & 0 \\ * & * & * & -\check{R} \end{bmatrix} < 0 \tag{24}$$

where

$$\begin{aligned} \bar{\Psi} &= \Xi_1 + \Xi_2 - \Xi_3 + \bar{\Xi}_4 + \bar{\Xi}_4^T \\ \bar{\Xi}_4 &= \Pi_1^T[(A + BK)\bar{\varepsilon}_1 + A_d\bar{\varepsilon}_2 - \bar{\varepsilon}_6] \\ \Pi_1 &= N_1\bar{\varepsilon}_1 + N_2\bar{\varepsilon}_2 + N_3\bar{\varepsilon}_6, \\ \Pi_2 &= (C + DK)\bar{\varepsilon}_1 + C_d\bar{\varepsilon}_2 \\ \bar{E}_i &= \left[E_i, 0_{2n\times(6n+r+m)}\right], i = 1, 2 \end{aligned} \tag{25}$$

and other notations are same as the one defined in Theorem 1.

Proof Construct the LKF as shown in (10), based on Theorem 1, it can be yielded that:

$$\dot{V}(t) + z^T(t)z(t) - \gamma^2\varpi^T(t)\varpi(t) \le \zeta_2^T(t)(\bar{\Psi} + \Xi_a)\zeta_2(t) + \Xi_b \tag{26}$$

where $\bar{\Psi}$ is defined in (25), Ξ_a is defined in (17), and

$$\Xi_b = \zeta_2^T(t)\Pi_1^T B_w\varpi(t) + \varpi^T(t)B_w^T\Pi_1\zeta_2(t) - \gamma^2\varpi^T(t)\varpi(t) + z^T(t)z(t)$$

It follows from Schur complement and convex combination technique that, when conditions (23) and (24) are satisfied, $\zeta_2^T(t)(\bar{\Psi} + \Xi_a)\zeta_2(t) + \Xi_b \le 0$ holds for $\forall\rho(t) \times \dot{\rho}(t) \in [0, h] \times [\tau_1, \tau_2]$, which in turn leads to

$$\dot{V}(t) + z^T(t)z(t) - \gamma^2\varpi^T(t)\varpi(t) \le 0 \tag{27}$$

By taking zero initial condition into account, it can be obtained from (27) that

$$V(t) - V(0) + \int_0^\infty z^T(t)z(t) - \gamma^2\int_0^\infty \varpi^T(t)\varpi(t) \le 0 \tag{28}$$

which implies that

$$\int_0^\infty z^T(t)z(t) \le \gamma^2\int_0^\infty \varpi^T(t)\varpi(t) \tag{29}$$

which indicates $\|z(t)\|_2 \le \gamma\|\varpi(t)\|_2$.

Based on the above analysis, if conditions (23) and (24) hold, then system (21) is asymptotically stable with a prescribed H_∞ performance index. The proof is finished. ■

Based on Theorem 2, we will give a method to design the H_∞ controller, whose conditions are summarized as the following criterion.

Theorem 3 *For given scalars h, $\tau_1 \le 0 \le \tau_2$, α_1 and α_2, system (19) can be stabilized with $\varpi(t) \equiv 0$ and satisfies a prescribed H_∞ performance index γ under zero initial conditions, if there exist a $3n \times 3n$-dimensional matrix $\hat{P} > 0$, $n \times n$-dimensional matrices $\hat{Q} > 0$, $\hat{R} > 0$, $\hat{Z} > 0$, $\mathscr{M}$, $\hat{K}$ and a $2n \times 2n$-dimensional matrix $\hat{S}$, such that the following LMIs hold for $\dot{\rho}(t) \in \{\tau_1, \tau_2\}$:*

$$\begin{bmatrix} \hat{\Psi}|_{\rho(t)=0} & \hat{\Pi}_1^T B_w & \hat{\Pi}_2^T & \bar{E}_1^T\hat{S} \\ * & -\gamma^2 I_r & D_w^T & 0 \\ * & * & -I_m & 0 \\ * & * & * & -\check{R} \end{bmatrix} < 0 \tag{30}$$

$$\begin{bmatrix} \hat{\Psi}|_{\rho(t)=h} & \hat{\Pi}_1^T B_w & \hat{\Pi}_2^T & \bar{E}_2^T\hat{S}^T \\ * & -\gamma^2 I_r & D_w^T & 0 \\ * & * & -I_m & 0 \\ * & * & * & -\check{R} \end{bmatrix} < 0 \tag{31}$$

where

$$\begin{aligned}
\hat{\Psi} &= \hat{\Xi}_1 + \hat{\Xi}_2 - \hat{\Xi}_3 + \hat{\Xi}_4 + \hat{\Xi}_4^T \\
\hat{\Xi}_1 &= \mathfrak{E}_1^T \hat{P} \mathfrak{E}_2 + \mathfrak{E}_2^T \hat{P} \mathfrak{E}_1 \\
\hat{\Xi}_2 &= \bar{\varepsilon}_1^T (\hat{Q} + \hat{Z}) \bar{\varepsilon}_1 - (1 - \dot{\rho}(t)) \bar{\varepsilon}_2^T \hat{Q} \bar{\varepsilon}_2 - \bar{\varepsilon}_3^T \hat{Z} \bar{\varepsilon}_3 + h^2 \bar{\varepsilon}_6^T \hat{R} \bar{\varepsilon}_6 \\
\hat{\Xi}_3 &= \begin{bmatrix} E_1 \\ E_2 \end{bmatrix}^T \begin{bmatrix} \frac{2h-\rho(t)}{h} \check{R} & \hat{S} \\ * & \frac{h+\rho(t)}{h} \check{R} \end{bmatrix} \begin{bmatrix} E_1 \\ E_2 \end{bmatrix}, \quad \check{R} = \begin{bmatrix} \hat{R} & 0 \\ 0 & 3\hat{R} \end{bmatrix} \\
\hat{\Xi}_4 &= \hat{\Pi}_1^T [(A\mathscr{M} + B\hat{K}) \bar{\varepsilon}_1 + A_d \mathscr{M} \bar{\varepsilon}_2 - \mathscr{M} \bar{\varepsilon}_6] \\
\hat{\Pi}_1 &= \bar{\varepsilon}_1 + \alpha_1 \bar{\varepsilon}_2 + \alpha_2 \bar{\varepsilon}_6, \\
\hat{\Pi}_2 &= (C\mathscr{M} + D\hat{K}) \bar{\varepsilon}_1 + C_d \mathscr{M} \bar{\varepsilon}_2
\end{aligned} \tag{32}$$

with other notations are same as the one defined in Theorem 1, and the H_∞ controller gain can be calculated by:

$$K = \hat{K} \mathscr{M}^{-1} \tag{33}$$

Proof Denote

$$\begin{aligned}
&\hat{P} = \mathscr{M}^T P \mathscr{M}, \quad \hat{Q} = \mathscr{M}^T Q \mathscr{M}, \quad \hat{Z} = \mathscr{M}^T Z \mathscr{M}, \quad \hat{R} = \mathscr{M}^T R \mathscr{M}, \quad \hat{S} = \mathscr{M}^T S \mathscr{M} \\
&\hat{K} = K \mathscr{M}, \quad N_1 = \mathscr{M}^{-1}, \quad N_2 = \alpha_1 \mathscr{M}^{-1}, \quad N_3 = \alpha_2 \mathscr{M}^{-1}
\end{aligned}$$

and define

$$\mathscr{X} = \mathrm{diag}\{\mathscr{M}, \mathscr{M}, \mathscr{M}, \mathscr{M}, \mathscr{M}, \mathscr{M}, I_r, I_m, \mathscr{M}, \mathscr{M}\}$$

Pre- and post-multiplying $\bar{\Phi}_1$ and $\bar{\Phi}_2$ with $\mathscr{X}^T$ and $\mathscr{X}$, respectively, we can get LMIs (30) and (31). The proof is finished. ■

3.3 A Numerical Example

A brief numerical example is introduced to illustrate the advantages of the proposed method.

Example 4 Consider time-delay system shown in (19) with the following parameters:

$$\begin{aligned}
&A = \begin{bmatrix} 0 & 0 \\ 0 & -1 \end{bmatrix}, \quad A_d = \begin{bmatrix} -1 & -1 \\ 0 & -0.9 \end{bmatrix}, \quad B = \begin{bmatrix} 0 \\ 1 \end{bmatrix}, \quad B_w = \begin{bmatrix} 1 \\ 0 \end{bmatrix} \\
&C = [0 \ \ 1], \quad C_d = \begin{bmatrix} 0 & 0 \\ 0 & 0 \end{bmatrix}, \quad D = 0.1, \quad D_w = 0
\end{aligned}$$

Table 2 MADUPs for various $\mu = -\tau_1 = \tau_2$ (Example 1.4)

Methods	$\mu = -\tau_1 = \tau_2$					
	1	2	3	5	10	Unknown
Theorem 2 [23]	–	–	–	–	–	1.38
Corollary 2 [24]	–	–	–	–	–	1.4075
Corollary 2 [25]	–	–	–	–	–	1.4142
Theorem 3	2.6457	2.5864	2.5762	2.5729	2.5729	2.5706

For given $\gamma = 0.1287, \alpha_1 = 1.2$ and $\alpha_2 = 3.2$, the MADUPs for various μ calculated by using Theorem 3, together with the ones recalled from [23–25], are listed in Table 2. It is found, from Table 2, that Theorem 3 can design the controller satisfying the preset performance requirement under a bigger delay, which shows the merits of the proposed method.

4 Load Frequency Control for A Delayed One-Area Power System

In this section, the proposed methods of stability analysis and robust H_∞ control are employed in LFC for power system considering communication delays.

4.1 *Dynamic Model of the LFC Scheme*

The structure of delay-dependent LFC scheme for a one-area power system is shown in Fig. 1, which includes the governor, the turbine, the rotating mass and load, and the control center.

Time delays will arise from the communication networks during the transmission of measurement signals from remote sensors to control center and control signals from the control center to plants [26].

To simplify the analysis, the delays in different parts are considered as a single one, which is located in the forward loop and is expressed by an exponential block $e^{-s\rho(t)}$ as shown in Fig. 1 [27] and $\rho(t)$ is the communication delay satisfying (3) and (4).

Considering the system with non-reheated turbines, the state-space model of LFC scheme can be expressed as [28]:

$$\begin{cases} \dot{x}(t) = Ax(t) + Bu(t) + F\varpi(t) \\ y(t) = Cx(t) \end{cases} \tag{34}$$

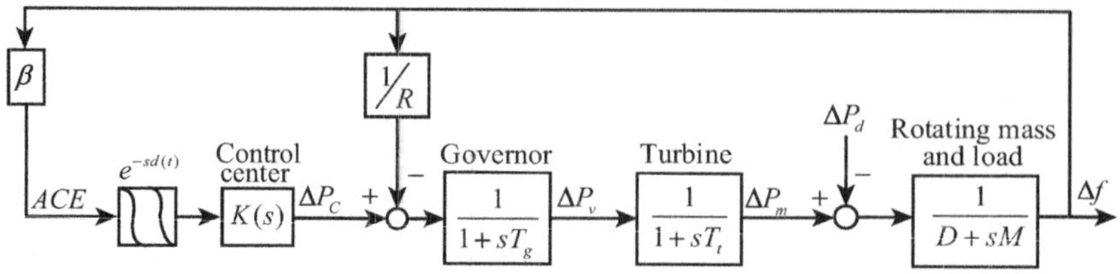

Fig. 1 Dynamic model of LFC for a one-area power system

where

$$x(t) = [\Delta f\ \Delta P_m\ \Delta P_v]^T,\ u(t) = \Delta P_c,\ y(t) = ACE,\ A = \begin{bmatrix} -\frac{D}{M} & \frac{1}{M} & 0 \\ 0 & -\frac{1}{T_t} & \frac{1}{T_t} \\ -\frac{1}{RT_g} & 0 & -\frac{1}{T_g} \end{bmatrix},$$

$$B = \begin{bmatrix} 0\ 0\ \frac{1}{T_g} \end{bmatrix}^T,\ F = \begin{bmatrix} -\frac{1}{M} & 0\ 0 \\ 0 & 0\ 0 \\ 0 & 0\ 0 \end{bmatrix},\ \varpi(t) = [\Delta P_d 00]^T,\ C = \begin{bmatrix} \beta\ 0\ 0 \end{bmatrix},$$

and Δf is the deviation of frequency; ΔP_m is the deviation of the generator mechanical output; ΔP_v is the deviation of valve position; ΔP_d is the deviation of load; M is the moment of inertia of the generator; D denotes generator damping coefficient; T_g denotes the time constant of the governor; T_t shows the time constant of the turbine; R indicates the speed drop; β the frequency bias factor; and ACE the area control error, which is defined as

$$ACE = \beta \Delta f \tag{35}$$

The control law $u(t)$ can be designed by different methods. In this section, two types of controllers, PI controller and state-feedback (SF) controller, are discussed.

4.2 Stability Assessment of PI-Based LFC Schemes

The following PI-type controller can be chosen as an LFC:

$$u(t) = -K_P ACE(t) - K_I \int ACE(t)dt \tag{36}$$

where K_P and K_I are proportional and integral gains, respectively.

Defining virtual state and measurement output vectors be $\bar{x}(t) = \begin{bmatrix} x^T(t) & \int y^T(t)dt \end{bmatrix}^T$ and $\bar{y}(t) = \begin{bmatrix} y^T(t) & \int y^T(t)dt \end{bmatrix}^T$, the model and controller are rewritten as

$$\begin{cases} \dot{\bar{x}}(t) = \bar{A}\bar{x}(t) + \bar{B}u(t) + \bar{F}\bar{\varpi}(t) \\ \bar{y}(t) = \bar{C}\bar{x}(t) \end{cases} \tag{37}$$

and

$$u(t) = -K\bar{y}(t - \rho(t)) \tag{38}$$

where $K = [K_P K_I]$ and

$$\bar{A} = \begin{bmatrix} A & 0 \\ C & 0 \end{bmatrix}, \ \bar{B} = \begin{bmatrix} B \\ 0 \end{bmatrix}^T, \ \bar{F} = \begin{bmatrix} F & 0 \\ 0 & 0 \end{bmatrix}, \ \bar{\varpi}(t) = \begin{bmatrix} \varpi(t) \\ 0 \end{bmatrix}, \ \bar{C} = \begin{bmatrix} C & 0 \\ 0 & 1 \end{bmatrix}.$$

Substituting (38) into (37), the closed-loop system is obtained:

$$\dot{\bar{x}}(t) = \bar{A}\bar{x}(t) + \bar{A}_d\bar{x}(t - \rho(t)) + \bar{F}\bar{\varpi}(t) \tag{39}$$

where $\bar{A}_d = -\bar{B}K\bar{C}$. Based on the above derivation, the LFC model for the power system considering communication delay is transformed into a linear time-delay system model.

For a designed LFC scheme, it is important to assess whether it can still work well when the time delay arises in the communication channels. For this case, the external disturbance is omitted, namely, the model of closed-loop LFC scheme is rewritten as follows

$$\dot{\bar{x}}(t) = \bar{A}\bar{x}(t) + \bar{A}_d\bar{x}(t - \rho(t)) \tag{40}$$

Obviously, based on the above disturbance-free model, Theorem 1 developed in Sect. 1 provides a theoretical approach to explain the impact of the communication delay on the LFC system's stability. Additionally, it can be applied to derive the delay stability region of the LFC system. Such region, which is a range of the time-varying communication delay that the LFC system can withstand for maintain its stability, comprises a lower bound and an upper bound. Then the delay margin of the LFC system is defined as the maximally acceptable upper bound of the time-varying communication delay, when defaulting the lower bound as zero.

The detailed procedure for calculating the delay margin is shown in Fig. 2.

Based on the above discussion, the steps to develop the delay margin to maintain the LFC system's stability are briefly listed, as follows.

Step 1: Modeling. The closed-loop LFC system is first established by combining the open-loop LFC in Sect. 4.1 with the PI-type control law. Then this system is transformed into a classic time-delay system by introducing the communication delay in LFC.

Step 2: Calculating the delay margin. First, preset the system parameters. The communication delay margin is then solved by the algorithm shown in Fig. 2.

Step 3: Analysis and verification of simulation tests. First, build the LFC system model in the Simulink platform in the MATLAB. Then the delay margin obtained in

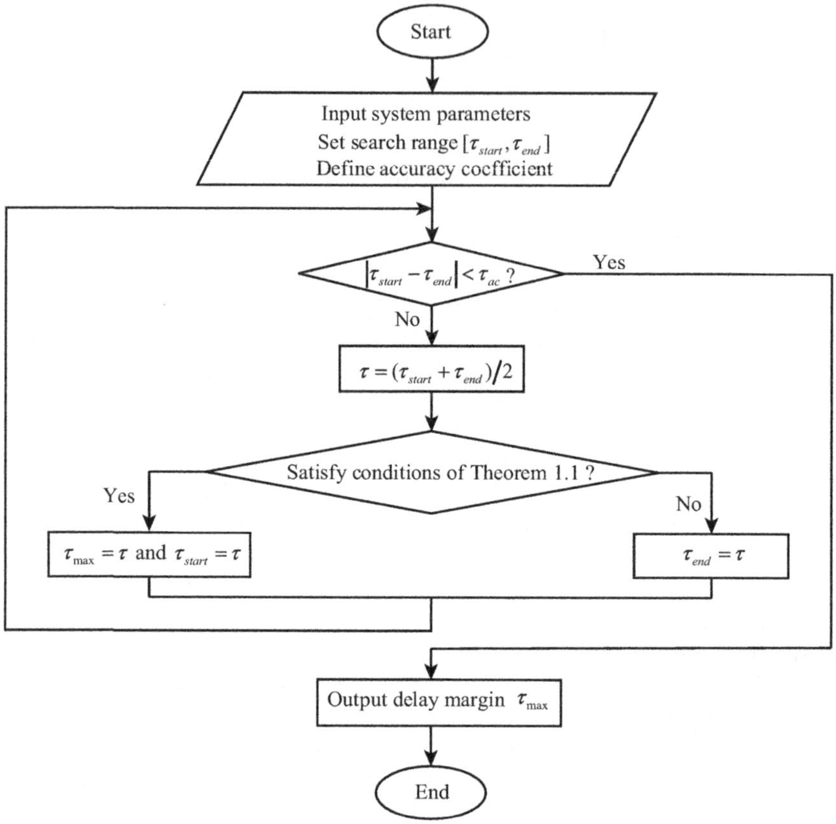

Fig. 2 Detailed calculation procedure of delay margins of LFC

Step 2 is tested by introducing the communication delay into the LFC process in the Simulink platform.

4.3 Design of the SF-Based LFC Scheme

When the designed LFC scheme does not work well due to the existence of time delays, one should redesign the scheme by taking into account the time delays. To guarantee the steady state of Δf to tend to zero, which can be guaranteed by an SF controller, the integral of Δf is used as a new state of the closed-loop LFC system. It is defined as $\Delta E = K_e \int \Delta f dt$, where K_e is similar to the gain of integral controller. By redefining the state vector as $\breve{x}^T = [\Delta f \ \Delta P_m \ \Delta P_v \ \Delta E]$, system (34) is rewritten as

$$\begin{cases} \dot{\breve{x}}(t) = \breve{A}\breve{x}(t) + \breve{B}u(t) + \breve{F}\breve{\varpi}(t) \\ \breve{z}(t) = \breve{C}\breve{x}(t) \end{cases} \tag{41}$$

where $\breve{z}(t)$ is controlled output and

$$\breve{A} = \begin{bmatrix} A & 0 \\ K_e & 0 \end{bmatrix}, \; \breve{B} = \begin{bmatrix} B \\ 0 \end{bmatrix}^T, \; \breve{F} = \begin{bmatrix} F & 0 \\ 0 & 0 \end{bmatrix}^T, \; \breve{\varpi}(t) = \begin{bmatrix} \varpi(t) \\ 0 \end{bmatrix}, \; \breve{C} = [1\,0\,0\,0].$$

Then the SF controller to be designed is given as

$$u(t) = K_s\breve{x}(t - \rho(t)) \tag{42}$$

where K_s represents the controller gains and $\rho(t)$ is the communication delay. By combining (41) and (42), the closed-loop LFC system model can be written as follows

$$\begin{cases} \dot{\breve{x}}(t) = \breve{A}\breve{x}(t) + \breve{B}K_s\breve{x}(t - \rho(t)) + \breve{F}\breve{\varpi}(t) \\ \breve{z}(t) = \breve{C}\breve{x}(t) \end{cases} \tag{43}$$

Aiming at making the LFC scheme can work well, the designed controller is required to have robustness against to both time delay and the disturbance. Thus, such a problem can be solved by using the delay-dependent H_∞ control method introduced in Sect. 3.

Based on the method of Sect. 3.2, the following criterion is given.

Theorem 4 *For three scalars h and $\tau_1 \le 0 \le \tau_2$, system (43) is asymptotically stable with $\breve{\varpi}(t) \equiv 0$ and satisfies a prescribed H_∞ performance index γ under zero initial conditions, if there exist a $3n \times 3n$-dimensional matrix $P > 0$, $n \times n$-dimensional matrices $Q > 0$, $R > 0$, $Z > 0, N_1$, N_2, N_3, a $2n \times 2n$-dimensional matrix S, and controller gains K_s such that the following conditions hold for both $\dot{\rho}(t) = \tau_1$ and $\dot{\rho}(t) = \tau_2$:*

$$\begin{bmatrix} \breve{\Psi}|_{\rho(t)=0} & E_1^T S \\ * & -\breve{R} \end{bmatrix} < 0 \tag{44}$$

$$\begin{bmatrix} \breve{\Psi}|_{\rho(t)=h} & E_2^T S^T \\ * & -\breve{R} \end{bmatrix} < 0 \tag{45}$$

where

$$\begin{aligned} \breve{\Psi} &= \Xi_1 + \Xi_2 - \Xi_3 + \breve{\Xi}_4 + \breve{\Xi}_4^T + \bar{\varepsilon}_1^T \breve{C}^T \breve{C} \bar{\varepsilon}_1 - \gamma^2 \bar{\varepsilon}_7^T \bar{\varepsilon}_7 \\ \breve{\Xi}_4 &= \left[\bar{\varepsilon}_1^T N_1 + \bar{\varepsilon}_2^T N_2 + \bar{\varepsilon}_6^T N_3\right] [\breve{A}\bar{\varepsilon}_1 + \breve{B}K_s\bar{\varepsilon}_2 + \breve{F}\bar{\varepsilon}_7 - \bar{\varepsilon}_6] \\ \bar{\varepsilon}_i &= [0_{n\times(i-1)n}, I, 0_{n\times(7-i)n}], \; i = 1, 2, \ldots, 7 \end{aligned}$$

and other notations are same as the one defined in Theorem 1.

Table 3 Parameters for stability analysis

β	D	M	R	T_g	T_t
21 pu/Hz	1.0 pu/Hz	10 pu·s	0.05 Hz/pu	0.1 s	0.3 s

Table 4 Parameters for control design

K_e	D	R	T_g	T_t	M
1	0.0830 pu/Hz	2.40 Hz/pu	0.08 s	0.30 s	0.1667 pu·s

Proof Redefine $\zeta_2(t) = [\breve{x}^T(t), \breve{x}^T(t-\rho(t)), \breve{x}^T(t-h), \breve{v}_1^T(t), \breve{v}_2^T(t), \dot{\breve{x}}^T(t), \breve{\varpi}(t)]^T$, where $\breve{v}_1(t) = \int_{t-\rho(t)}^{t} \frac{\breve{x}(\upsilon)}{\rho(t)} ds$, $\breve{v}_2(t) = \int_{t-h}^{t-\rho(t)} \frac{\breve{x}(\upsilon)}{h-\rho(t)} ds$. Then, the proof procedure is similar to the proof in Theorem 2 and is omitted here.

Note that, for unknown K_s, inequalities (44) and (45) are no longer the LMI-based conditions due to the product of $N_1^T K_s$, $N_2^T K_s$, and $N_3^T K_s$. A particle swarm optimization (PSO) method is applied to design an SF controller K_s for LFC scheme with a preset H_∞ performance as well as a good dynamical performance. The specific design procedure is shown as follows.

Step 1: Determine the system parameters; set communication delay parameters h_M and τ_M; preset H_∞ performance parameter γ.

Step 2: Initialize parameters of the PSO method and set the objective function "$Max(\varepsilon)$".

Step 3: Start PSO and repeat the following steps for $i = 1, 2, \ldots, n$ and $j = 1, 2, \ldots, m$. (1) Set $K_{si-j} = -[SW_{i-j}(1), SW_{i-j}(2), SW_{i-j}(3), SW_{i-j}(4)]$, $\varepsilon = 0$ and $\varepsilon_i = 0$. (2) Check the feasibility of inequalities (44) and (45). If (44) and (45) are feasible, set $\varepsilon_{i-j} = SW_{i-j}(4)$; else, set $\varepsilon_{i-j} = 0$. (3) If $\varepsilon_{i-j} > \varepsilon_i$, set $\varepsilon_i = \varepsilon_{i-j}$ and $K_{si} = K_{si-j}$. (4) If $j = m$ and $\varepsilon_i > \varepsilon$, set $\varepsilon = \varepsilon_i$ and $K_s = K_{si}$; and then if $i = n$, goto Step 4. where SW_i represents the particle swarm in the ith evolution and SW_{i-j} denotes the jth particle.

Step 4: Output optimal feedback gains K_s.

4.4 Case Studies

Case studies are carried out in this section, including the stability of the system with parameters shown in Table 3 and control design for the system with parameters given in Table 4.

A. Stability analysis. Delay margins with respect to different PI gains are calculated and listed in Tables 5 and 6. It is found that h decreases sharply with the increase of K_I for a fixed K_P and that h continues to decrease for some K_I (0.05) while it increases at first and then decreases for others K_I with the increasing of K_P. Moreover, the obtained results are less conservative in comparison to those of [27,

Table 5 Upper bound for constant delay ($\mu = 0$)

K_P	K_I	Theorem 1	[29]	[27]	[30]
0	0.05	30.85	30.03	27.92	27.92
0	0.1	15.17	14.85	13.77	13.77
0	0.2	7.32	7.23	6.69	6.69
0	0.4	3.38	3.36	3.12	3.12
0	0.6	2.04	2.03	1.91	1.91
0.1	0.05	30.13	29.40	27.03	27.03
0.1	0.1	15.50	15.20	13.68	13.69
0.1	0.2	7.78	7.60	6.94	6.94
0.1	0.4	3.60	3.58	3.29	3.29
0.1	0.6	2.22	2.21	2.02	2.02

Table 6 Upper bound for time-varying delay ($\mu = 0.5$)

K_P	K_I					
	0.05	0.1	0.2	0.4	0.6	1.0
0	27.27	13.40	6.46	2.95	1.75	0.73
0.05	27.18	13.79	6.66	3.06	1.82	0.77
0.1	25.71	13.27	6.78	3.14	1.88	0.80
0.2	22.43	11.68	6.25	3.18	1.93	0.84
0.4	14.78	7.96	4.51	2.54	1.66	0.75
0.6	5.46	3.14	1.86	1.08	0.82	0.54
1.0	0.43	0.42	0.41	0.39	0.36	0.29

29, 30]. A simulation test is given for the case that a positive load disturbance of 0.1 pu is added to the system at $t = 10$ s and $K_P = 0.1$, $K_I = 0.2$. The responses of the LFC scheme under different delays are given in Fig. 3. It is found from the figures that the responses are convergent when the communication delay is 7.78 s, which is closed to the delay margin listed in the table. That is, the delay margin calculated is accurate.

B. Controller design. By setting $h = 10$, $\mu = 1$, and $\gamma = 10$, and following the procedure described in Sect. 4.3, the gain of controller is obtained as

$$K_s = -[0.0180 \;\; 0.0393 \;\; 0.0117 \;\; 0.0435]. \qquad (46)$$

Simulation tests are carried out to show the effectiveness and the advantage of the proposed controllers (compared with the ones recalled from literature, including the state-feedback controller with $K_1 = -[2.7321\ 4.0167\ 0.8506\ 0.4318]$ in [31], the PID controller with $K_2 = -[0.4036\ 0.6356\ 0.1832]$ in [32], and the sampled-data controller with $K_3 = -[0.0622\ 0.1231\ 0.0226\ 0.1723]$ in [33]). Assume that 0.01 pu

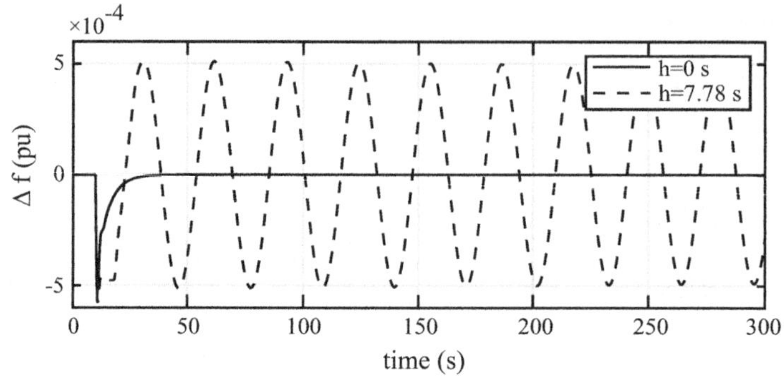

Fig. 3 Frequency responses with $h_{max} = 7.78$ ($\mu = 0$) and without delay

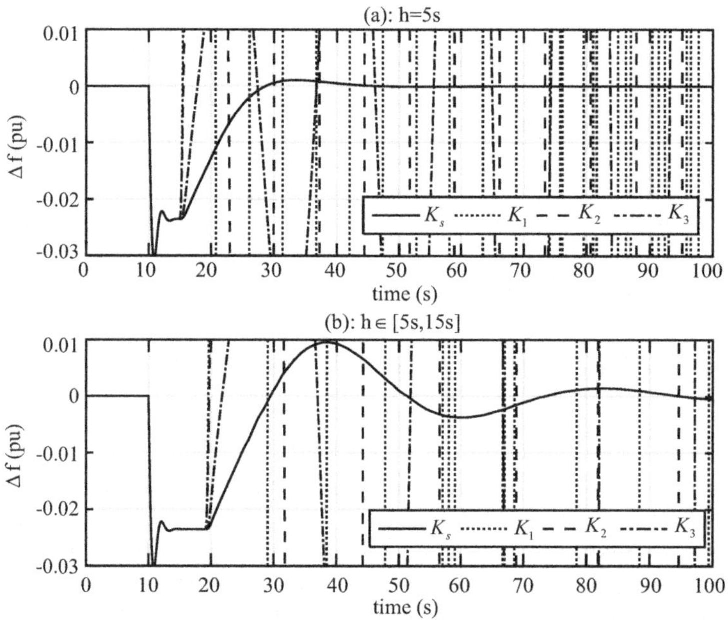

Fig. 4 Frequency deviation responses under controllers K_s, K_1,K_2 and K_3

step load disturbance is added at 10 s, and the communication delay is set to 5 s and random values between [5 s,15 s], respectively. Responses of frequency deviation are shown in Fig. 4, which shows that the proposed controller K_s provides robustness against the communication delays.

References

1. Xiong, J.L., Lam, J.: Stabilization of networked control systems with a logic ZOH. IEEE Trans. Autom. Control **54**(2), 358–363 (2009)
2. Gu, K.Q., Kharitonov, V.L., Chen, J.: Stability of Time-Delay Systems. Birkhauser, Basel (2003)
3. Fridman, E.: Introduction to Time-Delay Systems: Analysis and Control. Birkhauser, Basel (2014)
4. Seuret, A., Gouaisbaut, F.: Wirtinger-based integral inequality: application to time-delay systems. Automatica **49**(9), 2860–2866 (2013)
5. Fridman, E., Shaked, U.: Delay-dependent stability and H_∞ control: constant and time-varying delays. Int. J. Control **76**, 48–60 (2003)
6. Briat, C.: Linear Parameter-Varying and Time-Delay Systems: Analysis, Observation, Filtering & Control. Springer, Berlin (2015)
7. He, Y., Wu, M., She, J.H., Liu, G.P.: Delay-dependent robust stability criteria for uncertain neutral systems with mixed delays. Syst. Control Lett. **51**(1), 57–65 (2004)
8. Wu, M., He, Y., She, J.H., Liu, G.P.: Delay-dependent criteria for robust stability of time-varying delay systems. Automatica **40**(8), 1435–1439 (2004)
9. He, Y., Wang, Q.G., Lin, C., Wu, M.: Delay-range-dependent stability for systems with time-varying delay. Automatica **43**(2), 371–376 (2007)
10. He, Y., Wang, Q.G., Xie, L., Lin, C.: Further improvement of free-weighting matrices technique for systems with time-varying delay. IEEE Trans. Autom. Control **52**(2), 293–299 (2007)
11. Zeng, H.B., He, Y., Wu, M., She, J.H.: Free-matrix-based integral inequality for stability analysis of systems with time-varying delay. IEEE Trans. Autom. Control **60**(10), 2768–2772 (2015)
12. Zhang, C.K., He, Y., Jiang, L., Wu, Q.H., Wu, M.: Delay-dependent stability criteria for generalized neural networks with two delay components. IEEE Trans. Neural Netw. Learn. Syst. **25**(7), 1263–1276 (2014)
13. Gu, K.: An integral inequality in the stability problem of time-delay systems. In: Proceedings of the 39th IEEE Conference on Decision and Control, Sydney, Australia (2010)
14. Park, P.G., Lee, W.I., Lee, S.Y.: Auxiliary function-based integral inequalities for quadratic functions and their applications to time-delay systems. J. Frankl. Inst. **352**(4), 1378–1396 (2015)
15. Seuret, A., Gouaisbaut, F.: Hierarchy of LMI conditions for the stability analysis of time-delay systems. Syst. Control Lett. **81**, 1–7 (2015)
16. Zeng, H.B., He, Y., Wu, M., She, J.H.: New results on stability analysis for systems with discrete distributed delay. Automatica **60**, 189–192 (2015)
17. Seuret, A., Gouaisbaut, F., Ariba, Y.: Complete quadratic Lyapunov functionals for distributed delay systems. Automatica **62**, 168–176 (2015)
18. Hien, L.V., Trinh, H.: An enhanced stability criterion for time-delay systems via a new bounding technique. J. Frankl. Inst. **352**(10), 4407–4422 (2015)
19. Hien, L.V., Trinh, H.: Refined Jensen-based inequality approach to stability analysis of time-delay systems. IET Control Theory Appl. **9**(14), 2188–2194 (2015)
20. Park, P., Ko, J.: Stability and robust stability for systems with a time-varying delay. Automatica **43**(10), 1855–1858 (2007)
21. Kim, J.H.: Note on stability of linear systems with time-varying delay. Automatica **47**(9), 2118–2121 (2011)
22. Ariba, Y., Gouaisbaut, F.: An augmented model for robust stability analysis of time-varying delay systems. Int. J. Control **82**(9), 1616–1626 (2009)
23. Xu, S.Y., Lam, J., Zou, Y.: New results on delay-dependent robust H_∞ control for systems with time-varying delays. Automatica **42**(2), 343–348 (2006)
24. Wu, J., Chen, T.W., Wang, L.: Delay-dependent robust stability and H_∞ control for jump linear systems with delays. Syst. Control Lett. **55**, 939–948 (2006)

25. Tian, E., Yue, D., Zhang, Y.: On improved delay-dependent robust H_∞ control for systems with interval time-varying delay. J. Frankl. Inst. **348**(4), 555–567 (2011)
26. Yu, X., Kevin, T.: Application of linear matrix inequalities for load frequency control with communication delays. IEEE Trans. Power Syst. **19**(3), 1508–1515 (2004)
27. Jiang, L., Yao, W., Wu, Q.H., Wen, J.Y., Cheng, S.J.: Delay-dependent stability for load frequency control with constant and time-varying delays. IEEE Trans. Power Syst. **27**(2), 932–941 (2012)
28. Zhang, C.K., Jiang, L., Wu, Q.H., He, Y., Wu, M.: Delay-dependent robust load frequency control for time delay power systems. IEEE Trans. Power Syst. **28**(3), 2192–2201 (2013)
29. Yang, F., He, J., Pan, Q.: Further improvement on delay-dependent load frequency control of power systems via truncated B-L inequality. IEEE Trans. Power Syst. **33**(5), 5062–5071 (2018)
30. Ramakrishnan, K., Ray, G.: Stability criteria for nonlinearly perturbed load frequency systems with time-delay. IEEE J. Emerg. Sel. Top. Circuits Syst. **5**(3), 383–392 (2015)
31. Wang , Z.Q., Sznaier, M.: Robust control design for load frequency control using μ-synthesis. In: Conference Record Southcon, pp. 186–190 (1994)
32. Tan, W.: Unified tuning of PID load frequency controller for power systems via IMC. IEEE Trans. Power Syst. **25**(1), 341–350 (2010)
33. Shang-Guan, X.C., He, Y., Zhang, C.K., Jiang, L., Spencer, J.W., Wu, M.: Sampled-data based discrete and fast load frequency control for power systems with wind power. Appl. Energy. **259**, 114202 (2020)

Active Disturbance Rejection in Repetitive Control Systems

Jinhua She, Min Wu, Weihua Cao, Seiichi Kawata, and Kangzhi Liu

Abstract As a high-precision control method for periodic signals, repetitive control has been widely investigated from both theoretical and practical sides. Since a repetitive control system contains two completely different actions: continuous control within each repetition period and discrete learning between periods, exploring the best combination of these two actions may provide us with a potential to achieve higher levels of performance. For this reason, we devised a two-dimensional repetitive control method that features preferential adjustment of control and learning actions. On the other hand, there are aperiodic disturbances in a repetitive control system. It is a challenge to reject such kinds of disturbances. To solve this problem, the equivalent-input-disturbance approach is integrated into a two-dimensional repetitive control system to improve disturbance-rejection performance. This section explains the concepts of two-dimensional repetitive control and an equivalent input disturbance and then shows how these two concepts are combined to construct a high-precision control system.

Keywords Equivalent input disturbance (EID) · Linear matrix inequality (LMI) · Lyapunov stability theory · Repetitive control · Two-dimensional (2D) system

---

J. She (✉)
School of Engineering, Tokyo University of Technology, Hachioji,
Tokyo 192-0982, Japan
e-mail: she@stf.teu.ac.jp

J. She · M. Wu · W. Cao
School of Automation, China University of Geosciences, Wuhan 430074, Hubei, China

Hubei Key Laboratory of Advanced Control and Intelligent Automation
for Complex Systems, Wuhan 430074, China

S. Kawata
Graduate School of Industrial Technology, Advanced Institute of Industrial Technology,
Tokyo 140-0011, Japan

K. Liu
Department of Electrical and Electronic Engineering, Chiba University,
Chiba 263-8522, Japan

M. Wu et al. (eds.), *Developments in Advanced Control and Intelligent Automation for Complex Systems*, Studies in Systems, Decision and Control 329,
https://doi.org/10.1007/978-3-030-62147-6_2

Abbreviations

2D Two-dimensional
EID Equivalent input disturbance
LMI Linear matrix inequality
RC Repetitive control
RCS Repetitive-control system

1 Introduction

In control engineering practice, high accuracy tracking and/or rejection of periodic signals are important issues. Repetitive control (RC) is an effective approach to dealing with those problems. Due to its simple structure and high control accuracy, RC has been widely used in many control systems and has drawn much attention in various engineering fields. A repetitive-control system (RCS) contains two types of actions: continuous control within each repetition period and discrete learning between periods. Unlike a one-dimensional model, a two-dimensional (2D) continuous-discrete hybrid model is able to describe this nature of RC, which enables better control performance. On the other hand, a disturbance not only degrades control performance but also may cause vibration and environmental pollution problems. To achieve satisfactory system performance, we need to take into consideration of disturbance rejection in control systems. While an RCS provides perfect steady-state control performance for periodic signals, it is hard to reject aperiodic disturbances. In this chapter, we explain an equivalent-input-disturbance (EID)-based RCS. In this system, aperiodic disturbances and uncertainties are treated as a disturbance on the control input channel, which is called an EID. The EID is estimated by an EID estimator and is incorporated into an RC law to suppress their effects.

In this chapter, we first give a brief review of RC and build a 2D model for an RCS. Next, we explain the concept of an EID. Then, we present the configuration of a generalized EID (GEID)-based RCS and the design method of the system. Finally, we use an example to illustrate the design procedure and show the effectiveness of the method.

2 Repetitive Control

Repetitive control (RC) is a servomechanism for periodic signals [1–3]. A repetitive controller, which is

$$C_R(s) = \frac{1}{1 - e^{-Ts}}, \tag{1}$$

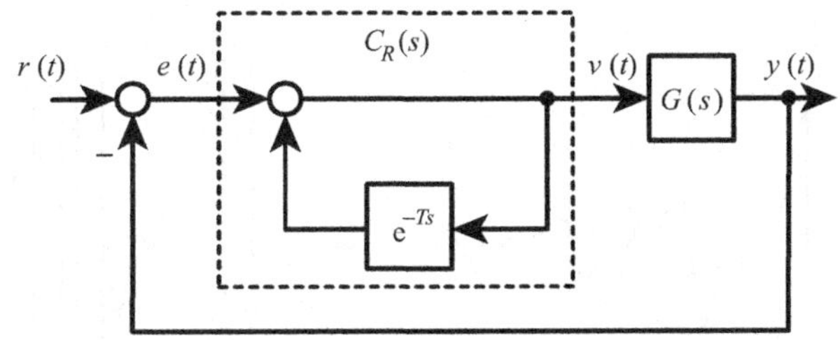

Fig. 1 Configuration of basic RCS

where T is the period of a periodic signal, in an RCS (Fig. 1) is an internal model of a periodic signal. This guarantees asymptotic tracking and/or rejection of a periodic reference input or a periodic disturbance, respectively. RC provides us a practical, effective solution to high-precision and high-performance control, and has been widely used in many engineering fields.

In this section, $\|x(t)\|$ $[= \sqrt{x^{\mathrm{T}}(t)x(t)}]$ is the Euclidean norm (2-norm); $\mathscr{L}\{x(t)\}$ $[= X(s)]$ and $\mathscr{L}^{-1}\{X(s)\}$ $[= x(t)]$ are the Laplace and inverse Laplace transforms, respectively.

2.1 *Two-Dimensional Property of Repetitive Control*

The output of the basic repetitive controller (Fig. 1) is

$$v(t) = e(t) + v(t - T). \tag{2}$$

Note that, in the above equation, while the tracking error, $e(t)$, is a signal in the present period, the control input, $v(t - T)$, is the one in the previous period. Since we can view the past control input as a kind of experience, $C_R(s)$ can be interpreted as a unit that produces the present control input by combining two different kinds of information: control $[e(t)]$ and learning $[v(t - T)]$. Focusing on the fact that RC involves continuous control within each repetition period and discrete learning between periods, we devise a design method that enables independent adjustment of the two actions so as to achieve better transient and tracking performance [4, 5].

To exploit the 2D characteristic in RC, we apply a lifting technique, $\mathscr{L}_{\mathscr{C}}$ (Fig. 2, [6]), to build a 2D model for an RCS. $\mathscr{L}_{\mathscr{C}}$ is used to slice the time axis, $[0, +\infty)$, into intervals of a length T and convert a vector-valued continuous-time signal $\xi(t)$ into a function-valued discrete-time sequence $\{\xi_k(\tau)\}$. Its element is denoted by $\xi(k, \tau)$

$$\xi(k, \tau) = \xi_k(\tau) := \mathscr{L}_{\mathscr{C}}[\xi(t)],\ t = kT + \tau,\ \tau \in [0,\ T],\ k \in \mathbb{Z}_+, \tag{3}$$

where $\mathscr{L}_{\mathscr{C}}$ is an isometric and isomorphic transformation between $L_2(\mathbb{R}_+,\ \mathbb{C}^p)$ and $\ell_2(\mathbb{Z}_+,\ \aleph)$. Clearly, τ and k are variables with respect to control and learning, respectively.

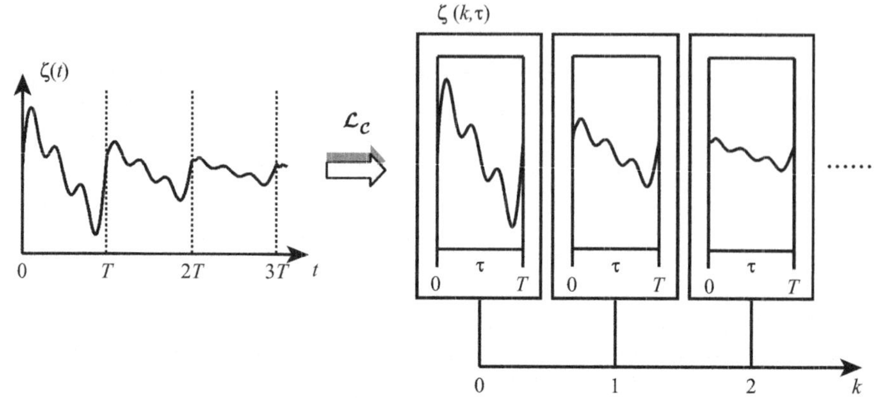

Fig. 2 Lifting

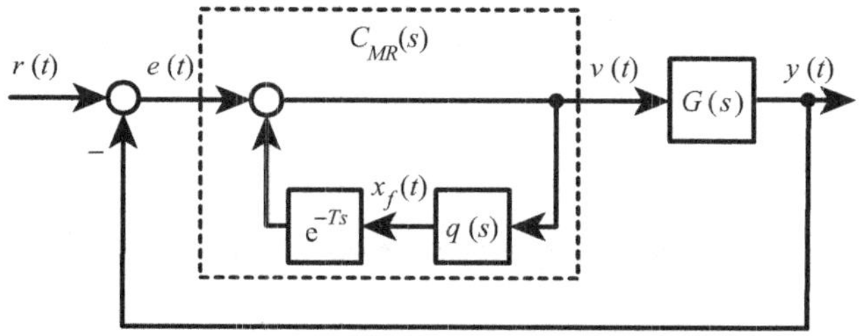

Fig. 3 Configuration of MRCS

2.2 *Modified Repetitive-Control System*

Since a repetitive controller, $C_R(s)$, has an infinite number of poles on the imaginary axis, an RCS is a neutral-type delay system, and the stabilization of such a system is a difficult task. As pointed out in [2], this type of system can be stabilized only for a plant with a relative degree of zero. However, this condition is quite strong for control engineering applications because many actual plants do not satisfy it. To apply RC to a plant with a relative degree that is larger than zero, that is, a strictly proper plant, the repetitive controller is modified by inserting a low-pass filter $q(s)$ into the time-delay feedback line:

$$C_{MR}(s) = \frac{1}{1 - q(s)\mathrm{e}^{-Ts}}. \tag{4}$$

This controller is called a modified repetitive controller (MRC) and the resulting system is called a modified RCS (MRCS) (Fig. 3). This modification makes the modified repetitive controller only be an approximate generator of a periodic signal. While an MRCS is easy to stabilize, it is impossible to perform perfect tracking.

Thus, we have to find a way in system design to solve the trade-off problem between tracking/rejection performance for periodic signals and other control performance.

The control and learning actions can easily be preferentially adjusted for an RCS [4]. However, the insertion of a low-pass filter in an MRC mixes the two actions together. An iterative algorithm was presented in [5] that searches for the best combination of the two actions.

3 Equivalent-Input-Disturbance Approach

The equivalent-input-disturbance (EID) approach is an active disturbance-rejection method. This section first explains the basic concept of the EID. Then, it presents the configuration of an EID-based control system. Finally, it performs the analysis of disturbance-rejection mechanism.

3.1 Basic Concept of Equivalent Input Disturbance

Consider the following linear time-invariant plant

$$\begin{cases} \dot{x}_p(t) = Ax_p(t) + Bu(t) + B_d d(t), \\ y(t) = Cx_p(t), \end{cases} \tag{5}$$

where $x_p(t)$ $(\in \mathbb{R}^n)$, $u(t)$ $(\in \mathbb{R}^{n_u})$, and $y(t)$ $(\in \mathbb{R}^{n_y})$ are the state, the input, and the output of the plant, respectively; $d(t)$ $(\in \mathbb{R}^{n_d})$ is an unknown disturbance; A $(\in \mathbb{R}^{n\times n})$, B $(\in \mathbb{R}^{n\times n_u})$, B_d $(\in \mathbb{R}^{n\times n_d})$, and C $(\in \mathbb{R}^{n_y\times n})$ are constant matrices. Two assumptions are made for the plant:

Assumption 1 (A, B, C) is controllable and observable.

Assumption 2 (A, B, C) has no zeros on the imaginary axis.

Assumptions 1 and 2 are standard. They ensure the internal stabilizability of the plant (5) for reference tracking [7]. Note that B and B_d in (5) may have different dimensions, that is, the disturbance may be imposed on a channel other than the control input channel. We can use stable-inversion approach [8] to construct a signal, $d_e(t)$ $(\in \mathbb{R}^{n_u})$, on the control input channel that produces the same effect on the output as the disturbance does (This signal is called an EID [9]). This shows the existence of an EID, and $d_e(t)$ is called an EID of $d(t)$.

Since we focus on the effect of the disturbance on the system output rather than the disturbance itself, we rewrite the plant using $d_e(t)$:

$$\begin{cases} \dot{x}_p(t) = Ax_p(t) + B[u(t) + d_e(t)], \\ y(t) = Cx_p(t). \end{cases} \tag{6}$$

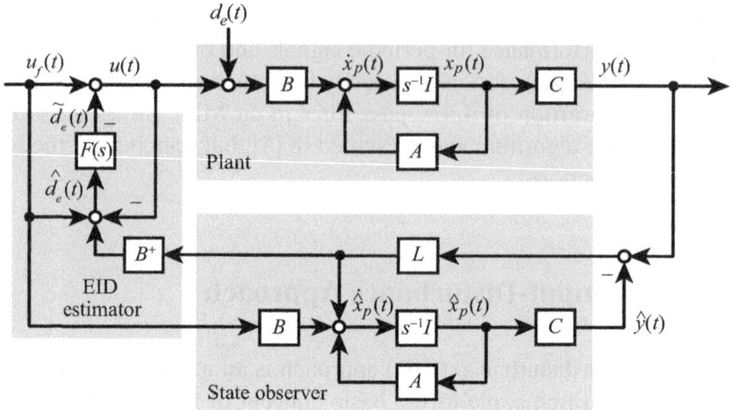

Fig. 4 EID estimator

We abused the notation a bit by using the same variable $x_p(t)$ for the state of the plant in both (5) and (6). This should not cause confusion.

While $d_e(t)$ can be exactly calculated using the stable-inversion approach, it is noncausal and needs the information of $y(t)$ in the future for calculation. This is not practical for control practice. For this reason, we present a causal method that produces an estimate of $d_e(t)$ in a real-time fashion in the next subsection.

In the rest of this section, we only consider the single-input single-output (SISO) case ($n_u = 1$ and $n_y = 1$) for simplicity. Note that the results obtained are easy to extend to the multi-input multi-output (MIMO) case.

3.2 Equivalent-Input-Disturbance Estimation

An EID estimator (Fig. 4) elaborately integrates the control input and output to produce an estimate of $d_e(t)$. The mechanism of EID estimation is explained below.

A state observer is used to estimate the state of the plant

$$\begin{cases} \dot{\hat{x}}_p(t) = A\hat{x}_p(t) + Bu_f(t) + L[y(t) - \hat{y}(t)], \\ \hat{y}(t) = C\hat{x}_p(t), \end{cases} \tag{7}$$

where $\hat{x}_p(t) \in \mathbb{R}^n$, $u_f(t) \in \mathbb{R}$, $\hat{y}(t) \in \mathbb{R}$, and L ($\in \mathbb{R}^{n\times 1}$) is the gain of the observer. To explain the disturbance estimation in Fig. 4, we define

$$\Delta x_p(t) = x_p(t) - \hat{x}_p(t). \tag{8}$$

Substituting (8) into (6) yields

$$\dot{\hat{x}}_p(t) = A\hat{x}_p(t) + B[u(t) + d_e(t)] + [A\Delta x_p(t) - \Delta\dot{x}_p(t)]. \tag{9}$$

Since (A, B) is controllable, there is a signal $\Delta d_e(t)$ that satisfies

$$\Delta\dot{x}_p(t) = A\Delta x_p(t) + B\Delta d_e(t). \tag{10}$$

Let an estimate of the EID be

$$\hat{d}_e(t) = d_e(t) - \Delta d_e(t). \tag{11}$$

Substituting (10) and (11) into (9), we have

$$\dot{\hat{x}}_p(t) = A\hat{x}_p(t) + B[u(t) + \hat{d}_e(t)]. \tag{12}$$

Comparing (7) and (12) gives

$$B\left[u(t) + \hat{d}_e(t) - u_f(t)\right] = LC\Delta x_p(t). \tag{13}$$

A least-squares optimal solution to (13) for $\hat{d}_e(t)$ is

$$\hat{d}_e(t) = B^{+}LC\Delta x_p(t) + u_f(t) - u(t), \tag{14}$$

where

$$B^{+} = \left(B^{\mathrm{T}}B\right)^{-1}B^{\mathrm{T}}. \tag{15}$$

A low-pass filter $F(s)$

$$\begin{cases} \dot{x}_F(t) = A_F x_F(t) + B_F\hat{d}_e(t), \\ \tilde{d}_e(t) = C_F x_F(t) \end{cases} \tag{16}$$

is used to choose an angular-frequency range for disturbance estimation [9]. Thus, the filtered disturbance estimate is given by

$$\tilde{d}_e(t) = \mathscr{L}^{-1}\left\{F(s)\mathscr{L}\left\{\hat{d}_e(t)\right\}\right\}. \tag{17}$$

The following condition has to be satisfied for the selection of $F(s)$:

$$|F(j\omega)| \approx 1,\ \forall\omega \in [0, \omega_d], \tag{18}$$

where ω_d is the highest angular frequency for disturbance estimation. As a result, we obtain a new control law

$$u(t) = u_f(t) - \tilde{d}_e(t) \tag{19}$$

that incorporates $\tilde{d}_e(t)$ on the control input channel to compensate for $d_e(t)$.

Combining (6), (7), (14), (17), and (19) and describing their relationship in a block diagram yield Fig. 4.

Remark 1 Equations (6) and (12) reveal that the difference between the state of the plant and that of the observer is equivalent to the difference between the exact value and the estimate of the EID. A suitable design of the observer guarantees that $\|\hat{x}_p(t) - x_p(t)\|$ and $\|\hat{d}_e(t) - d_e(t)\|$ are small enough. And Condition (18) for $F(s)$ ensures that $\tilde{d}_e(t)$ is a good approximation of $d_e(t)$.

3.3 Analysis of Disturbance Rejection

From (6), (7), (8), and (19), we have

$$\Delta\dot{x}_p(t) = (A - LC)\Delta x_p(t) + B[d_e(t) - \tilde{d}_e(t)]. \tag{20}$$

To analyze disturbance rejection performance, we redraw Fig. 4 from the EID to the output taking into consideration of (20) [10] and obtain Fig. 5.

Define

$$\begin{cases} W(s) = [sI - (A - LC)]^{-1}B, \\ P(s) = C(sI - A)^{-1}B, \\ K_L = B^{+}LC. \end{cases} \tag{21}$$

Using them to simplify Fig. 5 yields Fig. 6. The transfer function from $d_e(t)$ to $y(t)$ is

$$G_{yde}(s) = G_{FF}(s)P(s), \tag{22}$$

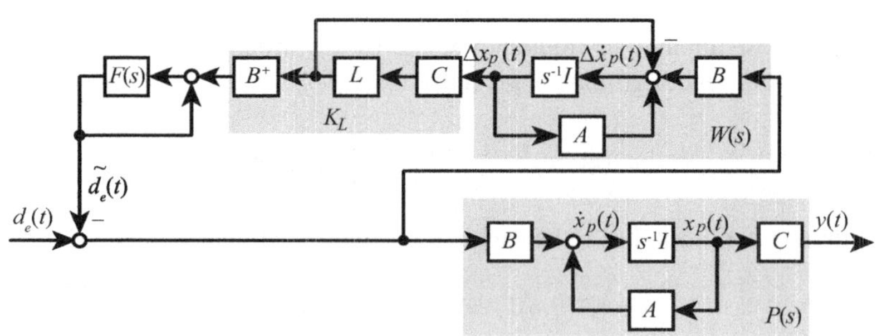

Fig. 5 Block diagram from EID to output

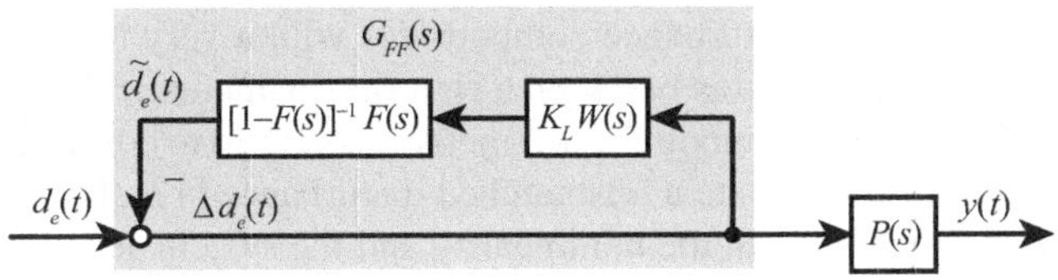

Fig. 6 Simplification of Fig. 5

where

$$G_{FF}(s) = \frac{1 - F(s)}{1 + F(s)[K_L W(s) - 1]}. \tag{23}$$

Since $F(s)$ is chosen such that (18) holds,

$$G_{FF}(j\omega) \approx 0, \ \forall \omega \in [0, \omega_d]. \tag{24}$$

Thus, the disturbance is blocked by $G_{FF}(s)$ to be transmitted to the output. The low-pass filter, $F(s)$, in $G_{FF}(s)$ plays a key role in disturbance rejection [10]. A first-order low-pass filter is usually selected for the following reasons:

(1) We require that a disturbance estimate passes through the filter without any gain reduction or phase lag in the required disturbance-rejection frequency band. Thus, we mainly concern about whether or not (18) is true in the band and do not care about the roll-off speed of the gain of $F(j\omega)$ outside the band.
(2) If a high-order filter is used to ensure (18) in the same angular-frequency band, phase lag usually begins at a lower angular frequency for the high-order filter than for the first-order one.
(3) A first-order filter makes the system design simple.

4 Disturbance Rejection for Repetitive Control System with Time-Varying Nonlinearity

Nonlinearities are common in mechanical devices, such as a manipulator [11], and a servo motor [12], and an offshore steel jacket platform [13]. They cause difficulty in control system design. Moreover, exogenous disturbances largely degrade the control performance of an RCS.

Various strategies have been adopted to enhance disturbance-rejection performance for nonlinear systems, for example, the disturbance-observer-based control (DOBC) method is used when a plant is known precisely, but the performance is degraded if the plant has uncertainty [14]. The DOBC was combined with the H_∞ control method to reject a disturbance [15]. It can deal with a special form of perturbations or uncertainties in the system. However, the above methods require that the state of the system is available. Moreover, for a disturbance with a complex form, these methods result in a disturbance compensator with a very high order because the exact linearization method is based on a rigorous mathematical model of a nonlinear system [16]. A sliding-mode control method was used to construct a nonlinear disturbance observer to estimate a mismatched disturbance [17]. However, It needs precise information about both the nonlinearity and the system state.

Recently, the EID approach was applied to deal with the nonlinearities in a nonlinear system. A conventional EID estimator was used to reject disturbances for a system with a state-dependent nonlinearity [18]. And the method was employed in an RCS to compensate for a nonlinearity using a conventional EID estimator [19]. However, it requires that the state of a plant is available and imposes conditions on system design, for example, a restrictive commutative condition [20] was needed for system analysis [21].

In this section, we combine the 2D RC and the EID approach to deal with a plant containing a time-varying nonlinearity, and present a new configuration that yields not only high precision tracking for periodic signals but also satisfactory rejection performance for aperiodic disturbances.

Consider the following nonlinear system

$$\begin{cases} \dot{x}_p(t) = Ax_p(t) + Bu(t) + K_f f[x_p(t), d(t), t] + B_d d(t), \\ y(t) = Cx_p(t). \end{cases} \tag{25}$$

A comparison between (25) and (5) show that the difference between these two models is the time-varying nonlinear term $f[x_p(t), d(t), t] \in \mathbb{R}^{n_f}$ in (25).

The following assumptions are made for (25).

Assumption 3 $d(t)$ is unknown and bounded, that is,

$$\|d(t)\| \leq d_M, \ \forall t \geq 0, \tag{26}$$

where d_M is a positive number.

Assumption 4 For $t \geq 0$, and any $x_p(t)$ and $z_p(t)$ in $\mathbb{R}^n$, the nonlinear function $f[x_p(t), d(t), t]$ satisfies

(1)

$$f[0, d(t), t] = 0; \tag{27}$$

(2)

$$\left\| f[x_p(t), d(t), t] - f[z_p(t), d(t), t] \right\| \leq \left\| U[x_p(t) - z_p(t)] \right\|, \tag{28}$$

where U is a constant matrix.

Many nonlinear systems can be described using the model (25) [14]. Assumption 3 is satisfied in practice. Note that, for (28),

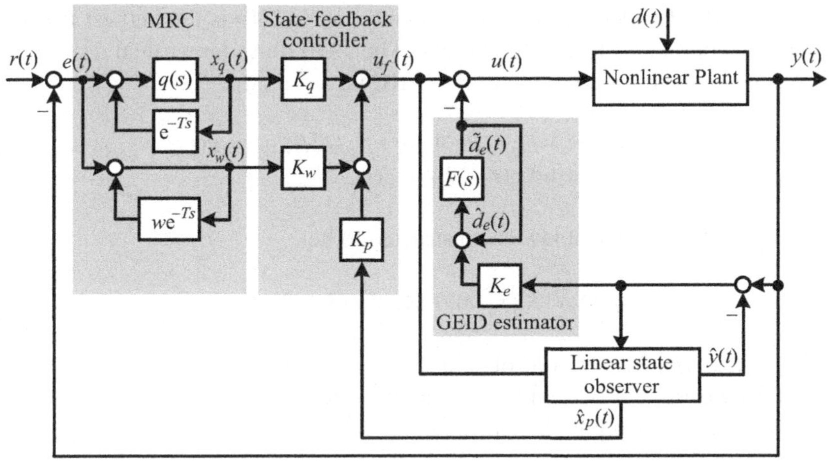

Fig. 7 Configuration of GEID-based nonlinear MRCS

$$\left\| f[x_p(t), d(t), t] - f[z_p(t), d(t), t] \right\| \leq \left\| U[x_p(t) - z_p(t)] \right\| \leq \|U\| \left\| x_p(t) - z_p(t) \right\|.$$

Thus, if we choose $\|U\|$ to be a Lipschitz constant, the Lipschitz condition is satisfied if Condition (28) holds.

The devised nonlinear MRCS (Fig. 7) has five parts: the plant, a linear state observer, a generalized EID (GEID) estimator, a state-feedback controller, and an MRC. The time delay T in the MRC is exactly the period of a reference input. The GEID inserts a gain K_e in the EID estimator to ease the adjustment of disturbance-rejection performance. The MRC has two positive feedback loops: One has a low-pass filter $q(s)$ and the other has a constant gain w [22]. $q(s)$ in the feed-forward path relaxes the stability condition of the system. The loop containing w provides us with a relaxing variable K_w to increase the flexibility in design.

We divide the nonlinear plant (25) into two parts: linear and nonlinear. Then, we take the linear part as an ideal system and the nonlinear part to be an artificial disturbance, and use the EID approach to estimate and compensate for the artificial and actual disturbances. Based on this strategy, we used a linear state observer that has exactly the form (7) in this study.

Without loss of generality a low-pass filter $q(s)$ in the MRC is chosen to be

$$q(s) = \frac{\omega_q}{s + \omega_q}, \tag{29}$$

where ω_q is the cutoff angular frequency for reference tracking. It is selected such that

$$|q(j\omega)| \approx 1, \ \forall \omega \in [0, \omega_r] \tag{30}$$

to ensure good reference-tracking performance. ω_r in (30) is the highest frequency of a reference input. ω_q is usually chosen to be 5–10 times larger than ω_r.

The state-space representation of the MRC is

$$\begin{cases} \dot{x}_q(t) = -\omega_q x_q(t) + \omega_q x_q(t-T) + \omega_q e(t), \\ x_w(t) = w x_w(t-T) + e(t). \end{cases} \tag{31}$$

A state-feedback control law is constructed to be

$$u_f(t) = K_q x_q(t) + K_w x_w(t) + K_p \hat{x}_p(t), \tag{32}$$

where K_q, K_p, and K_w are control gains.

The GEID estimator is used to estimate the effect of the nonlinearity and disturbance on the system output. An EID estimate is

$$\hat{d}_e(t) = K_e[y(t) - \hat{y}(t)] + u_f(t) - u(t), \tag{33}$$

where K_e is the gain of the GEID estimator.

The low-pass filter, $F(s)$, in the GEID is used to select the frequency band for the estimation of the nonlinearity and disturbance. Similar to the design of $q(s)$, the cutoff frequency of $F(s)$ is chosen to be 5–10 times larger than the highest frequency for the compensation of the nonlinearity and disturbance, ω_d. A first-order filter is usually a proper choice [10].

Incorporating $\tilde{d}_e(t)$ into the feedback control law yields the following improved control law:

$$u(t) = u_f(t) - \tilde{d}_e(t). \tag{34}$$

4.1 Analysis and Design of Nonlinear MRCS

Substituting (8) into (7) yields

$$\dot{\hat{x}}_p(t) = A\hat{x}_p(t) + LC\Delta x_p(t) + Bu_f(t). \tag{35}$$

Combining (25), (7), (16), and (34) yields

$$\Delta\dot{x}_p(t) = (A - LC)\Delta x_p(t) - BC_F x_F(t) + K_f f[x_p(t), d(t), t] + B_d d(t). \tag{36}$$

And combining (33), (16), and (34) gives

$$\dot{x}_F(t) = B_F K_e C\Delta x_p(t) + (A_F + B_F C_F)x_F(t). \tag{37}$$

Apply the lifting technique to the MRCS and define

$$\varphi(k,\tau) = [x_p^{\mathrm{T}}(k,\tau),\ \ x_q^{\mathrm{T}}(k,\tau),\ \ \Delta x_p^{\mathrm{T}}(k,\tau),\ \ x_F^{\mathrm{T}}(k,\tau)]^{\mathrm{T}}. \tag{38}$$

Taking (8) and (31) into consideration, we rewrite the 2D repetitive control law to be

$$u_f(k,\tau) = \tilde{K}_p x_p(k,\tau) + K_q x_q(k,\tau) - K_p \Delta x_p(k,\tau) + wK_w x_w(k-1,\tau) + K_w r(k,\tau), \tag{39}$$

where

$$\tilde{K}_p = K_p - K_w C. \tag{40}$$

Since the reference input, $r(k,\tau)$, is periodic, that is, $r(k,\tau) = r(k-1,\tau),\ \forall \tau \in [0,\ T],\ k \in \mathbb{Z}_+$, we can adjust the learning action by K_w and the control action by K_p and K_q. This shows the advantage of formulating the design problem in the 2D space. Although the control and learning actions cannot be adjusted completely independently due to the coupling between them, which is shown in (40), they can be preferentially adjusted by introducing two turning parameters. This provides us with a way of improving control performance.

Rearranging (31), (35)–(37), together with the 2D control law (39), we obtain a 2D model for the closed-loop MRCS in Fig. 7:

$$\begin{cases} \begin{bmatrix} \dot{\varphi}(k,\tau) \\ x_w(k,\tau) \end{bmatrix} = \begin{bmatrix} \overline{A}_{cl} & \overline{B}_w \\ -\overline{C} & w \end{bmatrix} \begin{bmatrix} \varphi(k,\tau) \\ x_w(k-1,\tau) \end{bmatrix} + \begin{bmatrix} h_r(k,\tau) \\ r(k,\tau) \end{bmatrix} + \begin{bmatrix} \overline{A}_d & 0 \\ 0 & 0 \end{bmatrix} \begin{bmatrix} \varphi(k-1,\tau) \\ x_w(k-2,\tau) \end{bmatrix} \\ \qquad + \begin{bmatrix} \overline{F}_f \\ 0 \end{bmatrix} f[x(k,\tau), d(k,\tau), k, \tau] + \begin{bmatrix} \overline{B}_d \\ 0 \end{bmatrix} d(k,\tau), \\ y(k,\tau) = \begin{bmatrix} \overline{C} & 0 \end{bmatrix} \begin{bmatrix} \varphi(k,\tau) \\ x_w(k-1,\tau) \end{bmatrix}, \end{cases} \tag{41}$$

where

$$h_r(k,\tau) = \begin{bmatrix} r^{\mathrm{T}}(k,\tau) K_w^{\mathrm{T}} B^{\mathrm{T}} & \omega_q r^{\mathrm{T}}(k,\tau) & 0 & 0 \end{bmatrix}^{\mathrm{T}},$$

$$\overline{A}_{cl} = \begin{bmatrix} A + B\tilde{K}_p & BK_q & -BK_p & -BC_F \\ -\omega_c C & -\omega_c & 0 & 0 \\ 0 & 0 & A - LC & -BC_F \\ 0 & 0 & B_F K_e C & A_F + B_F C_F \end{bmatrix},$$

$$\overline{K}_f = \begin{bmatrix} K_f \\ 0 \\ K_f \\ 0 \end{bmatrix},\quad \overline{B}_d = \begin{bmatrix} B_d \\ 0 \\ B_d \\ 0 \end{bmatrix},\quad \overline{A}_d = \begin{bmatrix} 0 & 0 & 0 & 0 \\ 0 & \omega_q & 0 & 0 \\ 0 & 0 & 0 & 0 \\ 0 & 0 & 0 & 0 \end{bmatrix},$$

$$\overline{B}_w = \begin{bmatrix} (wBK_w)^{\mathrm{T}} & 0 & 0 & 0 \end{bmatrix}^{\mathrm{T}},\quad \overline{C} = \begin{bmatrix} C & 0 & 0 & 0 \end{bmatrix}.$$

To make system design simple, we focus on the stability of the closed-loop MRCS so as to use the separation principle to design $\{K_q,\ K_w,\ K_p\}$ and $\{L,\ K_e\}$ independently. The linear-matrix-inequality (LMI) technique is used to derive two theorems for the design of those parameters: One is used to obtain $\{K_q,\ K_w,\ K_p\}$ and the other is used to obtain $\{L,\ K_e\}$ for known $\{K_q,\ K_w,\ K_p\}$.

Assuming that the nonlinearity and disturbance are completely suppressed, we can consider a simplified subsystem that only has $\left[x_p^{\mathrm{T}}(k,\tau)\ x_q^{\mathrm{T}}(k,\tau)\ x_w^{\mathrm{T}}(k,\tau)\right]^{\mathrm{T}}$:

$$\begin{cases} \dot{x}_p(k,\tau) = Ax_p(k,\tau) + Bu(k,\tau), \\ \dot{x}_q(k,\tau) = -\omega_q x_q(k,\tau) + \omega_q x_q(k-1,\tau) - \omega_q C x_p(k,\tau), \\ x_w(k,\tau) = w x_w(k-1,\tau) - C x_p(k,\tau), \end{cases} \tag{42}$$

and the control input is

$$u(k,\tau) = \tilde{K}_p x_p(k,\tau) + K_q x_q(k,\tau) + w K_w x_w(k-1,\tau). \tag{43}$$

The following theorem is devised for the design of $\{K_q,\ K_w,\ K_p\}$ for the system (42).

Theorem 5 *For a given cutoff frequency ω_q, a constant w (< 1), and positive scalars α and β, if there exist symmetric positive-definite matrices X, Y, and Z, and arbitrary matrices W_2 and W_3 of appropriate dimensions such that the following LMI holds:*

$$\begin{bmatrix} \Lambda_{11} & \alpha\hat{A}_d X & \beta w B_K W_3 - w\alpha X\hat{C}^{\mathrm{T}} & -\alpha X\hat{C}^{\mathrm{T}} \\ \star & -Y & 0 & 0 \\ \star & \star & \beta\left(w^2-1\right)Z & 0 \\ \star & \star & \star & -\beta Z \end{bmatrix} < 0, \tag{44}$$

where

$$\Lambda_{11} = \alpha\hat{A}X + \alpha X\hat{A} + \alpha B_K W_2 + \alpha W_2^{\mathrm{T}} B_K^{\mathrm{T}} + Y,$$
$$\hat{A} = \begin{bmatrix} A & 0 \\ -\omega_q C & -\omega_q \end{bmatrix},\quad B_K = \begin{bmatrix} B \\ 0 \end{bmatrix},\quad \hat{A}_d = \begin{bmatrix} 0 & 0 \\ 0 & \omega_q \end{bmatrix},\quad \hat{C} = \begin{bmatrix} C & 0 \end{bmatrix},$$

then the subsystem (42) is asymptotically stable. Furthermore, the gains are given by

$$K_q = W_2 X^{-1}\left[0\ 1\right]^{\mathrm{T}},\quad K_w = W_3 Z^{-1},\quad K_p = K_w C + W_2 X^{-1}\left[I_n\ 0\right]^{\mathrm{T}}. \tag{45}$$

The proof of the above Theorem 5 follows immediately by taking

$$\begin{aligned} V(k,\tau) =& \psi^{\mathrm{T}}(k,\tau)(\alpha X)^{-1}\psi(k,\tau) + \int_{\tau-T}^{\tau} \psi^{\mathrm{T}}(k,s)Y^{-1}\psi(k,s)\mathrm{d}s \\ &+ \int_{\tau-T}^{\tau} x_w^{\mathrm{T}}(k,s)(\beta Z)^{-1}x_w(k,s)\mathrm{d}s \end{aligned}$$

as a Lyapunov functional candidate for the subsystem (42). Thus, it is omitted.

For $\{K_q,\ K_w,\ K_p\}$ selected using Theorem 5, we present a stability condition for the closed-loop system in Fig. 7. For convenience, we rewrite

$$\overline{A}_{cl} = \overline{A}_L - \overline{L}\,\overline{C}_L, \tag{46}$$

where

$$\overline{A}_L = \begin{bmatrix} A + B\tilde{K}_p & BK_q & -BK_p & -BC_F \\ -\omega_q C & -\omega_q & 0 & 0 \\ 0 & 0 & A & -BC_F \\ 0 & 0 & 0 & A_F + B_F C_F \end{bmatrix},\ \overline{L} = \begin{bmatrix} 0 \\ \tilde{L} \end{bmatrix},\ \tilde{L} = \begin{bmatrix} L \\ -B_F K_e \end{bmatrix},$$
$$\overline{C}_L = \begin{bmatrix} 0\ 0\ C\ 0 \end{bmatrix}$$

and recall the following lemmas.

Lemma 1 ([23]) *Let D and E be real matrices with appropriate dimensions. For any $x,\ y \in \mathbb{R}^n$ and any $\varepsilon > 0$, the following is true:*

$$2x^{\mathrm{T}}DEy \le \varepsilon^{-1}x^{\mathrm{T}}DD^{\mathrm{T}}x + \varepsilon y^{\mathrm{T}}E^{\mathrm{T}}Ey. \tag{47}$$

Lemma 2 ([5]) *If there exists a semi-positive definite functional $V(k,\tau)$ for the MRCS in Fig. 7 that is continuous and decreases monotonically in every interval $[kT, (k+1)T)$, $k \in \{0, 1, 2, \ldots\}$, then the MRCS is asymptotically stable.*

Proof From the characteristics of repetitive control, we know that

$$V(k, T) = V(k+1, 0)$$

holds for the MRCS in Fig. 7. So,

$$V(t) = V(kT+\tau) := V(k,\tau)$$

is continuous and decreases monotonically in $[0, \infty)$. The MRCS in Fig. 7 is asymptotically stable [24]. This completes the proof. □

The following theorem is devised to design $\{L,\ K_e\}$.

Theorem 6 *For given cutoff frequency ω_q, positive scalars χ, γ, and μ, and $\{K_q,\ K_w,\ K_p\}$, if there exist symmetrical positive-definite matrices*

$$P = \begin{bmatrix} P_{11} & P_{12} & 0 & 0 \\ \star & P_{22} & 0 & 0 \\ \star & \star & \chi P_{33} & P_{34} \\ \star & \star & \star & \gamma P_{44} \end{bmatrix}, \tag{48}$$

Q, and R, positive numbers $\{\varepsilon_1, \ldots, \varepsilon_5\}$, and an appropriately dimensioned matrix W_1 such that the following LMI holds:

$$\begin{bmatrix} \Xi_{11} & \mu P\overline{A}_d & \Xi_{13} & -\overline{C}^{\mathrm{T}}R & \mu P\overline{K}_f & \mu P\overline{B}_d & \overline{C}^{\mathrm{T}}R & 0 & \mu P & \varepsilon_1\overline{U}^{\mathrm{T}} \\ \star & -Q & 0 & 0 & 0 & 0 & 0 & 0 & 0 & 0 \\ \star & \star & \Xi_{33} & 0 & 0 & 0 & 0 & wR & 0 & 0 \\ \star & \star & \star & -R & 0 & 0 & 0 & 0 & 0 & 0 \\ \star & \star & \star & \star & -\varepsilon_1 & 0 & 0 & 0 & 0 & 0 \\ \star & \star & \star & \star & \star & -\varepsilon_2 & 0 & 0 & 0 & 0 \\ \star & \star & \star & \star & \star & \star & -\varepsilon_3 & 0 & 0 & 0 \\ \star & \star & \star & \star & \star & \star & \star & -\varepsilon_4 & 0 & 0 \\ \star & \star & \star & \star & \star & \star & \star & \star & -\varepsilon_5 & 0 \\ \star & \star & \star & \star & \star & \star & \star & \star & \star & -\varepsilon_1 \end{bmatrix} < 0, \tag{49}$$

then the closed-loop MRCS (41) (Fig. 7) is bounded-input bounded-output (BIBO) stable for bounded reference input $r(t)$ as well as disturbance $d(t)$. Moreover, it is asymptotically stable for $r(t) = 0$ and $d(t) = 0$. Furthermore, the gains of state observer and GEID estimator are

$$L = \begin{bmatrix} I_n & 0 \end{bmatrix} P_2^{-1} W_1, \quad K_e = -B_F^{-1} \begin{bmatrix} 0 & 1 \end{bmatrix} P_2^{-1} W_1. \tag{50}$$

The matrices in (49) and (50) are

$$\begin{cases} \Xi_{11} = \mu P\overline{A}_L + \mu\overline{A}_L^{\mathrm{T}}P - \mu\overline{W}_1\overline{C}_L - \mu\overline{C}_L^{\mathrm{T}}\overline{W}_1^{\mathrm{T}} + Q, \quad \Xi_{13} = \mu P\overline{B}_w - w\overline{C}^{\mathrm{T}}R, \\ \Xi_{33} = (w^2-1)R, \quad \overline{U} = \begin{bmatrix} U^{\mathrm{T}} & 0 & 0 & 0 \end{bmatrix}^{\mathrm{T}}, \quad \overline{W}_1 = \begin{bmatrix} 0 & W_1^{\mathrm{T}} \end{bmatrix}^{\mathrm{T}}, \quad P_2 = \begin{bmatrix} 0 & I \end{bmatrix} P \begin{bmatrix} 0 & I \end{bmatrix}^{\mathrm{T}}. \end{cases} \tag{51}$$

Proof Choose a Lyapunov functional candidate to be

$$\begin{aligned} V(k,\tau) =& \varphi^{\mathrm{T}}(k,\tau)\mu P\varphi(k,\tau) + \int_{\tau-T}^{\tau} \varphi^{\mathrm{T}}(k,s)Q\varphi(k,s)\mathrm{d}s \\ &+ \varepsilon_1 \int_0^{\tau} \left[\|Ux(k,s)\|^2 - \|f(x(k,s),d(k,s),k,s)\|^2\right]\mathrm{d}s \\ &+ \int_{\tau-T}^{\tau} x_w^{\mathrm{T}}(k,s)Rx_w(k,s)\mathrm{d}s. \end{aligned} \tag{52}$$

The derivative of $V(k,\tau)$ along (41) is

$$\begin{aligned} \frac{dV(k,\tau)}{d\tau} \le& 2\mu\varphi^{\mathrm{T}}(k,\tau)P\dot{\varphi}(k,\tau) - \varphi^{\mathrm{T}}(k-1,\tau)Q\varphi(k-1,\tau) + \varphi^{\mathrm{T}}(k,\tau)Q\varphi(k,\tau) \\ &+ \varepsilon_1\left(\left\|Ux_p(k,\tau)\right\|^2 - \left\|f(x_p(k,\tau),d(k,\tau),k,\tau)\right\|^2\right) \\ &+ x_w^{\mathrm{T}}(k,\tau)Rx_w(k,\tau) - x_w^{\mathrm{T}}(k-1,\tau)Rx_w(k-1,\tau) \\ \le& \eta^{\mathrm{T}}(k,\tau)\Phi\eta(k,\tau) + \varepsilon_2 d^{\mathrm{T}}(k,\tau)d(k,\tau) \\ &+ (\varepsilon_3+\varepsilon_4)r^{\mathrm{T}}(k,\tau)r(k,\tau) + \varepsilon_5 h_r^{\mathrm{T}}(k,\tau)h_r(k,\tau) + r^{\mathrm{T}}(k,\tau)Rr(k,\tau), \end{aligned} \tag{53}$$

where

$$\eta(k,\tau) = \left[\varphi^{\mathrm{T}}(k,\tau)\ \varphi^{\mathrm{T}}(k-1,\tau)\ x_w^{\mathrm{T}}(k-1,\tau)\right]^{\mathrm{T}},$$

$$\Phi = \begin{bmatrix} \Phi_{11} & P\overline{A}_d & P\overline{B}_w - w\overline{C}^{\mathrm{T}}R \\ \star & -Q & 0 \\ \star & \star & (w^2-1)R + \frac{1}{\varepsilon_4}w^2R^{\mathrm{T}}R \end{bmatrix},$$

$$\begin{aligned}\Phi_{11} = &\mu P\overline{A}_L + \mu\overline{A}_L^{\mathrm{T}}P - \mu P\overline{L}\,\overline{C}_L - \mu\overline{C}_L^{\mathrm{T}}\overline{L}^{\mathrm{T}}P + Q + \varepsilon_1\overline{U}^{\mathrm{T}}\overline{U} + \frac{\mu^2}{\varepsilon_1}P\overline{K}_f\overline{K}_f^{\mathrm{T}}P \\ &+ \frac{\mu^2}{\varepsilon_2}P\overline{B}_d\overline{B}_d^{\mathrm{T}}P + \frac{1}{\varepsilon_3}\overline{C}^{\mathrm{T}}R^{\mathrm{T}}R\overline{C} + \frac{\mu^2}{\varepsilon_5}PP + \overline{C}^{\mathrm{T}}R\overline{C},\end{aligned} \tag{54}$$

$\{\varepsilon_2, \ldots, \varepsilon_5\}$ are positive scalars required for Lemma 1.

According to Lemma 2, if

$$\Phi < 0 \tag{55}$$

holds, and $r(k,\tau)$ and $d(k,\tau)$ are bounded, then the MRCS (41) is BIBO stable [25]. If $r(k,\tau) = 0$ and $d(k,\tau) = 0$, then the MRCS (41) is asymptotically stable.

Define

$$P_2 = \left[\,0\ I\,\right]P\left[\,0\ I\,\right]^{\mathrm{T}},\ \ \overline{W}_1 = P\overline{L},\ \ W_1 = P_2\tilde{L}. \tag{56}$$

Note that

$$\overline{W}_1 = \left[\,0\ W_1^{\mathrm{T}}\,\right]^{\mathrm{T}}. \tag{57}$$

Applying Schur complement argument [26] to inequality (55) yields (49). Finally, $\tilde{L}$, that is. L and K_e in (50), is obtained from (56).

This completes the proof. □

Regarding the relationship between Theorems 5 and 6, we have the follows.

Lemma 3 *Theorem 5 is a necessary condition of Theorem 6.*

Proof A necessary condition for LMI (49) to be true is that the following LMI holds:

$$\begin{bmatrix} \Xi_{11} & \mu P\overline{A}_d & \Xi_{13} & -\overline{C}^{\mathrm{T}}R \\ \star & -Q & 0 & 0 \\ \star & \star & \Xi_{33} & 0 \\ \star & \star & \star & -R \end{bmatrix} < 0. \tag{58}$$

Pre- and post-multiplying the matrix on the left side of (58) by

$$\mathrm{diag}\{(\mu P)^{-1}, (\mu P)^{-1}, R^{-1}, R^{-1}\}$$

yields

$$\begin{bmatrix} \hat{\Xi}_{11} & \overline{A}_d(\mu P)^{-1} & \hat{\Xi}_{13} & -(\mu P)^{-1}\overline{C}^{\mathrm{T}} \\ \star & -(\mu P)^{-1}Q(\mu P)^{-1} & 0 & 0 \\ \star & \star & \hat{\Xi}_{33} & 0 \\ \star & \star & 0 & R^{-1} \end{bmatrix} < 0, \tag{59}$$

where

$$\hat{\Xi}_{11} = \overline{A}_{cl}(\mu P)^{-1} + (\mu P)^{-1}\overline{A}_{cl}^{\mathrm{T}} + (\mu P)^{-1}Q(\mu P)^{-1},$$
$$\hat{\Xi}_{13} = \overline{B}_w R^{-1} - w(\mu P)^{-1}\overline{C}^{\mathrm{T}}, \quad \hat{\Xi}_{33} = (w^2 - 1)R^{-1}.$$

Then, pre- and post-multiply the matrix on the left side of (59) by Υ and Υ^{T}:

$$\Upsilon = \begin{bmatrix} I_{n+1} & 0 & 0 & 0 & 0 & 0 \\ 0 & 0 & I_{n+1} & 0 & 0 & 0 \\ 0 & 0 & 0 & 0 & 1 & 0 \\ 0 & 0 & 0 & 0 & 0 & 1 \\ 0 & I_{n+1} & 0 & 0 & 0 & 0 \\ 0 & 0 & 0 & I_{n+1} & 0 & 0 \end{bmatrix}.$$

Let $(\mu P)^{-1} = \mathrm{diag}\{\alpha X, \Sigma_1\}$, $(\mu P)^{-1}Q(\mu P)^{-1} = \begin{bmatrix} Y & \Sigma_2 \\ \star & \Sigma_3 \end{bmatrix}$, and $R^{-1} = \beta Z$, where Σ_i $(i = 1, 2, 3)$ denote compatible matrices. Extracting a $(n+1) \times (n+1)$ square matrix on the upper-left side of the resulting matrix gives LMI (44). This completes the proof. □

4.2 Design Algorithm for Nonlinear MRCS

A design algorithm for the GEID-based nonlinear MRCS, which is based on Theorems 5 and 6, is given as follows.

Algorithm of designing GEID-based nonlinear MRCS:

Step 1: Choose the cutoff frequency ω_q of the filter $q(s)$ according to the desired highest tracking frequency ω_r, and the constant w (< 1).

Step 2: Select ω_d according to the required highest estimation frequency of the overall effect of the nonlinearity and disturbance.

Step 3: Choose rough ranges of tuning parameters, χ, γ, μ, α, and β.

Step 4: Use the grid-search algorithm to perform a search in the ranges given at Step 3 to find a set of parameters and the corresponding $\{K_q,\ K_w,\ K_p\}$ [from (45)] and $\{L,\ K_e\}$ [from (50)] that ensure LMIs (49) and (44), respectively.

Step 5: Check whether or not a performance index (for example, ISE, IAE, ITSE and ITAE) is satisfied. If not, go to Step 3. Otherwise, finish.

Remark 2 α and β in LMI (44) are selected to ensure a feasible solution of LMI (44) and to enable satisfactory dynamical tracking performance. Too large K_q and K_w should be avoided because they may result in big measurement noise. On the other hand, χ and γ in LMI (49) are chosen to ensure a feasible solution of LMI (49) and provide satisfactory disturbance-rejection performance. The trade-off between estimation quality and noise sensitivity should also be taken into consideration in their selection.

4.3 Simulation Verification and Analysis

Assume that the parameters of the plant (25) are

$$\begin{cases} A = \begin{bmatrix} 0 & 1 \\ -1 & -1 \end{bmatrix}, \quad B = \begin{bmatrix} 0 \\ 1 \end{bmatrix}, \quad B_d = \begin{bmatrix} 1 \\ 1.2 \end{bmatrix}, \quad C = \begin{bmatrix} 1 & 1 \end{bmatrix}, \\ K_f = \begin{bmatrix} 0.7 & 0 \\ 0 & 0.8 \end{bmatrix}, \quad f[x_p(t), d(t), t] = \begin{bmatrix} x_{p2}(t) \sin 10t \\ x_{p1}(t) \sin 15t \end{bmatrix}, \quad U = \begin{bmatrix} 1 & 0 \\ 0 & 1 \end{bmatrix}. \end{cases} \tag{60}$$

Consider the problem of tracking the reference input (Fig. 8)

$$r(t) = \begin{cases} |2t + 0.5| - 0.5, & 0 \leq t \leq 0.5, \\ |2t - 2| - 2, & 0.5 \leq t \leq 2, \\ r(t - 2), & \text{otherwise}, \end{cases} \tag{61}$$

and rejecting the following disturbance (Fig. 8)

$$d = 8 \sin 1.5\pi t + 8 \tanh(t - 8) - 8 \tanh(t - 4), \ \ 2\text{ s} \leq t \leq 10\text{ s}. \tag{62}$$

The period of the reference is $T = 2$ s.

Choose

$$w = 0.99999, \quad \omega_c = 100 \text{ rad/s} \tag{63}$$

for reference tracking and

$$\omega_d = 10 \text{ rad/s}, \quad F(s) = \frac{100}{s + 100} \tag{64}$$

for the rejection of the overall effect of the nonlinearity and disturbance.

The grid search algorithm was used to choose the parameters. The search ranges and the lengths of steps for the search were

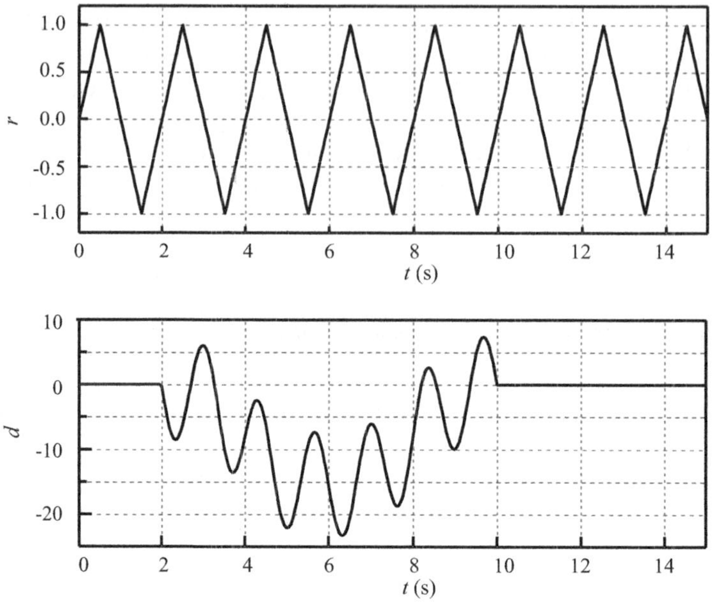

Fig. 8 Reference and disturbance signals

$$\begin{cases} \chi = 1, \quad \gamma \in [0.01 \quad 0.05], \quad \mu \in \left[1 \times 10^{-7} \quad 5 \times 10^{-7}\right], \\ \alpha \in [0.1 \quad 0.5], \quad \beta \in [0.015 \quad 0.025], \\ \Delta\chi = 0, \quad \Delta\gamma = 0.01, \quad \Delta\mu = 1 \times 10^{-7}, \quad \Delta\alpha = 0.1, \quad \Delta\beta = 0.001. \end{cases} \tag{65}$$

The state-feedback gains were restricted to be less than 1250 for feasibility. The integral of absolute error (IAE)

$$J_p = \int_0^t |e(t)| dt \tag{66}$$

is used as the performance index to assess the overall performance of the control system.

The resulting parameters and corresponding performance index were

$$\chi = 1, \ \ \gamma = 1 \times 10^{-7}, \ \ \mu = 0.01, \ \ \alpha = 0.3, \ \ \beta = 0.021, \ \ J_p = 9.4518 \times 10^{-3},$$

and the resulting control gains were

$$\begin{cases} K_q = 1.2218 \times 10^3, \quad K_w = 2.1812, \quad K_p = \left[-9.7188 \quad -7.7618\right], \\ L^{\mathrm{T}} = \left[-34.566 \; 61.357\right], \quad K_e = 1.8023 \times 10^2. \end{cases} \tag{67}$$

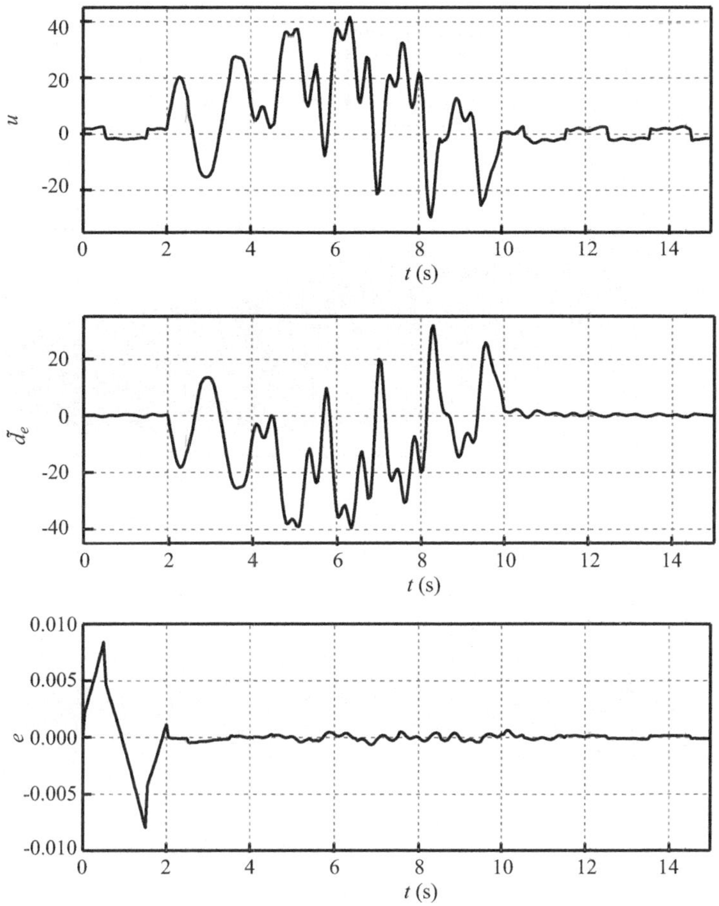

Fig. 9 Time response of GEID-based nonlinear MRCS

The time response of the system is shown in Fig. 9. It is clear from the tracking error $e(t)$ that the overall effect of the nonlinearity and disturbance is effectively estimated and suppressed (less than 1.1469×10^{-3}), and the system has good tracking control performance even in the transient state. Note that a nonlinearity in a system produce high-frequency components on the output of the system [11]. Since the GEID approach almost completely compensated for the nonlinearity in the low-frequency band, the high-frequency effect of the nonlinearity was suppressed to a very low level (Fig. 9). Moreover, the linear part of the closed-loop system, that is, the transfer function matrix from the output of the nonlinearity $K_f f[x_p(t), d(t), t]$ to the input, is

$$G_{\text{linear}}(s) = [sI - A + G_1(s) - G_2(s)G_3(s)]^{-1}, \tag{68}$$

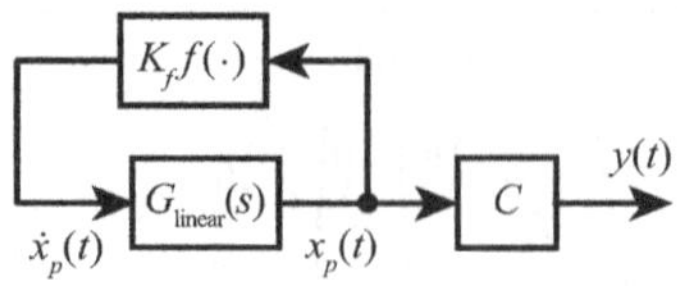

Fig. 10 Connection between $K_f f[x_p(t), d(t), t]$ and $CG_{\text{linear}}(s)$

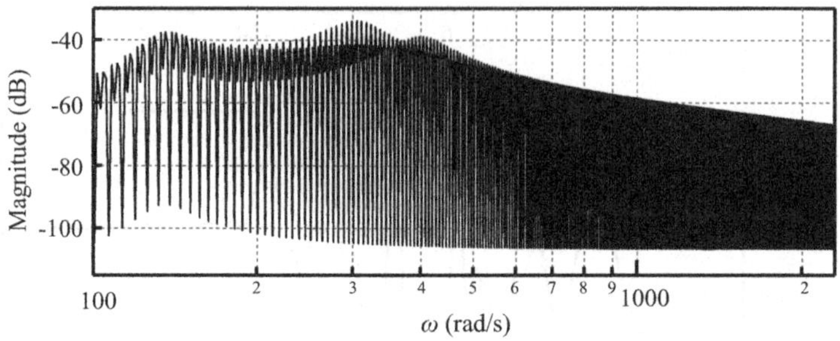

Fig. 11 Bode magnitude plot of (1, 1) entry of $CG_{\text{linear}}(s)$

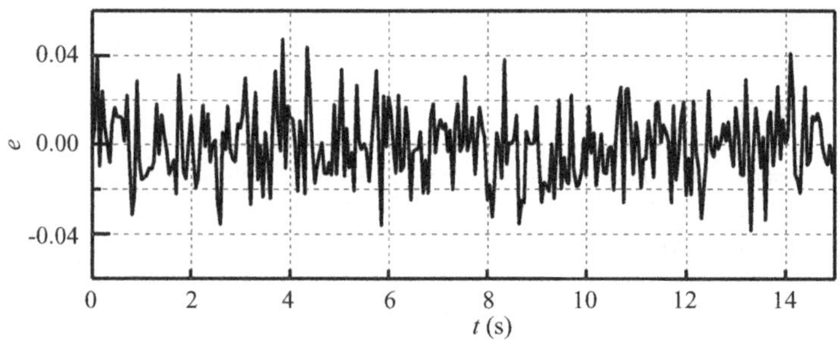

Fig. 12 Tracking error of GEID-based nonlinear MRCS for measurement noise

where

$$\begin{aligned}
&G_1(s) = B\left[G_{MRC}(s) + G_{GEID}(s)\right]C, \quad G_2(s) = B\left[K_p + G_{GEID}(s)C\right],\\
&G_3(s) = \left[sI - \left(A + BK_p\right) + LC\right]^{-1}\left[-BG_{MRC}(s)C + LC\right],\\
&G_{MRC}(s) = q(s)/[1 - q(s)\mathrm{e}^{-Ts}] + K_w/[1 - w\mathrm{e}^{-Ts}],\\
&G_{GEID}(s) = \left[B_F/(s - A_F - B_F)\right]K_e.
\end{aligned}$$

The connection between $K_f f[x_p(t), d(t), t]$ and $CG_{\text{linear}}(s)$ is shown in Fig. 10. Note that $f[x_p(t), d(t), t] \in \mathbb{R}^2$ and $y \in \mathbb{R}$. $CG_{\text{linear}}(s)$ is a 2×1 matrix. Since the two entries have similar frequency characteristics, the Bode magnitude plot of the (1, 1) entry of $CG_{\text{linear}}(s)$ is shown in Fig. 11 as example. Clearly, the gain of the linear part has low-pass nature and the gain in the high-frequency band is very low. Thus, the effect of the nonlinearity at high frequency is suppressed.

Simulations were carried out with band-limited white noise (peak value: ± 0.05) in the output measurement to show the effectiveness of the method in practice. The tracking error (Fig. 12) shows that the method has a good noise-suppression performance.

5 Conclusion

This chapter presented an active disturbance rejection method for an RCS that contains a time-varying nonlinearity. The 2D property of the RCS was explored and was made utilized in the system design to enable a preferential adjustment of control and learning actions. The GEID estimator is incorporated to estimate the overall effect of the nonlinearity and disturbance in the low-frequency band. The combination of the 2D repetitive control and the GEID approach enables both high reference-tracking and disturbance-rejection performance. The grid search was used to find suitable parameters of the control system. Simulation results demonstrate the validity of the method.

References

1. Inoue, T., Nakano, M., Iwai, S.: High accuracy control of a proton synchrotron magnet power supply. IFAC Proc. Vol. **14**(2), 3137C3142 (1981)
2. Hara, S., Yamamoto, Y., Omata, T., Nakano, M.: Repetitive control system: a new type servo system for periodic exogenous signals. IEEE Trans. Autom. Control **33**(7), 659–666 (1988)
3. Tomizuka, M., Tsao, T.-C., Chew, K.-K.: Analysis and synthesis of discrete-time repetitive controllers. J. Dyn. Syst. Meas. Control **111**(3), 353–358 (1988)
4. Wu, M., Zhou, L., She, J.H.: Design of observer-based H_∞ robust repetitive-control system. IEEE Trans. Autom. Control **56**(6), 1452–1457 (2011)
5. She, J.H., Zhou, L., Wu, M., Zhang, J., He, Y.: Design of a modified repetitive-control system based on a continuous-discrete 2D model. Automatica **48**(5), 844–850 (2012)
6. Chen, T., Francis, B.: Optimal Sampled-Data Control Systems. Springer, Berlin (2013)
7. Levine, W.S.: The Control Handbook. CRC Press, Boca Raton (1996)
8. Devasia, S., Chen, D., Paden, B.: Nonlinear inversion-based output tracking. IEEE Trans. Autom. Control **41**(7), 930–943 (1996)
9. She, J.H., Fang, M., Ohyama, Y., Hashimoto, H., Wu, M.: Improving disturbance-rejection performance based on an equivalent-input-disturbance approach. IEEE Trans. Ind. Electron. **55**(1), 380–389 (2008)
10. She, J.H., Xin, X., Pan, Y.: Equivalent-input-disturbance approach-analysis and application to disturbance rejection in dual-stage feed drive control system. IEEE/ASME Trans. Mechatron. **16**(2), 330–340 (2011)
11. Omata, T., Hara, S., Nakano, M.: Nonlinear repetitive control with application to trajectory control of manipulators. J. Robot. Syst. **4**(5), 631–652 (1987)
12. Han, S.H., Kim, Y.H., Ha, I.-J.: Iterative identification of state-dependent disturbance torque for high precision velocity control of servo motors. IEEE Trans. Autom. Control **43**(5), 724–729 (1998)
13. Terro, M.J., Mahmoud, M.S., Abdel-Rohman, M.: Multi-loop feedback control of offshore steel jacket platforms. Comput. Struct. **70**(2), 185–202 (1999)

14. Guo, L., Chen, W.-H.: Disturbance attenuation and rejection for systems with nonlinearity via DOBC approach. Int. J. Robust Nonlinear Control **15**(3), 109–125 (2005)
15. Wei, X.J., Guo, L.: Composite disturbance-observer-based control and H_∞ control for complex continuous models. Int. J. Robust Nonlinear Control **20**(1), 106–118 (2010)
16. Slotine, J.-J.E., Li, W.: Applied Nonlinear Control. Prentice Hall, Englewood Cliffs (1991)
17. Yang, J., Li, S., Yu, X.: Sliding-mode control for systems with mismatched uncertainties via a disturbance observer. IEEE Trans. Ind. Electron. **60**(1), 160–169 (2013)
18. Gao, F., Wu, M., She, J.H., Cao, W.H.: Disturbance rejection in nonlinear systems based on equivalent-input-disturbance approach. Appl. Math. Comput. **282**, 244–253 (2016)
19. Zhou, L., She, J.H., Zhou, S., Li, C.: Compensation for state-dependent nonlinearity in a modified repetitive control system. Int. J. Robust Nonlinear Control **28**(1), 213–226 (2018)
20. Ho, D.W.C., Lu, G.: Robust stabilization for a class of discrete-time nonlinear systems via output feedback: the unified LMI approach. Int. J. Control **76**(2), 105–115 (2003)
21. Yu, P., Wu, M., She, J.H., Liu, K.Z., Nakanishi, Y.: An improved equivalent-input-disturbance approach for repetitive control system with state delay and disturbance. IEEE Trans. Ind. Electron. **65**(1), 521–531 (2018)
22. Yu, P., Wu, M., She, J.H., Lei, Q.: Robust repetitive control and disturbance rejection based on two-dimensional model and equivalent-input-disturbance approach. Asian J. Control **18**(6), 2325–2335 (2016)
23. Wang, Y., Xie, L., de Souza, C.E.: Robust control of a class of uncertain nonlinear systems. Syst. Control Lett. **19**(2), 139–149 (1992)
24. Merkin, D.R.: Introduction to the Theory of Stability. Springer, New York (1997)
25. Khalil, H.K.: Nonlinear Systems. Macmillan, New York (1992)
26. Liu, K.-Z., Yao, Y.: Robust Control: Theory and Applications. Wiley, Hoboken (2016)

Intelligent Control of Underactuated Mechanical System

Xuzhi Lai, Yawu Wang, Chunyi Su, Jinhua She, and Min Wu

Abstract In this chapter, we present the control methods for the multi-link planar underactuated manipulator (PUM) with one passive joint. Firstly, we present a continuous control strategy based on the differential evolution algorithm for the system with passive first joint to quickly realize its position-posture control objective. The angular velocity constraint of the system is used to obtain the coupling equation in integral form between the angle of the passive joint and the states of the active joints. The step PD controllers are used to overcome the sudden change of the initial torques. The differential evolution algorithm is used to optimize the target angles of all joints and the design parameters of the step PD controllers corresponding to the target position-posture. Secondly, we present a unified trajectory planning and tracking control strategy for the multi-link PUMs with a passive joint at a different location. In addition, we employ intelligent optimization algorithm to optimize the parameters of the trajectories. The effectiveness of such strategies is demonstrated through simulation results.

Keywords Multi-link planar underactuated manipulator · Continuous control · Position and posture control · Trajectory planning · Tracking control · Intelligent optimization algorithm

---

X. Lai (✉) · Y. Wang · M. Wu
China University of Geosciences, Wuhan 430074, China
e-mail: laixz@cug.edu.cn

Hubei Key Laboratory of Advanced Control and Intelligent Automation for Complex Systems, Wuhan 430074, China

C. Su
Concordia University, Montreal, Canada

J. She
Tokyo University of Technology, Tokyo, Japan

M. Wu et al. (eds.), *Developments in Advanced Control and Intelligent Automation for Complex Systems*, Studies in Systems, Decision and Control 329,
https://doi.org/10.1007/978-3-030-62147-6_3

1 Introduction

Earlier mechanical systems were basically fully-actuated, that is each degree of freedom (DOF) corresponds to one actuator. Such a fully-actuated mechanical system (FMS) is relatively simple in motion control and scholars have proposed a variety of effective control methods [1–4]. However, when the FMS has one or more broken actuators, it will become an underactuated mechanical system (UMS) [5–7], and the control methods developed for it cannot be used for controlling the UMS. In this case, the control methods specially developed for the UMS can ensure the faulted FMS runs as usual. Besides, the UMS usually has the characteristics of increasing the flexibility of system and reducing costs and energy consumption. Thus, the research on the UMS is of great significance.

An underactuated manipulator (UM) is a typical class of the UMS, and more and more scholars devote themselves to the study of its motion control at present [8, 9]. Generally, the UM can be divided into two types: a vertical UM (VUM) affected by gravity, and a planar UM (PUM) unaffected by gravity.

The VUM has a straight-down equilibrium point and a straight-up one, and its movement space can be divided into a swing-up area and an attraction area. Due to the controllable linear approximation model of VUM at the straight-up equilibrium point [10], lots of scholars have proposed effective methods to achieve rapid swing and stable control of the system [11–14]. That is to say, the control objective in the swing-up area is to converge the states of the system to the attraction area and that in the attraction area is to stabilize the system at the straight-up equilibrium point. At present, the research on the VUM is relatively complete and mature. Since there is no gravity term in the dynamic model of the PUM, its control objective and controllability are different compared to the VUM. Therefore, the existing control methods applicable to the VUM cannot be used to achieve the stable control of the PUM.

For convenience, the PUM with one passive joint is represented to be the $\mathrm{A^mPA^n}$ PUM, where A denotes the active joint, P denotes the passive joint, $m \geq 0, n \geq 0$, $m + n$ is the number of the active joints, and $m + n \neq 0$. According to the integral characteristics, the $\mathrm{A^mPA^n}$ PUM is further divided into three types [15]: a holonomic system, which refers to the PUM with both angular constraint and angular velocity constraint, i.e., a PA ($m = 0, n = 1$) PUM; a first-order nonholonomic system, which refers to the PUM with only angular velocity constraint, i.e., a $\mathrm{PA^n}$ ($m = 0, n > 1$) PUM; a second-order nonholonomic system, which refers to the PUM neither angular velocity constraint nor angular constraint, i.e., an $\mathrm{A^mPA^n}$ ($m \geq 1, n \geq 0$) PUM. This lays a theoretical foundation for the realization of the control objective of the PUMs. The control objective is usually to move its end-point from an initial position to any target position.

For the PA PUM, the angular constraint shows that there is a certain nonlinear mathematical relationship between the passive angle and the active angle, which provides favorable conditions for the conjoint control of the passive angle. Using such a mathematical relationship, the position control of the PA PUM was achieved by controlling the active angle in [16].

For the PA^n ($n > 1$) PUM, the angular velocity constraint shows that the angular velocity of the passive joint can be controlled to zero by controlling that of the active joints to zero. Based on the constraint, [17] proposed a model reduction method for a PAA ($n = 2$) PUM, which reduced the original system to two PA PUMs, so as to achieve the position control via two stages. This method lays a foundation for the control of the PA^n ($n > 2$) PUM. A unified control method has been proposed to successful solve the position control problem of the PA^n ($n > 1$) PUM [18, 19]. The proposed method has very important theoretical value, and overcomes the problems of long control time and complex controller design in model reduction control.

The A^mPA^n ($m \geq 1, n \geq 0$) PUM has neither angular constraint nor angular velocity constraint [20]. Therefore, it is difficult to maintain the stability of the system, let alone controlling the passive angle to its target value. However, such a PUM has the different characteristics when the passive joint is at a different location. So, by analyzing and using the different characteristics, many scholars have proposed various control methods to realize the stable control.

Based on a nilpotent approximation model, [20, 21] proposed an open-loop contracting method to achieve the position control of an AP ($m = 1$, $n = 0$) PUM. Because such a method needs to be iterated, it takes a long time to complete the control. Moreover, using such a method to control the A^mPA^n ($m \geq 1, n \geq 1$) PUM is very complicated. For the AAP ($m = 2$, $n = 0$) PUM, [22] proposed a control method to transform its dynamic model to a model with a second-order chained structure, and used the transformed model to complete its position control. But the control method based on chain structure model is only applicable to the A^mP ($m > 1$, $n = 0$) PUM. Besides, the relationship among the actual input and the virtual control input of the transformed model is quite complicated.

For the APA^n ($n > 2$) PUM, [23] proposed a position control method, where keeping the first active angle being unchanged reduces the original system to a PA^n PUM. However, this control method reduces the reachable area of the APA^n PUM. Then, [24] also reduced an APAA PUM to the PAA PUM by adding a disturbance to attenuate the system energy to zero when the first active angle was controlled to the target value, so as to solve the position control problem. However, the first active angle cannot be guaranteed to converge to its target value when the system energy is attenuated to zero.

All the above control methods are proposed to achieve the position control of the PUMs. But for completing some particular tasks (such as welding and drilling) in practical applications, the posture control of the last link is also needed to be considered while realizing the position control of the end-point, that is the position-posture control objective. Obviously, the position-posture control is more hard than the pure position control because the number of the constraints on the joint angles is increased. Besides, we can see from the above reviews that most of the existing control methods depend on the specific characteristics of the PUM, and there is currently no unified control method that can quickly and effectively solve the control

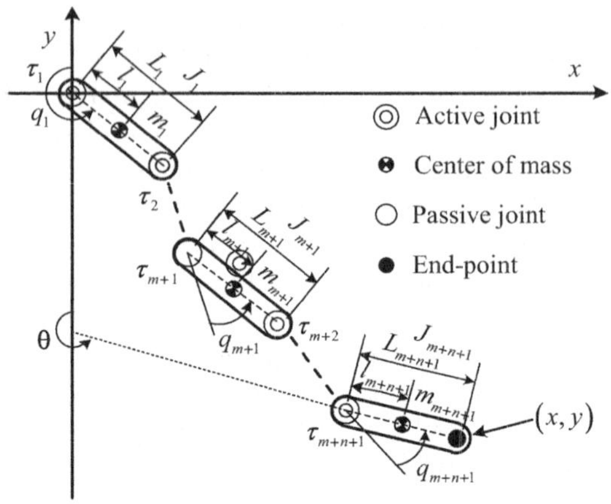

Fig. 1 Physical structure of PA^n ($n > 1$) PUM

problem of the PUM. Therefore, the authors are now working on the position-posture control method and the unified control method for the PUMs.

Next, the dynamic model and kinematic model of the A^mPA^n ($m \geq 0, n \geq 0$, and $m + n \neq 0$) PUM are introduced to prepare for developing the control methods. Then, a continuous position-posture control method is proposed based on the differential evolution (DE) algorithm for the PA^n ($n > 1$) PUM. Also, a unified trajectory planning and tracking control method is presented for the position control of the A^mPA^n ($m \geq 1, n \geq 0$) PUM. Finally, we analyze and summarize the current research results, as well as discuss the future work.

2 Preparations

The physical structure of the A^mPA^n ($m \geq 0, n \geq 0$, and $m + n \neq 0$) PUM is depicted in Fig. 1, where the parameters are
q_i: the angle of the ith joint, ($i = 1, 2, \ldots m + n + 1$)
L_i, m_i: the length and mass of the ith link,
τ_i: the applied torque to the ith joint,
J_i: the moment of inertia of ith link around its center of mass,
l_i: distance between the ith joint and the center of mass of the ith link,
(x, y): the coordinates of the manipulator end-point (MEP),
θ: the posture angle of the manipulator end-link (MEL).

Obviously, from Fig. 1, the position coordinates, (x, y), of the MEP are:

$$\begin{cases} x = -\sum_{i=1}^{\hat{n}} \left(L_i \sin \left(\sum_{j=1}^{i} q_j \right) \right) \\ y = \sum_{i=1}^{\hat{n}} \left(L_i \cos \left(\sum_{j=1}^{i} q_j \right) \right) \end{cases} \tag{1}$$

where $\hat{n} = m + n + 1$ is the number of all links/joints. Define (x_d, y_d) as the target position of the MEP, and then the position control is to stabilize the MEP at (x_d, y_d) from an initial position.

Then, according to Fig. 1, the posture angle, θ, of the MEL is:

$$\theta = \text{rem} \left(\sum_{i=1}^{\hat{n}} q_i, 2\pi \right) \tag{2}$$

where rem (A, B) represents the remainder when A is divided by B, and the sign of θ is same with B. In addition, rem $(A, 2\pi)$ can convert the range of A to $[0, 2\pi)$. Define θ_d as the target posture angle of the MEL, and then the posture control is to stabilize θ at θ_d from an initial posture angle.

According to (1) and (2), when the target position and posture angle of the end-point are given, we can convert the position/posture control objective into the angle control of each joint.

The system dynamic model is

$$M(q)\ddot{q} + H(q, \dot{q}) = \tau \tag{3}$$

where $q \in R^{\hat{n}\times 1}$ is the angle vector, and $\tau \in R^{\hat{n}\times 1}$ is the input torque vector. The passive joint is the $(m+1)$th joint, that is, $\tau_{m+1} = 0$. $M(q) \in R^{\hat{n}\times\hat{n}}$ is the symmetric positive definite matrix, and $H(q, \dot{q})$ is the combination of Coriolis and the centrifugal forces, where $H(q, \dot{q}) = \dot{M}(q)\dot{q} - (1/2)\,\partial\left(\dot{q}^T M(q)\dot{q}\right)/\partial q \in R^{n\times 1}$.

Let $X_1 = q$, $X_2 = \dot{q}$, and $X = \left[X_1^T \;\; X_2^T\right]^T$. Then we give the system state-space equation as follows

$$\begin{cases} \dot{X}_1 = X_2 \\ \dot{X}_2 = G(X)\tau + F(X) \end{cases} \tag{4}$$

where

$$G(X) = \begin{bmatrix} g_{11} & \cdots & g_{1\hat{n}} \\ g_{21} & \cdots & g_{2\hat{n}} \\ \vdots & \ddots & \vdots \\ g_{\hat{n}1} & \cdots & g_{\hat{n}\hat{n}} \end{bmatrix} = M^{-1}(q)$$

$$F(X) = \left[f_1 \; f_2 \cdots f_{\hat{n}} \right]^T = -M^{-1}(q) H(q, \dot{q})$$

We can see from (3) and (4) that the passive joint has no actuator, so it is difficult to converge the passive angle to its target value. The control of the passive angle only can be achieved by the constraint among the states of the active joints and passive one. Thus, next, we use the equation corresponding to the passive joint to analyse the constraint.

The passive part of (1) is

$$M_{p1}\ddot{q}_1 + M_{p2}\ddot{q}_2 + \cdots + M_{p\hat{n}}\ddot{q}_{\hat{n}} + H_p = 0 \tag{5}$$

where $p = m + 1$.

According to (5), the angular velocity and angle of the passive joint are obtained as follows:

$$\begin{cases} \dot{q}_p = -\dot{q}_{p0} - \int_0^T \left(\dfrac{\sum\limits_{r=1,r\neq p}^{\hat{n}} M_{pr}(q)\ddot{q}_r + H_p}{M_{pp}} \right) \mathrm{d}t \\ q_p = -q_{p0} + \int_0^T \dot{q}_p \mathrm{d}t \end{cases} \tag{6}$$

where $r = 1, 2, \ldots \hat{n}$ except p, and q_{p0} and $\dot{q}_{p0}$ are the initial angle and angular velocity of the passive joint, respectively. The constraint among the states of the active joints and passive one is given as (6). According to (6), the motion trajectories of all active joints and their final stable angles will affect the final stable angle of the passive joint.

Specially, for the A^mPA^n PUM, when $m = 0, n > 1$ (i.e., the PA^n PUM), and the initial state of system is static, we can get the angular velocity constraint from (5) as

$$\dot{q}_1 = -\sum_{i=2}^{\hat{n}} \left(\frac{M_{1i}(q)}{M_{11}(q)} \dot{q}_i \right) \tag{7}$$

Equation (7) shows when the states of active joints are static, the states of passive joint are also static.

Further, integrating (7) yields

$$q_1 = -\int_0^t \sum_{i=2}^{\hat{n}} \left(\frac{M_{1i}(q)}{M_{11}(q)} \dot{q}_i \right) \mathrm{d}t - q_{10} \tag{8}$$

Equation (8) shows a coupling relationship in the integral form between the passive angle and the states of all active joints. That is, the integral term of (8) is only affected by the angles and angular velocities of the active joints, but unaffected by the passive joint. Specifically, when all active joints rotate to different angles and finally maintain a stable state, the passive joint will be rotated to diverse angle and maintain a stable state, which provides the possibility of designing a continuous control strategy.

For the A^mPA^n PUM, when $m \neq 0$, although the constraint (6) is not integrable, we can indirectly control passive angle by controlling active joints with (6). The constraint (6) also provides the possibility of designing a unified control strategy for the PUMs.

3 A Continuous Control Method for Planar Underactuated Manipulator with Passive First Joint

In this section, a continuous control strategy is proposed to realize the position-posture control of the PA^n $(n > 1)$ PUM. The key problem to realize the control objective is how to converge all angles to the target values via one stage. Firstly, based on the constraint (8) and the control objective of all joints, a Lyapunov function is constructed and a PD controller is designed. In addition, in order to avoid the sudden changes of the control torques at initial time caused by the application of the PD controller, a step PD controller is used instead of the base PD controller. Next, the target angles of all joints are calculated according to the control objective and the parameters of the step PD controller are optimized by using the DE algorithm, so as to converge all angles to the target values via one stage. Finally, in order to explain the above continuous method more intuitively, we choose a PAAA $(m = 0, n = 3)$ as an example to do the simulation. The simulation comparison experiments indicate that the proposed control strategy is feasible and effective.

3.1 Continuous Controller Design

The control objective is to stabilize all active angles at $q_{2c}, q_{3c}, \ldots, q_{\hat{n}c}$, so we construct a Lyapunov function to be

$$V_1(X) = \sum_{i=2}^{\hat{n}} \left(\frac{P_i}{2} (x_i - x_{ic})^2 \right) + E \tag{9}$$

where P_i are positive constants, $x_{ic} = q_{ic}$. E is the system kinetic energy. Obviously, the angular velocity value of each active joint converges to zero when E attenuates to zero.

Taking the time derivative of (9), we can get

$$\dot{V}_1(X) = \sum_{i=2}^{\hat{n}} \left(x_{2i} \left(P_i (x_i - x_{ic}) + \tau_i \right) \right) \tag{10}$$

Let

$$\tau_i = -P_i(x_i - x_{ic}) - D_i x_{2i} \tag{11}$$

where D_i are positive constants. The PD controller (11) can guarantee

$$\dot{V}_1(X) = -\sum_{i=2}^{\hat{n}} D_i x_{2i}^2 \le 0 \tag{12}$$

We know that $V_1(X)$ monotonically decreases from (12), but there is no guarantee that x_i converges to x_{ic}. Therefore, Lasalle's invariance theorem is applied to the next stability analysis.

A closed-loop system is got by substituting (11) into (4) as follows

$$\dot{X} = F_a(X) \tag{13}$$

According to (12), we can know that $V_1(X)$ is bounded. Define

$$\Phi = \left\{ X \in R^{2\hat{n}} \middle| V_1(X) \le \delta \right\} \tag{14}$$

where δ represents a positive constant. Any solution X of (13) starting in Φ remains in Φ for all $t \ge 0$. We choose

$$\Psi = \left\{ X \in \Phi | \dot{V}_1(X) = 0 \right\} \tag{15}$$

as an invariant set of (13).

According to (12), we know when $\dot{V}_1(X) = 0$, $x_{22} = x_{23} \ldots = x_{2\hat{n}} = 0$. According to (7), $x_{21} = 0$. So, $X_2 = 0$. From (4), we obtain

$$G(X)\tau + F(X) = 0 \tag{16}$$

According to Appendix A of [25], $H(q, \dot{q}) = 0$ when $X_2 = 0$. From the characteristics of the positive definite matrix $M(q)$, we can get $M^{-1}(q) \ne 0$. We also can get $G(X) \ne 0$ and $F(X) = 0$ according to (4). Then,

$$\tau = 0 \tag{17}$$

Substituting (17) into (11), we can get $-P_i(x_i - x_{ic}) = 0$. That is, $x_i = x_{ic}$ $(i = 2, 3, \ldots, \hat{n})$. When all active angles converge to $x_{2c}, x_{3c}, \ldots, x_{\hat{n}c}$, according to (8), the passive angle will converge to x_{1c}. So, the maximal invariant set M of (13) is

$$\mathrm{M} = \left\{ X \in \Psi | x_i = x_{ic} \left(i = 1, 2, \ldots, \hat{n} \right), X_2 = 0 \right\} \tag{18}$$

According to LaSalle's invariance theorem, we can know that every solution X of the closed-loop system (13) starting in Φ approaches to M as $t \to +\infty$. That is, the

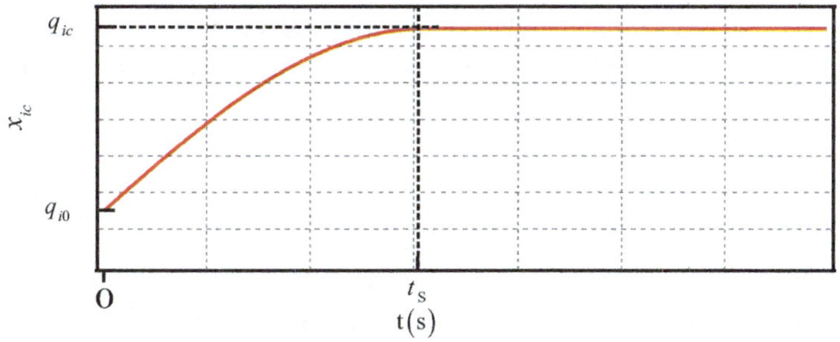

Fig. 2 Mathematical relationship of (19)

angular velocity values of all joints converge to zeroes, and the angles of all joints converge to x_{ic} $(i = 1, 2, \ldots, \hat{n})$.

In addition, the controllers (11) are the basic PD controllers that may cause sudden changes of the control torques at initial time. However, for the real PUM, the maximum output torque of the motor in each active joint is limited, and the initial torque is generally zero. So, in order to prevent such a problem, x_{ic} in (11) is improved to a step form, which is written as follows

$$\begin{cases} x_{ic} = (q_{ic} - q_{i0}) \cdot \sin(\pi t/(2t_s)) + q_{i0}, & 0 \le t \le t_s \\ x_{ic} = q_{ic}, & t > t_s \end{cases} \tag{19}$$

where q_{i0} represents the initial angle of ith joint $(i = 2, 3, \ldots, \hat{n})$. A picture describing the mathematical relationship of (19) is showed in Fig. 2. We can replace the basic PD controllers with the step PD controllers.

According to Fig. 2, it is easy to see that (19) is continuous. So, if we employ (19) in the system, the control torques can be guaranteed to zero at $t = 0$, that is to say, we can effectively overcome the sudden changes of the control torques at initial time.

3.2 Optimization of Target Angles and Design Parameters

In order to rapidly and efficiently realize the position-posture control of the PA^n $(n > 1)$ PUM, we should ensure that when the active angles converge to their target values, the passive angle also converges to its target value. It can be seen from (5) that when the design parameters of the controllers (11) and the target angle values of the active joints are selected differently, the motion trajectory of each active joint will be different, and also the motion trajectory of the passive joint will be different. So, the passive angle will converge to different value.

Considering the above problem, the design parameters of the controllers (11) and the target angle values of the active joints are optimized by employing DE algorithm.

In this way, it can be ensured that the passive joint can be stabilized at its target angle with the active joints stabilizing at their target angles, eventually achieving the control objective of the system.

The evaluation function is selected as

$$h_1 = \left|\text{rem}\left(e_{qz}, 2\pi\right)\right| + |e_x| + \left|e_y\right| \tag{20}$$

where $e_{qz} = \theta - \theta_d$, $e_x = x - x_d$, $e_y = y - y_d$.

The operation rules for mutation, crossover, and selection of the DE algorithm are shown below, respectively

$$\upsilon_k(g+1) = p_m\left[\chi_{r_2}(g) - \chi_{r_3}(g)\right] + \chi_{r_1}(g) \tag{21}$$

$$\varsigma_k(g+1) = \begin{cases} \upsilon_k(g+1), & \eta \le p_c \\ \chi_k(g), & \text{otherwise} \end{cases} \tag{22}$$

$$\chi_k(g+1) = \begin{cases} \varsigma_k(g+1), h(\varsigma_k(g+1)) < h(\chi_k(g)) \\ \chi_k(g), \quad \text{otherwise} \end{cases} \tag{23}$$

where $k = 1, 2, \ldots, N$, and N is the population size; g ($g = 0, 1, \ldots, G$) is the evolution number, and G is the maximum evolution number; p_m and p_c are the mutation and crossover rate, respectively; η is a random number with range [0, 1]; r_1, r_2 and r_3 represent the numbers of the individuals $\chi_{r_1}(g)$, $\chi_{r_2}(g)$ and $\chi_{r_3}(g)$ in $g-$th population, respectively, and $k \ne r_1 \ne r_2 \ne r_3$; $\upsilon_k(g+1)$ and $\varsigma_k(g+1)$ are the mutation result and crossover result, respectively; $h(\varsigma_k(g+1))$ are the evaluation values of $\varsigma_k(g+1)$; $\chi_k(g)$ and $\chi_k(g+1)$ are new individuals in $(g)-$th and $(g+1)-$th population, respectively; and $h(\chi_k(g))$ are the evaluation values of $\chi_k(g)$.

The search range of the target angle of each active joint is $\Omega_1 \in \left[\Omega_1^{\min}, \Omega_1^{\max}\right]$, and the search range of the design parameters of the controllers (11) is $\Omega_2 \in \left[\Omega_2^{\min}, \Omega_2^{\max}\right]$.

Then, the procedures of the DE algorithm is:

Step 1: Initialize the evolution number $g = 0$.

Step 2: Initialize $\chi_k(g)$ randomly to generate the population of N individuals, which contains $q_{ic}^k(g)$ $(i = 2, 3, 4)$, $P_{ic}^k(g)$ and $D_{ic}^k(g)$.

Step 3: Calculate $q_{1c}^k(g)$ by substituting $\chi_k(g)$ into (13).

Step 4: According to (1) and (2), we can use $q_{ic}^k(g)$ and $q_{1c}^k(g)$ to calculate (20).

Step 5: If the value of h_1 is less than ε, where ε is a small positive constant, we can obtain the optimization results: $q_{id} = q_{ic}^k(g)$, $q_{1d} = q_{1c}^k(g)$, $P_i = P_{ic}^k(g)$ and $D_i = D_{ic}^k(g)$, then the procedures end. Otherwise, the procedures continue.

Step 6: By sequentially performing the operation rules for mutation, crossover, and selection based on (21), (22) and (23), we can calculate $\chi_k(g+1)$. After that, $g = g + 1$.

Table 1 Model parameters of PAAA PUM

Link i	L_i/m	m_i/kg	J_i/kg·m^2	l_i/m
1	1.0000	1.0000	0.0833	0.5000
2	0.9500	0.9500	0.0752	0.4750
3	1.0500	1.0500	0.0919	0.5250
4	0.9000	0.9000	0.0675	0.4500

Step 7: If the number of evolutions reaches the maximum, that is $g = G$, stop. Otherwise, return to **Step 3**.

Based on the DE algorithm, not only the optimal target angle of each joint is obtained through calculation, but also the optimal parameters of (11) are obtained. After obtaining the optimization results, whether the control objective of the system can be effectively achieved is required to verify via simulation experiments, which is what we will do next.

3.3 Simulations

This section implements the simulations based on Matlab/Simulink software for verifying the proposed control strategy. The dynamic model and kinematic model of the PA^{n} ($n > 1$) PUM are given in Sect. 2, and we take the PAAA PUM as an example to do the following simulation. Moreover, to better reflect the effectiveness and superiority of the control method which is proposed in this chapter, we will compare the simulation results with those in [25]. At the same time, in order to make the simulation comparison results more convincing, the initial states, target position-posture and model parameters of the PAAA PUM are chosen to be the same as those of [25], and we show them respectively as follows.

$$\begin{cases} (q_{10}, q_{20}, q_{30}, q_{40}) = (0, 0, 0, 0) \text{ rad} \\ (\dot{q}_{10}, \dot{q}_{20}, \dot{q}_{30}, \dot{q}_{40}) = (0, 0, 0, 0) \text{ rad/ s} \\ (x_d, y_d) = (-2.5500, -2.1100) \text{ m} \\ \theta_d = 2.1000 \text{ rad} \end{cases} \tag{24}$$

Different parameter t_s in (19) will have a certain effect on the control performance of the PAAA PUM. If the value of t_s is very big, the system takes longer time to control. Conversely, if the value of t_s is very small, the control torques of the system will become large. Through a large number of simulation experiments, we find that when the value of t_s is in the range of $t_s \in [2, 5]$, the control objective of the PUM can be quickly realized, and at the same time, the control torque of the system is not too large. Therefore, in the simulations, we set $t_s = 4$.

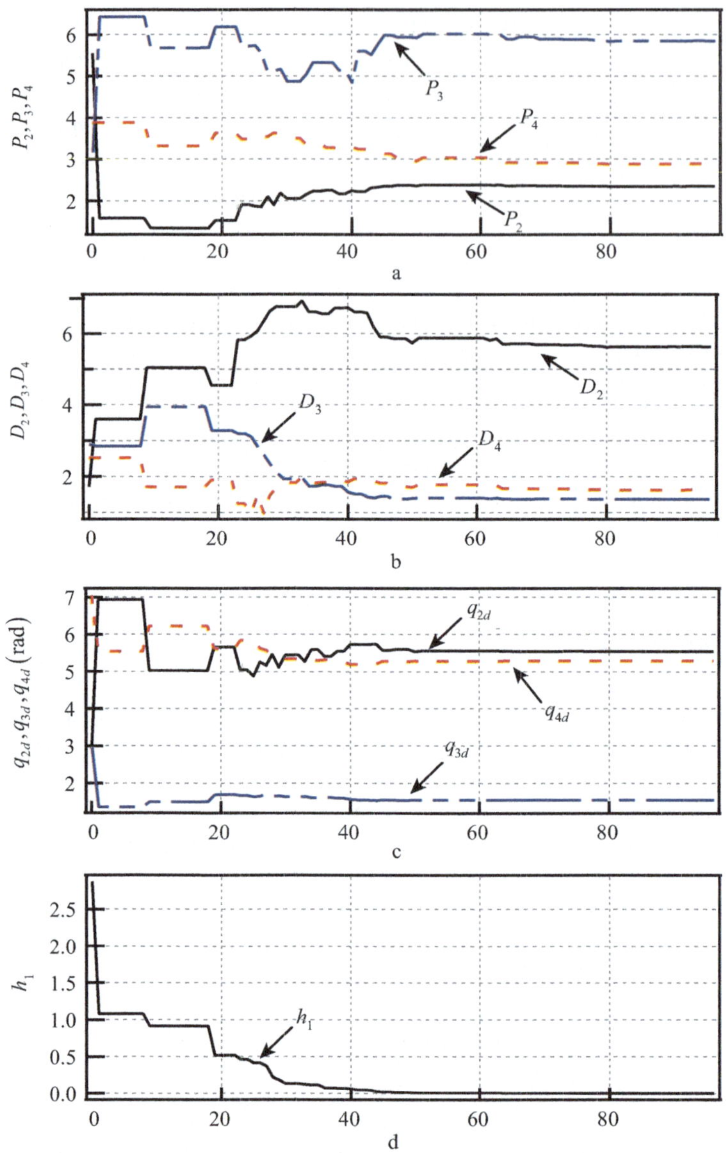

Fig. 3 Results of optimized variables and evaluation function

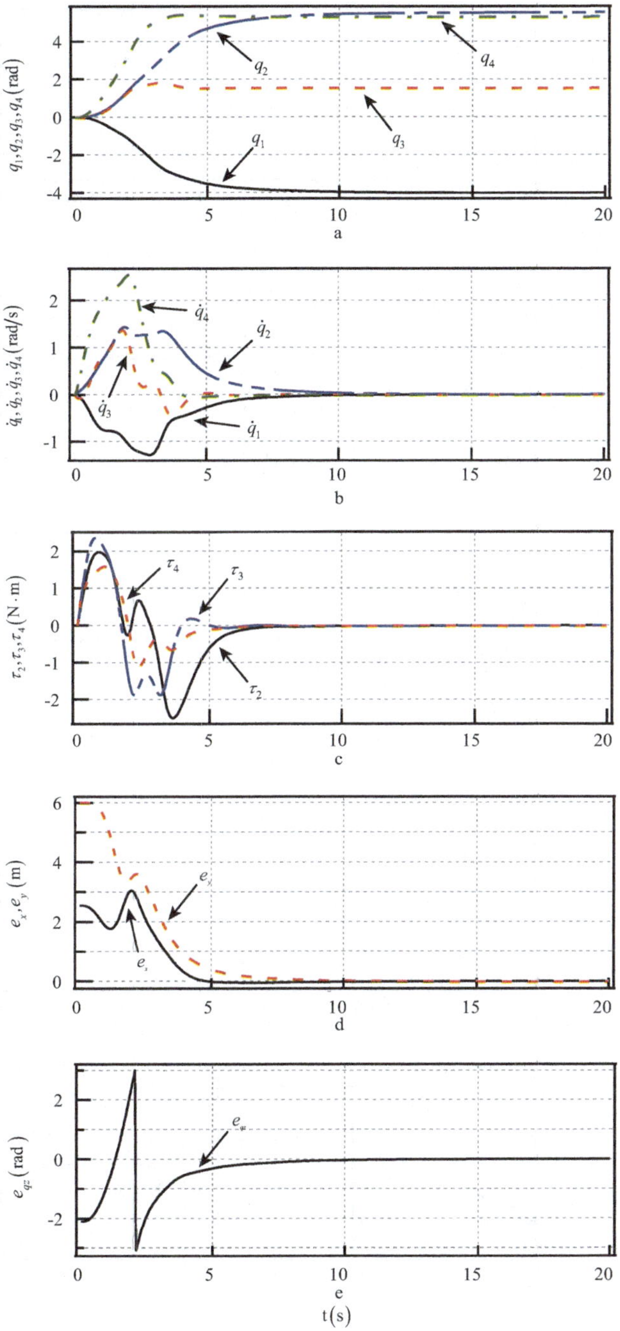

Fig. 4 Simulation results

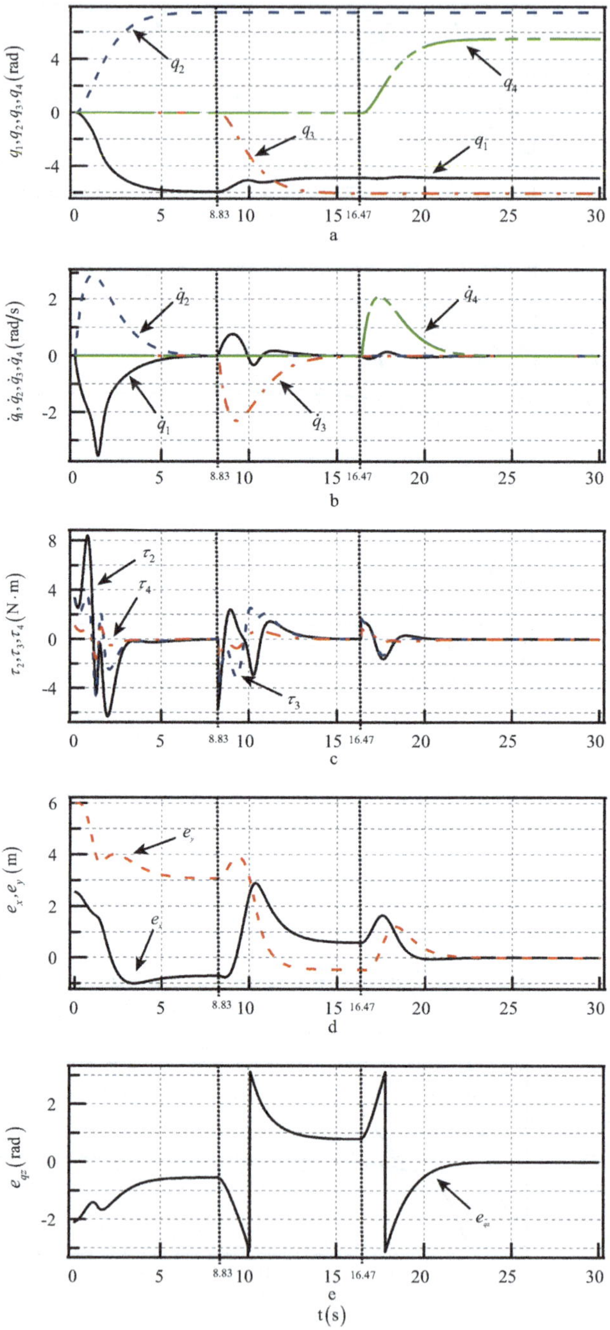

Fig. 5 Simulation results of [25]

The parameters of the DE algorithm are chosen to be $G = 150$, $N = 30$, $\Omega_1^{\min} = -5\pi$, $\Omega_1^{\max} = 5\pi$, $\Omega_2^{\min} = 1$, $\Omega_2^{\max} = 10$, $\varepsilon = 0.001$, $p_c = 0.7$, and $p_m = 0.4$.

In Fig. 3, we can clearly see the optimal target angle of each joint and parameters of (11) calculated by employing the DE algorithm, and changes of the evaluation function. We show the optimization results as follows:

$$\begin{cases} [q_{1d}, q_{2d}, q_{3d}, q_{4d}] = [-4.01, 5.54, 1.55, 5.30]\,\text{rad} \\ [P_2, P_3, P_4] = [2.35, 5.86, 2.89] \\ [D_2, D_3, D_4] = [5.63, 1.38, 1.64] \end{cases} \tag{25}$$

Based on the optimization results, we obtain the simulation results in Fig. 4. It can be seen from Fig. 4 that the angle of each joint smoothly converges to its target value. The angular velocity of each joint, the error of the end-point position, and the error of the posture angle converge to zero, respectively. Meantime, the torque of each active joint also converges to zero, and there is no sudden change in the control torque at initial time due to the application of step PD controllers. It can be seen from the simulation results that the proposed control method for the PAAA PUM is feasible and effective.

We show the simulation results of [25] in Fig. 5. Comparing Figs. 4 and 5, first we can see that each joint is controlled to the target position and stabilized at the target posture angle within 15 s, which is shorter than that in [25]. Secondly, the initial control torques in Fig. 5 are not zero, but the initial control torques in Fig. 4c are zero. This is because the problem of sudden changes of control torques is effectively solved by the application of step PD controllers. Also, the change of the torques is approximately from $-2.5\,\text{N}\cdot\text{m}$ to $2.4\,\text{N}\cdot\text{m}$, which is smaller than that in [25]. In addition, since the control objective in Fig. 4 is achieved by employing the continuous control strategy, the change of the torques is smaller than that in [25] which is achieved by employing the piecewise control strategy. Therefore, for the same position-posture control objective, the continuous control strategy proposed in this chapter requires smaller control torques and shorter time to achieve the control than that in [25]. In a word, this continuous control strategy has more advantages than that in [25].

4 A Unified Control Method for Planar Underactuated Manipulator with One Passive Joint

For the $\text{A}^\text{m}\text{PA}^\text{n}$ $(m \geq 0, n \geq 0, m + n \neq 1)$ PUM, this section presents a unified control approach using the trajectory planning and tracking control for all active joints to solve its position control problem. Firstly, according to the geometric constraint showed in (1), the position control can be transformed to the angle control of all joints. The target values of all angles can be calculated easily using the optimization algorithm (e.g. the DE algorithm in Sect. 3.2). The trajectory of each active joint consists of two parts, where the first part of the trajectory is designed from its initial

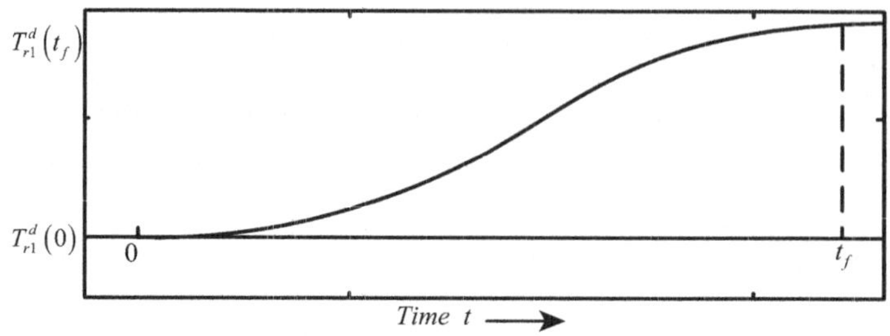

Fig. 6 The trajectory of T_{r1}^d

state to target state and the second one is designed for the realization of the target angle of the passive joint. At the same time, by means of the coupling relationship among the states of the active joints and passive one, the trajectory parameters are optimized to ensure that the angles of the passive joint and active joints reach to their target values simultaneously. Then, the controllers are developed to make the active joints track their trajectories. Simulation results show the effectiveness of the proposed unified strategy.

4.1 Trajectory Planning and Parameters Optimization

As mentioned above, the trajectory for each active joint includes two parts. Define the first part of the trajectory as T_{r1}^d, and

$$T_{r1}^d(t) = \begin{cases} q_{r0} + (q_{rd} - q_{r0})\left(\dfrac{t}{t_f} - \dfrac{1}{2\pi}\sin\left(\dfrac{2\pi t}{t_f}\right)\right) & , 0 \le t \le t_f \\ q_{rd} & , t > t_f \end{cases} \tag{26}$$

where q_{r0} is initial angle, q_{rd} is the target angle, t_f is the control time to achieve the control objective, $r = 1, 2, \ldots, \hat{n}$ except p. The trajectory T_{r1}^d is shown as Fig. 6.

The first and second derivatives of T_{r1}^d on time are

$$\dot{T}_{r1}^d(t) = \begin{cases} \dfrac{(q_{rd} - q_{r0})}{t_f}\left(1 - \cos\left(\dfrac{2\pi t}{t_f}\right)\right) & , 0 \le t \le t_f \\ 0 & , t > t_f \end{cases} \tag{27}$$

$$\ddot{T}_{r1}^d(t) = \begin{cases} \dfrac{(q_{rd} - q_{r0})}{t_f^2}\left(2\pi \sin\left(\dfrac{2\pi t}{t_f}\right)\right) & , 0 \le t \le t_f \\ 0 & , t > t_f \end{cases} \tag{28}$$

When $t = t_f$, we can see from (29) to (28) that

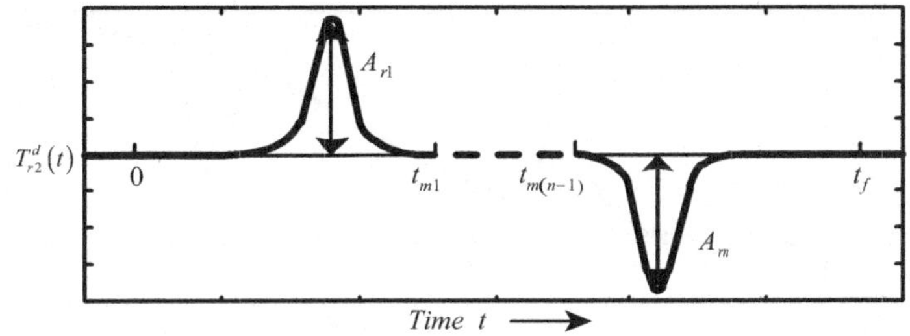

Fig. 7 The trajectory of T_{r2}^{d}

$$\begin{cases} T_{r1}^{d}\left(t_f\right) = q_{rd} \\ \dot{T}_{r1}^{d}\left(t_f\right) = 0 \\ \ddot{T}_{r1}^{d}\left(t_f\right) = 0 \end{cases} \tag{29}$$

which means the active angles are stabilized at their target values $t = t_f$. Then the next problem to be solved is how to stabilize the passive angle at its target value.

According to the constraint among the states of the active joints and passive one (5) and the corresponding analysis in Sect. 2, we know that the passive angle is determined by the motion trajectories of the active joints. Observed by (27) and (28), the state of the passive joint is different with the different t_f, which inspires us that the passive angle may be stabilized at its target value by increasing the adjustable parameters in (27) and optimizing them. So, the following second part of trajectory, defined to be T_{r2}^{d}, is planned.

$$T_{r2}^{d}(t) = A_{r1}\mathrm{sech}(\alpha_1) + A_{r2}\mathrm{sech}(\alpha_2) + \cdots + A_{rn}\mathrm{sech}(\alpha_k) \tag{30}$$

where

$$\begin{cases} \alpha_1 = \dfrac{\lambda\left(2t - t_{m_1}\right)}{t_{m1}} \\ \alpha_2 = \dfrac{\lambda\left(2t - t_{m2} - t_{m1}\right)}{t_{m2} - t_{m1}} \\ \vdots \\ \alpha_{k-1} = \dfrac{\lambda\left(2t - t_{m(k-1)} - t_{m(k-2)}\right)}{t_{m\left(k-1\right)} - t_{m(k-2)}} \\ \alpha_k = \dfrac{\lambda\left(2t - t_f - t_{m(k-1)}\right)}{t_f - t_{m(k-1)}} \end{cases} \tag{31}$$

$A_{r1}, \ldots, A_{rk}, t_{m1}, \ldots, t_{m(k-1)}$ are adjustable parameters and $0 < t_{mi} < t_f$; λ is a large positive constant. The trajectory T_{r2}^{d} is shown as Fig. 7.

The first and second derivatives of T_{r2}^{d} on time are

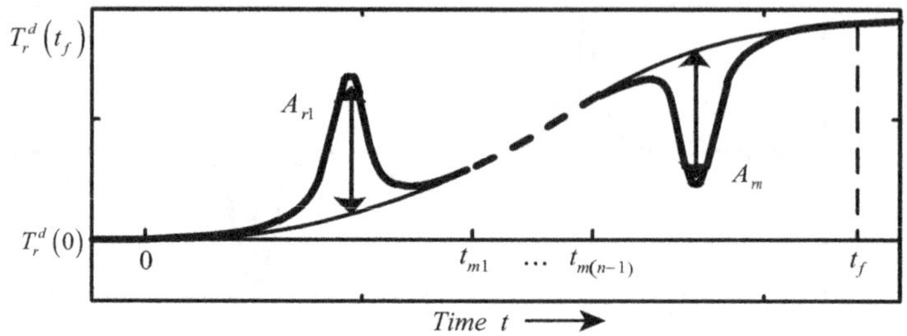

Fig. 8 The trajectory of T_r^d

$$\begin{aligned}\dot{T}_{r2}^d(t) = &-\frac{2\lambda A_{r1}\text{sech}(\alpha_1)\tanh(\alpha_1)}{t_{m1}} - \frac{2\lambda A_{r2}\text{sech}(\alpha_2)\tanh(\alpha_2)}{t_{m2}-t_{m1}} - \cdots \\ &- \frac{2\lambda A_{rn}\text{sech}(\alpha_k)\tanh(\alpha_k)}{t_f - t_{m(k-1)}}\end{aligned} \tag{32}$$

$$\begin{aligned}\ddot{T}_{r2}^d(t) = &\frac{4\lambda^2 A_{r1}\text{sech}(\alpha_1)(\tanh(\alpha_1))^2}{(t_{m1})^2} - \frac{4\lambda^2 A_{r1}(\text{sech}(\alpha_1))^3}{(t_{m1})^2} + \\ &\frac{4\lambda^2 A_{r2}\text{sech}(\alpha_2)(\tanh(\alpha_2))^2}{(t_{m2}-t_{m1})^2} - \frac{4\lambda^2 A_{r2}(\text{sech}(\alpha_2))^3}{(t_{m2}-t_{m1})^2} + \cdots \\ &+ \frac{4\lambda^2 A_{rn}\text{sech}(\alpha_k)(\tanh(\alpha_k))^2}{\left(t_f - t_{m(k-1)}\right)^2} - \frac{4\lambda^2 A_{rk}(\text{sech}(\alpha_k))^3}{\left(t_f - t_{m(k-1)}\right)^2}\end{aligned} \tag{33}$$

The trajectory for each active joint, defined as T_r^d, is superposed by T_{r1}^d and T_{r2}^d, that is

$$T_r^d = T_{r1}^d + T_{r2}^d \tag{34}$$

whose construction is shown as Fig. 8.

Obviously, combining the trajectory (34) and Fig. 8, the parameters t_f, A_{r1}, $A_{r2}, \ldots, A_{rk}$, $t_{m1}, t_{m2}, \ldots, t_{m(k-1)}$ are able to effect the curve of the planned trajectory from $(0, t_f)$, but have no effect on the states of the active joints at $t = t_f$. Thus, we employ the DE algorithm to optimize these parameters, so that the passive angle converges to its target value with the active angles stabilizing at their target values along the planned trajectory.

The evaluation function h_2 is designed to be

$$h_2 = \left|q_p(t_f) - q_{pd}\right| + \left|\dot{q}_p(t_f)\right| \tag{35}$$

where q_{pd} is the target angle of passive joint. The trajectory parameters' optimization is similar with the DE algorithm in the Sect. 3.2.

After we optimize the parameters $A_{r1}, A_{r2}, \ldots, A_{rk}, t_{m1}, t_{m2}, \ldots, t_{m(k-1)}$ and t_f, we can obtain the total optimized trajectory T_r^d. Then, the position control objective can be achieved by designing the controllers to make the active joints track the optimized trajectory T_r^d.

4.2 Trajectory Tracking Controllers Design

Next the tracking controllers are designed for the active joints tracking the optimized trajectory T_r^d. According to the control objectives, the Lyapunov function is designed as

$$V_2(X) = \sum_{r=1,r\neq p}^{\hat{n}} \left(\frac{P_r}{2}(x_r - T_{rd})^2 \right) + \frac{1}{2}x_{2r}^2 \tag{36}$$

where P_r is the positive constant.

Combining the state-space equation (4), $\dot{V}_2(X)$ is

$$\dot{V}_2(X) = \sum_{r=1,r\neq p}^{\hat{n}} \left(x_{2r} \left(P_r (x_r - T_{rd}) + f_r + G_r \tau_r \right) \right) \tag{37}$$

where $G_r = [g_{r1}\ g_{r2}\ \cdots\ g_{r\hat{n}}]$.

We choose

$$\tau_r = (-P_i(x_r - T_{rd}) - f_r - D_r x_{r+4} - T_j) g_{rr}^{-1} \tag{38}$$

where

$$T_j = \sum_{j=1,j\neq r,j\neq p}^{\hat{n}} g_{rj}\tau_j$$

to guarantee

$$\dot{V}_2(X) = -\sum_{r=1,r\neq p}^{\hat{n}} D_r x_{r+4}^2 \leq 0 \tag{39}$$

where D_r are positive constants.

Similar the stability analysis in Sect. 3.1, based on the LaSalle's invariance theorem, both the angles and angular velocities of all active joints can converge to the target states. Due to the parameters in trajectory are optimized based on the evaluation function (35), the passive angle can be stabilized at its target values simultaneously, thereby achieving the position control objective.

4.3 Simulations

Taking four-link PUMs (i.e., $\hat{n} = 4$) with a passive joint at a different position as examples, four simulations are performed to verify the effectiveness and uniformity of the proposed control strategy.

The model parameters of the four-link PUMs are showed in the Table 1.

The initial states and target states are showed in (24). And the target angles are the same as that in (25).

Considering the optimization calculation time and system stability accuracy after many simulation experiments, the parameters used in the trajectory are chosen as $\lambda = 9, k = 2, P_r = 1, D_r = 1.8, A_{r1}, A_{r2} \in [-1.8, 1.8], t_{m1} \in [1, 10]$ and $t_f \in [11, 20]$.

In the first simulation, take the PAAA PUM as example to do the experiment. Based on the target angles (25), the parameters of the planned trajectories, T_{22}^d, T_{32}^d, T_{42}^d, are optimized via the DE algorithm,

$$\begin{cases} A_{21} = -0.1269 \text{ rad} \\ A_{22} = -1.7611 \text{ rad} \\ A_{31} = -0.0526 \text{ rad} \\ A_{32} = 1.4677 \text{ rad} \\ A_{41} = 0.2265 \text{ rad} \\ A_{42} = 0.6482 \text{ rad} \\ t_{m1} = 3.6712 \text{ s} \\ t_f = 11.5937 \text{ s} \end{cases} \tag{40}$$

The simulation results are exhibited in Fig. 9. From it, we can see that the position control objective of the PAAA PUM is achieved at $t = 11.5937$ s.

In the second simulation, take the APAA PUM as example to do the experiment. Based on the target angles (25), the parameters of the planned trajectories, T_{12}^d, T_{32}^d, T_{42}^d, are optimized via the DE algorithm,

$$\begin{cases} A_{11} = -0.6871 \text{ rad} \\ A_{12} = 0.4637 \text{ rad} \\ A_{31} = -1.5414 \text{ rad} \\ A_{32} = 1.4755 \text{ rad} \\ A_{41} = 1.7903 \text{ rad} \\ A_{42} = 0.6175 \text{ rad} \\ t_{m1} = 8.9516 \text{ s} \\ t_f = 17.8045 \text{ s} \end{cases} \tag{41}$$

The simulation results are exhibited in Fig. 10. From it, we can see that the position control objective of the APAA PUM is achieved at $t = 17.8045$ s.

In the third simulation, take the AAPA PUM as example to do the experiment. Based on the target angles by (25), the parameters of the planned trajectories, T_{12}^d, T_{22}^d, T_{42}^d, are optimized via the DE algorithm,

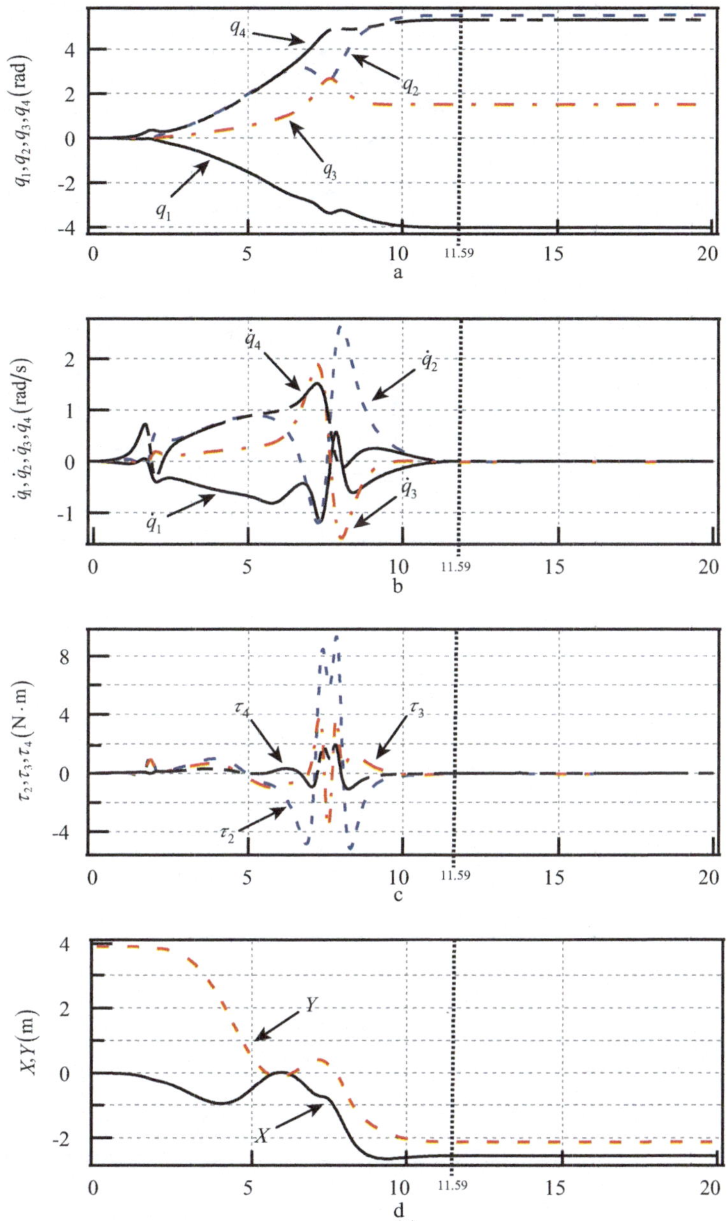

Fig. 9 Simulation results for first simulation

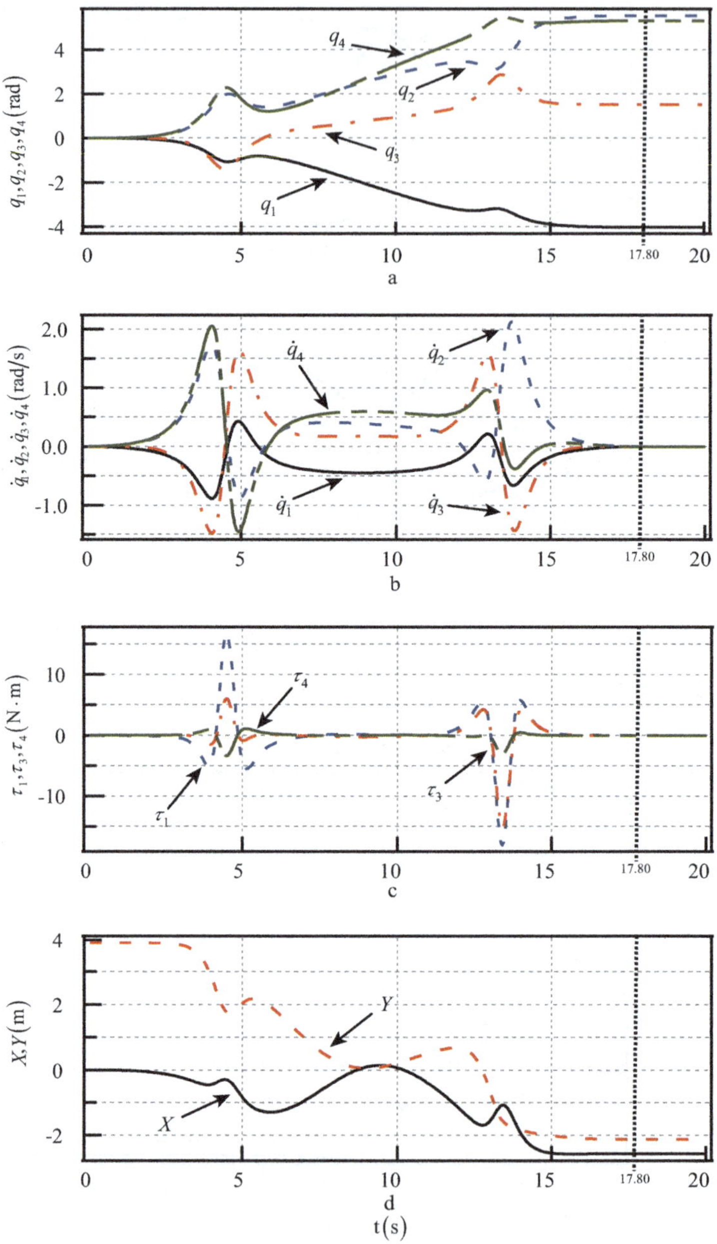

Fig. 10 Simulation results for second simulation

$$\begin{cases} A_{11} = 0.6069 \text{ rad} \\ A_{12} = 0.9054 \text{ rad} \\ A_{21} = 1.1852 \text{ rad} \\ A_{22} = -1.5655 \text{ rad} \\ A_{41} = -1.7418 \text{ rad} \\ A_{42} = 1.7132 \text{ rad} \\ t_{m1} = 7.5489 \text{ s} \\ t_f = 18.5660 \text{ s} \end{cases} \tag{42}$$

The simulation results are exhibited in Fig. 11. From it, we can see that the position control objective of the AAPA PUM is achieved at $t = 18.5660$ s.

In the fourth simulation, take the AAAP PUM as example to do the experiment. Based on the target angles (25), the parameters of the planned trajectories, T_{12}^d, T_{22}^d, T_{32}^d, are optimized via the DE algorithm,

$$\begin{cases} A_{11} = 0.4882 \text{ rad} \\ A_{12} = -1.3886 \text{ rad} \\ A_{21} = 0.0977 \text{ rad} \\ A_{22} = -0.6331 \text{ rad} \\ A_{31} = -1.3802 \text{ rad} \\ A_{32} = -1.4449 \text{ rad} \\ t_{m1} = 4.4128 \text{ s} \\ t_f = 17.9765 \text{ s} \end{cases} \tag{43}$$

The simulation results are exhibited in Fig. 12. From it, we can see that the position control objective of the AAAP PUM is achieved at $t = 17.9765$ s.

All the above simulation results confirm that the presented strategy is effective and uniformity for the $\mathrm{A^mPA^n}$ $(m \geq 0, n \geq 0, m + n \neq 0)$ PUM.

5 Conclusion

In this chapter, we present two continuous control strategies to effectively and quickly achieve the control of the $\mathrm{A^mPA^n}$ $(m \geq 0, n \geq 0, m + n \neq 0)$ PUM. First, using the coupling relationship between the passive joint and active joints and the DE algorithm, a position-posture control method is present for the $\mathrm{PA^n}$ $(m = 0, n > 1)$ PUM with the first-order nonholonomic constraint. Second, a unified position control method is presented based on the trajectory planning and tracking control for the $\mathrm{A^mPA^n}$ $(m \geq 0, n \geq 0, m + n \neq 0)$ PUM.

There are two points that are worth mentioning. On the one hand, the fact that the continuous position-posture control method is developed for the first-order nonholonomic system breaks the traditional viewpoint from most scholars that the stable control of the nonholonomic system cannot be achieved via a continuous method. The control strategy mentioned in this chapter provides a novel method to solve the

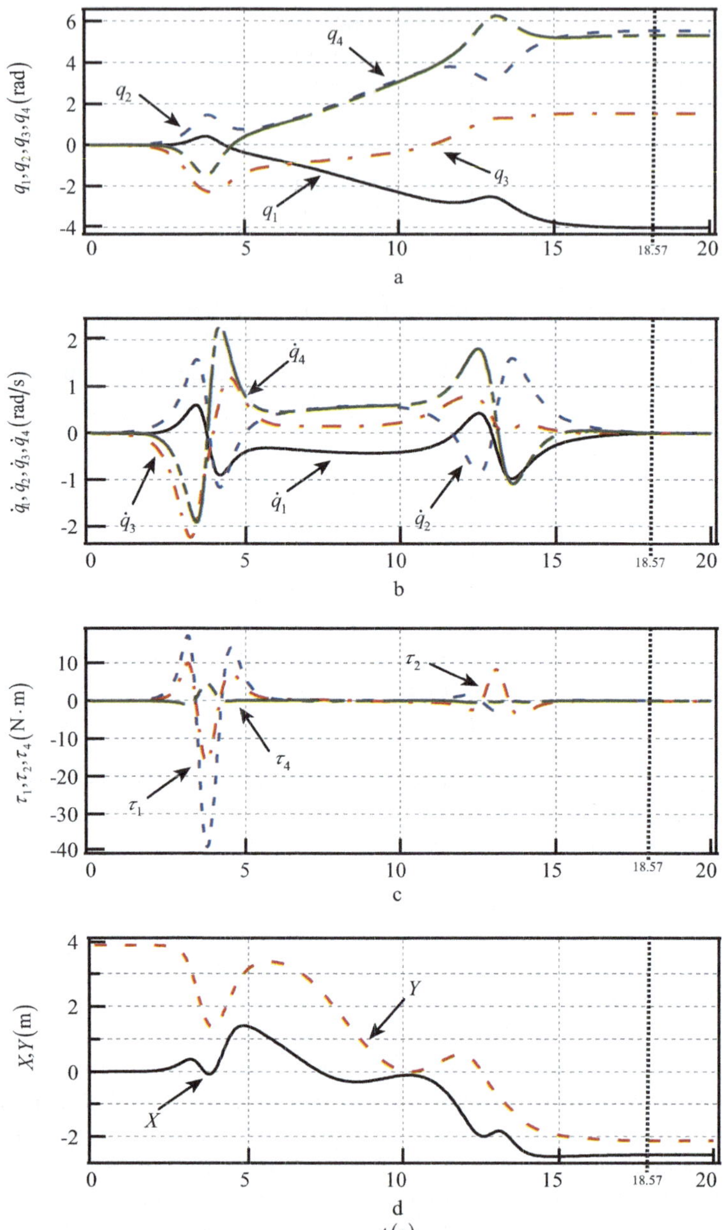

Fig. 11 Simulation results for third simulation

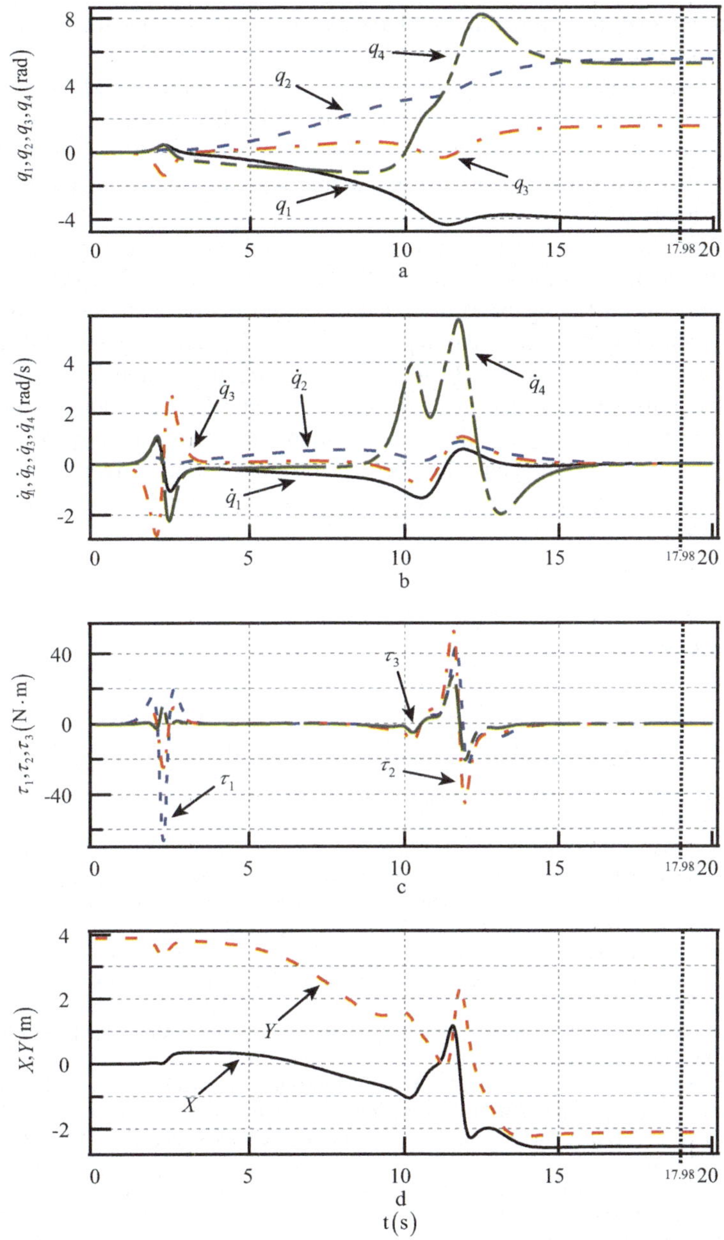

Fig. 12 Simulation results for fourth simulation

control problem of the multi-link PUM and a new research idea to the motion control of nonholonomic system.

On the other hand, the A^mPA^n ($m \geq 0, n \geq 0, m + n \neq 0$) PUM cannot be controlled to be stable via a unified control method. In this chapter, by using the DE algorithm to optimize the trajectory parameters, we successfully obtain the optimal trajectory and smoothly achieve the stable position control of the PUM via a one-stage method.

All in all, the control methods proposed in this chapter are of great significance to the fault operation strategy of the FMS, improve the control theory in the field of the UMS, and fill the gaps in the position-posture control strategy of the PUM. Also, the control methods have a rich practical application background and advanced practical application occasions. Besides, from the numerical simulation experiments using the control strategy we proposed, the good control results can be found, but a physical experimental result will be more convincing. Therefore, in order to further illustrate that the proposed control strategy is also feasible and effective in the actual environment, it is necessary to carry out actual experimental verification in future research work.

References

1. Spong, M.W., Vidyasagar, M.: Robot Dynamics and Control. Wiley, Hoboken (2008)
2. He, W., Huang, H., Ge, S.S.: Adaptive neural network control of a robotic manipulator with time-varying output constraints. IEEE Trans. Cybern. **47**(10), 3136–3147 (2017)
3. Xiao, B., Yin, S.: An intelligent actuator fault reconstruction scheme for robotic manipulators. IEEE Trans. Cybern. **48**(2), 639–647 (2018)
4. He, W., Huang, B., Dong, Y., Li, Z., Su, C.-Y.: Adaptive neural network control for robotic manipulators with unknown deadzone. IEEE Trans. Cybern. **48**(9), 2670–2682 (2018)
5. Yang, C., Li, Z., Li, J.: Trajectory planning and optimized adaptive control for a class of wheeled inverted pendulum vehicle models. IEEE Trans. Cybern. **43**(1), 24–36 (2013)
6. Moreno-Valenzuela, J., Aguilar-Avelar, C., Puga-Guzmán, S.A., Santibáñez, V.: Adaptive neural network control for the trajectory tracking of the Furuta pendulum. IEEE Trans. Cybern. **46**(12), 3439–3452 (2016)
7. Moreno-Valenzuela, J., Aguilar-Avelar, C.: Motion Control of Underactuated Mechanical Systems. Springer, Cham (2018)
8. Cao, V.K., Ngoc, S.N., Huy, A.H.P.: A stable Lyapunov approach of advanced sliding mode control for swing up and robust balancing implementation for the pendubot system. In: AETA 2015: Recent Advances in Electrical Engineering and Related Sciences, pp. 411–425. Springer, Cham (2016)
9. Azad, M., Featherstone, R.: Angular momentum based balance controller for an underactuated planar robot. Auton. Robot. **40**(1), 93–107 (2016)
10. Lai, X.Z., Pan, C.Z., Wu, M., Yang, S.X.: Unified control of n-link underactuated manipulator with single passive joint: a reduced order approach. Mech. Mach. Theory **56**, 170–185 (2012)
11. Brown, S.C., Passino, K.M.: Intelligent control for an acrobot. J. Intell. Robot. Syst. **18**(3), 209–248 (1997)
12. Zhang, A.C., Lai, X.Z., Wu, M.: Nonlinear stabilizing control for a class of underactuated mechanical systems with multi degree of freedoms. Nonlinear Dyn. **89**(3), 2241–2253 (2017)
13. Fantoni, I., Lozano, R., Spong, M.W.: Energy based control of the pendubot. IEEE Trans. Autom. Control **45**(4), 725–729 (2000)

14. Horibe, T., Sakamoto, N.: Swing up and stabilization of the acrobot via nonlinear optimal control based on stable manifold method. IFAC-Papersonline **49**(18), 374–379 (2016)
15. De Luca, A.: Underactuated manipulators: control properties and techniques. Mach. Intell. Robot. Control **4**(3), 113–126 (2002)
16. Lai, X.Z., She, J.H., Cao, W.H., Yang, S.X.: Stabilization of underactuated planar acrobot based on motion-state constraints. Int. J. Nonlinear Mech. **77**, 342–347 (2015)
17. Lai, X.Z., Wang, Y.W., Wu, M., Cao, W.H.: Stable control strategy for planar three-link underactuated mechanical system. IEEE/ASME Trans. Mechatron. **21**(3), 1345–1356 (2016)
18. Zhang, P., Lai, X.Z., Wang, Y.W., Su, C.-Y., Wu, M.: A quick position control strategy based on optimization algorithm for a class of first-order nonholonomic system. Inf. Sci. **460**, 264–278 (2018)
19. Lai, X.Z., Zhang, P., Wang, Y.W., Chen, L.F., Wu, M.: Continuous state feedback control based on intelligent optimization for first-order nonholonomic systems. IEEE Trans. Syst. Man Cybern.: Syst. **50**(7), 2534–2540 (2020)
20. He, G.P., Wang, Z.L., Zhang, J., Geng, Z.Y.: Characteristics analysis and stabilization of a planar 2R underactuated manipulator. Robotica **34**(3), 584–600 (2016)
21. De Luca, A., Mattone, R., Oriolo, G.: Stabilization of an underactuated planar 2R manipulator. Int. J. Robust Nonlinear Control **10**(4), 181–198 (2000)
22. Yoshikawa, T., Kobayashi, K., Watanabe, T.: Design of a desirable trajectory and convergent control for 3-DOF manipulator with a nonholonomic constraint. In: Proceedings of the IEEE International Conference on Robotics and Automation, San Francisco, USA, pp. 1805–1810 (2000)
23. Xiong, P.Y., Lai, X.Z., Wu, M.: Position and posture control for a class of second-order nonholonomic underactuated mechanical system. IMA J. Math. Control Inf. **35**(2), 523–533 (2016)
24. Xiong, P.Y., Lai, X.Z., Wu, M.: A stable control for second-order nonholonomic planar underactuated mechanical system: energy attenuation approach. Int. J. Control **91**(7), 1630–1639 (2018)
25. Lai, X.Z., Zhang, P., Wang, Y.W., Wu, M.: Position-posture control of a planar four-link underactuated manipulator based on genetic algorithm. IEEE Trans. Ind. Electron. **64**(6), 4781–4791 (2017)

Finite-Time Fault Detection and H_∞ State Estimation for Markov Jump Systems Under Dynamic Event-Triggered Mechanism

Xiongbo Wan, Min Wu, and Zidong Wang

Abstract This chapter addresses the problems of finite-time fault detection (FD) and H_∞ state estimation for event-triggered discrete-time Markov jump systems under hidden Markov mode observation. A new dynamic event-triggered mechanism is proposed to reduce unnecessary data transmissions so as to save limited network resources. The modes of the Markov jump system are observed by the FD filter (FDF) and the state estimator obeying a hidden Markov process. By constructing Markov-mode-dependent Lyapunov functions, sufficient conditions in terms of linear matrix inequalities (LMIs) are obtained under which the filtering error system of FD and the estimation error system of state estimation are stochastically H_∞ finite-time bounded. The parameters of the FDF and state estimator are designed when these LMIs are feasible. Two numerical examples are given to verify the effectiveness and merits of the designed FDF and state estimator.

Keywords Fault detection · Finite-time H_∞ state estimation · Markov jump systems · Dynamic event-triggered mechanism · Hidden Markov model

Abbreviations

FD	Fault detection
SE	State estimation

X. Wan (✉) · M. Wu
School of Automation, China University of Geosciences, Wuhan 430074, China
e-mail: xbwan23@cug.edu.cn

Hubei Key Laboratory of Advanced Control and Intelligent Automation for Complex Systems, Wuhan 430074, China

Engineering Research Center of Intelligent Technology for Geo-Exploration, Ministry of Education,Wuhan 430074, China

Z. Wang
Department of Computer Science, Brunel University London, Uxbridge UB8 3PH, UK

M. Wu et al. (eds.), *Developments in Advanced Control and Intelligent Automation for Complex Systems*, Studies in Systems, Decision and Control 329,
https://doi.org/10.1007/978-3-030-62147-6_4

DETM	Dynamic event-triggered mechanism
SETM	Static event-triggered mechanism
FDF	Fault detection filter
HMM	Hidden Markov model
SFTB	Stochastically finite-time bounded
SH_∞FTB	Stochastically H_∞ Finite-time bounded
TP	Transition probability
ZIC	Zero initial condition

1 Introduction

Networked systems have many advantages such as high reliability, easy maintenance, and resource sharing [1, 2], and therefore, they have been widely applied in wireless communications, intelligent buildings, and other related fields [3–5]. However, as the bandwidth of the communication network is so limited, the frequent data transmissions will unavoidably produce network-induced phenomena, which usually degrade the performance or even lead to instability of systems.

Due to increasing complexities and large scales of networked systems, they are prone to faults, and the fault detection (FD) is of great significance to reduce unnecessary losses. The FD of networked systems has attracted widespread attention with fruitful results available [6–8]. Note that these results are all on FD over an infinite-time interval. However, to achieve practical aims, people might be more interested in studying finite-time FD issues. Since the network resources are limited, how to implement finite-time FD with the consideration of saving network resources by avoiding unnecessary data transmissions is an issue that deserves further investigation.

As another research topic, state estimation (SE) becomes imperatively important for the analysis and design of networked systems [4]. In many situations, people might care more about the dynamical behaviors of systems over a finite-time interval than an infinite-time interval. Therefore, the finite-time SE issue of networked systems is of great significance.

An effective method of saving network resources is adopting the event-triggered mechanism under which the current data packet is transmitted only when a prespecified triggering condition is violated. The event-triggered mechanism is generally classified into two categories: the static event-triggered mechanism (SETM) [9–12] and the dynamic event-triggered mechanism (DETM) [13–16]. The DETM contains an additional internal dynamic variable, and the SETM has fixed threshold parameters. Generally speaking, the former yields a longer time interval between two successive transmissions than the latter does, and is more capable to save network resources.

When it comes to FD and SE problems under certain triggering mechanisms, fruitful results based on SETM have been reported. However, there have been very few results on DETM except [15–18]. Note that these DETMs in [15–18] can be further improved since there have some limitations regarding their parameters or dynamic characteristics. As such, one motivation is to study the FD and SE problems of networked systems under an improved DETM by introducing more flexible parameters with the hope to save more network resources.

For the existing results of FD of Markov jump systems, the fault detection filter (FDF) dependent on the mode is usually designed to generate the residual signals, where the mode of the FDF and that of the Markov jump system are assumed to be the same. In practical situations, however, there might exist significant differences between the modes of the FDF and the Markov jump system. Therefore, the situation is studied, where the mode of the Markov jump system is observed obeying a hidden Markov process. The relationship between the mode of the Markov jump system and its observed one is detailed by a hidden Markov model (HMM) with the form of conditional probabilities [19, 20]. Also, a state estimator is designed based on the HMM. Note that the HMM-based FD and SE problems of Markov jump systems have not been dealt with under the DETM. This is considered as the second motivation.

We focus on the finite-time FD and SE problems for discrete-time Markov jump systems under the DETM. An HMM-based FDF and an HMM-based state estimator are aimed to be designed ensuring that the error dynamics of FD and SE are stochastically H_∞ finite-time bounded (SH_∞FTB). *The main contributions are listed as follows*: *(1) a general DETM with flexible parameters is developed to reduce unnecessary data transmissions so as to avoid/alleviate data collisions and save limited network resources; (2) the DETM-based finite-time FD and SE problems are studied, for the first time, for the networked Markov jump systems; and (3) an HMM-based FDF and an HMM-based state estimator of Markov jump systems are devised with the consideration of mode mismatch caused by practical factors.*

2 Finite-Time Fault Detection

This section includes problem formulation, main results, detection threshold design, and a numerical example.

2.1 Problem Formulation

Consider the discrete-time Markov jump system:

$$\begin{cases} x(k+1) = A_{\delta(k)}x(k) + B_{\delta(k)}w(k) + E_{\delta(k)}f(k), \\ y(k) = C_{\delta(k)}x(k) + D_{\delta(k)}w(k), \end{cases} \tag{1}$$

where $x(k) \in R^{n_x}$ is the state vector, $w(k) \in R^{n_w}$ is the external disturbance belonging to $l_2[0, \infty)$, $f(k) \in R^{n_f}$ is the fault, $y(k) \in R^{n_y}$ is the measurement output, and $\delta(k)$ obeys a Markov chain which takes values in $\mathscr{S} = \{1, 2, \dots, s\}$ with s being a given integer. The mode TP matrix is $\Lambda = [\varepsilon_{ij}]_{s\times s}$, where the TP is described as follows:

$$\text{Prob}(\delta(k+1) = j|\delta(k) = i) = \varepsilon_{ij}, \tag{2}$$

where $\varepsilon_{ij} \geq 0$ $(i, j \in \mathscr{S})$ and $\sum_{j=1}^{s} \varepsilon_{ij} = 1$. Denote

$$y_g(k) = y(g_t), \ \forall k \in [g_t, g_{t+1}), \tag{3}$$

$$e_g(k) = y(k) - y_g(k), \tag{4}$$

where g_t denotes the tth triggering instant with $g_0 = 0$.

The following DETM is devised to avoid unnecessary data transmissions:

$$\begin{cases} g_{t+1} = \min_{k \in \mathrm{N}} \{k > g_t \mid a_1\sigma(k) + b_1 y^T(k)y(k) - b_2 e_g^T(k)e_g(k) \leq 0\}, \\ \sigma(k+1) = a_2\sigma(k) + b_3 y^T(k)y(k) - b_4 e_g^T(k)e_g(k), \end{cases} \tag{5}$$

where $\sigma(k)$ is an additional internal dynamic variable with $\sigma(0) \geq 0$, and $a_1 > 0$, $a_2 > 0$, $b_1 \geq 0$, $b_2 > 0$, $b_3 \geq 0$ and $b_4 > 0$ are given scalars.

From (3) and (4), we know that $e_g(k) = y(k) - y(g_t)$ holds for any $k \in [g_t, g_{t+1})$. In addition, we have $e_g(k) = y(k) - y(g_t)$ and $e_g(k) = y(k) - y_g(k)$ for any $k \in [0, \infty)$ as t takes values from 0 to $+\infty$ under the DETM (5).

Remark 1 By letting $a_1 > 0$ and $b_2 = b_4$, we rewrite the DETM (5) as follows:

$$\begin{cases} g_{t+1} = \min_{k \in \mathrm{N}} \{k > g_t \mid \sigma(k) + (b_1/a_1) y^T(k)y(k) - (b_2/a_1) e_g^T(k)e_g(k) \leq 0\}, \\ \sigma(k+1) = a_2\sigma(k) + b_3 y^T(k)y(k) - b_2 e_g^T(k)e_g(k). \end{cases} \tag{6}$$

Moreover, by setting $b_1 = \rho$, $b_2 = \vartheta$, $b_3 = \rho$, $a_1 = 1/\theta$, $a_2 = \kappa$, it can be seen that (6) becomes the DETM in [17]. Therefore, the DETM (5) is more general than that in [17].

Remark 2 When setting $a_1 = 0$, we have from (5) that

$$g_{t+1} = \min_{k \in \mathrm{N}} \{k > g_t \mid b_1 y^T(k)y(k) - b_2 e_g^T(k)e_g(k) \leq 0\}, \tag{7}$$

which is just the SETM in [21]. When $b_1 = 0$ and $\sigma(k) \equiv \bar{\sigma}$ with $\bar{\sigma}$ being a positive constant, the DETM (5) becomes

$$g_{t+1} = \min_{k \in \mathrm{N}} \{k > g_t \mid a_1\bar{\sigma} - b_2 e_g^T(k)e_g(k) \leq 0\}, \tag{8}$$

which is just the SETM adopted in [22]. Therefore, the DETM (5) is more general than those in [21, 22] concerning FD systems.

Construct the HMM-based FDF of the following form:

$$\begin{cases} \hat{x}(k+1) = A_{f,h(k)}\hat{x}(k) + B_{f,h(k)}y_g(k), \\ r(k) = C_{f,h(k)}\hat{x}(k) + D_{f,h(k)}y_g(k), \end{cases} \tag{9}$$

where $\hat{x}(k) \in R^{n_x}$ is the estimation of the system state, $r(k) \in R^{n_f}$ is the generated residual, $A_{f,h(k)}$, $B_{f,h(k)}$, $C_{f,h(k)}$ and $D_{f,h(k)}$ are mode-dependent FDF parameters. $h(k) \in \mathscr{H}$ is a switching parameter dependent on $\delta(k)$, where $\mathscr{H} = \{1, 2, \ldots, h\}$ with h being an integer. The conditional probability matrix is $\Delta = [\rho_{il}]_{s\times h}$, where the conditional probability is described as follows:

$$\text{Prob}(h(k) = l|\delta(k) = i) = \rho_{il}, \tag{10}$$

where $\rho_{il} \geq 0$ $(i \in \mathscr{S}, l \in \mathscr{H})$ and $\sum_{l=1}^{h} \rho_{il} = 1$.

Remark 3 An HMM described by (2) and (10) is introduced to reflect the mismatch between the mode of the FDF (9) and that of the Markov jump system (1). So far, for Markov jump systems, HMM-based filtering [19, 20] and control [23] problems have been addressed. However, the HMM-based FD problems have not been fully discussed yet, not to mention the consideration of the DETM-based data transmissions.

Let

$$\begin{aligned} e(k) &= x(k) - \hat{x}(k),\ \tilde{r}(k) = r(k) - f(k), \\ v(k) &= [w^T(k), f^T(k)]^T,\ \eta(k) = [x^T(k), e^T(k)]^T. \end{aligned}$$

From (1), (3), (4) and (9), we have

$$\begin{cases} \eta(k+1) = \bar{A}_{\delta(k),h(k)}\eta(k) + \bar{B}_{\delta(k),h(k)}e_g(k) + \bar{E}_{\delta(k),h(k)}v(k), \\ \tilde{r}(k) = \bar{C}_{\delta(k),h(k)}\eta(k) - D_{f,h(k)}e_g(k) + \bar{D}_{\delta(k),h(k)}v(k), \end{cases} \tag{11}$$

where

$$\bar{A}_{\delta(k),h(k)} = \begin{bmatrix} A_{\delta(k)} & 0 \\ A_{\delta(k)} - B_{f,h(k)}C_{\delta(k)} - A_{f,h(k)} & A_{f,h(k)} \end{bmatrix}, \bar{B}_{\delta(k),h(k)} = \begin{bmatrix} 0 \\ B_{f,h(k)} \end{bmatrix},$$

$$\bar{E}_{\delta(k),h(k)} = \begin{bmatrix} B_{\delta(k)} & E_{\delta(k)} \\ B_{\delta(k)} - B_{f,h(k)}D_{\delta(k)} & E_{\delta(k)} \end{bmatrix},$$

$$\bar{C}_{\delta(k),h(k)} = [C_{f,h(k)} + D_{f,h(k)}C_{\delta(k)},\ -C_{f,h(k)}],\ \bar{D}_{\delta(k),h(k)} = [D_{f,h(k)}D_{\delta(k)},\ -I].$$

Assumption 1 $v(k)$ satisfies

$$\sum_{k=0}^{N} v^T(k)v(k) < \theta, \tag{12}$$

where $N > 0$ and $\theta > 0$ are known integer and scalar.

Definition 1 ([4]) The filtering error dynamics (11) of FD is said to be stochastically finite-time bounded (SFTB) with respect to (c_1, c_2, U, N, θ), where c_1 and c_2 are two given scalars with $c_1 < c_2$, $U > 0$ is a given matrix, $N > 0$ is a given integer, if $\mathbf{E}\{\eta^T(0)U\eta(0)\} \le c_1 \Rightarrow \mathbf{E}\{\eta^T(k)U\eta(k)\} \le c_2,\ \forall k \in \{1, 2, \ldots, N\}$.

For a prescribed $\gamma > 0$, we will design an FDF (9) which ensures that the following (1) and (2) hold.

(1) The system (11) is SFTB.

(2) Under the ZIC, the following condition holds:

$$\mathbf{E}\left\{\sum_{k=0}^{N} \tilde{r}^T(k)\tilde{r}(k)\right\} \le \gamma^2 \mathbf{E}\left\{\sum_{k=0}^{N} v^T(k)v(k)\right\}. \tag{13}$$

The system (11) is SH_∞FTB if the above two conditions are met simultaneously.

2.2 Main Results

The following lemma is first presented.

Lemma 1 *Consider the DETM (5) with* $\sigma(0) \ge 0$. *The variable* $\sigma(k) \ge 0$, $\forall k \ge 0$ *if* a_1, a_2, b_1, b_2, b_3 *and* b_4 *in (5) satisfy* $0 < a_1 \le a_2$, $0 \le b_1 \le b_3$ *and* $0 < b_4 \le b_2$.

Proof It follows from (5) that

$$a_1\sigma(k) + b_1 y^T(k)y(k) - b_2 e_g^T(k)e_g(k) > 0 \tag{14}$$

holds for any $k \in [0, g_1)$. Equation (14) together with $\sigma(k+1) = a_2\sigma(k) + b_3 y^T(k)$ $y(k) - b_4 e_g^T(k)e_g(k)$ yields that

$$\sigma(k+1) > (a_2 - a_1)\sigma(k) + (b_3 - b_1)y^T(k)y(k) + (b_2 - b_4)e_g^T(k)e_g(k) \tag{15}$$

holds, $\forall k \in [0, g_1)$. As $0 \le b_1 \le b_3$ and $0 < b_4 \le b_2$, we obtain from (15) that

$$\sigma(k+1) > (a_2 - a_1)\sigma(k) \tag{16}$$

holds for any $k \in [0, g_1)$. As $0 < a_1 \le a_2$ and $\sigma(0) \ge 0$, we have from (16) that

$$\sigma(k+1) > (a_2 - a_1)\sigma(k) > (a_2 - a_1)^2\sigma(k-1) > \cdots > (a_2 - a_1)^{k+1}\sigma(0) \ge 0. \tag{17}$$

Therefore, we have from $\sigma(0) \ge 0$ and (17) that $\sigma(k) \ge 0$, $\forall k \in [0, g_1]$. Similarly, by considering the cases that $k \in [g_1, g_2)$, $k \in [g_2, g_3)$, ..., respectively, we have that $\sigma(k) \ge 0$, $\forall k \ge 0$.

Theorem 1 *For given scalars α, a_1, a_2, μ, b_1, b_2, b_3 and b_4 satisfying $\alpha > 1, 0 < a_1 \le a_2$, $\mu > 0$, $\mu a_1 + a_2 - \alpha < 0, 0 \le b_1 \le b_3$ and $0 < b_4 \le b_2$, the filtering error dynamics (11) of FD under the DETM (5) with $\sigma(0) = 0$ is SFTB with respect to (c_1, c_2, U, N, θ) if there exist matrices $P_i > 0$, $\Phi_i > 0$ and positive scalars β_1, β_2 such that*

$$\tilde{\Phi}_i = \tilde{\Pi}_i + (b_3 + \mu b_1)\Omega_i\Omega_i^T + \textstyle\sum_{l=1}^h \rho_{il}\Gamma_{i,l}\tilde{P}_i\Gamma_{i,l}^T < 0, \tag{18}$$

$$\beta_1 U < P_i < \beta_2 U, \tag{19}$$

$$\beta_2 c_1 + \sup\{\lambda_{\max}(\Phi_i)\}\theta \le \alpha^{-N} c_2 \beta_1, \tag{20}$$

hold for any $i \in \mathscr{S}$, where

$$\tilde{\Pi}_i = \mathrm{diag}\{-\alpha P_i, -(\mu b_2 + b_4)I, (\mu a_1 + a_2 - \alpha)I, -\Phi_i\}, \Gamma_{i,l} = [\bar{A}_{i,l}, \bar{B}_{i,l}, 0, \bar{E}_{i,l}]^T,$$
$$\tilde{P}_i = \textstyle\sum_{j=1}^s \varepsilon_{ij} P_j, \Omega_i = [\hat{C}_i, 0, 0, \hat{D}_i]^T, \ \hat{C}_i = [C_i, \ 0], \ \hat{D}_i = [D_i, \ 0].$$

Proof Choose the mode-dependent Lyapunov function

$$V(k) = \eta^T(k) P_{\delta(k)} \eta(k) + \sigma(k). \tag{21}$$

Let $\delta(k) = i$ and $\mathbf{E}\{\Delta V(k)\} = \mathbf{E}\{V(k+1)\} - \alpha\mathbf{E}\{V(k)\}$. Then, by noticing that (14) holds for any $k \ge 0$, we have

$$\begin{aligned}\mathbf{E}\{\Delta V(k)\} &< \mathbf{E}\{\Delta V(k)\} + \mu(a_1\sigma(k) + b_1 y^T(k)y(k) - b_2 e_g^T(k)e_g(k)) \\ &= \mathbf{E}\{\zeta^T(k)\bar{\Phi}_i\zeta(k)\},\end{aligned} \tag{22}$$

where

$$\begin{aligned}\bar{\Phi}_i &= \bar{\Pi}_i + (b_3 + \mu b_1)\Omega_i\Omega_i^T + \textstyle\sum_{l=1}^h \rho_{il}\Gamma_{i,l}(\sum_{j=1}^s \varepsilon_{ij}P_j)\Gamma_{i,l}^T \\ \bar{\Pi}_i &= \mathrm{diag}\{-\alpha P_i, -(\mu b_2 + b_4)I, (\mu a_1 + a_2 - \alpha)I, 0\} \\ \zeta(k) &= [\eta^T(k), e_g^T(k), \widetilde{\sigma}^T(k), v^T(k)]^T, \ \widetilde{\sigma}(k) = \sigma^{\frac{1}{2}}(k).\end{aligned}$$

Then, based on (22), we have

$$\mathbf{E}\{\Delta V(k) - v^T(k)\Phi_i v(k)\} < \mathbf{E}\{\zeta^T(k)\tilde{\Phi}_i\zeta(k)\}. \tag{23}$$

It follows from (18) and (23) that

$$\begin{aligned}\mathbf{E}\{V(k+1)\} &< \alpha\mathbf{E}\{V(k)\} + \mathbf{E}\{v^T(k)\Phi_i v(k)\} \\ &\le \alpha\mathbf{E}\{V(k)\} + \sup\{\lambda_{\max}(\Phi_i)\}\mathbf{E}\{v^T(k)v(k)\}.\end{aligned} \tag{24}$$

Circularly utilizing (24) yields

$$\mathbf{E}\{V(k)\} < \alpha^N\mathbf{E}\{V(0)\} + \sup\{\lambda_{\max}(\Phi_i)\}\alpha^N\theta. \tag{25}$$

We obtain from (19) and (21) that

$$\mathbf{E}\{V(0)\} \leq \beta_2 c_1. \tag{26}$$

Also, we have that

$$\mathbf{E}\{V(k)\} \geq \beta_1 \mathbf{E}\{\eta^T(k)U\eta(k)\}. \tag{27}$$

Then, from (20), (25)–(27), we have

$$\mathbf{E}\{\eta^T(k)U\eta(k)\} < \frac{\alpha^N}{\beta_1}[\beta_2 c_1 + \sup\{\lambda_{\max}(\Phi_i)\}\theta] \leq c_2. \tag{28}$$

Therefore, the system (11) is SFTB with respect to (c_1, c_2, U, N, θ).

Theorem 2 *Given scalars* α, a_1, a_2, μ, b_1, b_2, b_3 *and* b_4 *satisfying* $\alpha > 1, 0 < a_1 \leq a_2$, $\mu > 0$, $\mu a_1 + a_2 - \alpha < 0$, $0 \leq b_1 \leq b_3$ *and* $0 < b_4 \leq b_2$, *the system (11) under the DETM (5) with* $\sigma(0) = 0$ *is* $SH_\infty FTB$ *with respect to* $(c_1, c_2, U, N, \gamma, \theta)$ *if there exist matrices* $P_i > 0$ *and positive scalars* β_1, β_2 *such that*

$$\hat{\Phi}_i = \hat{\Pi}_i + (b_3 + \mu b_1)\Omega_i \Omega_i^T + \textstyle\sum_{l=1}^h \rho_{il} \Gamma_{i,l} \tilde{P}_i \Gamma_{i,l}^T + \sum_{l=1}^h \rho_{il} \Upsilon_{i,l} \Upsilon_{i,l}^T < 0, \tag{29}$$

$$\beta_2 c_1 + \alpha^{-N}\gamma^2\theta \leq \alpha^{-N} c_2 \beta_1, \tag{30}$$

and (19) hold, where

$$\hat{\Pi}_i = \mathrm{diag}\{-\alpha P_i, -(\mu b_2 + b_4)I, (\mu a_1 + a_2 - \alpha)I, -\alpha^{-N}\gamma^2 I\},$$
$$\Upsilon_{i,l} = [\bar{C}_{i,l}, -D_{f,l}, 0, \bar{D}_{i,l}]^T.$$

Proof Replace Φ_i in (18) and (20) with $\alpha^{-N}\gamma^2 I$. Then, we obtain that (18) holds if (29) is met. In addition, (30) implies (20). Then, we have from Theorem 1 that (11) is SFTB with respect to (c_1, c_2, U, N, θ) if (19), (29) and (30) hold.

Choosing the Lyapunov function (21) and adopting the similar techniques, we have

$$\mathbf{E}\{V(k+1) - \alpha V(k) + \tilde{r}^T(k)\tilde{r}(k) - \alpha^{-N}\gamma^2 v^T(k)v(k)\} \leq \mathbf{E}\{\zeta^T(k)\hat{\Phi}_i\zeta(k)\},$$

where $\zeta(k)$ is given in the proof of Theorem 1.

We have from (29) that

$$\mathbf{E}\{V(k+1) - \alpha V(k) + \tilde{r}^T(k)\tilde{r}(k) - \alpha^{-N}\gamma^2 v^T(k)v(k)\} < 0. \tag{31}$$

Under the ZIC, we obtain from (31) that

$$\begin{aligned}
&\mathbf{E}\{V(k+1)\} \\
&< \alpha^{k+1}\mathbf{E}\{V(0)\} - \sum_{d=0}^{k}\mathbf{E}\{\tilde{r}^T(d)\tilde{r}(d)\alpha^{k-d}\} + \gamma^2\alpha^{-N}\mathbf{E}\left\{\sum_{d=0}^{k}\alpha^{k-d}v^T(d)v(d)\right\} \\
&= -\sum_{d=0}^{k}\mathbf{E}\{\tilde{r}^T(d)\tilde{r}(d)\alpha^{k-d}\} + \gamma^2\alpha^{-N}\mathbf{E}\left\{\sum_{d=0}^{k}\alpha^{k-d}v^T(d)v(d)\right\} \qquad (32)
\end{aligned}$$

Then, we have

$$\mathbf{E}\left\{\sum_{d=0}^{N}\tilde{r}^T(d)\tilde{r}(d)\right\} < \gamma^2\mathbf{E}\left\{\sum_{d=0}^{N}v^T(d)v(d)\right\}. \qquad (33)$$

Theorem 3 *For given scalars α, a_1, a_2, μ, b_1, b_2, b_3 and b_4 satisfying $\alpha > 1, 0 < a_1 \le a_2$, $\mu > 0$, $\mu a_1 + a_2 - \alpha < 0$, $0 \le b_1 \le b_3$ and $0 < b_4 \le b_2$, the system (11) under the DETM (5) with $\sigma(0) = 0$ is $SH_\infty FTB$ with respect to $(c_1, c_2, U, N, \gamma, \theta)$ if there exist matrices $P_i > 0$, $Q_{A,l}$, $Q_{B,l}$, $G_l = \begin{bmatrix} G_l^{(1)} & G_l^{(2)} \\ \kappa_l G_l^{(3)} & \varsigma_l G_l^{(3)} \end{bmatrix}$ with given positive scalars κ_l, ς_l $(l \in \mathscr{H})$, and positive scalars β_1 and β_2 such that, $\forall i \in \mathscr{S}$, (19), (30) and the following LMI*

$$\begin{bmatrix} \hat{\Pi}_i & \tilde{\Xi}_i & \Omega_i & \tilde{\Upsilon}_i \\ * & \Psi_i & 0 & 0 \\ * & * & -(b_3+\mu b_1)^{-1}I & 0 \\ * & * & * & -I \end{bmatrix} < 0 \qquad (34)$$

hold, where

$$\begin{gathered}
\tilde{\Upsilon}_i = [\sqrt{\rho_{i,1}}\Upsilon_{i,1}, \sqrt{\rho_{i,2}}\Upsilon_{i,2}, \ldots, \sqrt{\rho_{i,h}}\Upsilon_{i,h}], \\
\Psi_i = I_h \otimes \tilde{P}_i - G^T - G, \ G = \mathrm{diag}\{G_1, G_2, \ldots, G_h\}, \\
\tilde{\Xi}_i = [\Xi_{i,1}, \Xi_{i,2}, \ldots, \Xi_{i,h}], \forall l \in \mathscr{H}, \ \Xi_{i,l} = \sqrt{\rho_{i,l}}[\mathscr{X}_{i,l}^T, \mathscr{Y}_{i,l}^T, 0, \mathscr{Z}_{i,l}^T]^T, \\
\mathscr{X}_{i,l} = \begin{bmatrix} \mathscr{X}_{i,l}^{(1)} & \mathscr{X}_{i,l}^{(2)} \\ \mathscr{X}_{i,l}^{(3)} & \mathscr{X}_{i,l}^{(4)} \end{bmatrix}, \ \mathscr{Z}_{i,l} = \begin{bmatrix} \mathscr{Z}_{i,l}^{(1)} & \mathscr{Z}_{i,l}^{(2)} \\ \mathscr{Z}_{i,l}^{(3)} & \mathscr{Z}_{i,l}^{(4)} \end{bmatrix}, \mathscr{Y}_{i,l} = [\kappa_l Q_{B,l}, \varsigma_l Q_{B,l}], \\
\mathscr{X}_{i,l}^{(1)} = A_i^T G_l^{(1)} + \kappa_l A_i^T G_l^{(3)} - \kappa_l C_i^T Q_{B,l} - \kappa_l Q_{A,l}, \\
\mathscr{X}_{i,l}^{(2)} = A_i^T G_l^{(2)} + \varsigma_l A_i^T G_l^{(3)} - \varsigma_l C_i^T Q_{B,l} - \varsigma_l Q_{A,l}, \mathscr{X}_{i,l}^{(3)} = \kappa_l Q_{A,l}, \ \mathscr{X}_{i,l}^{(4)} = \varsigma_l Q_{A,l}, \\
\mathscr{Z}_{i,l}^{(1)} = B_i^T G_l^{(1)} + \kappa_l B_i^T G_l^{(3)} - \kappa_l D_i^T Q_{B,l}, \mathscr{Z}_{i,l}^{(2)} = B_i^T G_l^{(2)} + \varsigma_l B_i^T G_l^{(3)} - \varsigma_l D_i^T Q_{B,l}, \\
\mathscr{Z}_{i,l}^{(3)} = E_i^T G_l^{(1)} + \kappa_l E_i^T G_l^{(3)}, \ \mathscr{Z}_{i,l}^{(4)} = E_i^T G_l^{(2)} + \varsigma_l E_i^T G_l^{(3)}.
\end{gathered}$$

Other notations not mentioned are presented in Theorems 1 and 2. In this case, the FDF parameters can be given as follows:

$$\begin{bmatrix} A_{f,l} & B_{f,l} \\ C_{f,l} & D_{f,l} \end{bmatrix} = \begin{bmatrix} G_l^{(3)^{-T}} Q_{A,l}^T & G_l^{(3)^{-T}} Q_{B,l}^T \\ C_{f,l} & D_{f,l} \end{bmatrix}, \forall l \in \mathscr{H}. \tag{35}$$

Proof The LMI (34) together with $Q_{A,l} = A_{f,l}^T G_l^{(3)}$, $Q_{B,l} = B_{f,l}^T G_l^{(3)}$ and $G_l = \begin{bmatrix} G_l^{(1)} & G_l^{(2)} \\ \kappa_l G_l^{(3)} & \varsigma_l G_l^{(3)} \end{bmatrix}$ $(l \in \mathscr{H})$ yields

$$\begin{bmatrix} \hat{\Pi}_i & \tilde{\Gamma}_i G & \Omega_i & \tilde{\Upsilon}_i \\ * & \Psi_i & 0 & 0 \\ * & * & -(b_3 + \mu b_1)^{-1} I & 0 \\ * & * & * & -I \end{bmatrix} < 0, \tag{36}$$

where

$$\tilde{\Gamma}_i = [\sqrt{\rho_{i,1}}\Gamma_{i,1}, \sqrt{\rho_{i,2}}\Gamma_{i,2}, \ldots, \sqrt{\rho_{i,h}}\Gamma_{i,h}]. \tag{37}$$

Here, $\Gamma_{i,l}$ $(l = 1, 2, \ldots, h)$ are defined in Theorem 1. Noticing that $-G^T(I_h \otimes \tilde{P}_i^{-1})G \le \Psi_i, \forall i \in \mathscr{S}$, we obtain from (36) that

$$\begin{bmatrix} \hat{\Pi}_i & \tilde{\Gamma}_i G & \Omega_i & \tilde{\Upsilon}_i \\ * & -G^T(I_h \otimes \tilde{P}_i^{-1})G & 0 & 0 \\ * & * & -(b_3 + \mu b_1)^{-1} I & 0 \\ * & * & * & -I \end{bmatrix} < 0. \tag{38}$$

In addition, the LMI (34) ensures that $\Psi_i = I_h \otimes \tilde{P}_i - G^T - G < 0$, which together with $\tilde{P}_i > 0$ yields that G is nonsingular. Let $\Sigma = \text{diag}\{I, G^{-1}, I, I\}$. Then, simple matrix manipulations using Σ^T and Σ yield

$$\begin{bmatrix} \hat{\Pi}_i & \tilde{\Gamma}_i & \Omega_i & \tilde{\Upsilon}_i \\ * & -I_h \otimes \tilde{P}_i^{-1} & 0 & 0 \\ * & * & -(b_3 + \mu b_1)^{-1} I & 0 \\ * & * & * & -I \end{bmatrix} < 0 \tag{39}$$

By the Schur complement [24], we obtain that (29) holds if and only if (39) is satisfied. According to Theorem 2, the system (11) is proved to be SH_∞FTB with respect to $(c_1, c_2, U, N, \gamma, \theta)$.

2.3 Detection Threshold Design

Select the residual evaluation function $J(k)$ and threshold J_{th}:

$$J(k) = \mathbf{E}\left\{\left(\sum_{s=l_0}^{l_0+k} r^T(s)r(s)\right)^{\frac{1}{2}}\right\}, \tag{40}$$

$$J_{\text{th}} = \sup_{w\in l_2, f=0}\mathbf{E}\left\{\left(\sum_{s=l_0}^{l_0+L} r^T(s)r(s)\right)^{\frac{1}{2}}\right\}, \tag{41}$$

where l_0 is the initial instant, and L denotes the maximum time step. The decision is made according to the rule:

$$\begin{cases} J(k) > J_{\text{th}} \Rightarrow \text{A fault is detected}, \\ J(k) \leq J_{\text{th}} \Rightarrow \text{No faults}. \end{cases}$$

2.4 Numerical Example

In this section, we consider the following numerical example.

The system (1) involves two subsystems and, therefore, $\mathscr{S} = \{1, 2\}$. The system parameters of the subsystem 1 and subsystem 2 are given as follows, respectively.

Subsystem 1:

$$A_1 = \begin{bmatrix} 0.7 & 0.1 \\ -0.8 & 0.1 \end{bmatrix},\ B_1 = \begin{bmatrix} 0.2 \\ -0.2 \end{bmatrix},\ E_1 = \begin{bmatrix} 0.3 \\ 0.35 \end{bmatrix},\ C_1 = [0.5,\ -1.3],\ D_1 = 0.2.$$

Subsystem 2:

$$A_2 = \begin{bmatrix} -0.5 & 0.15 \\ 0.75 & 0.2 \end{bmatrix},\ B_2 = \begin{bmatrix} -0.85 \\ 0.4 \end{bmatrix},\ E_2 = \begin{bmatrix} 1.3 \\ 0.4 \end{bmatrix},\ C_2 = [2,\ -2.1],\ D_2 = 0.3.$$

Assume that two modes of the system (1) can be observed and, therefore, $\mathscr{H} = \{1, 2\}$. The mode TP matrix Λ and conditional probability matrix Δ are as follows

$$\Lambda = \begin{bmatrix} \varepsilon_{11} & \varepsilon_{12} \\ \varepsilon_{21} & \varepsilon_{22} \end{bmatrix} = \begin{bmatrix} 0.1 & 0.9 \\ 0.25 & 0.75 \end{bmatrix},\quad \Delta = \begin{bmatrix} \rho_{11} & \rho_{12} \\ \rho_{21} & \rho_{22} \end{bmatrix} = \begin{bmatrix} 0.6 & 0.4 \\ 0.45 & 0.55 \end{bmatrix}.$$

The parameters in the DETM (5) are set as $a_1 = 0.4, a_2 = 0.5, b_1 = 0.02, b_2 = 4$, $b_3 = 0.1$, $b_4 = 0.2$. Let $\kappa_1 = \kappa_2 = 0.1$, $\varsigma_1 = \varsigma_2 = 1$, $\mu = 1$, $\alpha = 1.01$, $c_1 = 0.01$, $c_2 = 3$, $N = 30$, $\theta = 1$, $U = I_4$. It is found that the LMIs (19), (30) and (34) are feasible and the optimal H_∞ performance index is $\gamma_{\text{opt}} = 3.5728$. By solving these LMIs, we obtain the following FDF parameters:

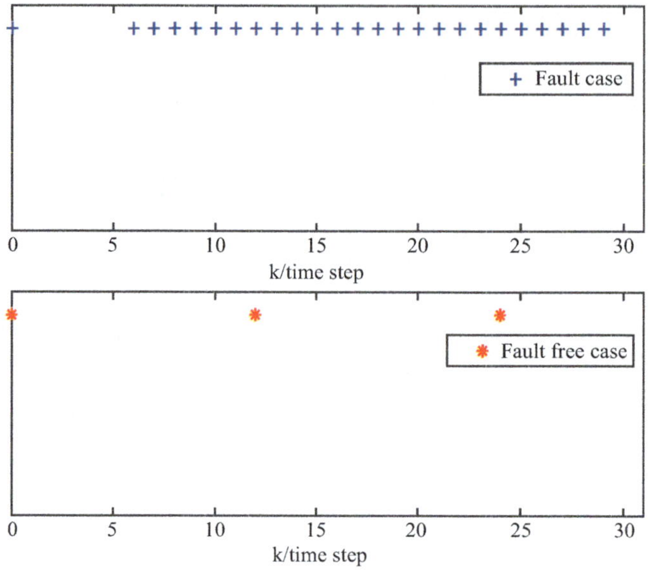

Fig. 1 Triggering instants

$$A_{f,1} = \begin{bmatrix} -0.0619 & -0.1020 \\ 0.0558 & 0.1967 \end{bmatrix},\ B_{f,1} = \begin{bmatrix} -0.2264 \\ 0.0342 \end{bmatrix},$$
$$C_{f,1} = [-0.0581,\ -0.2042],\ D_{f,1} = -0.0943,$$
$$A_{f,2} = \begin{bmatrix} 0.0264 & -0.2746 \\ -0.1326 & 0.2614 \end{bmatrix},\ B_{f,2} = \begin{bmatrix} -0.2502 \\ 0.0666 \end{bmatrix},$$
$$C_{f,2} = [0.0059,\ -0.2640],\ D_{f,2} = -0.1054.$$

According to Theorem 3, we obtain that the system (11) under the DETM (5) is SH_∞FTB. Next, set $x(0) = [-0.004,\ 0.004]^T$ and $\hat{x}(0) = [0,\ 0]^T$. Assume that $w(k) = 0.005e^{-1.1k}\sin(0.5k)$ and

$$f(k) = \begin{cases} 0.1, & k = 6, 7, \ldots, 11 \\ 0, & \text{otherwise} \end{cases}$$

In Fig. 1, the triggering instants under the DETM (5) are depicted. With given Λ and Δ, the Markov mode $\delta(k)$ and its observed mode $h(k)$ are illustrated in Fig. 2. In Figs. 3 and 4, the residual and its evaluation function are depicted, respectively. From Figs. 3 and 4, we find that the fault is easy to detect.

Select $J_{th} = \sup_{f=0} \mathrm{E}\left\{\sum_{k=0}^{30} r^T(k)r(k)\right\}^{\frac{1}{2}}$. Direct calculation yields J_{th}=0.0056. Furthermore, we have $J(6) < J_{th} < J(7)$, where $J(6) = 0.0039$ and $J(7) = 0.0138$. Then, we obtain that the fault is detected in one time step once it occurs.

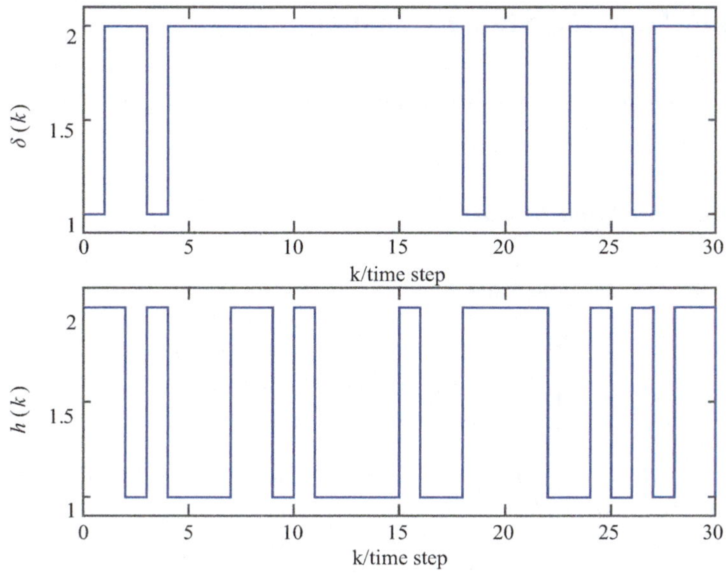

Fig. 2 Markov mode $\delta(k)$ and its observed mode $h(k)$

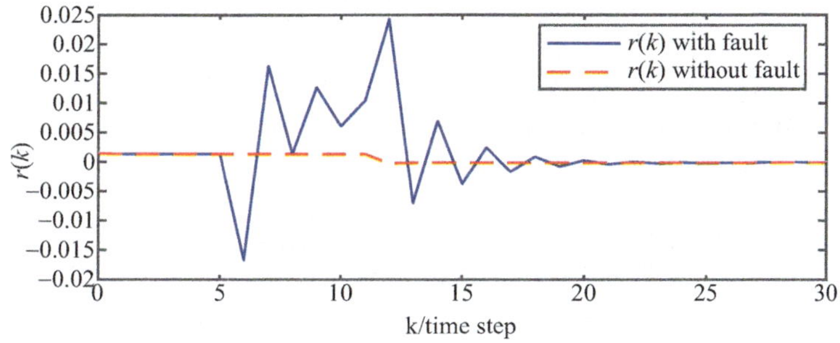

Fig. 3 Residual signal $r(k)$

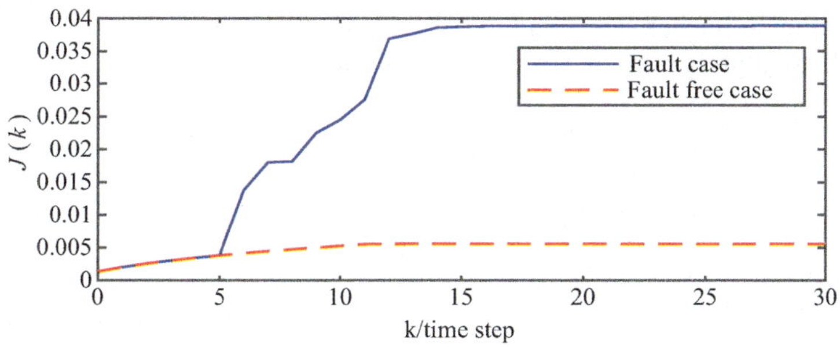

Fig. 4 Residual evaluation function $J(k)$

3 Finite-Time H_∞ State Estimation

This section includes problem formulation, main results, and a numerical example.

3.1 Problem Formulation

Consider the discrete-time Markov jump system:

$$\begin{cases} x(k+1) = A_{\delta(k)}x(k) + B_{\delta(k)}w(k), \\ y(k) = C_{\delta(k)}x(k) + D_{\delta(k)}w(k), \\ z(k) = L_{\delta(k)}x(k) + E_{\delta(k)}w(k), \end{cases} \tag{42}$$

where $x(k) \in R^{n_x}$ is the state vector, $w(k) \in R^{n_w}$ is the external disturbance belonging to $l_2[0, \infty)$, $y(k) \in R^{n_y}$ is the measurement output, $z(k) \in R^{n_z}$ is the signal to be estimated, and $\delta(k)$ is a Markov chain taking values in a finite state space $\mathscr{S} = \{1, 2, \ldots, s\}$ with s being a given integer. The mode TP matrix is $\Lambda = [\varepsilon_{ij}]_{s\times s}$, where the TP is described as follows:

$$\text{Prob}(\delta(k+1) = j|\delta(k) = i) = \varepsilon_{ij}, \tag{43}$$

where $\varepsilon_{ij} \geq 0$ $(i, j \in \mathscr{S})$ and $\sum_{j=1}^{s} \varepsilon_{ij} = 1$.

To avoid unnecessary data transmissions, we use the DETM (5). According to the DETM (5), we construct the state estimator as follows:

$$\begin{cases} \hat{x}(k+1) = \hat{A}_{h(k)}\hat{x}(k) + \hat{B}_{h(k)}y_g(k), \\ \hat{z}(k) = \hat{C}_{h(k)}\hat{x}(k) + \hat{D}_{h(k)}y_g(k), \end{cases} \tag{44}$$

where $\hat{x}(k) \in R^{n_x}$ is the estimation of the system state, $\hat{z}(k)$ is the estimation of the output $z(k)$, $\hat{A}_{h(k)}$, $\hat{B}_{h(k)}$, $\hat{C}_{h(k)}$ and $\hat{D}_{h(k)}$ are mode-dependent estimator parameters. $h(k) \in \mathscr{H}$ is a switching parameter dependent on $\delta(k)$, where $\mathscr{H} = \{1, 2, \ldots, h\}$ with h being an integer. The conditional probability matrix is $\Delta = [\rho_{il}]_{s\times h}$, where the conditional probability is described as follows:

$$\text{Prob}(h(k) = l|\delta(k) = i) = \rho_{il}, \tag{45}$$

where $\rho_{il} \geq 0$ $(i \in \mathscr{S}, l \in \mathscr{H})$ and $\sum_{l=1}^{h} \rho_{il} = 1$.

Define

$$e(k) = x(k) - \hat{x}(k),\ \tilde{z}(k) = z(k) - \hat{z}(k),\ \eta(k) = [x^T(k), e^T(k)]^T.$$

From (42) and (44), we have

$$\begin{cases}\eta(k+1) = \bar{A}_{\delta(k),h(k)}\eta(k) + \bar{B}_{\delta(k),h(k)}e_g(k) + \bar{E}_{\delta(k),h(k)}w(k), \\ \tilde{z}(k) = \bar{C}_{\delta(k),h(k)}\eta(k) + \hat{D}_{h(k)}e_g(k) + \bar{D}_{\delta(k),h(k)}w(k),\end{cases} \tag{46}$$

where

$$\bar{A}_{\delta(k),h(k)} = \begin{bmatrix} A_{\delta(k)} & 0 \\ A_{\delta(k)} - \hat{B}_{h(k)}C_{\delta(k)} - \hat{A}_{h(k)} & \hat{A}_{h(k)} \end{bmatrix}, \ \bar{B}_{\delta(k),h(k)} = \begin{bmatrix} 0 \\ \hat{B}_{h(k)} \end{bmatrix},$$

$$\bar{E}_{\delta(k),h(k)} = \begin{bmatrix} B_{\delta(k)} \\ B_{\delta(k)} - \hat{B}_{h(k)}D_{\delta(k)} \end{bmatrix},$$

$$\bar{C}_{\delta(k),h(k)} = [L_{\delta(k)} - \hat{C}_{h(k)} - \hat{D}_{h(k)}C_{\delta(k)}, \ \hat{C}_{h(k)}], \ \bar{D}_{\delta(k),h(k)} = E_{\delta(k)} - \hat{D}_{h(k)}D_{\delta(k)}.$$

Assumption 2 $w(k)$ satisfies

$$\sum_{k=0}^{N} w^T(k)w(k) < \theta, \tag{47}$$

where $N > 0$ is an integer and $\theta > 0$ is a scalar, which are both known.

The main purpose is to design a state estimator (44) such that (1) and (2) hold.
(1) The system (46) is SFTB.
(2) For a prescribed $\gamma > 0$, the following holds under the ZIC:

$$\mathbf{E}\left\{\sum_{k=0}^{N} \tilde{z}^T(k)\tilde{z}(k)\right\} \le \gamma^2 \mathbf{E}\left\{\sum_{k=0}^{N} w^T(k)w(k)\right\}. \tag{48}$$

3.2 *Main Results*

Theorem 4 *For given scalars α, a_1, a_2, μ, b_1, b_2, b_3 and b_4 satisfying $\alpha > 1, 0 < a_1 \le a_2$, $\mu > 0$, $\mu a_1 + a_2 - \alpha < 0$, $0 \le b_1 \le b_3$ and $0 < b_4 \le b_2$, the estimation error system (46) under the DETM (5) with $\sigma(0) = 0$ is SFTB with respect to (c_1, c_2, U, N, θ) if there exist matrices $P_i > 0$, $\Phi_i > 0$ and positive scalars β_1, β_2 such that*

$$\tilde{\Phi}_i = \tilde{\Pi}_i + (b_3 + \mu b_1)\Omega_i\Omega_i^T + \textstyle\sum_{l=1}^{h} \rho_{il}\Gamma_{i,l}\tilde{P}_i\Gamma_{i,l}^T < 0, \tag{49}$$

$$\beta_1 U < P_i < \beta_2 U, \tag{50}$$

$$\beta_2 c_1 + \sup\{\lambda_{\max}(\Phi_i)\}\theta \le \alpha^{-N} c_2 \beta_1, \tag{51}$$

hold for any $i \in \mathscr{S}$, where

$$\tilde{\Pi}_i = \text{diag}\{-\alpha P_i, -(\mu b_2 + b_4)I, (\mu a_1 + a_2 - \alpha)I, -\Phi_i\}, \ \Gamma_{i,l} = [\bar{A}_{i,l}, \bar{B}_{i,l}, 0, \bar{E}_{i,l}]^T,$$

$$\tilde{P}_i = \textstyle\sum_{j=1}^{s} \varepsilon_{ij} P_j, \ \Omega_i = [\hat{C}_i, 0, 0, D_i]^T, \ \hat{C}_i = [C_i, \ 0].$$

Proof Choose the mode-dependent Lyapunov function

$$V(k) = \eta^T(k)P_{\delta(k)}\eta(k) + \sigma(k). \tag{52}$$

Let $\delta(k) = i$ and $\mathbf{E}\{\Delta V(k)\} = \mathbf{E}\{V(k+1)\} - \alpha\mathbf{E}\{V(k)\}$. Then, by noticing that (14) holds for any $k \geq 0$, we have

$$\begin{aligned}\mathbf{E}\{\Delta V(k)\} &< \mathbf{E}\{\Delta V(k)\} + \mu(a_1\sigma(k) + b_1 y^T(k)y(k) - b_2 e_g^T(k)e_g(k)) \\ &= \mathbf{E}\{\zeta^T(k)\bar{\Phi}_i\zeta(k)\},\end{aligned} \tag{53}$$

where

$$\begin{aligned}\bar{\Phi}_i &= \bar{\Pi}_i + (b_3 + \mu b_1)\Omega_i\Omega_i^T + \textstyle\sum_{l=1}^{h} \rho_{il}\Gamma_{i,l}(\sum_{j=1}^{s} \varepsilon_{ij}P_j)\Gamma_{i,l}^T \\ \bar{\Pi}_i &= \mathrm{diag}\{-\alpha P_i, -(\mu b_2 + b_4)I, (\mu a_1 + a_2 - \alpha)I, 0\} \\ \zeta(k) &= [\eta^T(k), e_g^T(k), \widetilde{\sigma}^T(k), w^T(k)]^T,\ \widetilde{\sigma}(k) = \sigma^{\frac{1}{2}}(k).\end{aligned}$$

Then, based on (53), we have

$$\mathbf{E}\{\Delta V(k) - w^T(k)\Phi_i w(k)\} < \mathbf{E}\{\zeta^T(k)\tilde{\Phi}_i\zeta(k)\}. \tag{54}$$

It follows from (49) and (54) that

$$\begin{aligned}\mathbf{E}\{V(k+1)\} &< \alpha\mathbf{E}\{V(k)\} + \mathbf{E}\{w^T(k)\Phi_i w(k)\} \\ &\leq \alpha\mathbf{E}\{V(k)\} + \sup\{\lambda_{\max}(\Phi_i)\}\mathbf{E}\{w^T(k)w(k)\}.\end{aligned} \tag{55}$$

Circularly utilizing (55) yields that

$$\mathbf{E}\{V(k)\} < \alpha^N\mathbf{E}\{V(0)\} + \sup\{\lambda_{\max}(\Phi_i)\}\alpha^N\theta. \tag{56}$$

We obtain from (50) and (52) that

$$\mathbf{E}\{V(0)\} \leq \beta_2 c_1, \tag{57}$$

where β_2 is presented in Theorem 4. Also, we have that

$$\mathbf{E}\{V(k)\} \geq \beta_1\mathbf{E}\{\eta^T(k)U\eta(k)\}. \tag{58}$$

Then, from (51), (56)–(58), we have

$$\mathbf{E}\{\eta^T(k)U\eta(k)\} < \frac{\alpha^N}{\beta_1}[\beta_2 c_1 + \sup\{\lambda_{\max}(\Phi_i)\}\theta] \leq c_2. \tag{59}$$

Therefore, the system (46) is SFTB with respect to (c_1, c_2, U, N, θ).

Theorem 5 *For given scalars α, a_1, a_2, μ, b_1, b_2, b_3 and b_4 satisfying $\alpha > 1, 0 < a_1 \le a_2$, $\mu > 0$, $\mu a_1 + a_2 - \alpha < 0$, $0 \le b_1 \le b_3$ and $0 < b_4 \le b_2$, the estimation error system (46) under the DETM (5) with $\sigma(0) = 0$ is $SH_\infty FTB$ with respect to $(c_1, c_2, U, N, \gamma, \theta)$ if there exist matrices $P_i > 0$ and positive scalars β_1 and β_2 such that*

$$\hat{\Phi}_i = \hat{\Pi}_i + (b_3 + \mu b_1)\Omega_i\Omega_i^T + \textstyle\sum_{l=1}^h \rho_{il}\Gamma_{i,l}\tilde{P}_i\Gamma_{i,l}^T + \sum_{l=1}^h \rho_{il}\Upsilon_{i,l}\Upsilon_{i,l}^T < 0, \quad (60)$$

$$\beta_2 c_1 + \alpha^{-N}\gamma^2\theta \le \alpha^{-N}c_2\beta_1, \quad (61)$$

and (50) hold, where

$$\hat{\Pi}_i = \text{diag}\{-\alpha P_i, -(\mu b_2 + b_4)I, (\mu a_1 + a_2 - \alpha)I, -\alpha^{-N}\gamma^2 I\},$$
$$\Upsilon_{i,l} = [\bar{C}_{i,l}, \hat{D}_l, 0, \bar{D}_{i,l}]^T.$$

Proof Replace Φ_i in (49) and (51) with $\alpha^{-N}\gamma^2 I$. Then, we obtain that (49) holds if (60) is satisfied. Moreover, (61) implies (51). Then, we know that (46) is SFTB with respect to (c_1, c_2, U, N, θ) if (50), (60) and (61) hold.

Choosing the Lyapunov function (52) and adopting the similar techniques, we have

$$\mathbf{E}\{V(k+1) - \alpha V(k) + \tilde{z}^T(k)\tilde{z}(k) - \alpha^{-N}\gamma^2 w^T(k)w(k)\} \le \mathbf{E}\{\zeta^T(k)\hat{\Phi}_i\zeta(k)\},$$

where $\zeta(k)$ is presented in Theorem 4.

It follows from (60) that

$$\mathbf{E}\{V(k+1) - \alpha V(k) + \tilde{z}^T(k)\tilde{z}(k) - \alpha^{-N}\gamma^2 w^T(k)w(k)\} < 0. \quad (62)$$

Under the ZIC, we obtain from (62) that

$$\begin{aligned}
&\mathbf{E}\{V(k+1)\} \\
&< \alpha^{k+1}\mathbf{E}\{V(0)\} - \sum_{d=0}^{k}\mathbf{E}\{\tilde{z}^T(d)\tilde{z}(d)\alpha^{k-d}\} + \gamma^2\alpha^{-N}\mathbf{E}\left\{\sum_{d=0}^{k}\alpha^{k-d}w^T(d)w(d)\right\} \\
&= -\sum_{d=0}^{k}\mathbf{E}\{\tilde{z}^T(d)\tilde{z}(d)\alpha^{k-d}\} + \gamma^2\alpha^{-N}\mathbf{E}\left\{\sum_{d=0}^{k}\alpha^{k-d}w^T(d)w(d)\right\} \quad (63)
\end{aligned}$$

Then, we have

$$\mathbf{E}\left\{\sum_{d=0}^{N}\tilde{z}^T(d)\tilde{z}(d)\right\} < \gamma^2\mathbf{E}\left\{\sum_{d=0}^{N}w^T(d)w(d)\right\}. \quad (64)$$

Theorem 6 *For given scalars α, a_1, a_2, μ, b_1, b_2, b_3 and b_4 satisfying $\alpha > 1, 0 < a_1 \le a_2$, $\mu > 0$, $\mu a_1 + a_2 - \alpha < 0$, $0 \le b_1 \le b_3$ and $0 < b_4 \le b_2$, the estimation* *error system (46) under the DETM (5) with $\sigma(0) = 0$ is $SH_\infty FTB$ with respect to $(c_1, c_2, U, N, \gamma, \theta)$ if there exist matrices $P_i > 0$, $Q_{A,l}$, $Q_{B,l}$, $G_l = \begin{bmatrix} G_l^{(1)} & G_l^{(2)} \\ \kappa_l G_l^{(3)} & \varsigma_l G_l^{(3)} \end{bmatrix}$ with given positive scalars κ_l and ς_l ($l \in \mathscr{H}$), and positive scalars β_1 and β_2 such that, $\forall i \in \mathscr{S}$, (50), (61) and the following LMI*

$$\begin{bmatrix} \hat{\Pi}_i & \tilde{\Xi}_i & \Omega_i & \tilde{\Upsilon}_i \\ * & \Psi_i & 0 & 0 \\ * & * & -(b_3 + \mu b_1)^{-1} I & 0 \\ * & * & * & -I \end{bmatrix} < 0 \tag{65}$$

hold, where

$$\begin{aligned} \tilde{\Upsilon}_i &= [\sqrt{\rho_{i,1}}\Upsilon_{i,1}, \sqrt{\rho_{i,2}}\Upsilon_{i,2}, \ldots, \sqrt{\rho_{i,h}}\Upsilon_{i,h}], \\ \Psi_i &= I_h \otimes \tilde{P}_i - G^T - G, \ G = \text{diag}\{G_1, G_2, \ldots, G_h\}, \\ \tilde{\Xi}_i &= [\Xi_{i,1}, \Xi_{i,2}, \ldots, \Xi_{i,h}], \forall l \in \mathscr{H}, \ \Xi_{i,l} = \sqrt{\rho_{i,l}}[\mathscr{X}_{i,l}^T, \mathscr{Y}_{i,l}^T, 0, \mathscr{Z}_{i,l}^T]^T, \\ \mathscr{X}_{i,l} &= \begin{bmatrix} \mathscr{X}_{i,l}^{(1)} & \mathscr{X}_{i,l}^{(2)} \\ \mathscr{X}_{i,l}^{(3)} & \mathscr{X}_{i,l}^{(4)} \end{bmatrix}, \ \mathscr{Z}_{i,l} = \begin{bmatrix} \mathscr{Z}_{i,l}^{(1)} & \mathscr{Z}_{i,l}^{(2)} \end{bmatrix}, \mathscr{Y}_{i,l} = [\kappa_l Q_{B,l}, \varsigma_l Q_{B,l}], \\ \mathscr{X}_{i,l}^{(1)} &= A_i^T G_l^{(1)} + \kappa_l A_i^T G_l^{(3)} - \kappa_l C_i^T Q_{B,l} - \kappa_l Q_{A,l}, \\ \mathscr{X}_{i,l}^{(2)} &= A_i^T G_l^{(2)} + \varsigma_l A_i^T G_l^{(3)} - \varsigma_l C_i^T Q_{B,l} - \varsigma_l Q_{A,l}, \mathscr{X}_{i,l}^{(3)} = \kappa_l Q_{A,l}, \ \mathscr{X}_{i,l}^{(4)} = \varsigma_l Q_{A,l}, \\ \mathscr{Z}_{i,l}^{(1)} &= B_i^T G_l^{(1)} + \kappa_l B_i^T G_l^{(3)} - \kappa_l D_i^T Q_{B,l}, \mathscr{Z}_{i,l}^{(2)} = B_i^T G_l^{(2)} + \varsigma_l B_i^T G_l^{(3)} - \varsigma_l D_i^T Q_{B,l}. \end{aligned} \tag{66}$$

Other notations are presented in Theorems 4 and 5. In this case, the estimator parameters can be given as follows:

$$\begin{bmatrix} \hat{A}_l & \hat{B}_l \\ \hat{C}_l & \hat{D}_l \end{bmatrix} = \begin{bmatrix} G_l^{(3)^{-T}} Q_{A,l}^T & G_l^{(3)^{-T}} Q_{B,l}^T \\ \hat{C}_l & \hat{D}_l \end{bmatrix}, \forall l \in \mathscr{H}. \tag{67}$$

Proof The LMI (65) together with $Q_{A,l} = \hat{A}_l^T G_l^{(3)}$, $Q_{B,l} = \hat{B}_l^T G_l^{(3)}$ and $G_l = \begin{bmatrix} G_l^{(1)} & G_l^{(2)} \\ \kappa_l G_l^{(3)} & \varsigma_l G_l^{(3)} \end{bmatrix}$ ($l \in \mathscr{H}$) yields

$$\begin{bmatrix} \hat{\Pi}_i & \tilde{\Gamma}_i G & \Omega_i & \tilde{\Upsilon}_i \\ * & \Psi_i & 0 & 0 \\ * & * & -(b_3 + \mu b_1)^{-1} I & 0 \\ * & * & * & -I \end{bmatrix} < 0, \tag{68}$$

where

$$\tilde{\Gamma}_i = [\sqrt{\rho_{i,1}}\Gamma_{i,1}, \sqrt{\rho_{i,2}}\Gamma_{i,2}, \ldots, \sqrt{\rho_{i,h}}\Gamma_{i,h}]. \tag{69}$$

Here, $\Gamma_{i,l}$ $(l = 1, 2, \ldots, h)$ are defined in Theorem 4. Noting that $-G^T(I_h \otimes \tilde{P}_i^{-1}) G \leq \Psi_i, \forall i \in \mathscr{S}$, we obtain from (68) that

$$\begin{bmatrix} \hat{\Pi}_i & \tilde{\Gamma}_i G & \Omega_i & \tilde{\Upsilon}_i \\ * & -G^T(I_h \otimes \tilde{P}_i^{-1})G & 0 & 0 \\ * & * & -(b_3+\mu b_1)^{-1}I & 0 \\ * & * & * & -I \end{bmatrix} < 0. \tag{70}$$

In addition, the LMI (65) ensures that $\Psi_i = I_h \otimes \tilde{P}_i - G^T - G < 0$, which together with $\tilde{P}_i > 0$ yields that G is nonsingular. Let $\Sigma = \text{diag}\{I, G^{-1}, I, I\}$. Simple matrix manipulations using Σ^T and Σ yield

$$\begin{bmatrix} \hat{\Pi}_i & \tilde{\Gamma}_i & \Omega_i & \tilde{\Upsilon}_i \\ * & -I_h \otimes \tilde{P}_i^{-1} & 0 & 0 \\ * & * & -(b_3+\mu b_1)^{-1}I & 0 \\ * & * & * & -I \end{bmatrix} < 0 \tag{71}$$

By the Schur complement [24], we know that (60) holds if and only if (71) is satisfied. According to Theorem 5, the system (46) is proved to be SH_∞FTB with respect to $(c_1, c_2, U, N, \gamma, \theta)$.

3.3 Numerical Example

In this section, we consider the following numerical example.

The system (42) involves two subsystems and, therefore, $\mathscr{S} = \{1, 2\}$. The system parameters of the subsystem 1 and subsystem 2 are given as follows, respectively.

Subsystem 1:

$$A_1 = \begin{bmatrix} 0.5 & 0.2 \\ -0.7 & 0.2 \end{bmatrix}, \quad B_1 = \begin{bmatrix} 0.1 \\ -0.1 \end{bmatrix}, C_1 = [0.6, \ -1.4],$$
$$D_1 = 0.1, L_1 = [0.1, \ 0.1], \ E_1 = 0.1.$$

Subsystem 2:

$$A_2 = \begin{bmatrix} -0.4 & 0.25 \\ 0.65 & 0.3 \end{bmatrix}, \quad B_2 = \begin{bmatrix} -0.75 \\ 0.5 \end{bmatrix}, C_2 = [2.1, \ -2],$$
$$D_2 = 0.2, \ L_2 = [0.2, \ 0.2], \ E_2 = 0.2.$$

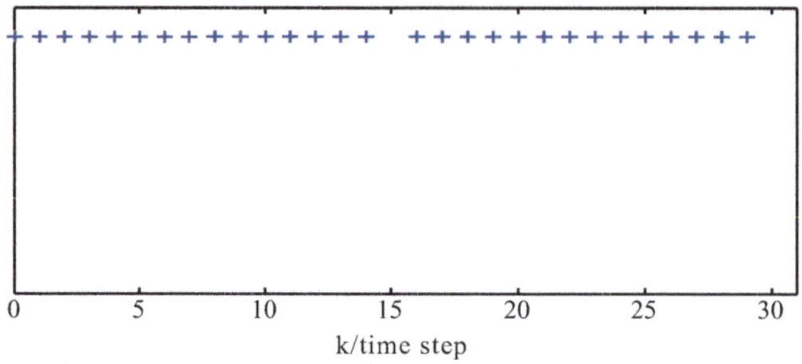

Fig. 5 Triggering instants

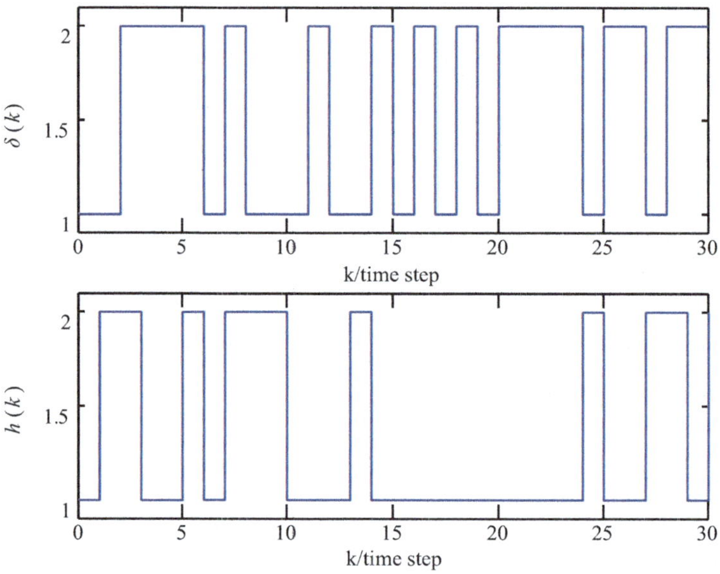

Fig. 6 Markov mode $\delta(k)$ and its observed mode $h(k)$

Assume that two modes of the system (42) can be observed and, therefore, $\mathscr{H} = \{1, 2\}$. The matrices Λ and Δ are as follows:

$$\Lambda = \begin{bmatrix} \varepsilon_{11} & \varepsilon_{12} \\ \varepsilon_{21} & \varepsilon_{22} \end{bmatrix} = \begin{bmatrix} 0.2 & 0.8 \\ 0.35 & 0.65 \end{bmatrix}, \quad \Delta = \begin{bmatrix} \rho_{11} & \rho_{12} \\ \rho_{21} & \rho_{22} \end{bmatrix} = \begin{bmatrix} 0.5 & 0.5 \\ 0.75 & 0.25 \end{bmatrix}.$$

The parameters in the DETM (5) are set as $a_1 = 0.1$, $a_2 = 0.5$, $b_1 = 0.002$, $b_2 = 4$, $b_3 = 1$, $b_4 = 0.2$. Let $\kappa_1 = \kappa_2 = 0.1$, $\varsigma_1 = \varsigma_2 = 1$, $\mu = 1$, $\alpha = 1.01$, $c_1 = 0.01$, $c_2 = 2$, $N = 30$, $\theta = 0.02$, $U = I_4$. It is found that the LMIs (50), (61) and (65) are feasible and the optimal H_∞ performance index is $\gamma^* = 3.9526$. By solving these LMIs in Theorem 6, we obtain the following state estimator parameters:

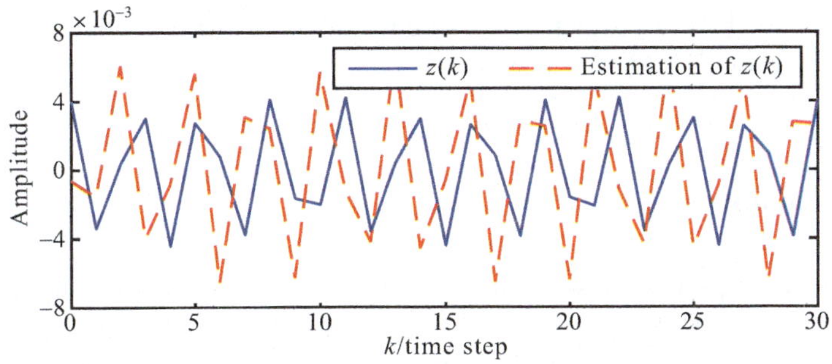

Fig. 7 Trajectories of $z(k)$ and its estimation

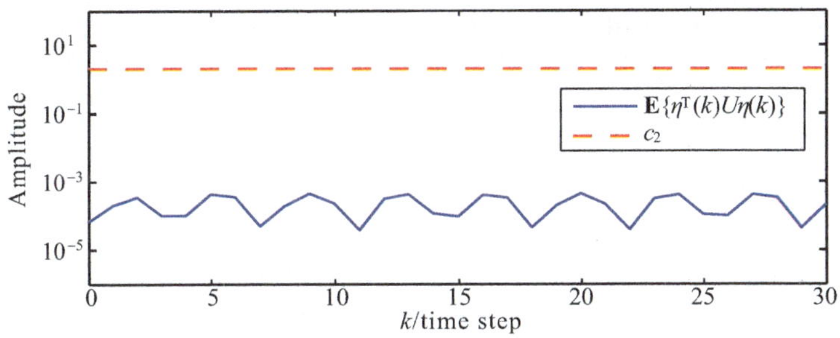

Fig. 8 $\mathbf{E}\{\eta^T(k)U\eta(k)\}$ and the bound c_2

$$\hat{A}_1 = \begin{bmatrix} 0.4296 & -0.2421 \\ -0.1281 & 0.4327 \end{bmatrix},\ \hat{B}_1 = \begin{bmatrix} -0.3962 \\ 0.1462 \end{bmatrix},$$
$$\hat{C}_1 = [0.2826,\ 0.3352],\ \hat{D}_1 = 0.0502,$$
$$\hat{A}_2 = \begin{bmatrix} 0.4816 & -0.3433 \\ -0.3494 & 0.1535 \end{bmatrix},\ \hat{B}_2 = \begin{bmatrix} -0.4032 \\ 0.1107 \end{bmatrix},$$
$$\hat{C}_2 = [0.1276, 0.3454],\ \hat{D}_2 = 0.0817.$$

According to Theorem 6, we know that the system (46) under the DETM (5) is SH_∞FTB. Next, set $x(0) = [-0.004,\ 0.004]^T$ and $\hat{x}(0) = [0,\ 0]^T$. Assume that $w(k) = 0.02\cos(4k)$. Figure 5 shows the triggering instants under the DETM (5). With given matrices Λ and Δ, the Markov mode $\delta(k)$ and its observed mode $h(k)$ are illustrated in Fig. 6. By the designed estimator, the trajectories of $z(k)$ and its estimation are given in Fig. 7. In Fig. 8, the trajectories of $\mathbf{E}\{\eta^T(k)U\eta(k)\}$ and the bound c_2 are depicted. From Fig. 8, we know that the system (46) is SFTB with respect to (c_1, c_2, U, N, θ).

Under the ZIC, we have

$$\sqrt{\frac{\mathbb{E}\left\{\sum_{k=0}^{30} \tilde{z}^T(k)\tilde{z}(k)\right\}}{\sum_{k=0}^{30} w^T(k)w(k)}} = 0.3127 < \gamma^*. \tag{72}$$

Then, we know that (48) is met. The above results verify the effectiveness of the SE scheme under the DETM.

4 Conclusion

The problems of finite-time FD and H_∞ SE have been addressed for Markov jump systems under a DETM. To reduce unnecessary data transmissions, an improved DETM has been devised, which includes some existing ones as special cases. An HMM has been utilized to describe the phenomenon of mode mismatch. Sufficient conditions have been established such that the filtering error system of FD and the estimation error system of SE are SH_∞FTB. Two numerical examples have been provided to demonstrate the effectiveness and merits of the FD and SE schemes.

References

1. Song, J., Wang, Z., Niu, Y.: On H_∞ sliding mode control under stochastic communication protocol. IEEE Trans. Autom. Control **64**(5), 2174–2181 (2019)
2. Wang, L., Wang, Z., Han, Q.-L., Wei, G.: Synchronization control for a class of discrete-time dynamical networks with packet dropouts: a coding-decoding-based approach. IEEE Trans. Cybern. **48**(8), 2437–2448 (2018)
3. Wan, X., Wang, Z., Wu, M., Liu, X.: H_∞ state estimation for discrete-time nonlinear singularly perturbed complex networks under the Round-Robin protocol. IEEE Trans. Neural Netw. Learn. Syst. **30**(2), 415–426 (2019)
4. Wan, X., Wang, Z., Han, Q.-L., Wu, M.: Finite-time H_∞ state estimation for discrete time-delayed genetic regulatory networks under stochastic communication protocols. IEEE Trans. Circuits Syst.-I: Regul. Pap. **65**(10), 3481–3491 (2018)
5. Wan, X., Wang, Z., Wu, M., Liu, X.: State estimation for discrete time-delayed genetic regulatory networks with stochastic noises under the Round-Robin protocols. IEEE Trans. Nanobioscience **17**(2), 145–154 (2018)
6. Wan, X., Fang, H., Fu, S.: Observer-based fault detection for networked discrete-time infinite-distributed delay systems with packet dropouts. Appl. Math. Model. **36**(1), 270–278 (2012)
7. He, X., Wang, Z., Zhou, D.H.: Robust fault detection for networked systems with communication delay and data missing. Automatica **45**(11), 2634–2639 (2009)
8. Li, F., Shi, P., Wang, X., Agarwal, R.: Fault detection for networked control systems with quantization and Markovian packet dropouts. Signal Process. **111**, 106–112 (2015)
9. Liu, H., Wang, Z., Shen, B., Liu, X.: Event-triggered H_∞ state estimation for delayed stochastic memristive neural networks with missing measurements: the discrete time case. IEEE Trans. Neural Netw. Learn. Syst. **29**(8), 3726–3737 (2018)
10. Wang, L., Wang, Z., Wei, G., Alsaadi, F.E.: Finite-time state estimation for recurrent delayed neural networks with component-based event-triggering protocol. IEEE Trans. Neural Netw. Learn. Syst. **29**(4), 1046–1057 (2018)
11. Jin, Z., Hu, Y., Sun, C.: Event-triggered state estimation for stochastic hybrid systems with missing measurements. IET Control Theory Appl. **12**(18), 2551–2561 (2018)
12. Li, L., Yu, D., Xia, Y., Yang, H.: Remote nonlinear state estimation with stochastic event-triggered sensor schedule. IEEE Trans. Cybern. **49**(3), 734–745 (2019)

13. Li, Q., Shen, B., Wang, Z., Huang, T., Luo, J.: Synchronization control for a class of discrete time-delay complex dynamical networks: a dynamic event-triggered approach. IEEE Trans. Cybern. **49**(5), 1979–1986 (2019)
14. Girard, A.: Dynamic triggering mechanisms for event-triggered control. IEEE Trans. Autom. Control **60**(7), 1992–1997 (2015)
15. Ma, L., Wang, Z., Cai, C., Alsaadi, F.E.: Dynamic event-triggered state estimation for discrete-time singularly perturbed systems with distributed time-delays. IEEE Trans. Syst. Man Cybern.: Syst. (2018). https://doi.org/10.1109/TSMC.2018.2876203
16. Ge, X., Han, Q.-L., Wang, Z.: A dynamic event-triggered transmission scheme for distributed set-membership estimation over wireless sensor networks. IEEE Trans. Cybern. **49**(1), 171–183 (2019)
17. Hajshirmohamadi, S., Sheikholeslam, F., Davoodi, M., Meskin, N.: Event-triggered simultaneous fault detection and tracking control for multi-agent systems. Int. J. Control (2019). https://doi.org/10.1080/00207179.2017.14202
18. Xiao, S., Zhang, Y., Zhang, B.: Event-triggered networked fault detection for positive Markovian systems. Signal Process. **157**, 161–169 (2018)
19. Dong, S., Wu, Z.-G., Pan, Y.-J., Su, H., Liu, Y.: Hidden-Markov-model-based asynchronous filter design of nonlinear Markov jump systems in continuous-time domain. IEEE Trans. Cybern. **49**(6), 2294–2304 (2019)
20. Zhu, Y., Zhong, Z., Zheng, W.X., Zhou, D.: HMM-based H_∞ filtering for discrete-time Markov jump LPV systems over unreliable communication channels. IEEE Trans. Syst. Man Cybern.: Syst. **48**(12), 2035–2046 (2018)
21. Hajshirmohamadi, S., Davoodi, M., Meskin, N., Sheikholeslam, F.: Event-triggered fault detection and isolation for discrete-time linear systems. IET Control. Theory Appl. **10**(5), 526–533 (2016)
22. Yang, L., He, X., Wang, Z., Zhou, D.-H.: Fault detection and diagnosis for a class of nonlinear systems with decentralized event-triggered transmissions. IFAC-PapersOnLine **48**(21), 1134–1139 (2015)
23. Li, F., Xu, S., Zhang, B.: Resilient asynchronous H_∞ control for discrete-time Markov jump singularly perturbed systems based on hidden Markov model. IEEE Trans. Syst. Man Cybern.: Syst. (2018). https://doi.org/10.1109/TSMC.2018.2837888
24. Wan, X., Fang, H.: Fault detection for networked nonlinear systems with time delays and packet dropouts. Circuits Syst. Signal Process. **31**(1), 329–345 (2012)

Intelligent Control and Decision-Making of Complex Metallurgical Processes

Intelligent Control of Sintering Process

Min Wu, Sheng Du, Jie Hu, Xin Chen, Weihua Cao, and Witold Pedrycz

Abstract The sintering process produces sinters as blast furnace materials. The quality of the sinter and the stability of the sintering process are related to the efficiency, energy consumption, and cost of the blast furnace production. The sintering thermal state parameters directly affect the quality of the sinter and the stability of the sintering process. The stability of the sintering process is based on the stable control of sintering thermal parameters. This chapter introduces the sintering process characteristics, presents a carbon efficiency optimization scheme, and describes the intelligent control of burn-through point and sintering ignition temperature.

Keywords Sintering process · Intelligent control · Sintering ignition temperature · Burn-through point · Carbon efficiency

Abbreviations

SIT	Sintering ignition temperature
TSD	Time series trend
PID	Proportional-integral-derivative
LS-SVM	Least-squares support vector machine
OCSCE	Optimization and control system of the carbon efficiency
HMI	Human machine interface

M. Wu (✉) · S. Du · J. Hu · X. Chen · W. Cao
School of Automation, China University of Geosciences, Wuhan 430074, China
e-mail: wumin@cug.edu.cn

Hubei Key Laboratory of Advanced Control and Intelligent Automation for Complex Systems, Wuhan 430074, China

S. Du · J. Hu · W. Pedrycz
Department of Electrical & Computer Engineering, University of Alberta, Edmonton AB T6R 2V4, Canada

M. Wu et al. (eds.), *Developments in Advanced Control and Intelligent Automation for Complex Systems*, Studies in Systems, Decision and Control 329,
https://doi.org/10.1007/978-3-030-62147-6_5

1 Sintering Process and Characteristics Analysis

This section describes sintering process, analyzes its characteristics, and introduces the carbon efficiency and thermal state parameters.

1.1 Iron Ore Sintering Process

Dwight-Lloyd sintering machine is adopted in sintering process, as shown in Fig. 1. The qualified sinter produced by the sintering process is utilized in the blast furnace process [1]. In this process, the ore is mixed with coke, returning sinter, and fine particles of limestone; and the heat generated by coke combustion can cause agglomeration. Finally, sintering is completed at the burn-through point.

The second most energy-intensive process in steel smelting is the sintering process [2]. The burning of carbon affects the process energy consumption. Reducing the energy consumption of carbon is of major importance to the realization of energy saving and consumption reduction in the steel industry. Thus, it is advantageous to realize carbon efficiency optimization and improve energy utilization. In the sintering process, the comprehensive coke ratio is used as an indicator of carbon efficiency [3–5]. The key to improving carbon efficiency is to first propose the hybrid prediction model of comprehensive coke ratio and then design an optimization strategy based on that model. In addition, on this basis, the quality of the sinter must also be guaranteed, which is directly related to the thermal state parameters of the process.

There are two important thermal parameters in the sintering process, namely the burn-through point and the ignition temperature. The ignition temperature greatly influences the initial melting state of the mixture and thus the quality of sinter. The stability of the burn-through point and the ignition temperature has a significant effect

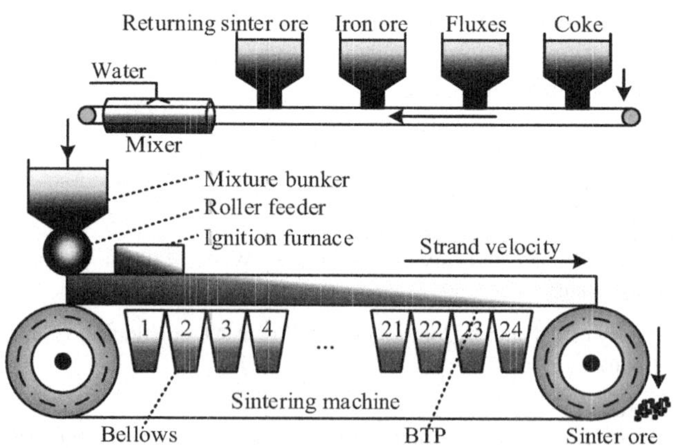

Fig. 1 Sintering process

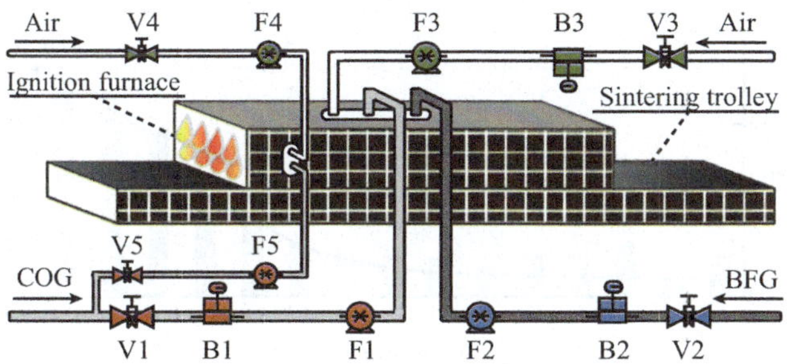

Fig. 2 Sintering ignition process

on the sinter's the quality and quantity. If the burn-through point occurs before the optimal position, the utilization factor is too low and the area of the sintering machine is not utilized efficiently. If the burn-through point occurs after the optimum position, part of the ore is still not sintered and the agglomeration rate is too slow. This decreases the quality and quantity of sinters and increases the amount of returning sinter. Besides, frequent changes in the burn-through point can shorten the life of sintering machine. Therefore, a key issue in sintering process control is to successfully follow the burn-through point and maintain it at an acceptable value.

The ignition furnace is the principal thermal transfer device in the sintering ignition process. Take the double inclined belt sintering furnace as an example. Figure 2 illustrates the structure of ignition furnace, which is located at the front of sintering machine and is fueled by a mixture of coke oven gas and blast furnace gas. There are two types of combustion reactions in an ignition furnace. One is the ignition reaction of air and coke oven gas, and the other is the ignition reaction between air and gas mixture. The ignition reaction stops when the sintering machine is in normal operation. The gas mixture stream in the ignition furnace is ignited at a negative pressure when the gas mixture is placed on the sintering machine. There is a direct transfer of heat through the hot flame to the gas mixture on the surface. The gas mixture then starts to fire.

Sintering is a nonlinear process with parametric coupling. Currently, there is no instrument that can measure burn-through point directly online. Image monitoring, exhaust gas (T_{Bel}) composition analysis, and exhaust gas pressure detection have been presented to measure the burn-through point, but the application of such approaches has been unsatisfactory due to poor real-time performance. Soft-sensing models are extensively employed to achieve the burn-through point measurement. 24 bellows are located on the sintering machine, and the temperature sensors are located only

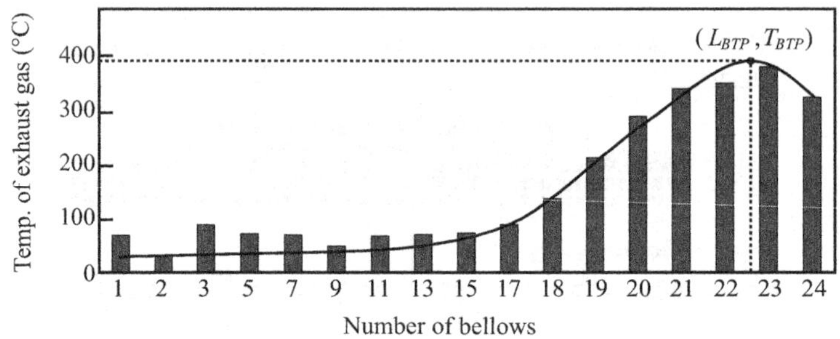

Fig. 3 Burn-through point soft-sensing model

on bellows NO. 1, 2, 3, 5, 7, 9, 11, 13, 15, 17, 18, 19, 20, 21, 22, 23, and 24. Burn-through point soft-sensing model works by taking the location on the sintering strand corresponding to the highest T_{Bel} as burn-through point (see Fig. 3). The burn-through point and burn-through point temperature can be fitted to the curve. A quadratic or higher-order polynomial function is employed to fit the allocation of T_{Bel}.

As shown in Fig. 3, the burn-through point is a variable that is strongly associated with the temperature change or combustion reaction. At the same time, thermal transfer is a time-dependent process. The present burn-through point is influenced by the earlier phase of it. Thus, analyzing the change in burn-through point from a time series perspective is feasible.

Sintering ignition is the beginning of the entire combustion process [6]. As intelligent control has evolved, including fuzzy control, expert control, neural network control, etc., many intelligent control techniques have been adopted for ignition temperature control. Zhang et al. [7] proposed an ignition temperature intelligent control method, which uses a field integrated bus to control air flow and gas flow. Ying et al. [8] introduced a proportional-integral-derivative (PID) neural network for controlling the sintering ignition temperature (SIT), overcoming the drawbacks of traditional PID. Endiyarov et al. proposed a sintering ignition adaptive control method, which depends on an adaptive model and continually improves the functionality of the model according to real-time knowledge of the sintering ignition process [9]. Cao et al. developed a continuous sintering furnace ignition temperature intelligent control system [10, 11]. Chen et al. [12] proposed an advanced SIT control method with improved equivalent input disturbance method.

The aforementioned methods of controlling the SIT provide a valid theory base of implementation on an industrial basis. But, the majority of these methods are dependent on data and operational experience. These methods may not be applicable when operating conditions change, such as fluctuations in gas calorific value and pressure.

The gas used in a steel plant has a problem with pressure fluctuations. This problem can be effectively solved if the gas is processed in the pressurization station. However, many steel plant do not pressurize the gas because of the additional production costs that may be incurred. Thus, an intelligent control method of SIT is studied, which has important scientific and economic value.

An intelligent control method according to ignition temperature prediction is proposed to address the instability of gas pressure in steel plant. The gas pressure is taken as the constraint, the gas valve as the actuator, and the ignition temperature as control target. At the end, the validity of the proposed method is experimentally demonstrated.

In burn-through point control, the problem is that sintering is a complex and dynamic process that is nonlinear, coupled, and time-delayed, making it it hard to establish a precise numerical model. Conventional methods depend heavily on operator experience, resulting in burn-through point control that still depends on individual judgment, experience, psychology, responsibility, and other human factors. When handling complex operating modes, these methods, in addition to increasing labor intensity, are no longer sufficient. So, how to stabilize the burn-through point remains an ongoing concern.

In recent years, in addition to the sintering process [13, 14], a wide range of fuzzy controls are already in use in various fields, such as industrial robots [15], wastewater treatment process [16], cleaning process [17], drilling process [18], food industry [19], agricultural production process [20], and other chemical processes [21–24]. Besides, fuzzy control has changed from traditional fuzzy control to stable adaptive fuzzy control, evolutionary fuzzy control, gray predictive fuzzy control, optimal fuzzy control, network-based fuzzy control, etc. These methods facilitate the integration of fuzzy control with other methods and expand the scope of application of fuzzy control.

Time series forecasting exploits the continuously changing nature of an object by using data from previous series for predicting its future trend. Most industrial processes are continuous, a fact that provides great chances for the implementation of time series forecasting. For example, Majidpour et al. adapted time series forecasting to demand forecasting for charging stations [25]. In addition, there are many successful applications in predicting the overflow of combined sewers [26] and the failure probability of tool condition monitoring [27]. Besides, several time series models such as the improved gray model [28], the multiple models [29], the higher-order time correlation model [30] have also been developed. These studies have demonstrated the validity of time series forecasting.

Time series trend analysis is a vital way for analyzing monotonic trends in time series and has applications in a number of areas, such as meteorology, climate, temperature, and precipitation [31–33]. In recent years, the prediction of fuel leaks from underground storage tanks [34] and the assessment of land degradation [35] have also benefited from this method. Thus, it is anticipated that this method to analyze time series trends will also be useful in applying to the sintering process.

The sintering process exhibits multiple parameters, nonlinearity, strong coupling, and multiple operating conditions. To come true the smooth and steady operation of the process, the complex characteristics of the process must be analyzed. Furthermore, intelligent control of the thermal state parameters of the process can be achieved under the premise of achieving optimal carbon efficiency.

1.2 Characteristics Analysis for Sintering Process

The sintering process not only has strong nonlinearities and large time constants, but also many factors are strongly coupled with each other. The analysis of the sintering technique reveals the following characteristics.

(1) Multiple types of parameters:

   (a) Ignition process parameters: the blast furnace gas valve (V_B), coke oven gas valve (V_C), air valve (V_A), and blast furnace gas flow (Q_B), coke oven gas flow (Q_C), air flow (Q_A), blast furnace gas pressure (P_B), coke oven gas pressure (P_C), air pressure (P_A), air-fuel ratio of the coke oven gas (r_C), and air-fuel ratio of the blast furnace gas (r_B), the SIT (T_I).
   (b) Raw material parameters: the proportion of coke (C_C), the returning sinter (C_{RS}), the CaO content (C_{CaO}), the SiO_2 content (C_{SiO_2}), the MgO content (C_{MgO}), and the TFe content (C_{TFe}).
   (c) Operating parameters: the strand velocity (V_S) and the height of the sinter material layer (H).
   (d) State parameters: the bellows negative pressure (N_P), the average vertical burning speed (V_{avs}), the burning-rising point temperature (T_{BRP}), the burning-rising point (L_{BRP}), the burn-through point temperature (T_{BTP}), and the burn-through point (L_{BTP}).

(2) Strong Nonlinearity: the sintering process is a complex physical and chemical process involving oxidation-reduction and solid-state reactions, decomposition of raw materials, etc. The stability of the process is influenced by ignition temperature, chemical composition and particle size distribution of the feedstock, burn-through point, and many other factors. These parameters change over time, and the relationship between them is highly nonlinear. As a result, an accurate numerical model is difficult to develop.
(3) Strong coupling: mixing, blending, feeding, granulation, and sintering are closely related in sintering process; and the coupling between sintering and feeding is particularly strong.
(4) Multiple operating conditions: sintering employs multiple ingredient lists to guide production, each representing one operating condition. Different characteristics of the sintering production data exist for different operating conditions.
(5) Large time constant: the process from charging to discharging requires approximately one hour. Large time constants indicate that a long lag time elapses between the result and the control action. Thus, it is often too late to execute another control action if the results are unsatisfactory. It leads to unsatisfactory control results and the inability to adjust the burn-through point in time when problems occur.

1.3 Control Objectives

To ensure stable operation and energy-saving and emission reduction of the sintering process, it is essential to realize intelligent control of the thermal state parameters of the process under the premise of carbon efficiency optimization.

A. Control Objective of Sintering Ignition Temperature

The ignition effect of sintering is typically evaluated in terms of SIT and the ignition intensity. The SIT is the temperature reached by the gas as it burns in the ignition furnace. The ignition intensity is the amount of heat or gas supplied to the mixture per unit area during the ignition process. The SIT and ignition intensity are not independent of each other. Usually, a specific relationship exists between SIT and ignition intensity. If the SIT is too low or the ignition intensity is not strong enough, the sintered surface will be poorly formed and may even be impossible to form sinter. If the SIT or intensity is too high, the sintered body on the surface will be over-melted and of undesirable quality, resulting in unwanted wastage of fuel.

The ignition effect of sintering is controlled mainly via the control of SIT, which depends on an ignition point of a solid fuel in a sintering mixture. In general, the SIT has not changed much. In the case of the unstable gas pressure, the optimal ignition temperature is preferably 1150 °C, and the SIT variation should be ±50 °C.

B. Control Objective of Burn-Through Point

The thermal state is a direct reflection of the combustion of sintering process. As an important parameter of the thermal state, the burn-through point is the first location on the sintering strand where the mixed material is completely burned. In other words, if the burn-through point is at the front of the target position, the efficient area of the sintering strand is under utilized and the sintering strand will be less utilized. Conversely, sintering strand are not fully burned during unloading, leading to an increases in returning sinter and a reduction in output. Thus, stabilizing the burn-through point plays an important role in improving economic efficiency.

The previous methods show the excellent properties of fuzzy control to control the burn-through point. Nonetheless, some approaches were used to study the sintering process from a time series perspective. Because of the temperature influence of the combustion reaction, the time series data of the burn-through point are autocorrelated. Therefore, it is very interesting to study the time series trends based on the burn-through point.

In a steel plant, the sintering process employs a large sintering bed with high and low strand velocity (i.e., H is approximately 860 mm, and V_S is approximately 2 m/min). In this case, the required burn-through point is the located position at the bellows No. 22.5, with an allowable variation of 0.5 bellows.

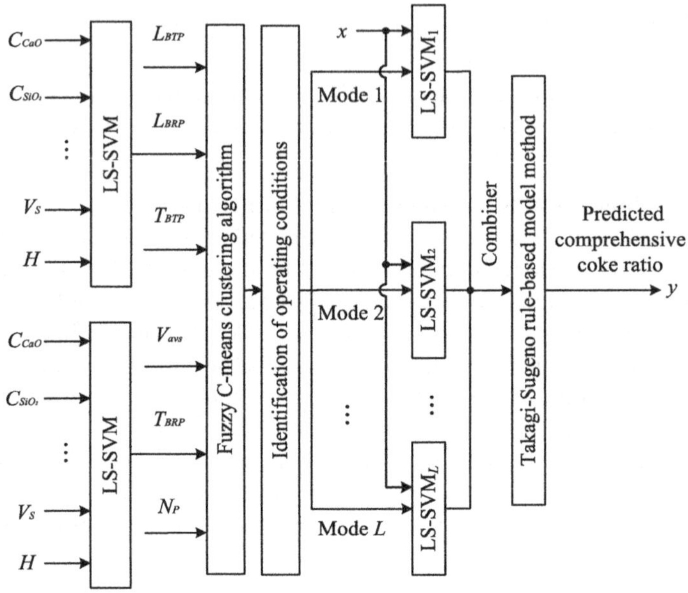

Fig. 4 Configuration of comprehensive coke ratio prediction model

2 Carbon Efficiency Prediction and Optimization

This section develops a configuration of a hybrid prediction model for comprehensive coke ratio and presents an optimization scheme for comprehensive coke ratio.

2.1 Carbon Efficiency Hybrid Prediction Model

The hybrid prediction model is proposed to realize an effective forecasting of comprehensive coke ratio (see Fig. 4). It consists of three points: identification of operating conditions, individual modeling, and aggregation.

The sintering parameters that affect the comprehensive coke ratio should be determined before the identification of operating conditions. According to the sintering mechanism, the state parameters $(N_P, V_{avs}, T_{BRP}, L_{BRP}, T_{BTP}, L_{BTP})$ have a direct impact on the comprehensive coke ratio. To ascertain the correlation between comprehensive coke and various process parameters, there is a Pearson's correlation coefficient employed here.

Pearson's correlation coefficient equals to the product of the covariance of two variables divided by their standard deviations. Assume that two variables are $X = [X_1, \dots, X_i, \dots, X_N]^T$ and $Y = [Y_1, \dots, Y_i, \dots, Y_N]^T$. The Pearson's correlation coefficient of X and Y is

Table 1 Pearson's correlation coefficients for the comprehensive coke ratio and state parameters

	N_P	T_{BTP}	L_{BTP}	L_{BRP}	T_{BTP}	V_{avs}
Comprehensive coke ratio	−0.284	0.307	−0.259	−0.267	0.214	−0.329

$$r = \frac{\sum_{i=1}^{N}(X_i - \bar{X})(Y_i - \bar{Y})}{\sqrt{\sum_{i=1}^{N}(X_i - \bar{X})^2}\sqrt{\sum_{i=1}^{N}(Y_i - \bar{Y})^2}}, \quad (1)$$

where $\bar{X}$ and $\bar{Y}$ are the average of X and Y, respectively. r is in the range $[-1, 1]$.

Table 1 shows the Pearson's correlation coefficients for the comprehensive coke ratio and the state parameters. A strong correlation is represented by an absolute value of Pearson's correlation coefficient that is larger or equal to 0.2. Therefore, these parameters in Table 1 are available as inputs to a comprehensive coke ratio prediction model.

Since the sintering process exhibits multiple operating conditions, it is difficult to accurately predict the process with a single model. It makes sense to identify different operating conditions through clustering methods. Fuzzy C-means clustering is adopted here to split data into different clusters.

There are nonlinearities in the sintering process. Least-squares support vector machine (LS-SVM) is the preferred option for building a comprehensive coke ratio prediction model due to the advantages of LS-SVM in solving nonlinear problems, and the LS-SVM is available as a individual modeling approach that then builds LS-SVM model for different operating conditions. In order to enhance the forecasting precision of comprehensive coke ratio, all models are integrated using a model approach based on the Takagi–Sugeno rule as an aggregation strategy. The comprehensive coke ratio prediction model is then established. The comprehensive coke ratio prediction model is explained in the following subsections.

A. Identification of Operating Conditions

Since L_{BRP} and T_{BRP} are not recorded in daily report, it is important to develop their predictive models, which are directly influenced by raw material parameters (C_{RS}, C_{MgO}, C_{SiO_2}, C_C, C_{TFe}, C_{CaO}) and operating parameters (V_S, H). The LS-SVM is then used to build the L_{BRP} and T_{BRP} prediction models. Comprehensive coke ratio prediction model is built using the prediction results of L_{BRP} and T_{BRP}.

Before building the comprehensive coke ratio prediction model, multiple operating conditions should be identified through fuzzy C-means clustering [36]. The state parameters (N_P, V_{avs}, T_{BRP}, L_{BRP}, T_{BTP}, L_{BTP}) serve as inputs to this comprehensive coke ratio model. The input sample data constitute $S = \{x_i \in \mathbb{R}^6 | i = 1, 2, \ldots, N\}$, where N is the number of sample.

Fuzzy C-means algorithm splits S into L clusters (i.e., L operating conditions) by minimizing J_m,

$$\min J_m = \sum_{k=1}^{L}\sum_{i=1}^{N} (\mu_{ki})^m d_{ki}^2,$$
$$\text{s.t.} \begin{cases} \sum_{k=1}^{L} \mu_{ki} = 1, \\ 0 < \mu_{ki} < 1, \end{cases} \tag{2}$$

where L is the number of clusters, m is the fuzzification coefficient, μ_{ki} is the membership degree of the ith sample in the kth cluster, $C = [c_1, c_2, \dots, c_L]$ is the prototypes matrix, and $d_{ki} = \|x_i - c_k\|$.

Fuzzy C-means has been the iterative algorithm for creating prototypes and partition matrices.

$$\mu_{ki} = \frac{1}{\sum_{f=1}^{L} (d_{ki}/d_{fi})^{\frac{2}{m-1}}}, \tag{3}$$

$$c_k = \frac{\sum_{i=1}^{N} (\mu_{ki})^m x_i}{\sum_{i=1}^{N} (\mu_{ki})^m}. \tag{4}$$

B. Individual Modeling

Because of its low computational complexity and versatility, LS-SVM is successfully applied in industrial processes modeling. Therefore, it is favoured as an individual modeling approach. Set the input and output sample as $(x_i, y_i),\ i = 1, 2, \dots, N$. The kth operating condition is represented by $\{(x_i, y_i),\ i = 1, 2, \dots, N_k\}$, and $N = \sum_{k=1}^{L} N_k$. The linear regression function of the LS-SVM for kth operating condition is

$$y^k = \omega^T \varphi(x_i) + b, \tag{5}$$

where ω is a vector of weights, b is a offset, and $\varphi(\cdot)$ is a nonlinear function.

Equation (5) translates into a constrained optimization problem based on the structural risk minimization principle.

$$\begin{cases} \min\ J = \frac{1}{2}\omega^T\omega + \frac{\gamma}{2}\sum_{i=1}^{N_k} (e_i)^2, \\ \text{s.t.}\ y^k = \omega^T \varphi(x_i) + b + e_i, \quad i = 1, 2, \dots, N_k, \end{cases} \tag{6}$$

where e_i is a modeling error, and γ ($\gamma > 0$) is a regularization parameter.

The optimization problem (6) is resolved using the Karush–Kuhn–Tucker condition, Lagrange function, and Mercer condition. The LS-SVM model is then expressed as

$$\left\{ y^k = \sum_{i=1}^{N_k} a_i K(x, x_i) + b, \right. \tag{7}$$

where $K(x, x_i)$ is a radial basis function, i.e., $K(x, x_i) = \exp\left\{\frac{-\|x-x_i\|^2}{2\sigma^2}\right\}$, σ is a kernel width parameter, and a_i is a Lagrange multiplier.

C. Individual Models Aggregation

After using the fuzzy C-means, different operating conditions are identified with corresponding membership degrees. Then, all models for different operating conditions are aggregated using the membership degrees to form the comprehensive coke ratio prediction model,

$$y = \sum\nolimits_{k=1}^{L} \mu_{ki} y^k = f(x). \tag{8}$$

In order to verify the validity of comprehensive coke ratio prediction model, the root-mean-square error ($RMSE$), average relative error (ARE), maximum relative error ($E_{\max}$), and standard deviation (σ) are selected as the evaluation metrics.

$$ARE = \frac{1}{M}\sum_{t=1}^{M}\frac{|y^*(t) - y(t)|}{y^*(t)} \times 100\%, \tag{9}$$

where M is the number of the testing sample, $\hat{y}(t)$ is the actual comprehensive coke ratio, and $y(t)$ is the predicted comprehensive coke ratio.

$$RMSE = \sqrt{\frac{1}{M}\sum_{t=1}^{M}[y^*(t) - y(t)]^2}. \tag{10}$$

$$\sigma = \sqrt{\frac{1}{M}\sum_{t=1}^{M}[y(t) - \bar{y}]^2}, \quad \bar{y} = \frac{1}{M}\sum_{t=1}^{M} y^*(t). \tag{11}$$

$$E_{\max} = \max_{1\le t\le M}\left\{\frac{|y^*(t) - y(t)|}{y^*(t)}\right\} \times 100\%. \tag{12}$$

The comprehensive coke ratio prediction model is layered and reflects well the sintering system dynamics. Therefore, the model is suitable for carbon efficiency optimization.

2.2 *Carbon Efficiency Intelligent Optimization*

The goal of intelligent optimization is to achieve the minimization of the comprehensive coke ratio. On the basis of the comprehensive coke ratio model, an optimization scheme is proposed by adjusting the operating parameters (see Fig. 5).

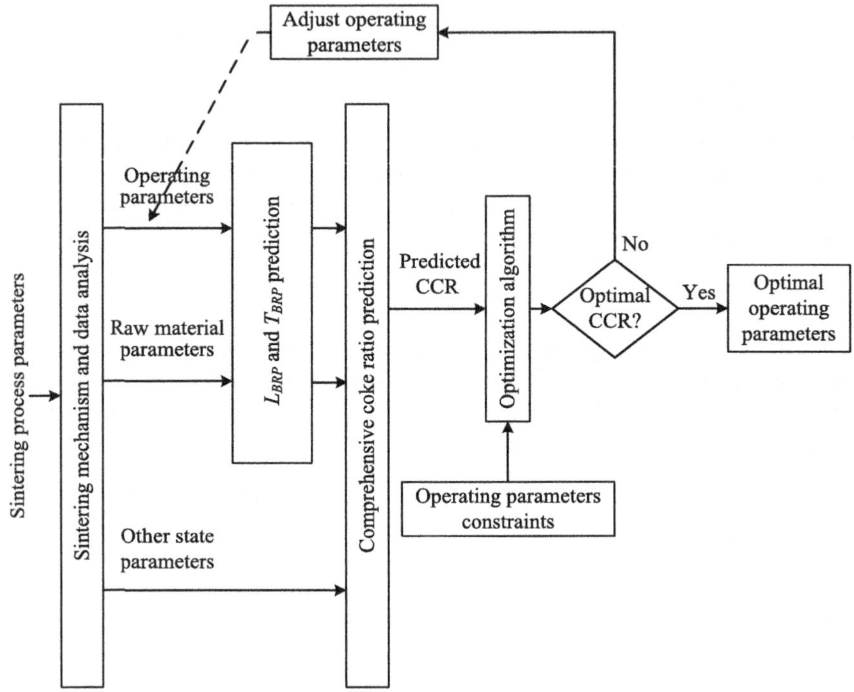

Fig. 5 Carbon efficiency optimization scheme. (CCR: comprehensive coke ratio)

$$\begin{cases} \min \eta, \\ \eta = f(x), \\ x = [V_{\text{avs}}, N_P, L_{BRP}, T_{BRP}, L_{BTP}, T_{BTP}], \\ L_{BRP} = \text{LS–SVM}(C_{CaO}, C_{SiO_2}, \ldots, V_S, H), \\ T_{BRP} = \text{LS–SVM}(C_{CaO}, C_{SiO_2}, \ldots, V_S, H), \end{cases} \tag{13}$$

and the constraints are

$$\begin{cases} V_{S_{\min}} \leq V_S \leq V_{S_{\max}}, \\ H_{\min} \leq H \leq H_{\max}, \end{cases} \tag{14}$$

where $H_{\max}$ are the bottom and top bounds of H and $V_{S_{\min}}$ and $V_{S_{\max}}$ are the bottom and top bounds of V_S, $H_{\min}$.

To optimize comprehensive coke ratio, it is crucial to find an easy-to-implement optimization algorithm, and the particle swarm optimization algorithm is a commonly used method in industrial processes. However, particle swarm optimization is easily caught up in a local optimum. Chaos dynamics is a popular approach used in particle swarm optimization to solve this problem. Then, the chaotic particle swarm optimization is presented, which is characterized by the ergodicity and randomness of the chaotic motion. Therefore, chaotic particle swarm optimization as an optimization algorithm for intelligent optimization strategy is a better choice to deal with the current problem than particle swarm optimization.

Use the following expression to iteratively refresh the position and velocity of the particle swarm optimization.

$$V_d^{j+1} = w(j)V_d^j + \alpha_1 Rand_1({P_{best}}^j - Z_d^j) + \alpha_2 Rand_2(G_{best} - Z_d^j), \tag{15}$$

$$Z_d^{j+1} = Z_d^j + V_d^{j+1}, \tag{16}$$

where $w(j)$ is the inertial weight, j is the iteration index, V_d^j is the particle velocity, $Rand_1$ and $Rand_2$ denote two random numbers within $[0, 1]$, α_1 and α_2 represent the acceleration coefficients within $[1, 2]$, G_{best} is global best position, ${P_{best}}^j$ is personal best position, and Z_d^j is a position of particle d.

In order to prevent being caught in a local optimum, a chaotic local search is used for globally optimal individuals. The chaotic queue can be obtained using logistic equations.

$$W_{o+1} = \mu W_o(1 - W_o), \quad o = 0, 1, 2, \cdots, z, \tag{17}$$

where W is the variable, $0 \leq W_0 \leq 1$ and μ is the control parameter. When $\mu = 4$ and $W_0 \notin \{0.25, 0.5, 0.75\}$, (17) is in the complete state of chaos. A chaotic local search is

$$p_o = p_o + (2W_o - 1), \tag{18}$$

where p_o is the oth variable of G_{best}.

The fitness function of chaotic particle swarm optimization is

$$fitness = f(x). \tag{19}$$

The intelligent optimization with chaotic particle swarm optimization is used to achieve the minimization of comprehensive coke ratio of (13). In the hybrid prediction model of comprehensive coke ratio, L_{BRP} and T_{BRP} take the predicted values, and other state parameters take the actual values of the present state. By optimizing the operating parameters using chaotic particle swarm optimization, the predicted values of L_{BRP} and T_{BRP} are dynamically changed. Then, the model's output is changed. Conclusively, the minimum comprehensive coke ratio and the optimal operating parameters are obtained, which provides instructions for the workers during the sintering process.

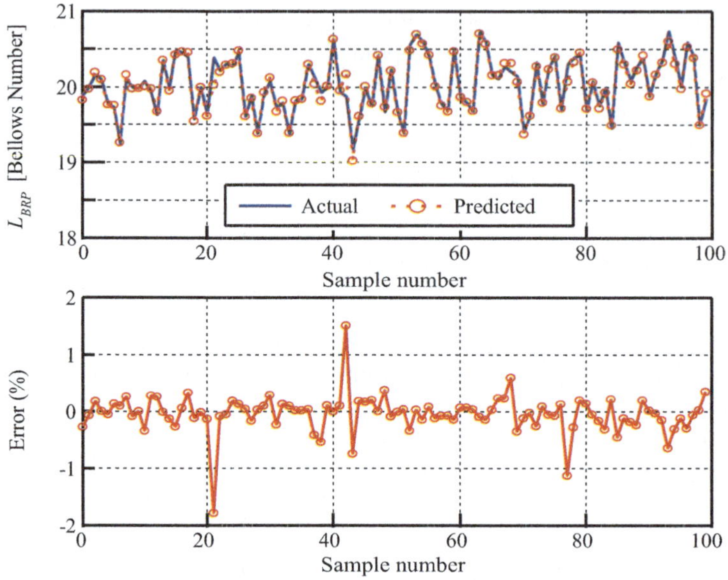

Fig. 6 Prediction results of L_{BRP} model

2.3 *Experimental Results and Analysis*

This subsection conducts simulation experiments using actual run data for the comprehensive coke ratio prediction and optimization.

A. Comprehensive Coke Ratio Prediction

The data of 3000 sets of experimental samples were received from the existing sintering plant with a sampling period of 2 h. These data were subjected to preprocessing and time series registration. At the same time, 2000 sets of data were used for training, and 10 trials were conducted, with 100 sets of data tested in each trial.

Figures 6, 7 denote the prediction results of L_{BRP} and T_{BRP} models. The prediction error of L_{BRP} model is within $[-2\%, 2\%]$, and the prediction error of T_{BRP} model is within $[-3\%, 3\%]$. Thus, the L_{BRP} and T_{BRP} models have high prediction accuracy and these predictions can serve in the comprehensive coke ratio prediction model.

Meanwhile, LS-SVM models are built for 6 operating conditions identified by fuzzy C-means. Then, the comprehensive coke ratio model is built by combining all models using the model approach based on the Takagi–Sugeno rule. To demonstrate the validity of the comprehensive coke ratio model, it is compared with a back propagation neural network model and a multilevel prediction model [3]. The weights of the multilevel prediction model obtained using differential evolution algorithm are: $w_1 = 0.22, w_2 = 0.15, w_3 = 0.17, w_4 = 0.14, w_5 = 0.21$, and $w_6 = 0.11$.

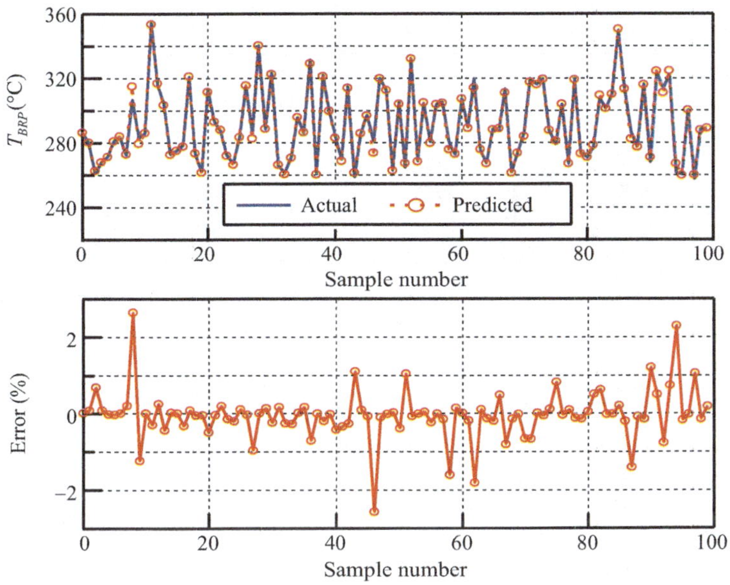

Fig. 7 Prediction results of T_{BRP} model

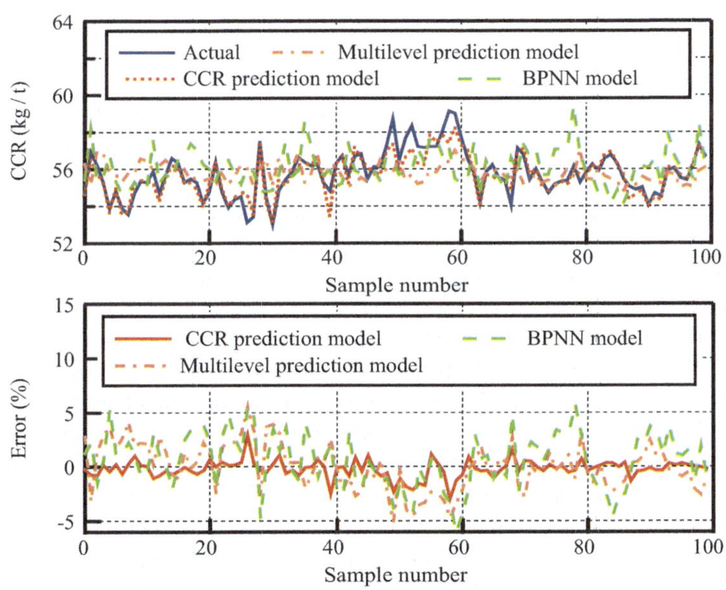

Fig. 8 Comparison of different models. (CCR: comprehensive coke ratio, BPNN: back propagation neural network)

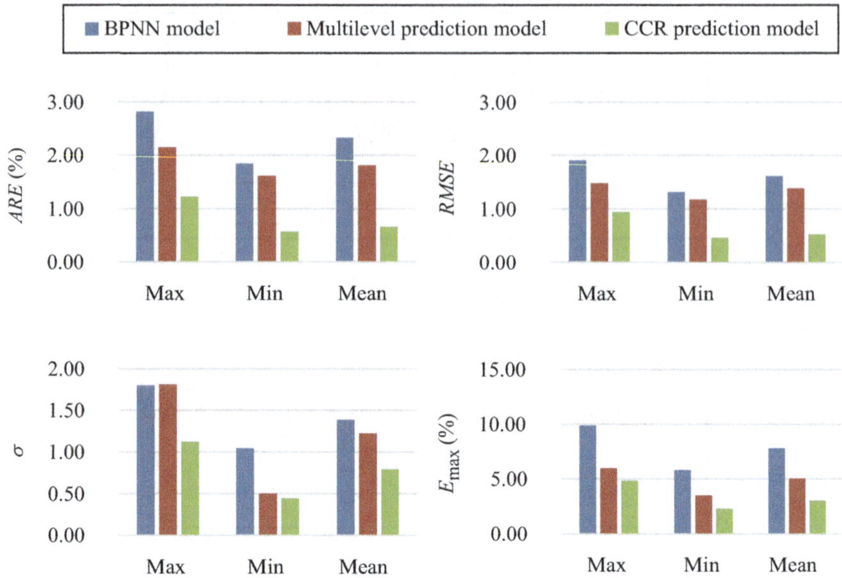

Fig. 9 Statistical results of ARE, $RMSE$, $E_{\max}$, and σ for different models in 10 trials. (CCR: comprehensive coke ratio, BPNN: back propagation neural network)

Figure 8 represents the prediction results of different models. The fitting performance of comprehensive coke ratio model outperforms other models. From the error curves (see Fig. 8), the prediction error of comprehensive coke ratio model is within $[-3\%, 3\%]$, whereas, the prediction error of back propagation neural network model is within $[-6\%, 6\%]$ and the prediction error of multilevel prediction model is within $[-5\%, 5.5\%]$. Therefore, the comprehensive coke ratio model provides higher performance than other models and is suitable as a carbon efficiency optimization model.

For the evaluation metrics described in (10)–(11), smaller $RMSE$ implies smaller error fluctuations, as does ARE and $E_{\max}$. Figure 9 shows the statistical results of ARE, $RMSE$, $E_{\max}$, and σ for different models in 10 trials. It demonstrates that the comprehensive coke ratio model outperforms other models in all evaluation metrics.

$RMSE$, ARE, and $E_{\max}$ of the comprehensive coke ratio prediction model are less than those of the back propagation neural network model and the multilevel prediction model. In addition, the average value of σ for comprehensive coke ratio model is 0.80, which is less than 1.23 for the multilevel prediction model and 1.39 for the back propagation neural network model. The maximum value of $E_{\max}$ for comprehensive coke ratio model is 4.88%, which is also less than the multilevel prediction model's 6.01% and the back propagation neural network model's 9.91%.

B. Comprehensive Coke Ratio Optimization

The higher H or faster V_S means that the sinter does not maintain a consistent quality, while lower H or slower V_S will result in premature burn through of the sintered layer, which in turn will damage the sintering machine. To maintain a stable sintering

Table 2 Results of carbon efficiency optimization. (CCR: comprehensive coke ratio)

No.	Actual CCR	Optimized CCR	Actual H	Actual V_S	Optimized H	Optimized V_S	Computational time (min)
1	54.76	53.60	691.26	2.78	858.10	1.96	36.12
2	54.38	53.45	693.24	2.75	737.99	2.93	35.39
3	57.73	55.73	696.71	2.68	886.61	2.09	35.74
4	55.82	54.88	675.07	2.49	850.80	2.68	36.59
5	52.48	50.79	716.35	2.29	748.41	2.37	35.95
6	53.49	51.80	723.26	2.68	823.12	2.00	36.72
7	54.45	52.59	737.17	2.48	799.84	2.92	35.48
8	53.41	52.38	724.15	2.09	895.39	1.67	35.18
9	55.17	52.84	785.19	2.87	857.10	2.09	35.60
10	52.79	52.44	754.14	2.46	790.53	1.93	35.79
11	54.00	52.01	701.54	2.71	821.65	2.04	36.17
12	52.79	52.19	719.32	2.47	681.78	2.97	36.29
13	53.58	52.27	734.59	2.38	836.13	2.37	35.44
14	53.73	52.36	705.49	2.65	907.42	1.89	35.85
15	53.54	52.67	698.45	2.28	709.46	1.57	36.37
16	54.75	51.89	714.79	2.61	869.27	2.13	36.48
17	55.20	52.37	754.16	2.82	805.66	1.51	35.99
18	53.92	52.69	729.49	2.54	860.45	1.54	35.82
19	54.99	52.55	768.35	2.79	873.29	2.32	35.77
20	53.13	52.16	689.59	2.12	878.48	1.71	36.25

process, H is maintained within [680, 920] mm and V_S is maintained within [1.5, 3.0] m/min, i.e., $H_{\min} = 680\,\text{mm}$, $H_{\max} = 920\,\text{mm}$, $V_{S_{\min}} = 1.5\,\text{m/min}$, and $V_{S_{\max}} = 3.0\,\text{m/min}$.

By adopting the intelligent optimization strategy with chaotic particle swarm optimization, the carbon efficiency optimization results are presented in Table 2. It can be seen that the optimized comprehensive coke ratio is on average 1.52 kg/t lower than the actual comprehensive coke ratio and the average computational time is 35.95 min. In addition, in actual sintering production, the optimization results are available within 60 min, which guarantees the real-time control of this process. The increased computation time does not affect the normal operation of sintering system, and the better optimization result can be obtained. Thus, the proposed optimization strategy decreases the consumption of energy, thus reducing the cost of sintering production which benefits the development of the steel industry.

3 Intelligent Control of Sintering Ignition Based on the Prediction of Ignition Temperature

The beginning spot in the whole burning process is the sintering ignition. Under fluctuating gas pressure, stable the SIT has essential application prospects. An intelligent control approach is presented via the prediction of the SIT. Firstly, an SIT prediction model is developed with a mechanism study approach as well as a data-driven approach. Next, the switch mode is given with the real operating practice. The SIP intelligent controller is created to get the expected gas flow. Lastly, a case match technique is involved, and the control experiment is achieved utilizing the raw data. The result shows that the proposal controls SIT well with fluctuating gas pressure, which has a sound application view.

3.1 Control Requirements and Control Structure

The actuators are the coke oven gas valve as well as the blast furnace gas valve. The constraints are the coke oven gas pressure as well as the blast furnace gas pressure. The control target is the SIT. An intelligent control structure with the SIT prediction is presented, which is displayed in Fig. 10. (The dotted line means the variable requires to be refreshed.)

The structure involves the intelligent controller for the SIT, the blast furnace gas flow controller, as well as the coke oven gas flow controller. the variation (ΔT, °C) within the expected SIT (T_D, °C) and the present SIT is assessed by the intelligent switching controller, and the switch mode is given. Meantime, considering the constraints of P_C and P_B, the intelligent controller for the SIT gets the expected blast furnace gas flow (Q'_B, m^3/h) as well as the expected coke oven gas flow (Q'_C, m^3/h) for T_D. The coke oven gas flow controller as well as the blast furnace gas flow controller get V_C and V_B, making Q_C and Q_B close to Q'_C and Q'_B.

3.2 Prediction Model of Ignition Temperature

A useful SIT prediction model represents a foundation to realize the control approach. The ignition is a fast & fierce burning process. All burning process obeys the energy conservation principle, which is all energy input is equivalent to all energy output. The input energy in the burning process involves,

(a) Chemical energy of the blast furnace gas and the coke oven gas (E_1);
(b) Intrinsic energy of the blast furnace gas and the coke oven gas (E_2);
(c) Intrinsic energy of the additional gas (E_3);
(d) Intrinsic energy of the stuff inside the previous furnace (E_4).

The output energy involves,

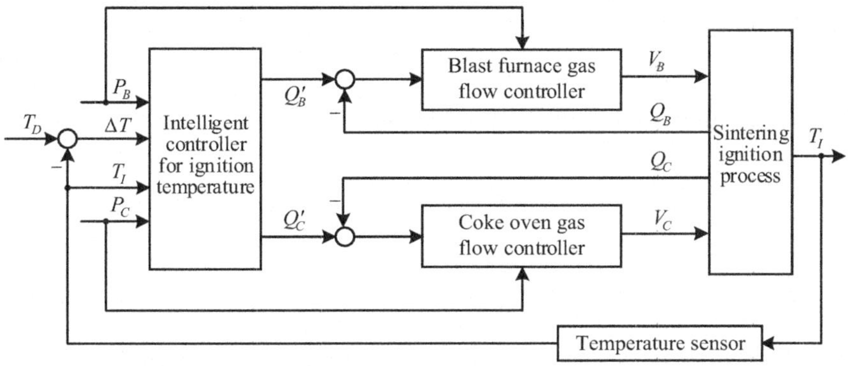

Fig. 10 Intelligent control structure for the SIT

(a) Intrinsic energy of the burning products (E_i);
(b) Burning energy loss (E_{ii});
(c) Energy loss caused by the incomplete burning (E_{iii});
(d) Other energy loss (E_{iv}).
The energy conservation during the burning process can be expressed as,

$$E_1 + E_2 + E_3 + E_4 = E_i + E_{ii} + E_{iii} + E_{iv}, \tag{20}$$

$$\begin{aligned}
E_1 &= m_B q_B + m_C q_C, \\
E_2 &= m_B c_B T_0 + m_C c_C T_0, \\
E_3 &= m_A c_A T_0, \\
E_4 &= M_h c_h T_0', \\
E_i &= M_H c_H T, \\
E_{iii} &= (1 - \eta_B) m_B q_B + (1 - \eta_C) m_C q_C,
\end{aligned} \tag{21}$$

where η_C is the coke oven gas combustion rate, and η_B is the blast furnace gas combustion rate. q_C is the coke oven gas calorific value, and q_B is the blast furnace gas calorific value. c_C is the coke oven gas specific energy capacity, and c_B is the blast furnace gas specific energy capacity. c_h is the mean specific energy capacity of the stuff during the furnace before burning, and c_H is the mean specific energy capacity of the stuff during the furnace after burning. m_A is the quantity of the air, m_B is the quantity of the blast furnace gas. m_C is the quantity of the coke oven gas quantity. M_h is the quantity of the stuff during the furnace before burning, and M_H is the quantity of the stuff during the furnace after burning. T_0 means the original gas temperature, T_0' means the original furnace temperature, and T represents the furnace temperature after burning.

The relation between density, ρ, volume, V, and mass, m can be expressed as,

$$m = \rho V. \tag{22}$$

The gas flow (Q) is,

$$V = Qt. \tag{23}$$

where t is the time. The Ideal-Gas Equation denotes,

$$PM = \rho RT, \tag{24}$$

so

$$\rho = \frac{PM}{RT} = \lambda P, \lambda = \frac{M}{RT}, \tag{25}$$

where R denotes the gas constant, M denotes the gas molar mass, T denotes the temperature, and P means the gas pressure. The mass of the coke oven gas as well as the blast furnace gas is,

$$\begin{aligned} m_C &= \rho_C V_C = \lambda_C P_C Q_C t, \lambda_C = \tfrac{M_C}{RT_0}, \\ m_B &= \rho_B V_B = \lambda_B P_B Q_B t, \lambda_B = \tfrac{M_B}{RT_0}. \end{aligned} \tag{26}$$

Substituting (26) and (21) into (20) produces,

$$\begin{aligned} T = \tfrac{1}{M_H c_H} & (\lambda_B P_B Q_B t(\eta_B q_B + c_B T_0) - Q_{ii} - Q_{iv} \\ & + \lambda_C P_C Q_C t(\eta_C q_C + c_C T_0) + m_A c_A T_0 + M_h c_h T_0'\big). \end{aligned} \tag{27}$$

Based on earlier research, a relation exists in the air-fuel ratio, r, and the combustion rate, η. Next, assume η_B and η_C is,

$$\begin{aligned} \eta_B &= k_1 r_B^2 + k_2 r_B + k_3, \\ \eta_C &= h_1 r_C^2 + h_2 r_C + h_3, \end{aligned} \tag{28}$$

where k_1, k_2, k_3, h_1, h_2 and h_3 are constants. r_B is the blast furnace gas air-fuel ratio, and r_C means the coke oven gas air-fuel ratio. As T_0' could not be diametrically caught, let

$$T_0' = g_1 T_1 + g_2 \Delta T_1, \tag{29}$$

where g_1 and g_2 are constants; T_1 means the sensing value of the SIP during the last instant; ΔT_1 denotes the change of T_1.

As t, T_0, q_B, q_C, c_A, c_B, and c_C are constants, besides λ_B and λ_C are constants. Assume c_h, c_H, M_h, M_H, Q_{ii}, and Q_{iv} are constant. Taking (28) and (29) into (27) produces,

$$\begin{aligned} T = \tfrac{1}{M_H c_H} & \big(\lambda_B P_B Q_B t((k_1 r_B^2 + k_2 r_B + k_3) q_B + c_B T_0) \\ & + \lambda_C P_C Q_C t((h_1 r_C^2 + h_2 r_C + h_3) q_C + c_C T_0) \\ & + m_A c_A T_0 + M_h c_h (g_1 T_1 + g_2 \Delta T_1) - Q_{ii} - Q_{iv}\big). \end{aligned} \tag{30}$$

Let,

$$T = A^{\mathrm{T}} X = a_1 x_1 + a_2 x_2 + a_3 x_3 + a_4 x_4 + a_5 x_5 + a_6 x_6 + a_7 x_7 + a_8 x_8 + a_9 x_9, \tag{31}$$

in which,

$$A = \begin{bmatrix} a_1 \\ a_2 \\ a_3 \\ a_4 \\ a_5 \\ a_6 \\ a_7 \\ a_8 \\ a_9 \end{bmatrix} = \begin{bmatrix} \frac{1}{M_{HCH}} \lambda_B t (k_3 q_B + c_B T_0) \\ \frac{1}{M_{HCH}} \lambda_B t k_2 q_B \\ \frac{1}{M_{HCH}} \lambda_B t k_1 q_B \\ \frac{1}{M_{HCH}} \lambda_C t (h_3 q_C + c_C T_0) \\ \frac{1}{M_{HCH}} \lambda_C t h_2 q_C \\ \frac{1}{M_{HCH}} \lambda_C t h_1 q_C \\ \frac{1}{M_{HCH}} M_h c_h g_1 \\ \frac{1}{M_{HCH}} M_h c_h g_2 \\ \frac{1}{M_{HCH}} (m_A c_A T_0 - Q_{ii} - Q_{iv}) \end{bmatrix}, X = \begin{bmatrix} x_1 \\ x_2 \\ x_3 \\ x_4 \\ x_5 \\ x_6 \\ x_7 \\ x_8 \\ x_9 \end{bmatrix} = \begin{bmatrix} Q_B P_B \\ r_B Q_B P_B \\ r_B^2 Q_B P_B \\ Q_C P_C \\ r_C Q_C P_C \\ r_C^2 Q_C P_C \\ T_1 \\ \Delta T \\ 1 \end{bmatrix}. \tag{32}$$

The parameter matrix ($\hat{A}$) is identified by using the least-squares approach.

3.3 *Design of Intelligent Controller*

The intelligent switching controller and two flow controllers compose the intelligent controller of the SIT.

A. Intelligent Switching Controller

The blast furnace gas calorific value equals approximately one-third of the coke oven gas calorific value. Therefore, the SIT is easier to change with changes in the coke oven gas flow. In the real site, operators control the SIT in the light of this feature. For instance, if the change, ΔT between T_I and T_D stays small, regulate the blast furnace gas flow. If ΔT is big, regulate the coke oven gas flow. If T_I severely varies from T_D, regulate the coke oven gas flow as well as the blast furnace gas flow in common.

In light of the real operating experience, we define four switch modes, and its switching rule shows below:

R1: IF $|\Delta T| \leq T_{S1}$, keep Q_C and Q_B;
R2: IF $T_{S1} \leq |\Delta T| \leq T_{S2}$, keep Q_C, and regulate Q_B;
R3: IF $T_{S2} \leq |\Delta T| \leq T_{S3}$, regulate Q_C, and keep Q_B;
R4: IF $|\Delta T| \geq T_{S3}$, regulate Q_C and Q_B.

Above rules, T_{S1}, T_{S2}, and T_{S3} denote thresholds of ΔT. T_{S1} means the expected top boundary of the SIT variation. T_{S2} denotes the top boundary of the SIT variation with the effect of the blast furnace gas flow. T_{S3} denotes the top boundary of the maximum SIT variation that operators could allow.

Because of the weakness of the mechanical construction, the regulation of V_B (or V_C) must be an integral number. For example, the gas valve opening is able to 2% or 3%, but not 2.6%. This has led to the gas flow only changes in a specific limit like V_B (or V_C). Thus, the control of gas flow needs not too accurate, and a special range is enough.

Based on the burning principle, when other states remain unchanging, the gas flow increase will increase the SIT. Therefore, Q'_B and Q'_C that match the expected SIT are found using the binary search approach. The limit of Q_B and Q_C are in 5000 m^3/h. Suppose the expected precision Q'_B or Q'_C is in 5 m^3/h. As $2^{10} > 1000$, just ten times estimation is required to get the expected Q'_B or Q'_C. Though Q'_B and Q'_C explore together, just a hundred times estimation is needed.

B. Gas Flow Controller

As long-time employment in the real site, the butterfly valve exists a severe nonlinearity on the flow and the valve opening. Besides, the fluctuating gas pressure results in the valve opening will not be linearly related to the gas flow. Nonetheless, the pipeline's gas flow needs to serve the fluid mechanical energy conservation. The gas flow, Q is,

$$Q = A_p v, \tag{33}$$

where A_p means the cross-sectional pipe area and v denotes the gas flow speed. The fluid mechanical energy conservation equation is,

$$C = P + \frac{1}{2}\rho v^2 + \rho g h, \tag{34}$$

where C denotes a constant, P denotes a spot fluid pressure, ρ represents the fluid density, v means the instant spot fluid flow speed, h is the spot height, and g means the gravity acceleration.

Take the face where the pipe is placed as the reference face (In the real site, all pipes are chiefly spread in one horizontal face or spread in multiple horizontal faces with little vertical height differences. Thus, assume all pipes are placed in one horizontal face.), set $h = 0$, taking (24), (25) and (33) into (34) produces,

$$C = P + \frac{1}{2}\rho v^2 = P + \frac{1}{2}\lambda P\left(\frac{Q}{A_p}\right)^2, \tag{35}$$

On the flow detection spot, $C = P_a + \frac{1}{2}\lambda P_a\left(\frac{Q_a}{A_p}\right)^2$.

On the pressure detection spot, $C = P_b + \frac{1}{2}\lambda P_b\left(\frac{Q_b}{A_p}\right)^2$.

As the flow detection spot occurs near the pipe exit, P_a is regarded as the external gas pressure, P_0. The pressure detection point occurs near the pipe entrance. Suppose the gas flow speed, v_b is related to the gas flow speed in the chief pipe, v_0,

$$v_b = k v_0, \tag{36}$$

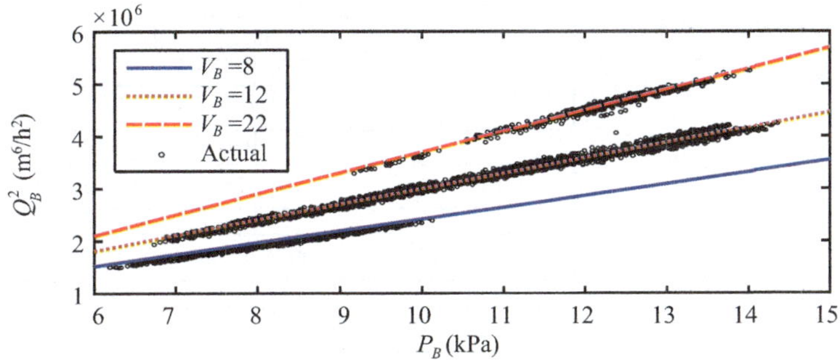

Fig. 11 Fit curve of Q_B^2 versus P_B

where k is determined by the valve. Taking $P_a = P_0$, and $Q_b = kA_p v_0$, produces,

$$Q_a^2 = \left(\frac{2A_p^2}{\lambda P_0} + \frac{k^2 A_p^2 v_0^2}{P_0}\right) P_b - \frac{2A_p^2}{\lambda} = \alpha P_b + \beta, \tag{37}$$

where $\alpha = \frac{2A_p^2}{\lambda P_0} + \frac{k^2 A_p^2 v_0^2}{P_0}$, $\beta = -\frac{2A_p^2}{\lambda}$. If other states remain unchanging, Q_a^2 is linear related to P_b. We fit the original data to obtain α and β.

Some representatively fit curves about Q_B^2 versus P_B are displayed in Fig. 11. The result validates our assume (neglect extra factors). So, if $V_B = i$, then

$$\hat{Q}_B^2 = \hat{\alpha}_i P_B + \hat{\beta}_i, \tag{38}$$

where i means the blast furnace gas valve regulatable value. We fit the original data to obtain $\hat{\alpha}_i$ and $\hat{\beta}_i$. Similarly, such a relation exists amongst Q_C, P_C, and V_C.

Q'_B and P_B are the blast furnace gas flow controller inputs. V_B is regarded as the output. Firstly, P_B is used to estimate $\hat{Q}_B$ with different V_B. Then, estimate $|\hat{Q}_B - Q'_B|$ to get V_B with the minimum $|\hat{Q}_B - Q'_B|$, and it is the blast furnace gas flow controller output. Similarly, the coke oven gas flow controller output is produced.

3.4 Experiment and Result Analysis

According to the SIT control system in the real site, experiments are produced to test the presented intelligent control approach with the SIT prediction.

A. Research Object Model

The research object model represents the foundation of the experimental implementation. A case library is built by collecting the plentiful original data from the real site. The controlled object model is built with the case matching approach. If

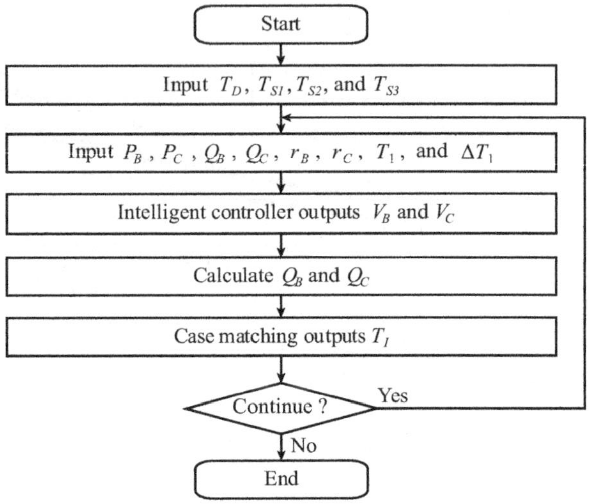

Fig. 12 Experimental flow chart

the intelligent control is performed, V_B and V_C vary, which leads to Q_B and Q_C to vary. P_B, P_C, and the SIT on the last instant (T_1, °C) form the constraint. Thus, the case matching inputs are Q_B, Q_C, P_B, P_C and T_1, and its output is T_I.

Many case match approaches exist. After careful thought, we select the extended cosine theory as the case match approach, which is,

$$F(X, Y) = \frac{x_1 y_1 + x_2 y_2 + \ldots + x_5 y_5}{\sqrt{x_1^2 + x_2^2 + \ldots + x_5^2} \cdot \sqrt{y_1^2 + y_2^2 + \ldots + y_5^2}}, \tag{39}$$

where $X = [x_1, x_2, \ldots, x_5]$ denotes Q_B, Q_C, P_B, P_C, and T_1 of Case A, and $Y = [y_1, y_2, \ldots, y_5]$ denotes Q_B, Q_C, P_B, P_C, and T_1 of Case B. As all variables are greater than zero, the scope of $F(X, Y)$ is zero to one. Supposing Case A and Case B are alike, $F(X, Y)$ equals one. Set $Fmax$ as the greatest amount of $F(X, Y)$. If the $Fmax$ is produced by the marked case and Case j, then T_I in Case j is the case match output. To enhance the case match's authenticity, the mean value of T_I with $F(X, Y)$ greater than 0.9 represents the ultimate output.

B. Analysis of Result

The experiment is developed. The flow diagram is displayed in Fig. 12. The gas flow is computed by (38). Based on the operating experience, T_D is 1150 °C, T_{S1} is 20 °C , T_{S2} is 40 °C, and T_{S3} is 60 °C.

We performed a hundred minutes experiment with two hundred groups of data. The result is displayed in Fig. 13. The gas pressure variations are in line with the initial property analysis. The blast furnace gas pressure varies more considerable than the coke oven gas pressure. Meanwhile, according to the SIT and the valve variations, the experiment is conducted chiefly in R1 mode. The blast furnace gas

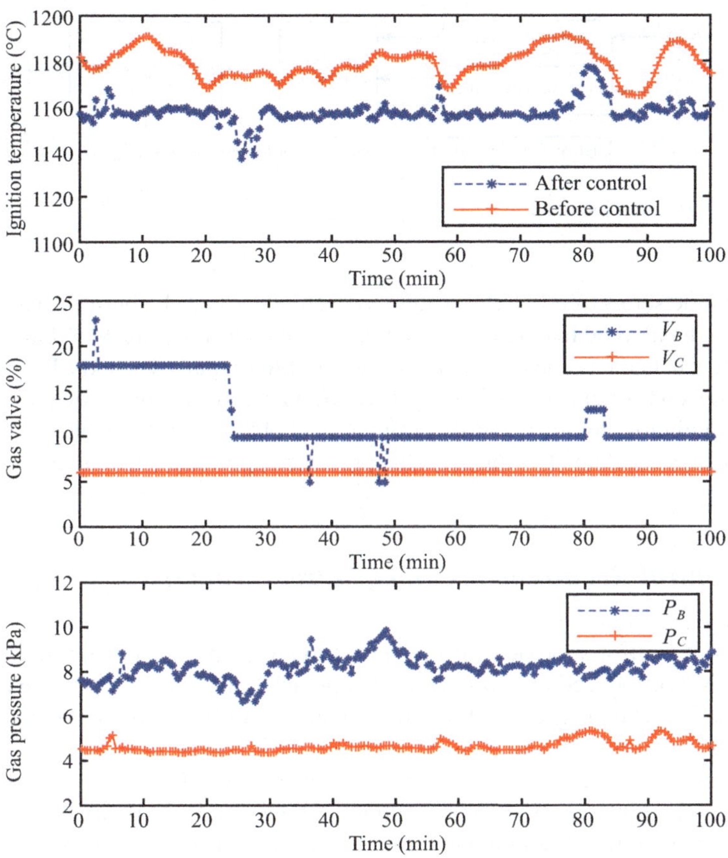

Fig. 13 Experimental results

valve exists two massive variations and is kept for a while, leading to an enormous SIT variation. The blast furnace gas pressure varies almost severely in this period.

It exists a meaningful development in the SIT after using the intelligent control approach, and the SIT is maintained in ±20 °C of the expected SIT with the fluctuating gas pressure. The result shows the intelligent control approach's usefulness with the SIT prediction and signifies a well-applied landscape.

4 Fuzzy Control of Burn-Through Point Based on the Feature Extraction of Time Series Trend

The burn-through point is a meaningful representation for measuring the sintering process security. A fuzzy control approach is developed for the burn-through point

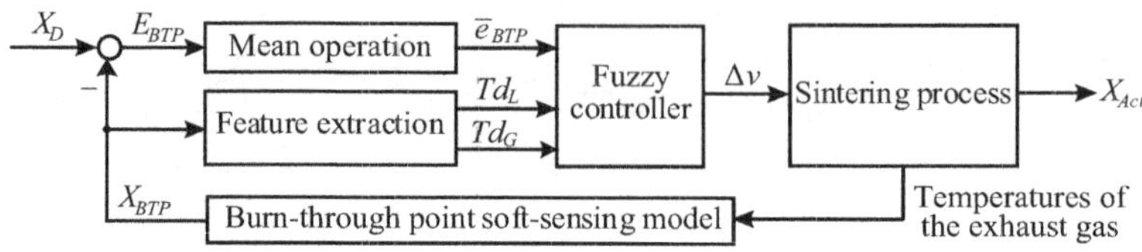

Fig. 14 Fuzzy control structure

with the time series trend (TSD) feature extraction. Firstly, the burn-through point's global and local trend feature parameters are obtained by adopting the Mann-Kendall test approach, and it is considered the fuzzy controller inputs. Then, the burn-through point fuzzy controller is produced to get the strand speed adjustment quantity. Lastly, with a simulation platform and the original data obtained in the real site, a control experiment is conducted to illustrate the presented approach's usefulness.

4.1 Control Requirements and Control Structure

The present control approach for the burn-through point depends on the operating practice. We introduced the fuzzy control approach to fusing the TSD with the operating practice. The fuzzy control structure for the burn-through point is demonstrated in Fig. 14.

The target is to keep the burn-through point in a specific scope of the expected burn-through point, L_D. The control output is strand speed. This structure is formed by the TSD feature extraction and a fuzzy controller of the burn-through point. As the real burn-through point time series, X_{Act}, could not be immediately sensing, the sensing value, X_{BTP}, is got through the burn-through point soft-sensing model. The variation within the expected burn-through point time series, X_D, and X_{BTP} is the error, E_{BTP}. The mean value of E_{BTP} is $\bar{e}_{BTP}$. Furthermore, the global TSD, Td_G, and the local TSD, Td_L, of X_{BTP} are acquired through the TSD feature extraction. Lastly, the burn-through point's fuzzy controller inputs are $\bar{e}_{BTP}$, Td_G, and Td_L, and the strand speed adjustment quantity, Δv, is acquired.

4.2 Feature Extraction of Time Series Trend

Two TSD feature parameters of the burn-through point are described, and the Mann-Kendall test approach extracts the TSD feature parameters.

A. Feature Extraction Method

A generally utilized approach for testing the TSD is the Mann-Kendall test [37, 38]. It tests how TSD varies and illustrates strength. We use the Mann-Kendall test approach to feature extraction. The burn-through point time series, $X_{BTP} =$

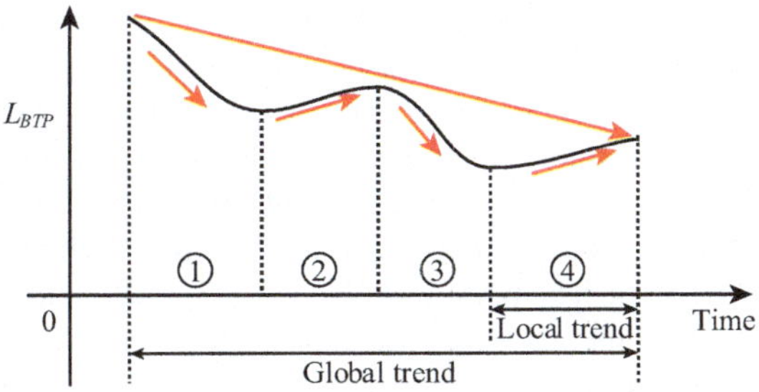

Fig. 15 Global TSD with many local TSDs. (① and ③ are decreasing TSDs. ② and ④ are increasing TSDs. The global TSD is decreasing one.)

$\{x_1, x_2, ..., x_N\}$, the statistic (H) means the count of positive variations minus the count of negative variations [39],

$$H = \sum_{i=1}^{N-1} \sum_{j=i+1}^{N} sign(x_j - x_i), \tag{40}$$

in which sign() denotes a sign function,

$$sign(x_j - x_i) = \begin{cases} 1, & x_j - x_i > 0 \\ 0, & x_j - x_i = 0 \\ -1, & x_j - x_i < 0 \end{cases}. \tag{41}$$

To great samples (N greater than ten), the test adopts the Gaussian distribution,

$$\mathrm{Var}(H) = \frac{N(N-1)(2N+5)}{18}. \tag{42}$$

The standard deviation (Y-statistic) means,

$$Y = \begin{cases} \frac{(H-1)}{\sqrt{\mathrm{Var}(H)}}, & H > 0 \\ 0, & H = 0 \\ \frac{(H+1)}{\sqrt{\mathrm{Var}(H)}}, & H < 0 \end{cases}. \tag{43}$$

During a double-sided trend test, if $|Y| \geq |Y_{1-\beta/2}|$ (β is a confidence degree, $\beta \in [0,\ 100\%]$), the null presumption is intolerable. With β, the data shows a notable increasing or decreasing trend. If $Y > 0$, it means the increasing trend, and if $Y < 0$, it is a decreasing trend. Further, the amount of Y denotes the trend strength. Thus, concerning Y being the feature parameter is feasible for the TSD.

B. Feature Parameter

A global TSD feature parameter describes the entire variation existing during the time series. Nevertheless, if multiple local TSDs exist during the time series, how the time series varies just with one global TSD feature parameter is hard to display. From Fig. 15, the global TSD denotes decreasing and it including four local TSDs.

The last local TSD denotes increasing, which is different from the global TSD. The last local TSD could not be expressed by the global TSD feature parameter. Thus, the local TSD feature parameter is used to explain the exhaustive variations.

Furthermore, the time series length represents an essential part of the extraction. In light of the burn-through point soft-sensing model's operating procedure, the burn-through point is linked with the last three bellows temperatures. The sinter strand is 90 m long. The bellows of No. 1, 2, 3, 4, 23, and 24 are three meters long, and others are four meters long. The bellows from No. 22 to No. 24 is ten meters long. In light of the original data, the mean strand speed remains approximately 2 m/min. Thus, the sinter strand moves from the bellows No. 22 to No. 24 within five minutes, then set t_G as five minutes. The sinter strand moves through the bellows No. 22 within two minutes, then set t_L as two minutes.

By Eqs. (42) and (43), Y stays linked with the time series length (N). Therefore, N in Td_G and Td_L must be the same to get the same value scope. The original data sampling interval holds five seconds. Five minutes will get 60 groups of data, and two minutes will get 24 groups of data. By the largest common divisor, set N_G and N_L as 12, which meets $N > 10$.

4.3 Design of Fuzzy Controller

Figure 14 offers the fuzzy controller of the burn-through point. Td_G and Td_L predicts the future variation of the burn-through point. $\bar{e}_{BTP}$ assesses if the present operating state satisfies the need, and Δv is used to regulate the burn-through point.

Firstly, the uniform fuzzification method is used to the physical parameter ($\bar{e}_{BTP}$ and Δv). A novel fuzzification method is required for the imaginary parameter (Td_G and Td_L). β is a physical parameter, Td_G, Td_L could be fuzzified by β. From the standard Gaussian distribution function [40], the different $\left|Y_{1-\beta/2}\right|$ with β is got in Table 3. $\left|Y_{1-\beta/2}\right|$ is not varied uniformly with β. Thence, the fuzzy partition is conducted on $\left|Y_{1-\beta/2}\right|$ with β. In light of β, four TSDs are listed below.

(a) If $\beta \in [0,\ 30\%)$, the TSD remains smooth.

(b) If $\beta \in [30\%,\ 60\%)$, the TSD increases/decreases slightly.

(c) If $\beta \in [60\%,\ 90\%)$, the TSD increases/decreases moderately.

(d) If $\beta \in [90\%,\ 100\%]$, the TSD increases/decreases rapidly.

Distinct Y results in the time series changing obviously. Particularly if $|Y|$ rises, the TSD grows more evident. Next, 7 fuzzy subsets are defined for Td_G and Td_L, and the fuzzy variables are NB, NM, NS, O, PS, PM, and PB. The triangular function is fuzzified Td_G or Td_L in Fig. 16.

In real site, the scope of the burn-through point is $L_{BTP} \in [20,\ 24]$. if L_D is 22.5, $\bar{e}_{BTP}$ is $[-2.5,\ 1.5]$. The burn-through point frequently varies in ± 1 bellows. Meanwhile, the burn-through point is steady with $\bar{e}_{BTP}$ keeping in ± 0.5 bellows. Thence, $\bar{e}_{BTP}$ is partition as 5 fuzzy subsets described as E involving NB, NS, O, PS, and PB. Likewise, the triangular function is also used to fuzzify $\bar{e}_{BTP}$ in Fig. 17.

Table 3 Different $|Y_{1-\beta/2}|$ of β

β	0	10%	20%	30%	40%	50%	60%	70%	80%	90%	100%
$\lvert Y_{1-\beta/2}\rvert$	0	0.13	0.26	0.39	0.53	0.68	0.84	1.04	1.28	1.65	$+\infty$

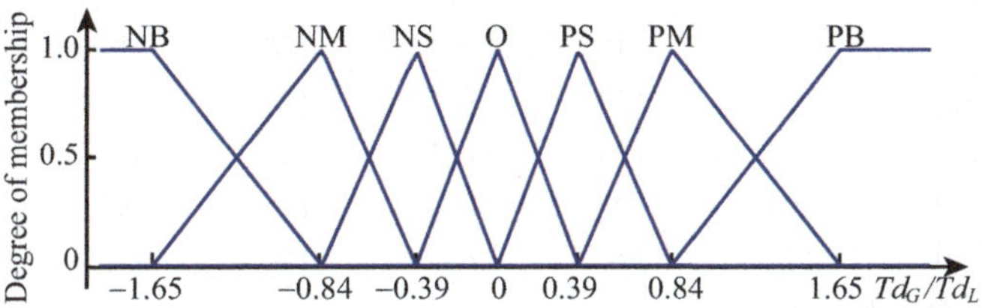

Fig. 16 Membership function of Td_G or Td_L

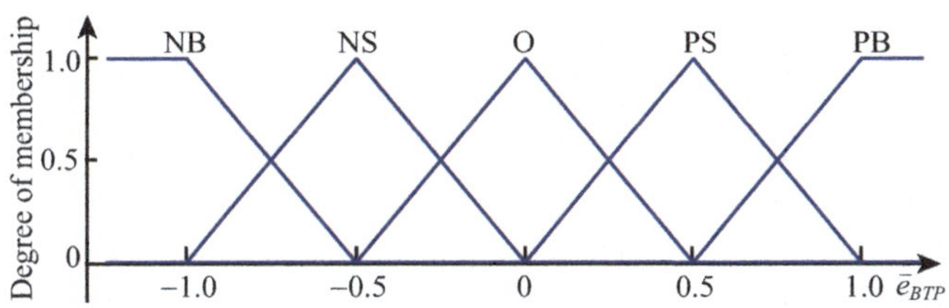

Fig. 17 Membership function of $\bar{e}_{BTP}$

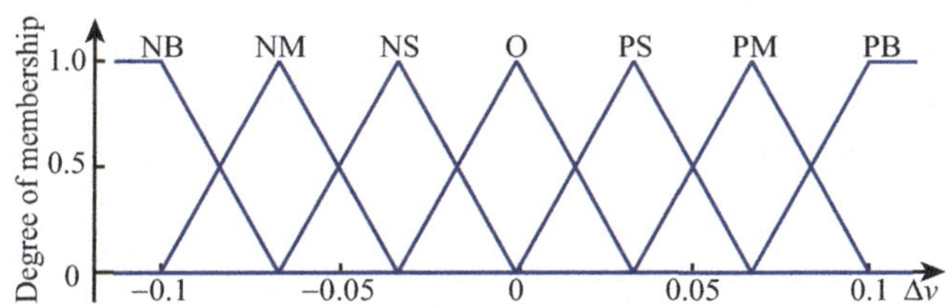

Fig. 18 Membership function of Δv

The strand speed is regulated by mechanical restriction and operating needs. Thence, the regulatable scope is $\Delta v \in [-0.1,\ 0.1]$ m/min after operating starting. To further accurate control, Δv is partition as 7 fuzzy subsets described as $\Delta v\ (U)$ including NB, NM, NS, O, PS, PM, and PB. The triangular function is also applied to fuzzify Δv in Fig. 18.

The presented approach's target keeps the burn-through point in a specific field of the expected burn-through point. By the operating experience included in the real site, the fuzzy rule is posted in Table 4. To reduce $\bar{e}_{BTP}$, the present $\bar{e}_{BTP}$ regards as the beginning spot. Td_G and Td_L provides the reference trend and strength for the variation of $\bar{e}_{BTP}$.

Table 4 Fuzzy Rules

TD_L	TD_G (E is NB)						
	NB	NM	NS	O	PS	PM	PB
NB	PB	PB	PB	PB	PB	PM	PM
NM	PB	PB	PB	PB	PM	PM	PM
NS	PB	PB	PB	PM	PM	PM	PS
O	PB	PB	PM	PM	PM	PS	PS
PS	PB	PM	PM	PM	PS	PS	PS
PM	PM	PM	PM	PS	PS	PS	O
PB	PM	PM	PS	PS	PS	O	O
TD_L	TD_G (E is NS)						
	NB	NM	NS	O	PS	PM	PB
NB	PB	PB	PM	PM	PM	PS	PS
NM	PB	PM	PM	PM	PS	PS	PS
NS	PM	PM	PM	PS	PS	PS	O
O	PM	PM	PS	PS	PS	O	O
PS	PM	PS	PS	PS	O	O	O
PM	PS	PS	PS	O	O	O	NS
PB	PS	PS	O	O	O	NS	NS
TD_L	TD_G (E is O)						
	NB	NM	NS	O	PS	PM	PB
NB	PM	PM	PS	PS	PS	O	O
NM	PM	PS	PS	PS	O	O	O
NS	PS	PS	PS	O	O	O	NS
O	PS	PS	O	O	O	NS	NS
PS	PS	O	O	O	NS	NS	NS
PM	O	O	O	NS	NS	NS	NM
PB	O	O	NS	NS	NS	NM	NM
TD_L	TD_G (E is PS)						
	NB	NM	NS	O	PS	PM	PB
NB	PS	PS	O	O	O	NS	NS
NM	PS	O	O	O	NS	NS	NS
NS	O	O	O	NS	NS	NS	NM
O	O	O	NS	NS	NS	NM	NM
PS	O	NS	NS	NS	NM	NM	NM
PM	NS	NS	NS	NM	NM	NM	NB
PB	NS	NS	NM	NM	NM	NB	NB

(continued)

Table 4 (continued)

TD_L	TD_G (E is PB)						
	NB	NM	NS	O	PS	PM	PB
NB	O	O	NS	NS	NS	NM	NM
NM	O	NS	NS	NS	NM	NM	NM
NS	NS	NS	NS	NM	NM	NM	NB
O	NS	NS	NM	NM	NM	NB	NB
PS	NS	NM	NM	NM	NB	NB	NB
PM	NM	NM	NM	NB	NB	NB	NB
PB	NM	NM	NB	NB	NB	NB	NB

4.4 Experimental Study and Result Analysis

By the simulation platform and the original data, we experimented with verifying the presented approach's usefulness.

A. Simulation Platform

A simulation platform is produced in our lab to revert the real site, as displayed in Fig. 19. The necessary hardware is in line with that in the real site. Hence, the simulation platform is cooperative with the real site. If the controller could work on the simulation platform, it could be transferred to the real site's control system.

The simulation platform is consists of five levels.

(a) Monitoring level (① and ②). It is used to monitor the online situation.

(b) Basic automation level (④ and ⑤). The programmable logic controller (PLC) collects sensing data and drives the actuator.

(c) Optimal control level (⑥ and ⑦). Operator stations give the operator a working console. Engineer stations provide the engineer with a console to promote and manage software.

(d) Process simulation level (③ and ⑨). It reverts the real site and refreshes several essential parameters.

(e) Service level (⑧, ⑩ and ⑪). The switch connects the whole platform with the local region network. Servers afford data assistance involving sharing and storing.

The divergence between the simulation platform and the real site's control system is that the platform has no actual machine. ③ replaces the actual machine, and it completes the essential parameters refreshing. ⑨ replaces the sensor and the driver, and it creates analog outputs and inputs.

With the original data, the simulation platform's data stream is given in Fig. 20. Switches transfer the production data to engineer stations and the PLC. The controller produces the control quantity, and it is transferred to the PLC through switches. The PLC converts the production data and the control quantity to an electronic signal, and it is sent to the simulation computer by a data conversion card. The simulation computer gets the simulation value, and it is converted to an electronic signal being

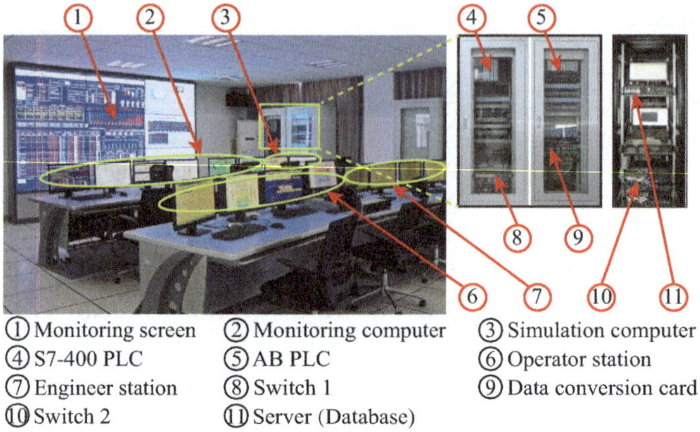

Fig. 19 Simulation platform

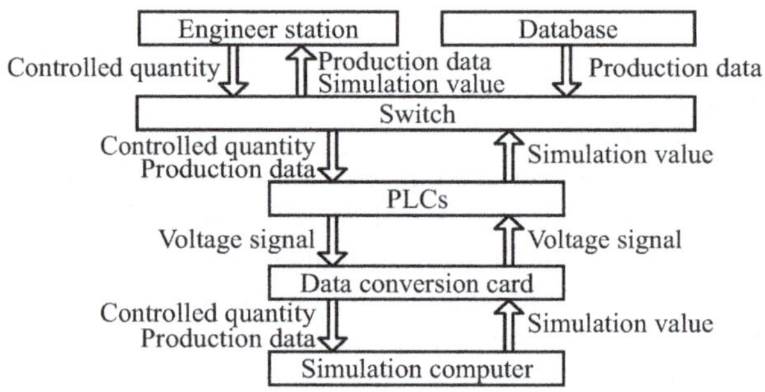

Fig. 20 Data stream of the simulation platform

transferred to the PLC. The operating premise of the simulation platform is a simulation procedure. Thus, a satisfactory burn-through point simulation procedure is needed. The procedure in [14] is chosen. This specific procedure can enhance the experiment's reliability.

B. Result and Analysis

The presented approach inserts into engineer stations. The test for the presented approach is conducted per five minutes with the original data. While four comparative tests are manual control, expert-fuzzy control [41], hierarchical intelligent control [42], and fuzzy predictive control [43], sequentially. Besides, let L_D be 22.5. The result is displayed in Figs. 21, 22, 23, 24, 25.

Three criteria are outlined to analyze the result acquired by five control approaches.

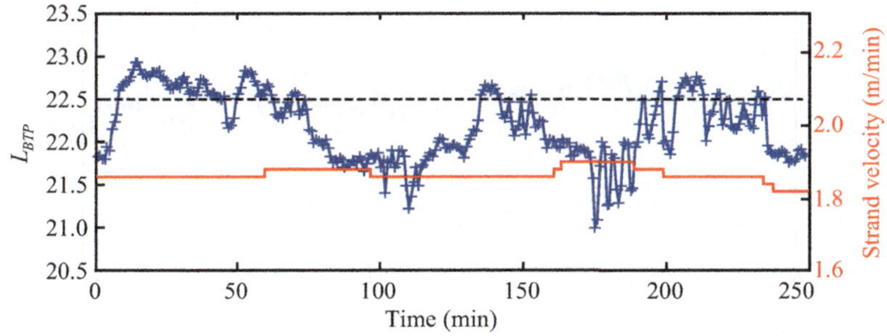

Fig. 21 Result of manual control

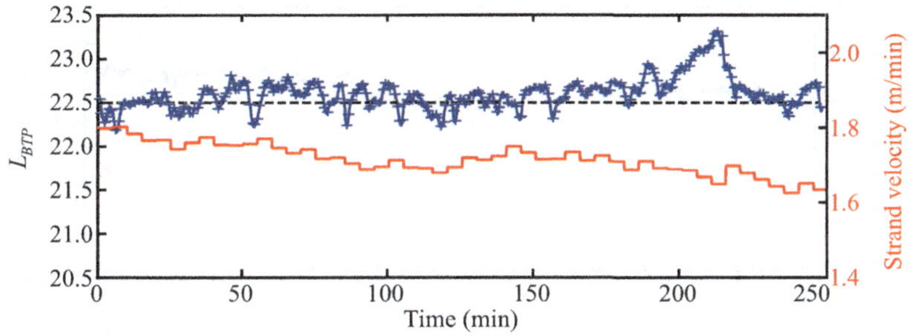

Fig. 22 Result of expert-fuzzy control [41]

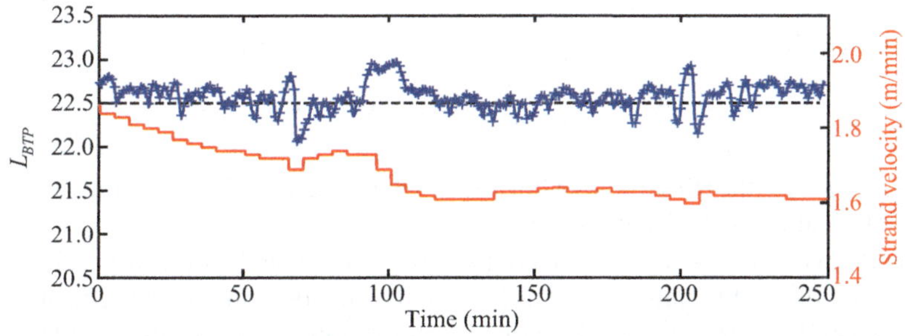

Fig. 23 Result of hierarchical intelligent control [42]

(a) Variation index,

$$\varepsilon_{BTP}=\frac{1}{L_D}\left(\max_{1\le i\le M}(L_{BTP}(i))-\min_{1\le i\le M}(L_{BTP}(i))\right)\times 100\%, \tag{44}$$

(b) Mean error,

$$\bar{L}_{BTP}=\frac{1}{M}\sum\nolimits_{i=1}^{M}L_{BTP}(i), \tag{45}$$

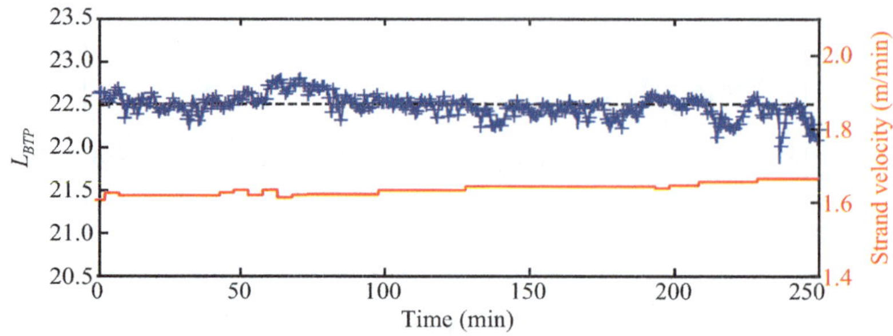

Fig. 24 Result of fuzzy predictive control [43]

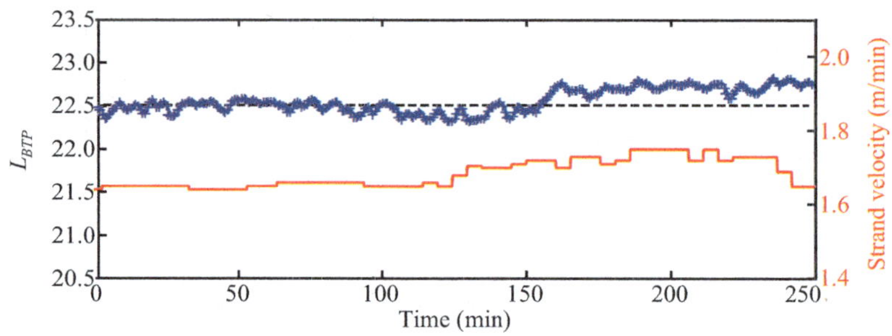

Fig. 25 Result of our method

(c) Variance,

$$\delta^2 = \frac{1}{M-1} \sum\nolimits_{i=1}^{M} \left(L_{BTP}(i) - \bar{L}_{BTP}\right)^2, \tag{46}$$

where $L_{BTP}(i)$ is the ith sample of burn-through point, and M is the count of samples. The analytical result is listed in Table 5.

Clearly, with employing manual control, the burn-through point varies violently in Fig. 21, and this is fundamentally hard to satisfy the operating needs. The operator's bodily-mental health affects the manual control approach, and they seldom regulate the strand speed. Moreover, changeable and unpredictable factors are included. Nevertheless, the remaining approaches fusing intelligent control and operating experience to regulate the strand speed. From Figs. 22, 23, 24, 25, the three control approaches massively satisfy the operating needs.

From Table 5, our approach is better than others on the variation index. Corresponded to fuzzy predictive control, hierarchical intelligent control, expert-fuzzy control, and manual control, our approach reduces ε_{BTP} by 2.39%, 1.83%, 2.78%, and 5.50%, sequentially. Besides, apart from fuzzy predictive control, $\bar{L}_{BTP}$ produced by our approach is nearest to L_D, and our approach reaches the least δ^2. Though the fuzzy predictive control seems victorious on $\bar{L}_{BTP}$ and δ^2, our approach is yet superior for a best ε_{BTP}.

Table 5 Statistical result of experiments ($L_D = 22.5$)

Method	$\varepsilon_{BTP}(\%)$	$\bar{L}_{BTP}$	δ^2
Manual control	8.80	22.21	0.1605
Expert-fuzzy control [41]	5.08	22.61	0.0338
Hierarchical intelligent control [42]	4.13	22.58	0.0234
Fuzzy predictive control [43]	4.69	22.48	0.0165
Our method	2.30	22.57	0.0188

5 Optimization and Control System of Carbon Efficiency

The optimization and control system of carbon efficiency (OCSCE) [33] in sinter green manufacturing is designed to maximize the use of carbon and best protect the environment in the sintering process. The OCSCE has two parts: the optimization for the carbon efficiency and the coordinated control of thermal states. The optimization for carbon efficiency is employed to optimize the state parameters. The coordinated control of thermal states is applied to control the sintering ignition temperature and the burn-through point. Finally, an implementation scheme of the OCSCE is put forward for the industrial site. The results of preliminary tests show that the OCSCE is in line with the needs of the industrial site, and it will have a significant performance after the OCSCE is put into the industrial site.

5.1 Architecture of OCSCE

The architecture of the OCSCE is shown in Fig. 26. It is a hierarchical structure, which has three parts. The L1 is the basic automation system, which converts the optimized operating parameters to the control quantities for the control of the sintering process, and it sends the process data to L2. The L2 is the coordinated control of thermal states, which realizes the intelligent control of the thermal state parameters, and it gets the optimized operating parameters. The L3 is the optimization of carbon efficiency to obtain the optimized sintering parameters.

A. Optimization for Carbon Efficiency

The carbon efficiency optimization includes its prediction and intelligent optimization. This gets the key parameters of the carbon efficiency, which serve as a reference for the comprehensive performance evaluation for the carbon efficiency. And then, according to the comprehensive performance evaluation result, the state parameters are optimized to restrict the coordinated optimization and control for the production phases.

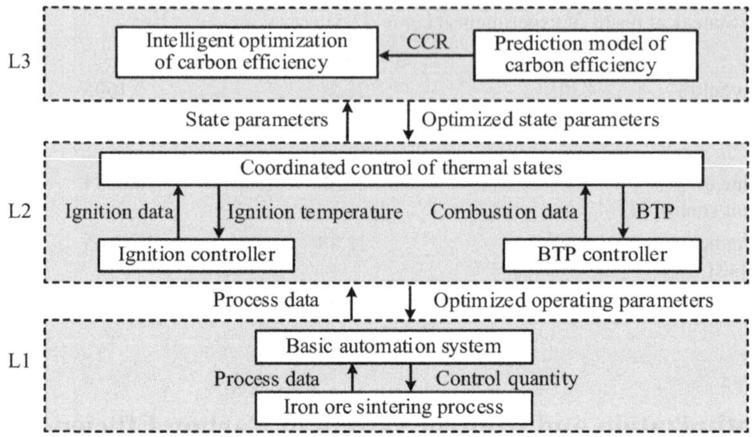

Fig. 26 Architecture of OCSCE. (CCR: comprehensive coke ratio, BTP: burn-through point)

The premise of achieving carbon efficiency optimization is its prediction. The correlation of carbon efficiency is analyzed from the perspective of energy flow. According to the state parameters, the prediction model of the carbon efficiency is presented using data-driven modeling and mechanism modeling. And the comprehensive coke ratio is employed as the carbon efficiency measurement standard.

An intelligent optimization strategy for carbon efficiency is designed according to the prediction model of carbon efficiency. The operating parameters are the comprehensive reflection of the operation state, which serve as a reference of the intelligent optimization for the comprehensive coke ratio. The goal of intelligent optimization is to minimize the comprehensive coke ratio, which can be influenced by adjusting the operating parameters to influence the state parameters and ultimately achieve the optimal carbon efficiency. So, from the perspective of material flow, the intelligent optimization model of carbon efficiency is formed.

B. Coordinated Control of Thermal States

The optimized state parameters obtained by the optimization for the carbon efficiency, which is aimed at the optimal carbon efficiency, not considering the current production conditions. So, there is a coordinated controller needs to comprehensively consider the current production conditions and carbon efficiency. According to the optimized state parameters obtained by the optimization of the carbon efficiency and the process data, the coordinated control of thermal states is realized. Finally, after coordinated control of thermal states, the control of state parameters is realized by two controllers: the ignition controller and the burn-through point controller. As a result, the optimized operating parameters obtained by the two controllers are sent to the basic automation system for execution.

Ignition control. The ignition is the beginning of combustion, and its stability affects the whole combustion process. The optimized parameters need to be executed by the control, so the ignition control is a means to realize the optimization of carbon

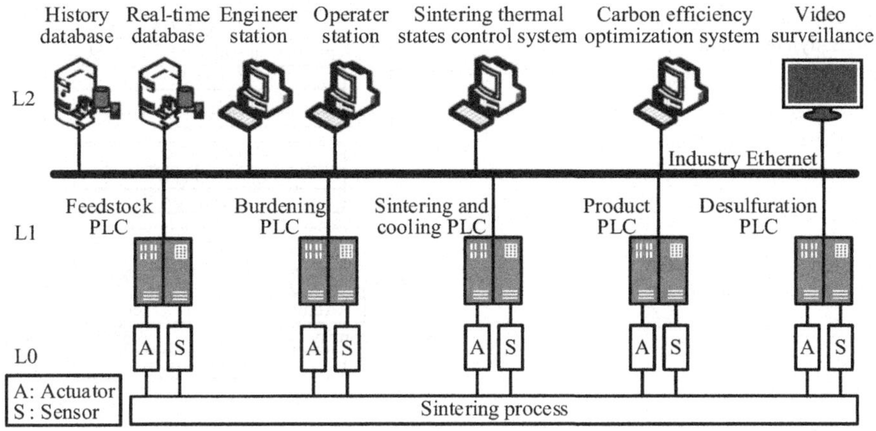

Fig. 27 Hardware structure

efficiency. The coal gas for the ignition includes blast furnace coal gas and coke oven gas. The ignition controller takes the ignition temperature obtained by the coordinated controller of thermal states as a control objective. The control objective is realized by the regulation of the coke oven gas valve and the blast furnace coal gas valve.

Burn-through point control. Burn-through point is a critical parameter of the thermal states, and its stability is associated with the energy consumption and the quality and yield of the sinter, so the burn-through point control is an important means to realize the optimization of the carbon efficiency. Changing the strand velocity is the main means of controlling the burn-through point. A given burn-through point is obtained by the coordinated controller of thermal states, which serves as the objective of the controller.

5.2 Implementation Scheme

An implementation scheme for OCSCE is designed according to the sintering process control system of an industrial site of a steel plant. The software data flow and hardware structure of the scheme are presented and it is demonstrated that the scheme is executable.

A. System Hardware Structure

Figure 27 shows the hardware structure of the sintering system. It is a network architecture, which has a good performance with the combination of the OCSCE. The OCSCE can be implemented on this hardware system. It mainly includes three layers: L0, L1, and L2, and the L1 and L2 are connected by industrial Ethernet. L0 includes the actuator and sensor. L1 consists of 5 PCLs, which are feedstock PLC, burdening PLC, sintering and cooling PLC, product PLC, and denaturation

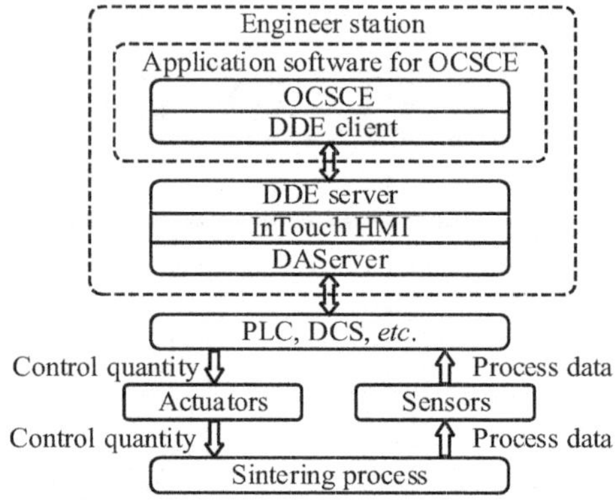

Fig. 28 Data flow diagram. (DDE: dynamic data exchange)

PLC. In the L2, the core of the system is the carbon efficiency optimization system and the sintering thermal state control system. Engineer and Operator stations are built by InTouch configuration software. The engineer station is aimed at maintainer and developer for the modification and generation of the control interface. Operator station is aimed at the operator provides real-time monitoring of the sintering process. In the end, the operation data are shown on the video surveillance, and the real-time database and history database keeps the operation data sequentially.

B. Data Flaw of System

Depending on the sintering process control system at the industrial site, the application software of OCSCE is added to the main control system in the engineer station. The human-machine interface (HMI) of the whole sintering process is designed by using InTouch configuration software. This application software is developed in Visual C++, and it adopts dynamic data exchange communication protocol to realize real-time data exchange with the InTouch through the dynamic data exchange client and dynamic data exchange server, and the communication between PLC and InTouch configuration software is implemented using DAServer. The data flow diagram is shown in Fig. 28.

6 Conclusion

Increasing carbon efficiency shows valuable to decrease operating expenses and power loss in the sintering process. A smart optimization scheme for Increasing carbon efficiency was invented. An intelligent control approach for the SIT was displayed, in which the SIT prediction treats the fluctuating gas pressure. A fuzzy control approach with the TSD feature extraction was provided to stabilizing the burn-through point. The test result with the original data illustrates the usefulness of the designed control approach and optimization scheme. An OCSCE with a hierarchical structure is developed, which realizes the optimization of carbon efficiency and the intelligent control of the thermal states. The OCSCE is in line with the industrial site's requirements, and it has great utilization prospects.

References

1. Kwon, W.H., Kim, Y.H., Lee, S.J., Paek, K.N.: Event-based modeling and control for the Burnthrough Point in sintering processes. IEEE Trans. Control Syst. Technol. **7**(1), 31–41 (1999)
2. Zhou, K., Chen, X., Wu, M., Cao, W., Hu, J.: A new hybrid modeling and optimization algorithm for improving carbon efficiency based on different time scales in sintering process. Control Eng. Pract. (2019). https://doi.org/10.1016/j.conengprac.2019.104104
3. Hu, J., Wu, M., Chen, X., Du, S., Zhang, P., Cao, W., She, J.: A multilevel prediction model of carbon efficiency based on differential evolution algorithm for iron ore sintering process. IEEE Trans. Ind. Electron. **65**(11), 8778–8787 (2018)
4. Hu, J., Wu, M., Chen, X., Du, S., Cao, W., She, J.: Hybrid modeling and online optimization strategy for improving carbon efficiency in iron ore sintering process. Inf. Sci. **483**, 232–246 (2019)
5. Hu, J., Wu, M., Chen, X., Cao, W., Pedrycz, W.: Multi-model ensemble prediction model for carbon efficiency with application to iron ore sintering process. Control Eng. Pract. **88**, 141–151 (2019)
6. Du, S., Wu, M., Chen, X., Cao, W.: An intelligent control strategy for iron ore sintering ignition process based on the prediction of ignition temperature. IEEE Trans. Ind. Electron. **67**(2), 1233–1241 (2020)
7. Zhang, C., Jing, H., Long, Y., Cai, Y.: Intelligent temperature control of ignition furnace in sintering machine. In: Proceedings of the 2004 IEEE Conference on Cybernetics and Intelligent Systems, pp. 224–228 (2005)
8. Ying, Y.Q., Lu, J.G., Chen, J.S., Sun, Y.X.: PIDNN based Intelligent control of ignition oven. Adv. Mater. Res. **396–398**, 493–497 (2012)
9. Endiyarov, S.V.: Adaptive control of the ignition of sintering batch. Steel Transl. **46**(10), 728–732 (2016)
10. Cao, S., Shi, L., Ze, X., Zhang, H.: Continuously sintering furnace temperature control system based on intelligent PID adjustment. IEEE, Phuket, pp. 190–193 (2008)
11. Cao, S., Ze, X., Xu, J., Shi, L.: Intelligent control system of multi-segments continuously sintering furnace. In: International Symposium on Knowledge Acquisition and Modeling, pp. 869-872 (2008)
12. Chen, X., Jiao, W., Wu, M., Cao, W.: EID-estimation-based periodic disturbance rejection for sintering ignition process with input time delay. Asian J. Control **20**(5), 1–14 (2018)
13. Du, S., Wu, M., Chen, L., Zhou, K., Hu, J., Cao, W., Pedrycz, W.: A fuzzy control strategy for burn-through point based on feature extraction of time series trend in iron ore sintering process. IEEE Trans. Ind. Inf.**16**(4), 2357–2368 (2020)
14. Chen, X., Hu, J., Wu, M., Cao, W.: T-S fuzzy logic based modeling and robust control for burning-through point in sintering process. IEEE Trans. Ind. Electron. **64**(12), 9378–9388 (2017)
15. Wang, W.J., Chou, H.G., Chen, Y.J., Lu, R.C.: Fuzzy control strategy for a hexapod robot walking on an incline. Int. J. Fuzzy Syst. **19**(6), 1703–1717 (2017)
16. Belchior, C.A.C., Rui, A.M.A., Landeck, J.A.C.: Dissolved oxygen control of the activated sludge wastewater treatment process using stable adaptive fuzzy control. Comput. Chem. Eng. **37**(4), 152–162 (2012)
17. Geert, C., Jossede, B., Bart, M., Wouter, S.: Fuzzy control of the cleaning process on a combine harvester. Biosyst. Eng. **106**(2), 103–111 (2010)
18. Haber, R.E., Toro, R.M.D., Gajate, A.: Optimal fuzzy control system using the cross-entropy method: a case study of a drilling process. Inf. Sci. **180**(14), 2777–2792 (2010)
19. Chung, C.C., Chen, H.H., Ting, C.H.: Grey prediction fuzzy control for pH processes in the food industry. J. Food Eng. **96**(4), 575–582 (2010)
20. Neugebauer, M., Sołowiej, P., Piechocki, J.: Fuzzy control for the process of heat removal during the composting of agricultural waste. J. Mater. Cycles Waste Manag. **16**(2), 291–297 (2014)

21. Wu, H.N., Li, H.X.: A multiobjective optimization based fuzzy control for nonlinear spatially distributed processes with application to a catalytic rod. IEEE Trans. Ind. Inf. **8**(4), 860–868 (2012)
22. Wang, T., Qiu, J., Yin, S., Gao, H., Fan, J., Chai, T.: Performance-based adaptive fuzzy tracking control for networked industrial processes. IEEE Trans. Cybern. **46**(8), 1760–1770 (2016)
23. Tong, W., Qiu, J., Gao, H., Wang, C.: Network-based fuzzy control for nonlinear industrial processes with predictive compensation strategy. IEEE Trans. Syst. Man Cybern.: Syst. **47**(8), 2137–2147 (2017)
24. Muthuramalingam, T., Mohan, B., Rajadurai, A., Devaraj, S.: Monitoring and fuzzy control approach for efficient electrical discharge machining process. Adv. Manuf. Process. **29**(3), 281–286 (2014)
25. Majidpour, M., Qiu, C., Chu, P., Gadh, R., Pota, H.R.: Fast prediction for sparse time series: demand forecast of EV charging stations for cell phone applications. IEEE Trans. Ind. Inf. **11**(1), 242–250 (2015)
26. Zhao, W., Beach, T.H., Rezgui, Y.: Automated model construction for combined sewer overflow prediction based on efficient LASSO algorithm. IEEE Trans. Syst. Man Cybern.: Syst. **49**(6), 1254–1269 (2019)
27. Yue, W., Hong, G.S., Wong, W.S.: Prognosis of the probability of failure in tool condition monitoring application-a time series based approach. Int. J. Adv. Manuf. Technol. **76**(1–4), 513–521 (2015)
28. Li, K., Lu, L., Zhai, J., Khoshgoftaar, T.M., Li, T.: The improved grey model based on particle swarm optimization algorithm for time series prediction. Eng. Appl. Artif. Intell. **55**, 285–291 (2016)
29. George, K., Mutalik, P.: A multiple model approach to time-series prediction using an online sequential learning algorithm. IEEE Trans. Syst. Man Cybern.: Syst. **49**(5), 976–990 (2019)
30. Jing, P., Su, Y., Jin, X., Zhang, C.: High-order temporal correlation model learning for time-series prediction. IEEE Trans. Cybern. **49**(6), 2385–2397 (2019)
31. Addisu, S., Selassie, Y.G., Fissha, G., Gedif, B.: Time series trend analysis of temperature and rainfall in lake Tana Sub-basin. Ethiopia. Environ. Syst. Res. **4**, 25 (2015)
32. Sayemuzzaman, M., Jha, M.K.: Seasonal and annual precipitation time series trend analysis in North Carolina, United States. Atmos. Res. **137**, 183–194 (2014)
33. Du, S., Wu, M., Chen, X., Cao, W., She, J.: Design of an optimization and control system for carbon efficiency in the green manufacturing of sinter ore. In: Proceedings of the 36th Chinese Control Conference, pp. 4470–4475 (2017)
34. Dai, X., Bikdash, M.: Trend analysis of fragmented time series for mHealth apps: hypothesis testing based adaptive spline filtering method with importance weighting. IEEE Access **5**, 27767–27776 (2017)
35. Wang, H.Y., Li, Z.Y., Gao, Z.H., Wu, J.J., Sun, B., Li, C.L.: Assessment of land degradation using time series trend analysis of vegetation indictors in Otindag Sandy land. IOP Conf. Ser.: Earth Environ. Sci. **17**, 012065 (2014)
36. Hu, J., Wu, M., Chen, X., She, J., Cao, W., Chen, L., Ding, H.: Hybrid prediction model of carbon efficiency for sintering process. IFAC PapersOnLine **50**(1), 10238–10243 (2017)
37. Mann, H.B.: Nonparametric tests against trend. Econometrica **13**(3), 245–259 (1945)
38. Kendall, M.G.: Rank correlation methods. Hafner Publishing Co., Oxford, England (1955)
39. Nourani, V., Danandeh Mehr, A., Azad, N.: Trend analysis of hydroclimatological variables in Urmia lake basin using hybrid wavelet Mann-Kendall and Şen tests. Environ. Earth Sci. **77**(207), 1–18 (2018)
40. Kisi, O., Ay, M.: Comparison of Mann-Kendall and innovative trend method for water quality parameters of the Kizilirmak River, Turkey. J. Hydrol. **513**, 362–375 (2014)
41. Du, S., Wu, M., Chen, X., Lai, X., Cao, W.: Intelligent coordinating control between burn-through point and mixture bunker level in an iron ore sintering process. J. Adv. Comput. Intell. Intell. Inf. **21**(1), 139–147 (2017)
42. Wang, C.S., Wu, M.: Hierarchical intelligent control system and its application to the sintering process. IEEE Trans. Ind. Inf. **9**(1), 190–197 (2013)

43. Wu, M., Duan, P., Cao, W., She, J., Xiang, J.: An intelligent control system based on prediction of the burn-through point for the sintering process of an iron and steel plant. Expert Syst. Appl. **39**(5), 5971–5981 (2012)

Decision-Making of Burden Distribution for Blast Furnace

Jianqi An, Min Wu, Jinhua She, Takao Terano, and Weihua Cao

Abstract Burden distribution affects the whole process of iron-making in a blast furnace (BF), including the distribution of the gas flow, the quality and production of the pig iron, and the consumption of the energy. Gas utilization rate (GUR) is one of the most important indictors to reflect the development of a BF. Designing a control strategy of the burden distribution for the GUR is necessary to improve the utilization of the gas flow, increase the production of the pig iron, and decrease the consumption of energy. Thus, this chapter introduces methods to predict and control the trend of the GUR. First, this chapter introduces the iron-making process of a BF. Based on this, this chapter introduces the development and the influencing factors of the GUR. Meanwhile, this chapter introduces the burden distribution and its impact on the gas flow and GUR. Then, this chapter introduces three different models to predict the GUR, which consider the chaotic characteristic of GUR, the qualitative influence of the operations on the development trend of the GUR, and the quantitative influence of the operations on the value of the GUR, respectively. Finally, this chapter introduces a decision-making strategy to improve the development trend of the GUR. According to some simulations, these methods predict the GUR accurately and change the development trend of the GUR qualitatively.

Keywords Iron-making process · Gas utilization rate · Prediction model · Decision-making strategy · Burden distribution

---

J. An (✉) · M. Wu · J. She · W. Cao
School of Automation, China University of Geosciences, Wuhan 430074, China
e-mail: anjianqi@cug.edu.cn

Hubei Key Laboratory of Advanced Control and Intelligent Automation for Complex Systems, Wuhan 430074, China

Engineering Research Center of Intelligent Technology for Geo-Exploration, Ministry of Education, Wuhan 430074, China

J. She
School of Engineering, Tokyo University of Technology, Tokyo 192-0982, Japan

T. Terano
Chiba University of Commerce, Ichikawa 272-8512, Japan

M. Wu et al. (eds.), *Developments in Advanced Control and Intelligent Automation for Complex Systems*, Studies in Systems, Decision and Control 329,
https://doi.org/10.1007/978-3-030-62147-6_6

Abbreviations

BD	Burden distribution
BF	Blast furnace
CO	Carbon monoxide
CO_2	Carbon dioxide
DGL	Distribution of the gas flow
DIGL	Distribution of the initial gas flow
EMD	Empirical mode decomposition
GL	Gas flow
GUR	Gas utilization rate
HBS	Hot-blast supply
H-L	Heat load
IMF	Intrinsic mode function
LTS	Long-time-scale
MFPM	Multi-time-scale fusion prediction model
O-E	Oxygen enrichment
P-C-I	Pulverized coal injection
P-I	Permeability index
RBF	Radial basis function
STS	Short-time-scale
SVR	Support vector regression
T-T	Top temperature

1 Analysis of Ironmaking and Burden Distribution

The metallurgical industry is the pillar of the national economy. The iron-making system is an important process in the metallurgical industry. A BF is the main part of the iron-making system. By physical changes and chemical reactions, a BF is used to change the iron ore to molten iron (Fig. 1) [1–3]. The GUR is an important indicator of a BF reflecting the DGL, the consumption of the energy, and the development state of a BF. BD and HBS are main operations to change the GUR.

1.1 Ironmaking Process

The ironmaking process is a highly difficult metallurgical production process with nonlinear and multi-time scale characteristics [4]. There are two main operations of a BF, one is the BD, and the other one is the HBS.

BD is set up to change the DGL in the upper part of a BF by adjusting the coke and iron-ore layers, thereby improving the GUR. The parameters of the BD are the

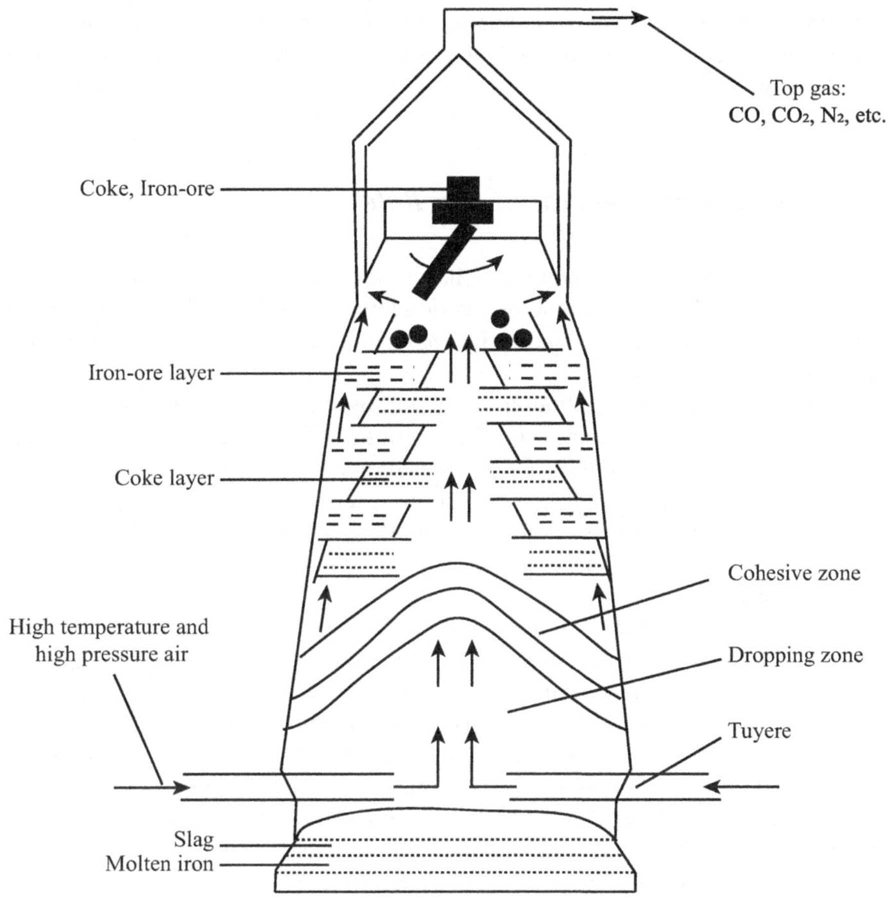

Fig. 1 Structure of a BF

angle of the distribution chute, the number of distribution circles, the weight of the batch, and other parameters [5]. The coke and iron ore fell into a BF from the top by the BD, which forms the coke layer and iron-ore layer. Then, the DGL is changed according to the shape of the layers.

The main purpose of the HBS is to adjust the DIGL. The main parameters of the HBS are air volume, air pressure, oxygen enrichment, and coal injection [6]. The hot blast discharged into BF by tuyeres reacts with the coke to form the amount of CO and heat, thereby forming the high-temperature initial GL [7, 8].

In the iron-making process, the iron ore reacts with the CO in the upward GL to form pig iron and CO_2 [9]. The pig iron flows from the bottom of a BF. The gas containing the left CO and CO_2 is discharged from the top of a BF, which is called top gas. The ratio of the volume of CO and the volume of the CO and CO_2 in the top gas is defined as the GUR [10].

In order to meet the production requirements, operators change the BD and HBS to adjust the smelting conditions and DGL and improve the redox reaction, thereby increasing the GUR [11].

1.2 Gas Flow and Gas Utilization Rate

The DGL is one of the important factors that cause fluctuations in furnace conditions. The DGL is closely related to the production target [12]. In a BF, the judgment of condition, shape management, and optimization of the distribution system all require extrapolation of the DGL.

The GL goes up from the bottom to the top by the pressure difference [13]. In this process, the iron ore reacts with the CO in the GL to form pig iron, slag, and the gas containing CO_2 and the left CO. The GL goes through the coke and iron-ore layers. The GL goes through three distributions in a BF: first, it diffuses upwards and toward the center from the air outlet; second, the gas passes through the drip zone and moves laterally in the coke layer of the reflow zone; finally, the gas stream twists up through the block [14].

In this process, there are two main types for the GL, one is the edge GL, and the other is the center GL. The characteristic of the edge GL is that the development intensity of the gas in the marginal area is relatively large. The development of the edge GL is not conducive to the long life of a BF. Therefore, it is generally necessary to avoid an overdevelopment of edge GL in actual production. The characteristics of the center GL are: the central airflow is strong, and the edge airflow is weaker. The center GL is mainly controlled by the BD that changes the coke ratio of the furnace throat to make the center GL stronger and try to suppress the edge GL.

The main oxidation Creduction reactions in a BF are

$$\begin{cases} C + O_2 = CO_2 \\ CO_2 + C = 2CO \\ 3Fe_2O_3 + CO = 2Fe_3O_4 + CO_2 \\ Fe_3O_4 + CO = 3FeO + CO_2 \\ FeO + CO = Fe + CO_2 \end{cases} . \tag{1}$$

According to the above equation, if CO_2 increases, the consumption of CO will increase and the reduction reactions will be more complete. In top GL, the definition of the GUR is shown as [15]

$$\eta_{CO} = \frac{V_{CO_2}}{V_{CO} + V_{CO_2}}. \tag{2}$$

The GUR is great of importance to reflect the overall states of a BF. Improving the GUR is a benefit for the consumption of the energy, and production of the pig iron [16].

The HBS affects the GUR by adjusting the DIGL [17]. The changes in the parameters of the HBS can adjust the volume of the gas, the distribution of the temperature, the difference of top pressure and bottom pressure, and GL rate, thereby changing the DIGL. Then, the GUR is changed by the DIGL. The BD affects the GUR by changing the DGL in the upper part of a BF [18]. The BD first changes the shape of the layers, which changes the gas permeability. Then, the DGL is changed.

1.3 Effect of Burden Distribution

A BF implements the BD through the charging equipment at the top, which is an important means for adjusting the upper part of a BF. In the process of BD, the ore and coke are fed in batches from the top of the furnace through a rotating chute, so that the ore layer and the coke layer form an ideal surface shape at the throat of the furnace, forming a specific ore and coke ratio distribution, and then affecting the rising channel of the GL, which affects the state of a BF. The charge is usually charged into a BF throat in batches [19]. Set the ore amount of a batch of material according to requirements, and determine the amount of coke composed of the batch by calculating the coke load, and load it into the furnace throat using a distributor. From the longitudinal section of the furnace throat, the ore and coke present a layered and overlapping structure [20].

The BD has an important influence on the GL and the shape and position of the melt zone [21]. By changing the wind capacity in the furnace, the development of the GL is adjusted to change the output level, improve operation, and significantly reduce fuel consumption. The BD can also control the edge GL through the special control of the edge GL, so as to prevent the erosion of the furnace wall and increase the service life of a BF [22].

Different state parameters of a BF are affected differently by the BD, so the characteristics related to the BD are also different. This feature is described by the matrix of the BD [23]. If the matrix of the the cloth matrix is

$$C_{32,2223}^{40,37.9,35.7,33.4,31,27} O_{3443.2}^{40,37.9,35,7,33.4,31} \tag{3}$$

it is defined as setting 3, 2, 2, 2, 2, and 3 turns of coke when the inclination of the chute is 40, 37.9, 35.7, 33.4, 31, 27, and setting 3, 4, 4, 3, 2 turns of ore when the chute inclination is 40, 37.9, 35.7, 33.4, 31, respectively.

The impact of the BD on the GUR is mainly due to the redistributed GL. The redistributed GL is affected by the ore-to-coke ratios in the radial direction. The ore-to-coke ratios can be changed by adjusting the parameters of the BD [24]. Thus, the BD makes a BF in an ideal working condition by redistributing the layers and GL to a reasonable state [25]. Then, the utilization of CO is maximized.

According to the real iron-making process, the DGL has three types, including edge-center balance type, edge overdevelopment type, and center overdevelopment type (Fig. 2). The edge-center balance type means that the loss of the beat is less

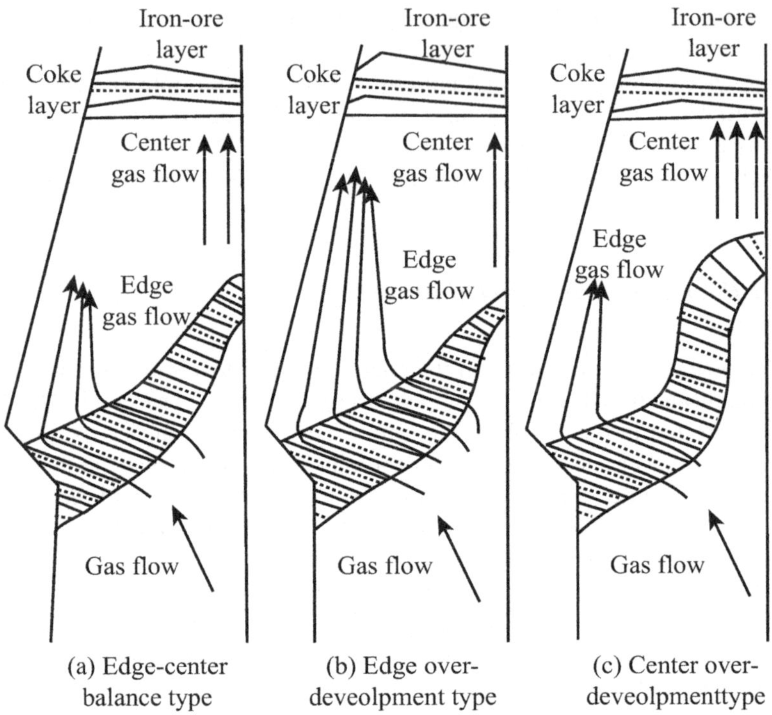

Fig. 2 Distribution of GL in different BD

and the utilization of CO is more completely, which is the best type. The edge overdevelopment type is formed by replenishing a lot of coke on the edge of a BF. And the center overdevelopment type is formed by replenishing coke in the center part of a BF [26].

2 Prediction Model of Gas Utilization Rate

Taking the GUR as an index, an accurate prediction model is a basis to establish a control method for the GUR. The development trend of the GUR is not only influenced by the operations of a BF (BD and HBS) but also determined by its own development trend. Based on the chaotic character of the GUR, a model considering the own developmental implications of the GUR is introduced to predict the GUR. Then, a qualitative model using the case-matching method is introduced to predict the changing trend of the GUR. Finally, a quantitative model considering the multi-time-scale relationships is introduced to predict the accurate value of the GUR.

2.1 Prediction Model of GUR Based on Chaos Theory

The development trend of the GUR affects its future development when the BD and HBS are stable. This characteristic of the GUR is called chaos. Then, this part introduces a GUR-predicted model using the chaos RBF algorithm, which considers the chaotic characteristics of the evolution process of the GUR.

A. Chaotic Characteristic

This part analyzes the chaotic character of the development of the GUR. The analysis method calculates the Kolmogorov entropies of the GUR according to the correlation integral algorithm. The Kolmogorov entropy is a key parameter to judge and describe the chaotic process. If the value of the Kolmogorov entropies is finite and positive, the development of the GUR is proved a chaotic process.

The Kolmogorov entropy, K, is calculated by

$$K = -\lim_{\tau\to 0}\lim_{\alpha\to 0}\lim_{d\to\infty}\frac{1}{d\tau}\ln C_d(a), \tag{4}$$

$$C_d(a) = \frac{1}{M^2}\sum_j\sum_i \theta(a - d_{ij}), \tag{5}$$

where $C_d(a)$ is the correlation integral; M, the number of points in reconstructed phase space; d, the dimension phase space can be divided into n boxes; τ, the interval; $\theta(\cdot)$, the Heaviside function; d_{ij}, the Euclidean distance of point i and point j; a, the random selective.

The simulation uses the real-word data of GUR from two BFs at different conditions as the sample data. The Kolmogorov entropies of these two BFs are shown in Figs. 3 and 4, respectively. The Kolmogorov entropy values of the two BFs are finite and positive, thus the development of the GUR is corroborated to be a chaotic process.

B. Prediction Model

The part introduces a prediction model based on the chaotic characteristic of the GUR. The method first reconstructs the phase space of the GUR. The phase space of the GUR is

$$G = \begin{Bmatrix} g(t)^1 \\ g(t)^2 \\ \vdots \\ g(t)^j \end{Bmatrix} = \begin{Bmatrix} g(t)^1 & g(t)^{1+\tau} & \cdots & g(t)^{1+(m-1)\tau} \\ g(t)^2 & g(t)^{2+\tau} & \cdots & g(t)^{2+(m-1)\tau} \\ \vdots & \vdots & \vdots & \vdots \\ g(t)^j & g(t)^{j+\tau} & \cdots & g(t)^{j+(m-1)\tau} \end{Bmatrix}, \tag{6}$$

where τ is the lag time and m is the embedded dimension. This part calculates these two variables by autocorrelation coefficient and G-P method.

The prediction model is built by the chaos RBF algorithm. The number of the inputs is the embedded dimension, and the interval of the samples is set as τ. Thus, the prediction model is shown as

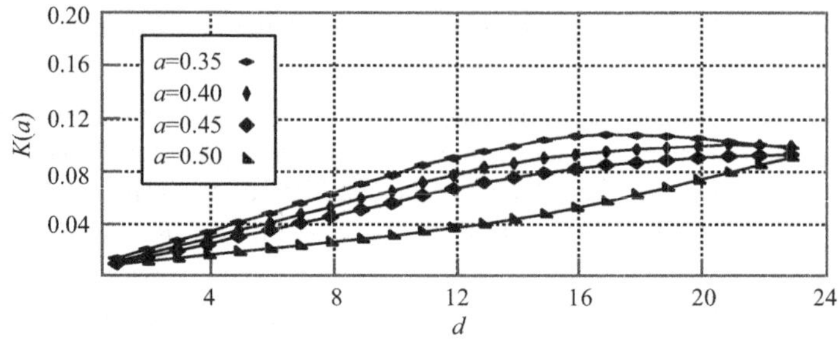

Fig. 3 Kolmogorov entropy of the $1100\,m^3$ BF

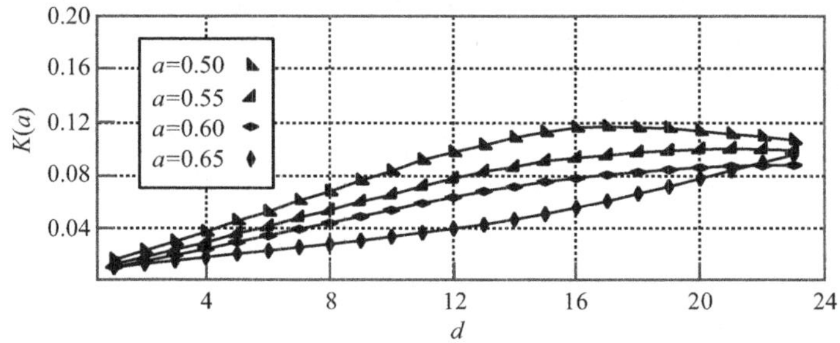

Fig. 4 Kolmogorov entropy of the the $3200\,m^3$ BF

$$
\begin{aligned}
g(t+r) &= f(g(t)) \\
&= \sum_{i=1}^{m} \omega_i \varphi_i(\|g(t) - r_i\|) \\
&= W^T \phi,
\end{aligned} \tag{7}
$$

$$
\varphi_i(\|g(t) - r_i\|) = e^{g\frac{(\|g(t)-r_i\|)^2}{2c_i^2}}, \; i = 1, 2, \ldots, m, \tag{8}
$$

where $g(t)$ is the input; f, the output; $\varphi = [\varphi_1, \varphi_2, \ldots, \varphi_m]^T$, the output vector of the hidden layer; m, the number of hidden layer; $\phi(\cdot)$, gaussian function; $W = [\omega_1, \omega_2, \ldots, \omega_m]^T$, the weight vector in the network output layer; r_i and c_i are the centre and width of gaussian function, respectively.

The results of the prediction models of the $1100\,m^3$ BF and $3200\,m^3$ BF are shown in Figs. 5 and 6, respectively.

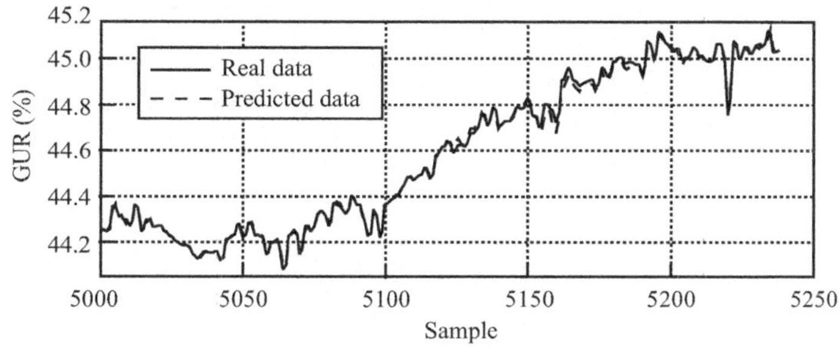

Fig. 5 Predicted results of GUR for the 1100 m^3 BF

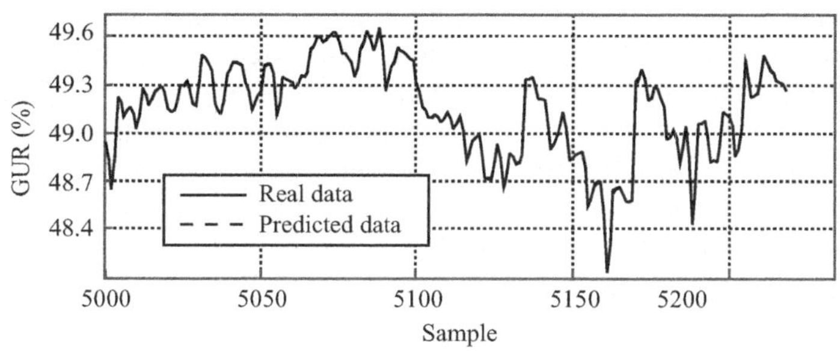

Fig. 6 Predicted results of GUR for the 3200 m^3 BF

2.2 *Prediction Model of GUR Based on Case-Matching*

The influence of the own development trend of GUR on its development is analyzed according to the chaos RBF prediction model mentioned above. However, the relationship between the GUR and the operations of a BF is not analyzed. Thus, this part introduces a case-matching model for the GUR to give the predicted change trend of GUR and the probability of its occurrence, which analyzes the influence of the BD on the changing trend of the GUR. In the iron-making process, the BD first adjusts some BF variables that are related to the GUR. Then, the adjustment of these BF variables results in the change of the trend for the GUR. Thus, a model that is used to predict these BF variables based on the BD parameters is established according to the SVR algorithm firstly. Then, a model using the case-matching method is established to predict the changing trend of the GUR according to the different states of the BF variables. The structure of this model is shown in Fig. 7.

A. Selection of Inputs and Outputs

In order to build an accurate prediction model, the inputs and outputs need to select carefully. This part selects the inputs and outputs according to the mechanism relationship and correlation.

In order to select the reasonable BF variables that have more impact on the GUR, the relations of the BF variables and GUR are calculated by the mutual information

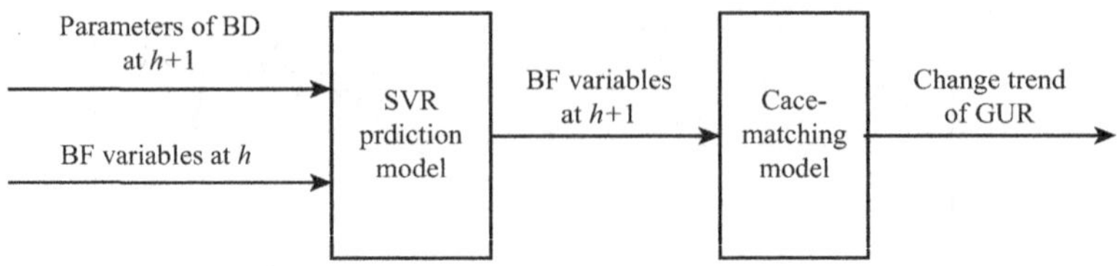

Fig. 7 Structure of case-matching model

algorithm. Then, the BF variables related to GUR are selected according to the relations of different cut-off frequencies.

According to the calculation results, the relations increases when the cut-off frequency decreases. Considering that the low cut-off frequency means more loss of the information, an appropriate cut-off frequency is determined. Then, this method takes H-L, T-T, P-I, O-E, and P-C-I as the inputs of the case-matching model. Among them, O-E and P-C-I are set by the operator. Thus, the next time BD and the current H-L, T-T, and P-I are taken as the inputs of the SVR prediction model. And the next time H-L, T-T, and P-I are the corresponding output, respectively.

B. SVR Prediction Model

A model using SVR algorithm is introduces to predict the BF variables in this part. The inputs are shown as

$$D_{In,i} = (d^{h}_{hl,i}, d^{h}_{tt,i}, d^{h}_{pi,i}, d^{h+1}_{oc,i}, d^{h+1}_{rc,i}), \tag{9}$$

and Output data

$$H_{Out,i} = d^{h+1}_{hl,i} / d^{h+1}_{tt,i} / d^{h+1}_{pi,i}, \tag{10}$$

where i is the sample sequence; h, the time sequence; D_{In} contains the next time BD parameters and the current H-L, T-T, and P-I; H_{Out}, one of the next time H-L, T-T, and P-I.

For a SVR prediction model

$$H_{Out,i} = w\phi(D_{In,i}) + b, \tag{11}$$

where $\phi(\cdot)$ is a nonlinear map; w, weight; b, bias. The training of a SVR model is to find the optimal w and b to minimize the difference of the actual value of H_{Out} and the fitted value of H_{Out}. By using the slack variables ξ_i and ξ_i^*, the problem of fitting the actual value becomes

$$\begin{aligned} &\min : \tfrac{1}{2}||w||^2 + C\sum_{i=1}^{n}(\xi_i + \xi_i^*), \\ &\text{s.t.} \begin{cases} H_{Out,i} - w\phi(D_{In,i}) - b \le \varepsilon + \xi_i \\ -H_{Out,i} + w\phi(D_{In,i}) + b \le \varepsilon + \xi_i{}^* \\ \xi_i \ge 0, \xi_i^* \ge 0 \end{cases}, \end{aligned} \tag{12}$$

where C is a penalty factor; $\varepsilon > 0$, a small number. C determines whether the prediction model is overfilling or underfitting. Thus, it is necessary to select an optimal C to ensure an optimal fit.

The simulation uses 200 samples to verify the effectiveness of these prediction models. The training process uses 180 samples and the testing process using 20 samples. Each test sample is selected at the specific time that the parameters of the BD are changed in the historical data. Thus, the sampling interval of the two testing samples is not fixed.

If the predicted trend of the BF variable is the same as the actual trend, the predicted value of the BF variable is taken as correct. Thus, according to the results in Fig. 8, three prediction models predict the H-L, T-T, and P-I within the acceptable range of the error, respectively.

C. Case-Matching Model

This part introduces the case-matching model that is used to predict the changing trend of the GUR and gives the probability of its occurrence.

As the combination change trend of the five BF variables determines the change trend of the GUR, these five BF variables are the basis of the prediction for the change trend of the GUR. The change trend of the GUR is divide into increase type and decrease type. The case-matching model uses the probabilities of different change trends to predict the GUR. The probabilities are calculated as

$$P_{inc}^{case_z} = \frac{N_{inc}^{case_z}}{N_{inc}^{case_z} + N_{dec}^{case_z}}; \; z = 1, \ldots, 32, \tag{13}$$

$$P_{dec}^{case_z} = \frac{N_{dec}^{case_z}}{N_{inc}^{case_z} + N_{dec}^{case_z}}; \; z = 1, \ldots, 32, \tag{14}$$

where $P_{inc}^{case_z}$ is the probability of increase and $P_{dec}^{case_z}$ is the probability of decrease when the combination is $case_z$. $N_{inc}^{case_z}$ is the number when the change trend of GUR is increase in $case_z$ and $N_{dec}^{case_z}$ is the number when the change trend of GUR is decrease. The change trend of GUR to increase or decrease depends on whether the probability of increase is less than the probability of decrease.

This part uses 20 samples to verify the effectiveness of the prediction model using the case-matching method. The predicted results are shown in Table 1.

In Table 1, P-GUR means the prediction for the changing trend of GUR, and A-GUR means the actual change trend. According to the results, the case-matching model can predict the changing trend of the GUR.

For the results, some samples can be explained by analyzing the mechanism. For instance, the trend of H-L and T-T in sample 1 is increasing. The trend of the P-I, O-E, and P-C-I is decreasing. The decrease of P-I means the decrease in air volume or the increase in pressure difference. When the pressure difference increases, the GL rate, and T-T increases, thereby reducing the reaction time between iron ore and the GL. The decrease of the P-C-I and O-E has a bad impact on the reduction reaction. Thus, the changing trend of GUR is decreasing.

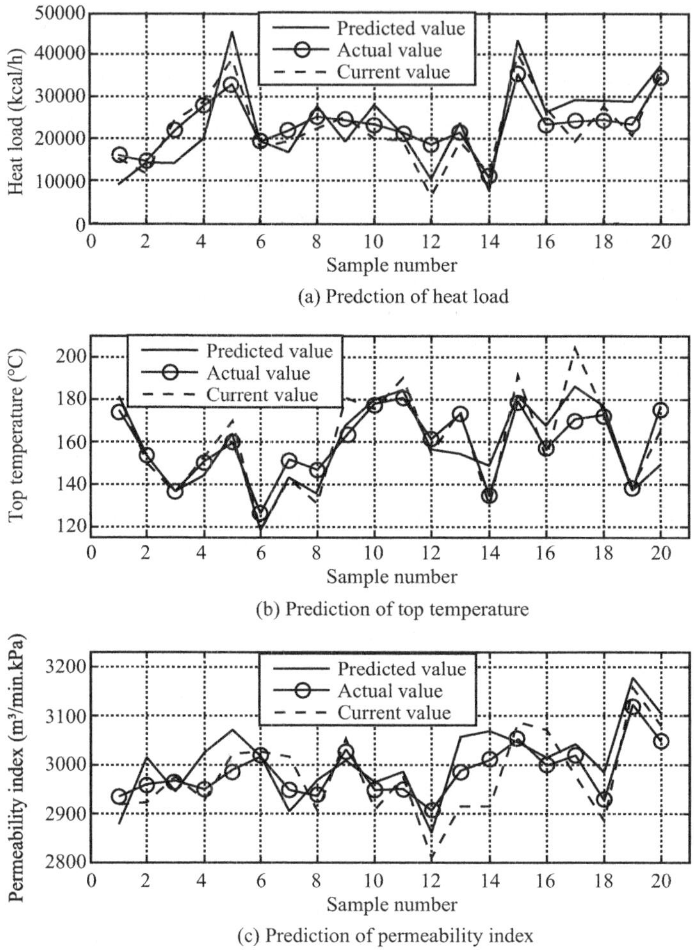

Fig. 8 Prediction results of SVR model

2.3 *Prediction Model of GUR Based on Multi-time-Scale*

The qualitative relation of BD and the changing trend of the GUR is analyzed by the model using the case-matching method. This model does not give the quantitative relationship between the operations of BF and the change of the GUR. Thus, this part introduces an MFPM to analyze the quantitative relationship between the operations of BF and the GUR.

A. Analysis of Multi-Time-Scale Characteristics

In the iron-making process, the changes for the parameters of the BD result in the different ore-to-coke and central-coke ratios. This part uses the ore-to-coke and

Table 1 The Prediction results of the change trend of GUR

Sample	H-L	T-T	P-I	O-E	P-C-I	P-GUR	A-GUR
1	↑	↑	↓	↓	↓	↓	↓
2	↑	↓	↓	↑	↑	↑	↑
3	↑	↑	↓	↑	↑	↓	↑
4	↑	↑	↑	↑	↓	↓	↓
5	↓	↓	↓	↓	↑	↑	↑
6	↑	↓	↓	↓	↓	↓	↓
7	↓	↑	↓	↑	↓	↓	↑
8	↓	↑	↓	↑	↓	↓	↓
9	↑	↑	↓	↓	↓	↓	↑
10	↑	↑	↓	↓	↑	↓	↓
11	↓	↑	↓	↓	↑	↓	↑
12	↑	↑	↓	↓	↓	↓	↓
13	↑	↓	↓	↑	↓	↑	↑
14	↑	↑	↑	↓	↓	↓	↑
15	↑	↓	↓	↓	↓	↑	↑
16	↑	↑	↓	↑	↑	↓	↓
17	↓	↑	↓	↓	↑	↓	↓
18	↓	↓	↓	↓	↑	↑	↑
19	↑	↓	↓	↓	↓	↑	↑
20	↓	↑	↓	↓	↓	↓	↓

central-coke ratios to represent the parameters of the BD. BD improves the GUR by adjusting the shape of the coke and iron-ore layers. As the complete change for the shape of the layers takes a long time, the BD affects the change of the GUR on an LTS.

The HBS improves the GUR by adjusting the DIGL. The changes for the parameters of the HBS result in the changes of the bosh gas volume, pressure difference, GL rate, and distribution of the temperature, which affect the DIGL. As the change of the initial GL takes a short time, the HBS affects the GUR on a short-time scale.

B. Decomposition and Reconstruction of GUR

The relation of the operations and the GUR can be understood deeply according to analyzing the different-time-sale parts of the GUR. The EMD decomposes the original time series to different-time-sale subseries according to the characteristics of the original signal. Taking the EMD method as the basis, a method is introduced to decompose the GUR.

The time series of the GUR is decomposed to subseries with different frequencies. The subseries contains the several IMFs and a residual term. Each IMF has two characteristics: (1) the maximum difference between the numbers of extreme value and the points passing through zero is one; (2) the mean value of the maximum and minimum envelope is zero. The decomposition of the GUR is shown as

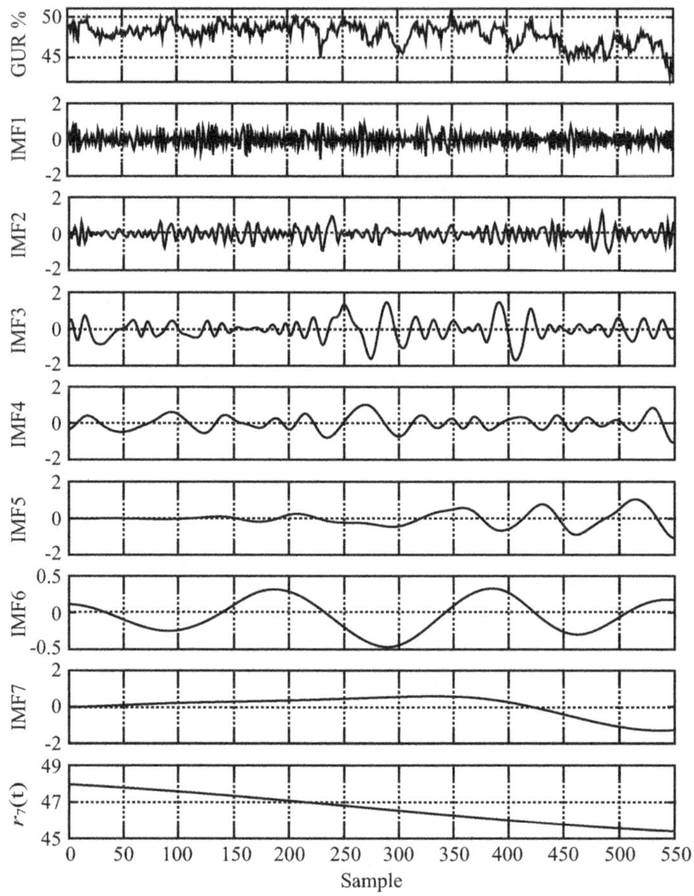

Fig. 9 GUR and its sub-series decomposed by EMD

$$G(t) = \sum_{j=1}^{n} \hat{c}_j(t) + r_n(t), \tag{15}$$

where $G(t)$ is the original signal of the GUR; $\hat{c}_j(t)$ $(j = 1, 2, \cdots, n)$, the jth IMF; $r_n(t)$, the residual term; n, the number of IMFs.

The decomposition results are shown in Fig. 9. The time series of the GUR is decomposed into 8 subseries, which contains 7 IMFs and a residual term. Each subseries has different time scales. The time scales decrease from IMF 1 to $r_7(t)$.

In order to find the relation of the operations and the GUR on different-time scales, the decomposition results of the GUR are reconstructed into an LTS part and an STS part. In order to reconstruct the GUR reasonably, this part analyzes the correlations between the operations and each sub-series of the GUR based on the Pearson coefficient method. According to the correlation, the GUR is reconstructed

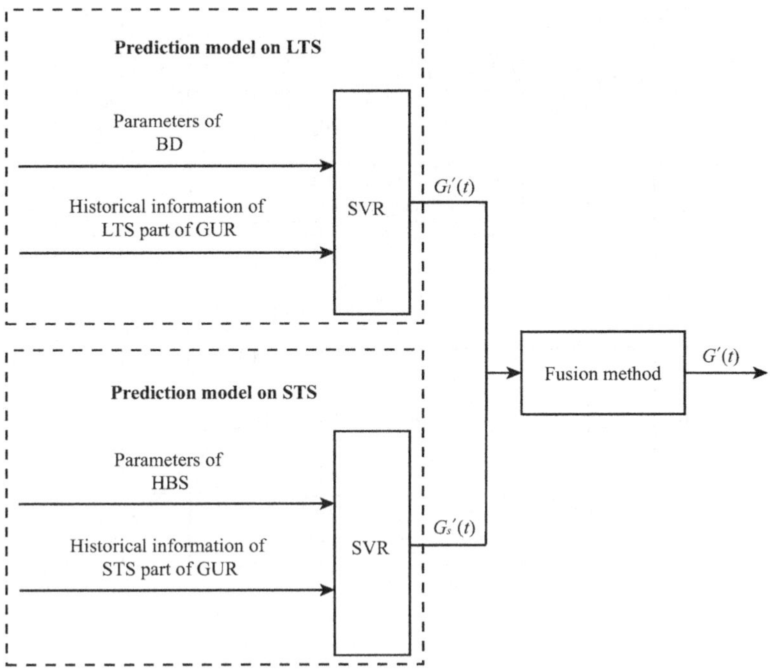

Fig. 10 Structure of the MFPM

as

$$G_h(t) = \hat{c}_1(t), \tag{16}$$

$$G_s(t) = \hat{c}_2(t) + \hat{c}_3(t) + \hat{c}_4(t) + \hat{c}_5(t), \tag{17}$$

$$G_l(t) = \hat{c}_6(t) + \hat{c}_7(t) + \hat{c}_8(t), \tag{18}$$

where $G_h(t)$ is the noise component of the GUR; $G_s(t)$, the STS part; $G_l(t)$, the LTS part.

C. Analysis of MFPM

According to the relations of operations and the multi-time-scale parts of the GUR, this part introduces an MFPM for the GUR. The structure of the model is shown as Fig. 10.

The prediction models are built by the SVR algorithm. According to the EMD method, the decomposition results add up to the original time series. Thus, the fusion method is

$$G'(t) = G_s'(t) + G_l'(t), \tag{19}$$

where $G_s'(t)$, $G_l'(t)$, and $G'(t)$ are the predictions of the STS part of the GUR, LTS part of the GUR, and the fused GUR, respectively.

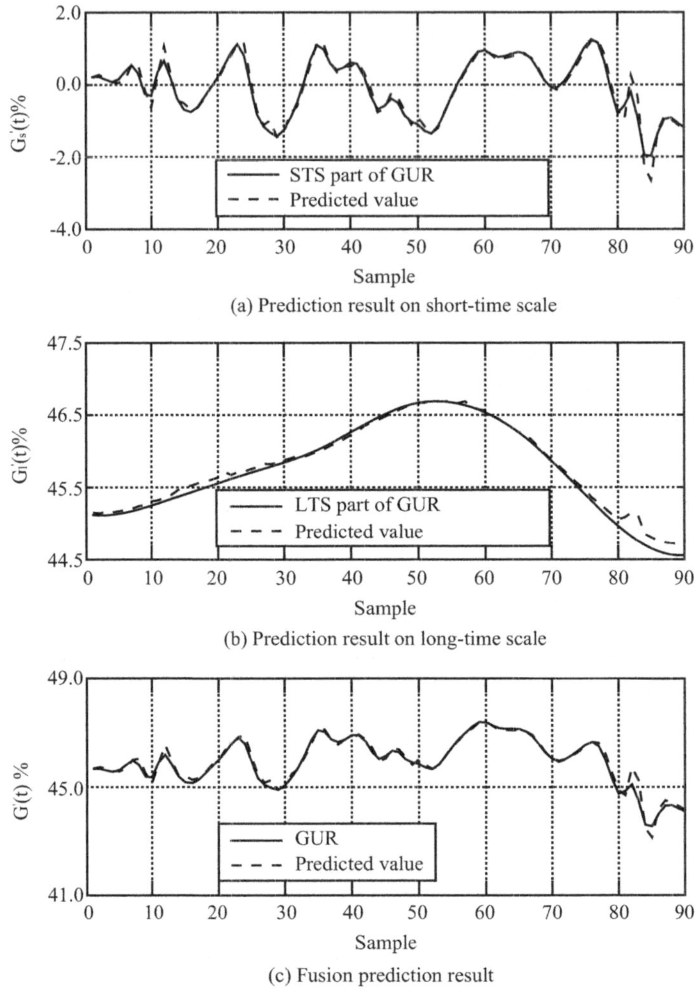

(a) Prediction result on short-time scale

(b) Prediction result on long-time scale

(c) Fusion prediction result

Fig. 11 Prediction results of the method

In order to verify the effectiveness of the fusion prediction model, the simulation predicts the STS part, LTS part, and fusion part of the GUR based on the corresponding prediction model, respectively.

Figure 11a shows that the prediction model on STS predicts the corresponding part of the GUR accurately. Figure 11b shows that the prediction model on LTS predicts the corresponding part of the GUR accurately. For Fig. 11c, the fused prediction value is calculated by the STS part and the LTS part of the GUR. According to the result, the fusion prediction model can obtain the predicted GUR accurately.

3 Decision-Making Strategy

A BF is a huge container with a complex mechanism environment. The means of regulating BF are both hot blast supply and BD. In this section, BD is chosen as the adjustment method. It means to adjust the BD of a BF by top charging equipment. The change of the distribution parameters is mainly to change the distribution matrix, including changing the angle of the distribution chute, the number of distribution circles, the weight of the batch, and other parameters. Different distribution strategies will have different effects on the GUR. Then, a variety of GUR prediction methods are mentioned, such as MFPM, case-matching model prediction model, and so on. GUR is an important indicator of BF operational status and BD is an important parameter that affects the DGL and GUR of BF. The decision-making strategy in this section is generated based on the previous forecast of GUR. Due to the complex mechanism relationship and long-time delay in the iron-making process, making a reasonable distribution matrix for the BD is difficult. The purpose of this section is to devise a BD decision method.

3.1 Structure of Decision-Making Strategy

In the iron-making process, the instantaneous value of the GUR is affected by the complex industrial environment. And the effect of the BD on the GUR has a long-time delay. Thus, The changing trend of the GUR is more concerned than the instantaneous value. The control strategy of the BD is based on the changing trend of the GUR. This part introduces the structure of the decision-making strategy for the BD, which is shown in Fig. 12. In the Fig. 12, h means the current time, and $h + 1$ mean the next time.

In order to design the decision-making strategy for the BD, a model is established to obtain the changing trend of the GUR. The structure of the model is shown in Fig. 13. First, a model is built to predict the BF variables highly related to the GUR. Then, the case-matching model based on the probability is built to predict the changing trend of the GUR. This prediction model is used in the decision-making strategy as a basis for evaluating the parameters of the BD. The following sections detail the decision model for allocating the BD.

3.2 Decision-Making Procedure

This part introduces the decision-making procedure for the parameters of the BD. In order to obtain the suitable parameters of the BD, the parameters of the BD are simplified into four modes. The current parameters are set as x_{oc}^{h} and x_{rc}^{h}. The next

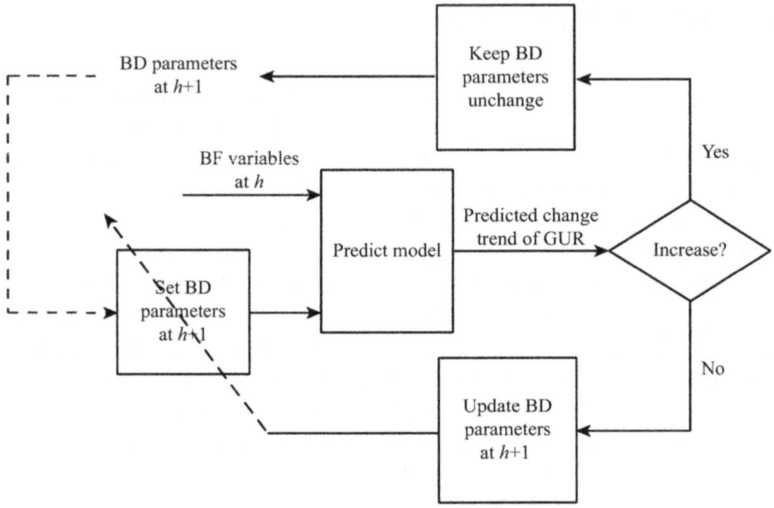

Fig. 12 Decision-making strategy

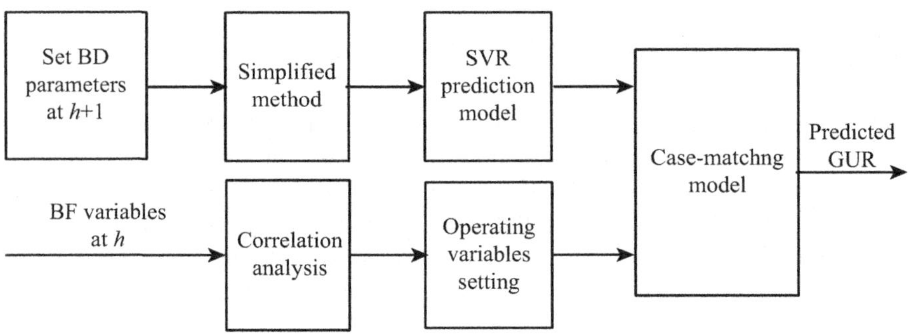

Fig. 13 GUR prediction model

Table 2 The modes of BD

Mode	Parameters of BD
1	$d_{oc}^{h+1} = d_{oc}^{h}, d_{rc}^{h+1} > d_{rc}^{h}$
2	$d_{oc}^{h+1} = d_{oc}^{h}, d_{rc}^{h+1} < d_{rc}^{h}$
3	$d_{rc}^{h+1} = d_{rc}^{h}, d_{oc}^{h+1} > d_{oc}^{h}$
4	$d_{rc}^{h+1} = d_{rc}^{h}, d_{oc}^{h+1} < d_{oc}^{h}$

time parameters are set as x_{oc}^{h+1} and x_{rc}^{h+1}. The simplified models of the parameters are shown in Table 2.

The modes are shown in Table 2 are the inputs of the model using the SVR algorithm to predict the BF variables. The corresponding BF variables are the outputs of this model. Then, the five BF variables are set as the inputs of the case-matching model to predict the changing trend of the GUR. Based on the above methods, the influence of the BD on the changing trend of the GUR is obtained. If the mode of the BD increases the corresponding GUR, this mode is suitable. The detailed decision-making steps are as follows.

(1) The current values of H-L, T-T, P-I, O-E, and P-C-I are set as d_{hl}^{m}, d_{tt}^{m}, x_{pi}^{m}, d_{oe}^{m}, and d_{pci}^{m}. The next time values of the parameters of the BD are d_{oc}^{h+1} and d_{rc}^{h+1}.

(2) The next time values of the H-L, T-T, P-I are set as d_{hl}^{h+1}, d_{tt}^{h+1}, and d_{pi}^{h+1}, which are predicted by the corresponding SVR prediction models, respectively. The next time values of O-E and P-C-I are set as d_{oe}^{h+1} and d_{pci}^{h+1}, which are obtained by the current states of a BF.

(3) The changes of the BF variables are calculated by $d_{hl}^{h+1} - d_{hl}^{h}$, $d_{pi}^{h+1} - d_{pi}^{h}$, $d_{pi}^{h+1} - d_{pi}^{h}$, $d_{oe}^{h+1} - d_{oe}^{h}$, and $d_{pci}^{h+1} - d_{pci}^{h}$.

(4) The changes of the BF variables are set as the inputs of the case-matching model to predict the corresponding change trend of the GUR.

(5) If the corresponding change trend of GUR decreases, the next time parameters of the BD need to be updated according to the above steps. The four simplified models of the parameters are listed in Table 2. The stop condition is that the mode of the parameters increases the GUR. The parameters corresponding to this mode is suitable.

3.3 Decision-Making Verification

In order to verify the effectiveness of the decision-making strategy, the simulation analyzes the effect of the modes of the BD parameters on the changing trend of the GUR. The results are shown in Table 3.

In Table 3, Modes 1 to 4 represent four types of BD modes corresponding to the Table 2. $P - mode$ is the predicted mode of parameters based on the decision-making strategy, and $A - mode$ is the actual mode of parameters. $P - GUR$ is the predicted change trend of the GUR, and $A - GUR$ is the actual change trend of the GUR.

For the results in Table 3, the predicted mode of the parameters in samples 2, 5, 7, 8, and 10 is the same as the actual mode. Among them, the predicted change trend of the GUR increase in sample 8, which is different from the actual change trend. Besides, the predicted change trend of the GUR decrease in samples 3 and 9 while the actual change trend of these two samples is increased. Analyzing the

Table 3 The decisions of BD

Sample	P-GUR of Mode 1	P-GUR of Mode 2	P-GUR of Mode 3	P-GUR of Mode 4	P-mode	A-mode	A-GUR
1	↓	↑	↓	↑	2/4	1	↓
2	↓	↑	↓	↑	2/4	2	↑
3	↓	↓	↓	↑	4	3	↑
4	↑	↑	↑	↓	1/2/3	4	↓
5	↓	↑	↓	↑	2/4	2	↑
6	↓	↓	↓	↑	4	1	↓
7	↓	↑	↑	↓	2/3	2	↑
8	↑	↑	↓	↑	1/2/4	1	↓
9	↓	↓	↓	↑	4	2	↑
10	↓	↑	↓	↓	2	2	↓

results of the 10 samples, the accuracy rate of the correct decisions is 70%. Thus, the decision-making strategy for the BD is effective.

In the actual iron-making process, this method can be used only when the parameters of the BD are provided. This method is a judgment on the rationality of the parameters.

4 Conclusion

This chapter introduces the strategy of BD to improve the GUR. The strategy contains two parts: the prediction model and control strategy. The chapter introduces the tree prediction model for the GUR. The first one is a chaotic RBF prediction model that analyzes the chaotic character of the development trend of the GUG. The second one is a case-matching prediction model that gives the influence of the BD on the changing trend of the GUR qualitatively. The last one is an MFPM that analyzes the relationship between the operations and the GUR quantitatively. Based on the case-matching prediction model, a decision-making strategy of BD is introduced. This strategy can provide guidance for setting suitable BD parameters.

References

1. An, J., Shen, X., Wu, M., She, J.: A multi-time-scale fusion prediction model for the gas utilization rate in a blast furnace. Control Eng. Pract. https://doi.org/10.1016/j.conengprac.2019.104120
2. An, J., Yang, J., Wu, M., She, J., Terano, T.: Decoupling control method with fuzzy theory for top pressure of blast furnace. IEEE Trans. Control Syst. Technol. **27**(6), 2735–2742 (2019)

3. An, J., Zhang, J., Wu, M., et al.: Soft-sensing method for slag-crust state of blast furnace based on two-dimensional decision fusion. Neurocomputing **315**(13), 405–411 (2018)
4. Helle, H., Helle, M., Saxn, H.: Nonlinear optimization of steel production using traditional and novel blast furnace operation strategies. Chem. Eng. Sci. **66**(24), 6470–6481 (2011)
5. Li, Z., Kuang, S., Liu, S., Gan, J., Yu, A., Li, Y., Mao, X.: Numerical investigation of burden distribution in ironmaking blast furnace. Powder Technol. **353**, 385–397 (2019)
6. Zhang, F., Mao, Q., Mei, C., Li, X., Hu, Z.: Dome combustion hot blast stove for huge blast furnace. J. Iron Steel Res. Int. **19**(9), 1–7 (2012)
7. Matino, I., Dettori, S., Colla, V., Weber, V., Salame, S.: Two innovative modelling approaches in order to forecast consumption of blast furnace gas by hot blast stoves. Energy Proc. **158**, 4043–4048 (2019)
8. Zetterholm, J., Ji, X., Sundelin, B., Martin, P., Wang, C.: Dynamic modelling for the hot blast stove. Appl. Energy **185**(2), 2142–2150 (2017)
9. Duan, W., Yu, Q., Liu, J., Hou, L., Xie, H., Wang, K., Qin, Q.: Characterizations of the hot blast furnace slag on coal gasification reaction. Appl. Thermal Eng. **98**, 936–943 (2016)
10. Ji, Y., Zhang, S., Yin, Y., Su, X.: Application of the improved the ELM algorithm for prediction of blast furnace gas utilization rate. IFAC-PapersOnLine **51**(21), 59–64 (2018)
11. Li, Y., Liu, X., Gao, S., Duan, X., Hu, Z., Chen, X., Shen, R., Guo, H., An W.: A generalized model for gas flow prediction in shale matrix with deduced coupling coefficients and its macroscopic form based on real shale pore size distribution experiments. J. Petrol. Sci. Eng. https://doi.org/10.1016/j.petrol.2019.106712
12. Dong, Z., Wang, J., Zuo, H., She, X., Xue, Q.: Analysis of gasCsolid flow and shaft-injected gas distribution in an oxygen blast furnace using a discrete element method and computational fluid dynamics coupled model. Particuology **32**, 63–72 (2017)
13. Shi, L., Zhao, G., Li, M., Ma, X.: A model for burden distribution and gas flow distribution of bell-less top blast furnace with parallel hoppers. Appl. Math. Modell. **40**(23–24), 10254–10273 (2016)
14. Wright, B., Zulli, P., Zhou, Z.Y., et al.: Gas-solid flow in an ironmaking blast furnace - I: Physical modelling. Powder Technol. **208**(1), 86–97 (2011)
15. Zhang, S., Jiang, H., Yin, Y., Xiao, W., Zhao, B.: The prediction of the gas utilization ratio based on TS fuzzy neural network and particle swarm optimization. Sensors (Basel) **18**(2), 625–644 (2018)
16. Li, Y., Zhang, S., Yin, Y., Xiao, W., Zhang, J.: A novel online sequential extreme learning machine for gas utilization ratio prediction in blast furnaces. Sensors (Basel) **17**(8), 1847–1870 (2017)
17. An, J., Yang, Y., Wu, M., Terano, T.: Analysis of influencing factors on carbon monoxide utilization rate of blast furnace based on multi-timescale characteristics. In: Proceedings of the 2017 36th Chinese Control Conference (CCC), pp. 4459–4463 (2017)
18. Zhang, Y., Zhou, P., Cui, G.: Multi-model based PSO method for burden distribution matrix optimization with expected burden distribution output behaviors. IEEE/CAA J. Autom. Sin. **6**(6), 1506–1512 (2019)
19. Wang, L., Zhang, B., Zhang, Y., et al.: Mathematical model of charging shape in bell-less blast furnace burden distribution. J. Iron Steel Res. **30**(9), 696–702 (2018)
20. Su, X., Zhang, S., Yin, Y., et al.: Data-driven prediction model for adjusting burden distribution matrix of blast furnace based on improved multilayer extreme learning machine. Soft Comput. **22**, 3575–3589 (2018)
21. Xiao, D., An, J., He, Y., et al.: The chaotic characteristic of the carbon-monoxide utilization ratio in the blast furnace. ISA Trans. **68**, 109–115 (2017)
22. Zhang, K., Wu, M., An, J., Cao, W., et al.: Relation model of burden operation and state variables of blast furnace based on low frequency feature extraction. IFAC-PapersOnLine **50**(1), 13796–13801 (2017)
23. Fu, D., Chen, Y., Zhou, Q.: Mathematical modeling of blast furnace burden distribution with non-uniform descending speed. Appl. Math. Modell. **39**(23–24), 7554–7567 (2015)

24. Shi, P., Zhou, P., Fu, D., et al.: Mathematical model for burden distribution in blast furnace. Ironmaki. Steelmak. **43**(1), 74–81 (2016)
25. Zhou, P., Shi, P., Song, Y., et al.: Evaluation of burden descent model for burden distribution in blast furnace. J. Iron Steel Res. Int. **23**(8), 765–771 (2016)
26. Wu, M., Zhang, K., An, J., She, J., Liu, K.: An energy efficient decision-making strategy of burden distribution for blast furnace. Control Eng. Pract. **78**, 186–195 (2018)

Intelligent System and Machine Learning

Granular Computing: Fundamentals and System Modeling

Witold Pedrycz

Abstract In the plethora of conceptual and algorithmic developments supporting system modeling, we encounter growing challenges associated with the complexity of systems, diversity of available data and a variety of requests as to the quality of the models. The accuracy of models is important. At the same time, the interpretability and explainability of models are equally important and of high practical relevance. We advocate that the level of abstraction at which models are constructed (and which could be flexibly adjusted), is conveniently realized through Granular Computing. Granular Computing is concerned with the development and processing information granules formal entities which facilitate a way of organizing and representing knowledge about the available data and relationships existing there. This study identifies the principles of Granular Computing, shows how information granules are constructed and subsequently used in describing relationships present among the data contributing to the realization of models.

Keywords Granular Computing · Information granules · Fuzzy sets · Design of information granules · Clustering · Principle of justifiable granularity · Aggregation

1 Introduction

The apparent reliance on data and experimental evidence in system modeling, decision-making, pattern recognition, and control engineering, just to enumerate several representative spheres of interest, entails the centrality of data and emphasizes their paramount role in data science. To capture the essence of data, facilitate building their essential descriptors and reveal key relationships, as well as having

W. Pedrycz (✉)
Department of Electrical & Computer Engineering, University of Alberta,
Edmonton, AB T6R 2V4, Canada
e-mail: wpedrycz@ualberta.ca

Systems Research Institute, Polish Academy of Sciences, 01-447 Warsaw, Poland

M. Wu et al. (eds.), *Developments in Advanced Control and Intelligent Automation for Complex Systems*, Studies in Systems, Decision and Control 329,
https://doi.org/10.1007/978-3-030-62147-6_7

all these faculties realized in an efficient manner as well as deliver transparent, comprehensive, and user-oriented results, we advocate a genuine need for transforming data into information granules. In the realized setting, information granules become regarded as conceptually sound knowledge tidbits over which various models could be developed and utilized.

A tendency, which is being witnessed more visibly nowadays, concerns human centricity. Data science and big data revolve around a two-way efficient interaction with users. Users interact with data analytics processes meaning that the terms such as data quality, actionability, transparency are of relevance and are provided in advance. With this regard, information granules emerge as a sound conceptual and algorithmic vehicle owing to their way of delivering a more general view at data, ignoring irrelevant details and supporting a suitable level of abstraction aligned with the nature of the problem at hand.

Our objective is to provide a general overview of Granular Computing, identify the main items on its agenda and associate their usage in the setting of data analytics. To organize our discussion in a coherent way and highlight the main trends as well as deliver a self-contained material, the study is structured in a top-down manner. Some introductory material offering some motivating insights and main concepts and objectives are presented in Sect. 2. The formal frameworks of information granules and the existing formalisms are covered in Sect. 3. Section 4 is devoted to the design of information granules; along with a general taxonomy discussed is a role clustering or fuzzy clustering and the principle of justifiable granularity. Several augmentations of the generic version of the principle are elaborated on in Sect. 5. A symbolic characterization of prototypes is covered in Sect. 6. An application of the principle to spatiotemporal data in the form of granular probes is presented in Sect. 7. Granular models with an emphasis on the concepts, methodologies and main design directions are the topic of Sect. 8. Some prospects and main conclusions are covered in Sect. 9.

2 Information Granules and Information Granularity

The framework of Granular Computing along with a diversity of its formal settings offers a critically needed conceptual and algorithmic environment. A suitable perspective built with the aid of information granules is advantageous in realizing a suitable level of abstraction. It also becomes instrumental when forming sound and pragmatic problem-oriented tradeoffs among precision of results, their easiness of interpretation, value, and stability (where all of these aspects contribute vividly to the general notion of actionability).

Information granules are intuitively appealing constructs, which play a pivotal role in human cognitive and decision-making activities [3, 4, 49, 51, 52, 71, 72]. We perceive complex phenomena by organizing existing knowledge along with available experimental evidence and structuring them in a form of some meaningful, semantically sound entities, which are central to all ensuing processes of describing the world, reasoning about the environment, and support decision-making activities.

The terms information granules and information granularity themselves have emerged in different contexts and numerous areas of application. Information granule carries various meanings. One can refer to Artificial Intelligence (AI) in which case information granularity is central to a way of problem solving through problem decomposition, where various subtasks could be formed and solved individually. Information granules and the area of intelligent computing revolving around them being termed Granular Computing are quite often presented with a direct association with the pioneering studies by Zadeh [71]. He coined an informal, yet highly descriptive and compelling concept of information granules. Generally, by information granules one regards a collection of elements drawn together by their closeness (resemblance, proximity, functionality, etc.) articulated in terms of some useful spatial, temporal, or functional relationships. Subsequently, Granular Computing is about representing, constructing, processing, and communicating information granules. The concept of information granules is omnipresent and this becomes well documented through a series of applications, cf. [27, 30, 45, 47, 75–77].

Granular Computing exhibits a variety of conceptual developments; one may refer here to selected and representative pursuits:

graphs [8, 38, 64]
information tables [7]
mappings [55]
knowledge representation [9]
micro and macro models [5]
association discovery and data mining [19, 66]
clustering [62] and rule clustering [65]
classification [31, 56]

There are numerous applications of Granular Computing, which are reported in recent publications:

Forecasting time series [20, 58]
Prediction tasks [17]
Manufacturing [27]
Concept learning [28]
Perception [21]
Optimization [33]
Credit scoring [54]
Analysis of microarray data [53, 61]
Linguistic modeling of time series [32]
Clustering granular data [59].

It is again worth emphasizing that information granules permeate almost all human endeavors. No matter which problem is taken into consideration, we usually set it up in a certain conceptual framework composed of some generic and conceptually meaningful entities-information granules, which we regard to be of relevance to the problem formulation, further problem solving, and a way in which the findings are communicated to the community. Information granules realize a framework in which we formulate generic concepts by adopting a certain level of generality.

Information granules naturally emerge when dealing with data, including those coming in the form of data streams. The ultimate objective is to describe the underlying phenomenon in an easily understood way and at a certain level of abstraction. This requires that we use a vocabulary of commonly encountered terms (concepts) and discover relationships between them and reveal possible linkages among the underlying concepts.

Information granules are examples of abstractions. As such, they naturally give rise to hierarchical structures: the same problem or system can be perceived at different levels of specificity (detail) depending on the complexity of the problem, available computing resources, and particular needs to be addressed. A hierarchy of information granules is inherently visible in processing of information granules. The level of captured details (which is represented in terms of the size of information granules) becomes an essential facet facilitating a way a hierarchical processing of information with different levels of hierarchy indexed by the size of information granules.

Even such commonly encountered and simple examples presented above are convincing enough to lead us to ascertain that (a) information granules are the key components of knowledge representation and processing, (b) the level of granularity of information granules (their size, to be more descriptive) becomes crucial to the problem description and an overall strategy of problem solving, (c) hierarchy of information granules supports an important aspect of perception of phenomena and deliver a tangible way of dealing with complexity by focusing on the most essential facets of the problem, (d) there is no universal level of granularity of information; commonly the size of granules is problem-oriented and user dependent.

Human-centricity comes as an inherent feature of intelligent systems. It is anticipated that a two-way effective human-machine communication is imperative. Human perceive the world, reason, and communicate at some level of abstraction. Abstraction comes hand in hand with non-numeric constructs, which embrace collections of entities characterized by some notions of closeness, proximity, resemblance, or similarity. These collections are referred to as information granules. Processing of information granules is a fundamental way in which people process such entities. Granular Computing has emerged as a framework in which information granules are represented and manipulated by intelligent systems. The two-way communication of such intelligent systems with the users becomes substantially facilitated because of the usage of information granules.

It brings together the existing plethora of formalisms of set theory (interval analysis) under the same banner by clearly visualizing that in spite of their visibly distinct underpinnings (and ensuing processing), they exhibit some fundamental commonalities. In this sense, Granular Computing establishes a stimulating environment of synergy between the individual approaches. By building upon the commonalities of the existing formal approaches, Granular Computing helps assemble heterogeneous and multifaceted models of processing of information granules by clearly recognizing the orthogonal nature of some of the existing and well established frameworks (say, probability theory coming with its probability density functions and fuzzy sets

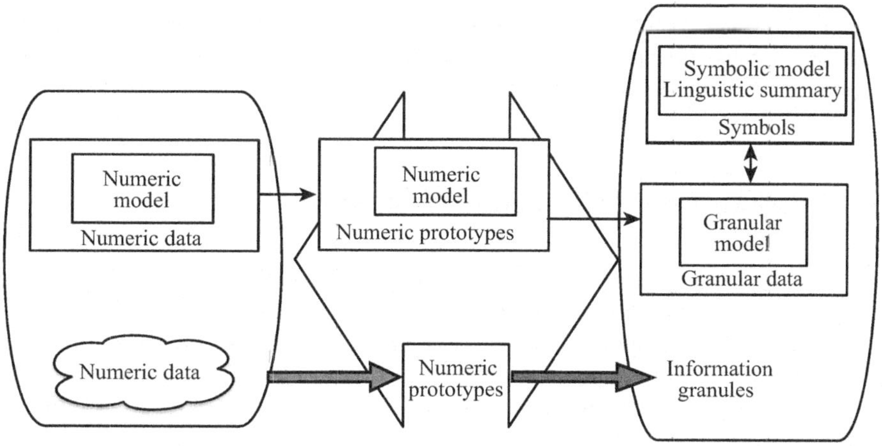

Fig. 1 Relationships among modeling environments and information granules: emphasized is the way of moving from experimental data to their representatives, information granules and linguistic summaries

with their membership functions). Granular Computing fully acknowledges a notion of variable granularity, whose range could cover detailed numeric entities and very abstract and general information granules. It looks at the aspects of compatibility of such information granules and ensuing communication mechanisms of the granular worlds. Granular Computing gives rise to processing that is less time demanding than the one required when dealing with detailed numeric processing.

In system modeling, information granules play at least two fundamental roles: (a) as building blocks using which a variety of models is built. The concept of granular models deals with models that establish mappings among information granules and realize tasks of prediction, classification and associations realized at a suitable level of abstraction implied by information granules. Granular models are constructed more efficiently than their numeric counterparts (the number of information granules is far smaller than the masses of numeric data), become more transparent and interpretable, (b) as a means to express the quality of the numeric models. In this case granular models incorporate the mechanisms of granular processing and the parameters of the model are made granular following the optimal allocation of information granularity. With this regard, it is instructive to link the developments of information granules with the way in which they support ways of system modeling as illustrated in Fig. 1. Numeric data and numeric prototypes associate with numeric models. Granular prototypes give rise to granular models. The symbolic manifestation of information granules entails symbolic (qualitative) models; the symbols used there are well-grounded in virtue of the construction scheme supporting the buildup of information granules.

There are two clearly visible layers of processing. The one is concerned with the abstraction of available data: we proceed with numeric data (commonly acquired experimentally as a manifestation of the system), determine their numeric representatives (prototypes) and build information granules. In parallel, the these activities give rise to particular processing realized in system modeling as portrayed at the upper portion of the figure.

3 Frameworks of Information Granules

There are numerous formal frameworks of information granules; for illustrative purposes, we recall some selected alternatives.

Sets (intervals) realize a concept of abstraction by introducing a notion of dichotomy: we admit element to belong to a given information granule or to be excluded from it. Along with the set theory comes a well-developed discipline of interval analysis [2, 35, 36].

Fuzzy sets deliver an important conceptual and algorithmic generalization of sets [10–12, 25, 37, 44, 49, 70, 73]. By admitting partial membership of an element to a given information granule, we bring an important feature which makes the concept to be in rapport with reality. It helps working with the notions, where the principle of dichotomy is neither justified, nor advantageous. Furthermore owing to the smooth nature of membership functions, fuzzy sets are helpful in realizing optimization tasks, in particular those engaging gradient-based optimization schemes. Fuzzy sets come with a spectrum of operations, usually realized in terms of triangular norms [24, 57].

Shadowed sets [46, 48] offer an interesting description of information granules by distinguishing among three categories of elements. Those are the elements, which (i) fully belong to the concept, (ii) are excluded from it (iii) their belongingness is completely *unknown*. Some design issues are discussed by [69].

Rough sets [39–43] are concerned with a roughness phenomenon, which arises when an object (pattern) is described in terms of a limited vocabulary of certain granularity. The description of this nature gives rise to a so-called lower and upper bounds forming the essence of a rough set.

The list of formal frameworks is quite extensive; as interesting examples, one can recall here probabilistic sets [18] and axiomatic fuzzy sets [29].

At the practical end, one should emphasize that while the existing approaches come with some conceptual motivation pointing at their relevance, it is of paramount importance to have them equipped with sound development mechanisms and estimation procedures; in several situations this is not the case.

Information Granules of Higher Type and Higher Order

There are two important directions of generalizations of information granules, namely information granules of higher type and information granules of higher order. The essence of information granules of higher type comes with a fact that the characterization (description) of information granules is described in terms of information granules rather than numeric entities. Well-known examples are fuzzy sets of type-2 [34, 60], granular intervals, or imprecise probabilities. For instance, a type-2 fuzzy set is a fuzzy set whose grades of membership are not single numeric values (membership grades in [0, 1]) but fuzzy sets, intervals or probability density functions truncated to the unit interval. There is a hierarchy of higher type information granules, which are defined in a recursive manner. Therefore we talk about type-0, type-1, type-2 fuzzy sets, etc. In this hierarchy, type-0 information granules are numeric entities, say, numeric measurements. With regard to higher order information granules, those are granules defined in some space whose elements are information granules themselves.

4 Information Granules and Their Two-Phase Development Process

Building information granules constitutes a central item on the agenda of Granular Computing with far-reaching implications on its applications. We present a way of moving from data to numeric representatives, information granules and their linguistic summarization. Overall the organization of the overall scheme and relationships among the resulting constructs are displayed in Fig. 2.

4.1 Clustering as a Prerequisite of Information Granules

Along with a truly remarkable diversity of detailed algorithms and optimization mechanisms of clustering, the paradigm itself delivers a viable prerequisite to the formation of information granules (associated with the ideas and terminology of fuzzy clustering, rough clustering, and others) and applies both to numeric data and information granules. Information granules built through clustering are predominantly data-driven, viz. clusters (either in the form of fuzzy sets, sets, or rough sets) are a manifestation of a structure encountered (discovered) in the data.

Numeric prototypes are formed through invoking clustering algorithms, which yield a partition matrix and a collection of the prototypes. Clustering realizes a certain process of abstraction producing a small number of the prototypes based on a large number of numeric data. Interestingly, clustering can be also completed in the

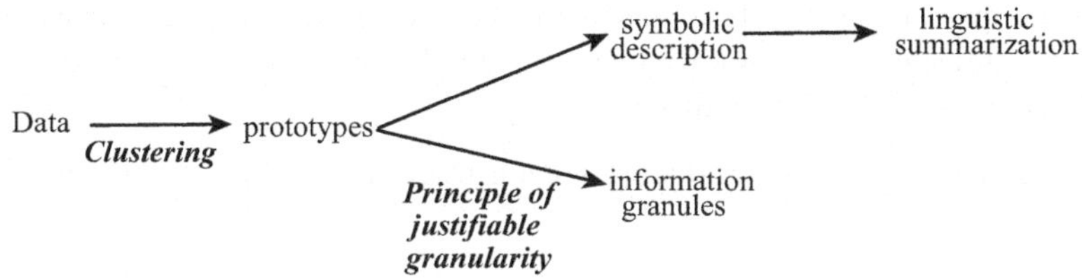

Fig. 2 From data to information granules and linguistic summarization

feature space. In this situation, the algorithm returns a small collection of abstracted features (groups of features) that might be referred to as meta-features.

Two ways of generalization of numeric prototypes treated as key descriptors of data and manageable chunks of knowledge are considered: (i) symbolic and (ii) granular. In the symbolic generalization, one moves away from the numeric values of the prototypes and regards them as sequences of integer indexes (labels). Along this line, developed are concepts of (symbolic) stability and (symbolic) resemblance of data structures. The second generalization motivates the construction of information granules (granular prototypes), which arise as a direct quest for delivering a more comprehensive representation of the data than the one delivered through numeric entities. This entails that information granules (including their associated level of abstraction), have to be prudently formed to achieve the required quality of the granular model.

As a consequence, the performance evaluation embraces the following sound alternatives: (i) evaluation of representation capabilities of numeric prototypes, (ii) evaluation of representation capabilities of granular prototypes, and (iii) evaluation of the quality of the granular model

A. Evaluation of Representation Capabilities of Numeric Prototypes

In the first situation, the representation capabilities of numeric prototypes are assessed with the aid of a so-called granulation-degranulation scheme yielding a certain reconstruction error. The essence of the scheme can be schematically portrayed as follows:

$$\boldsymbol{x} \longrightarrow \text{internal representation} \longrightarrow \text{reconstruction}$$

The formation of the internal representation is referred to as granulation (encoding) whereas the process of degranulation (decoding) can be sought as an inverse mechanism to the encoding scheme. In terms of detailed formulas, one encounters the following flow of computing:

- encoding leading to the degrees of activation of information granules by input $\boldsymbol{x}$, say $A_1(\boldsymbol{x}), A_2(\boldsymbol{x}), \ldots, A_c(\boldsymbol{x})$ with

$$A_i(\boldsymbol{x}) = \frac{1}{\sum_{j=1}^{c}\left(\frac{\|\boldsymbol{x}-\boldsymbol{v}_i\|}{\|\boldsymbol{x}-\boldsymbol{v}_j\|}\right)^{2/(m-1)}} \tag{1}$$

in case the prototypes are developed with the use of the Fuzzy C-Means (FCM) clustering algorithm, the parameter $m(>1)$ stands for the fuzzification coefficient and $\|.\|$ denotes the Euclidean distance.

- degranulation (decoding) producing a reconstruction of $\boldsymbol{x}$ via the following expression

$$\hat{\boldsymbol{x}} = \frac{\sum_{i=1}^{c} A_i^m(\boldsymbol{x})\boldsymbol{v}_i}{\sum_{i=1}^{c} A_i^m(\boldsymbol{x})} \tag{2}$$

It is worth stressing that the above stated formulas are a consequence of the underlying optimization problems. For any collection of numeric data, the reconstruction error is a sum of squared errors (distances) of the original data and their reconstructed versions.

4.2 The Principle of Justifiable Granularity

The principle of justifiable granularity guides a construction of an information granule based on available experimental evidence [50, 51]. In a nutshell, a resulting information granule becomes a summarization of data (viz. the available experimental evidence). The underlying rationale behind the principle is to deliver a concise and abstract characterization of the data such that (i) the produced granule is justified in light of the available experimental data, and (ii) the granule comes with a well-defined semantics meaning that it can be easily interpreted and becomes distinguishable from the others.

Formally speaking, these two intuitively appealing criteria are expressed by the criterion of coverage and the criterion of specificity. Coverage states how much data are positioned behind the constructed information granule. Put it differently C coverage quantifies an extent to which information granule is supported by available experimental evidence. Specificity, on the other hand, is concerned with the semantics of information granule stressing the semantics (meaning) of the granule.

A. One-Dimensional Case

The definition of coverage and specificity requires formalization and this depends upon the formal nature of information granule to be formed. As an illustration, consider an interval form of information granule A. In case of intervals built on a basis of one-dimensional numeric data (evidence) $x_1, x_2, \ldots, x_N$, the coverage measure is associated with a count of the number of data embraced by A, namely

$$\text{cov}(A) = \frac{1}{N} \text{card}\,\{x_k | x_k \in A\} \tag{3}$$

$card(.)$ denotes the cardinality of A, viz. the number (count) of elements x_k belonging to A. In essence, coverage has a visible probabilistic flavor. The specificity of A, $sp(A)$ is regarded as a decreasing function g of the size (length) of information granule. If the granule is composed of a single element, $sp(A)$ attains the highest value and returns 1. If A is included in some other information granule B, then $sp(A) > sp(B)$. In a limit case if A is an entire space $sp(A)$ returns zero. For an interval-valued information granule $A = [a, b]$, a simple implementation of specificity with g being a linearly decreasing function comes as

$$sp(A) = g(\text{length}(A)) = 1 - \frac{|b-a|}{range} \tag{4}$$

where range stands for an entire space over which intervals are defined.

If we consider a fuzzy set as a formal setting for information granules, the definitions of coverage and specificity are reformulated to take into account the nature of membership functions admitting a notion of partial membership. Here we invoke the fundamental representation theorem stating that any fuzzy set can be represented as a family of its α-cuts, namely

$$A(x) = \sup_{\alpha \in [0,1]} [\min(\alpha, A_\alpha(x))] \tag{5}$$

where

$$A_\alpha(x) = \{x | A(x) \geq \alpha\} \tag{6}$$

The supremum (sup) operation is taken over all values of α. In virtue of the theorem we have any fuzzy set represented as a collection of sets. Having this in mind and considering (3) as a point of departure for constructs of sets (intervals), we have the following relationships-coverage

$$\text{cov}(A) = \int_X A(x)dx/N \tag{7}$$

where X is a space over which A is defined; moreover, one assumes that A can be integrated. The discrete version of the coverage expression comes in the form of the sum of membership degrees. If each data point is associated with some weight, the calculations of the coverage involve these values

$$\text{cov}(A) = \int_X w(x)A(x)dx/\int_X w(x)dx \tag{8}$$

– specificity

$$sp(A) = \int_0^1 sp(A_\alpha)\, d\alpha \tag{9}$$

The criteria of coverage and specificity are in an obvious relationship, Fig. 3. We are interested in forecasting temperature: the more specific the statement about the prediction is, the lower the likelihood of its satisfaction.

Let us introduce the following product of the criteria

$$V = cov(A)sp(A) \tag{10}$$

It is apparent that the coverage and specificity are in conflict; the increase in coverage associates with the drop in the specificity. Thus the desired solution is the one where the value of V attains its maximum.

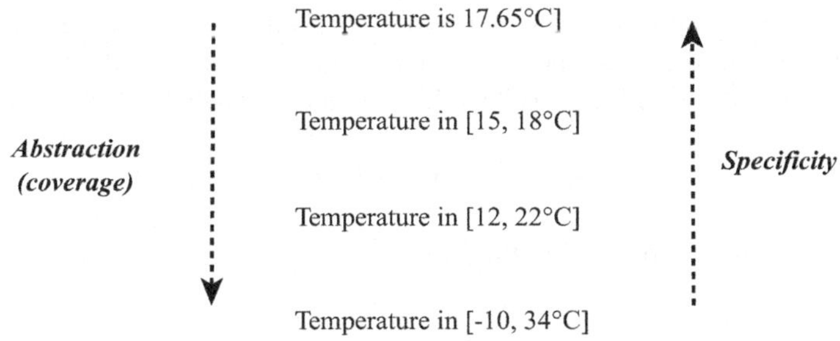

Fig. 3 Relationships between abstraction (coverage) and specificity of information granules of temperature

The design of information granule is accomplished by maximizing the above product of coverage and specificity. Formally speaking, consider that an information granule is described by a vector of parameters p, $V(p)$. The principle of justifiable granularity gives to an information granule that maximizes V, $p_{opt} = arg_p V(p)$.

To maximize the index V through the adjusting the parameters of the information granule, two different strategies are encountered:

(i) a two-phase development is considered. First a numeric representative (mean, median, mode, etc.) is determined. It can be regarded as an initial representation of the data. Next the parameters of the information granule are optimized by maximizing V. For instance, in case of an interval, one has two bounds (a and b) to be determined. These two parameters are determined separately, viz. the a and b are formed by maximizing $V(a)$ and $V(b)$. The data used in the maximization of $V(b)$ involves the data larger than the numeric representative. Likewise $V(a)$ is optimized on a basis of the data lower than this representative.

(ii) a single-phase procedure in which all parameters of information granule are determined at the same time.

B. Multi-dimensional Case

The results of clustering coming in the form of numeric prototypes $v_1, v_2, \ldots, v_c$ can be further augmented by forming information granules giving rise to so-called granular prototypes. This can be regarded as a result of an immediate usage of the principle of justifiable granularity and its algorithmic underpinning as elaborated earlier. Around the numeric prototype v_i, one spans an information granule V_i, $V_i = (v_i, \rho_i)$ whose optimal size is obtained as the result of the maximization of the well-known criterion

$$\rho_{\text{i,opt}} = \arg \text{Max}_{\rho_i} \left[\text{cov}\,(V_i)\,\text{sp}\,(V_i)\right] \tag{11}$$

where

$$\begin{aligned} \text{cov}\,(V_i) &= \frac{1}{N}\,\text{card}\,\{\boldsymbol{x}_k\ \|\boldsymbol{x}_k - \boldsymbol{v}_i\| \le \rho_i\} \\ \text{cov}\,(V_i) &= 1 - \rho_i \end{aligned} \tag{12}$$

assuming that we are concerned with normalized data. In case of the FCM method, the data come with their membership grades (entries of the partition matrix). The coverage criterion is modified to reflect this. Let us introduce the following notation

$$\Omega_i = \{\boldsymbol{x}_k | \, \|\boldsymbol{x}_k - \boldsymbol{v}_i\| \le \rho_i\} \tag{13}$$

Then the coverage is expressed in the form

$$\mathrm{cov}\,(V_i) = \frac{1}{N} \sum_{k_k \in \Omega_i} u_{ik} \tag{14}$$

C. Representation Aspects of Granular Prototypes
It is worth noting that having a collection of granular prototypes, one can conveniently assess their abilities to represent the original data (experimental evidence). The reconstruction problem, as outlined before for numeric data, can be formulated as follows: given x_k, complete its granulation and de-granulation using the granular prototypes V_i, $i = 1, 2, \ldots, c$. The detailed computing generalizes the reconstruction process completed for the numeric prototypes and for given x yields a granular result $\hat{X} = (\hat{v}, \hat{\rho})$ where

$$\hat{\boldsymbol{v}} = \frac{\sum_{i=1}^{c} A_i^m(x)\boldsymbol{v}_i}{\sum_{i=1}^{c} A_i^m(x)} \tag{15}$$

$$\hat{\rho} = \frac{\sum_{i=1}^{c} A_i^m(x)\rho_i}{\sum_{i=1}^{c} A_i^m(x)} \tag{16}$$

The quality of reconstruction uses the coverage criterion formed with the aid of the Boolean predicate

$$T\,(\boldsymbol{x}_k) = \begin{cases} 1 & \text{if } \left\|\hat{\boldsymbol{v}}_i - \boldsymbol{x}_k\right\| \le \hat{\rho} \\ 0, & \text{otherwise} \end{cases} \tag{17}$$

and for all data one takes the sum of (19) over them. It is worth noting that in addition to the global measure of quality of granular prototypes, one can associate with them their individual quality (taken as a product of the coverage and specificity computed in the formation of the corresponding information granule).

The principle of justifiable granularity highlights an important facet of elevation of the type of information granularity: the result of capturing a number of pieces of numeric experimental evidence comes as a single abstract entity—information granule. As various numeric data can be thought as information granule of type-0, the result becomes a single information granule of type-1. This is a general phenomenon of elevation of the type of information granularity. The increased level of abstraction is a direct consequence of the diversity present in the originally available granules. This elevation effect is of a general nature and can be emphasized by stating that when dealing with experimental evidence composed of information granules of type-n, the result becomes a single information granule of type (n+1). Some generalizations and their augmentations are reported by Zhongjie and Jian [74] and Wang et al. [63].

As a way of constructing information granules, the principle of justifiable granularity exhibits a significant level of generality in two essential ways. First, given the underlying requirements of coverage and specificity, different formalisms of information granules can be engaged. Second, experimental evidence could be expressed as information granules articulated in different formalisms and on this basis certain information granule is being formed.

It is worth stressing that there is a striking difference between clustering and the principle of justifiable granularity. First, clustering leads to the formation at least two information granules (clusters) whereas the principle of justifiable granularity produces a single information granule. Second, when positioning clustering and the principle vis-à-vis each other, the principle of justifiable granularity can be sought as a follow-up step facilitating an augmentation of the numeric representative of the cluster (such as e.g., a prototype) and yielding granular prototypes where the facet of information granularity is retained.

5 Augmentation of the Design Process of Information Granules

So far, the principle of justifiable granularity presented is concerned with a generic scenario meaning that experimental evidence gives rise to a single information granule. Several conceptual augmentations are considered where several sources of auxiliary information are supplied:

A. Involvement of Auxiliary Variable

Typically, these could be some dependent variable one encounters in regression and classification problems. An information granule is built on a basis of experimental evidence gathered for some input variable and now the associated dependent variable is engaged. In the formulation of the principle of justifiable granularity, this additional information impacts a way in which the coverage is determined. In more detail, we discount the coverage; in its calculations, one has to take into account the nature of experimental evidence assessed on a basis of some external source of knowledge. In regression problems (continuous output/dependent variable), in the calculations of specificity, we consider the variability of the dependent variable y falling within the realm of A. More precisely, the value of coverage is discounted by taking this variability into consideration. In more detail, the modified value of coverage is expressed as

$$\text{cov}'(A) = \text{cov}(A) \exp\left(-\beta\sigma_y^2\right) \tag{18}$$

where σ is a standard deviation of the output values associated with the inputs being involved in the calculations of the original coverage $cov(A)$. β is a certain calibration factor controlling an impact of the variability encountered in the output space. Obviously, the discount effect is noticeable, $\text{cov}'(A) < \text{cov}(A)$.

In case of a classification problem in which p classes are involved $\omega = \{\omega_1, \omega_2, \ldots, \omega_p\}$, the coverage is again modified (discounted) by the diversity of the data embraced by the information granule where this diversity is quantified in the form of the entropy function $h(\omega)$

$$\text{cov}'(A) = \text{cov}(A)(1 - h(\omega)) \tag{19}$$

This expression penalizes the diversity of the data contributing to the information granule and not being homogeneous in terms of class membership. The higher the entropy, the lower the coverage $\text{cov}'(A)$ reflecting the accumulated diversity of the data falling within the umbrella of A. If all data for which A has been formed belong to the same class, the entropy returns zero and the coverage is not reduced, $\text{cov}'(A) = \text{cov}(A)$.

B. Adversarial Information Granules

In untargeted adversarial attacks [15, 26], one considers x' such that it is close to the data coming from the training data $x_1, x_2, \ldots, x_N$ and producing a significantly different results than those reported for the neighboring data. The nature of the adversarial data x' can be quantified and generalized to the idea of the granular adversarial data. In light of the essence of the adversarial property, we determine x' such that it is close to x_k and $f(x')$ is different from $f(x_k)$ where $f(.)$ is a certain classifier or a model realizing this mapping $f.x'$ is sought as an adversarial example.

The granular adversarial data centered around x' and denoted by $A(x'; \rho)$ whose size ρ is the one which maximize the following ratio

$$V(\rho) = \frac{\sum_{x_k:\|x_k, x'| \le n_p^2} \left| f(\boldsymbol{x}_k) - f(\boldsymbol{x}') \right|}{\sum_{x_k:|x_k - x'\| \le n_p^2} \|x_k - \boldsymbol{x}'\|} \tag{20}$$

viz.

$$\rho_{\max} = \arg \text{Max}_\rho \, V(\rho) \tag{21}$$

where $\|.\|$ is a certain distance function, say the Euclidean one.

Having the training data, one can assess the adversarial nature of individual datum by picking up $x' = x_k$ and determining the maximum of V and reporting the associated size (radius) ρ thus producing $A(x_k; \rho_k)$. In this way, the data can be ranked with respect to their adversarial property by ordering the corresponding values of $V(\rho_k)$, and forming the resulting sequence starting from the highest value of this index.

6 Symbolic View at Information Granules and Their Symbolic Characterization and Summarization

Information granules are described through numeric parameters (or eventually granular parameters in case of information granules of higher type). There is an alternative view at a collection of information granules where we tend to move away from numeric details and instead look at the granules as symbols and engage them in further symbolic processing. Interestingly, symbolic processing is vividly manifested in Artificial Intelligence (AI) resulting in qualitative modeling [1, 6, 13, 14, 68]. Consider that a collection of the prototypes has been generated as a result of clustering. The prototypes are projected on the individual variables (features) and their projections are ordered linearly. At the same time, the distinguishability of the prototypes is evaluated: if two projected prototypes are close to each other they are deemed indistinguishable and collapsed. The merging condition involves the distance between the two close prototypes: If $|v_i C v_{i+1}| < range/c\varepsilon$ then the prototypes are deemed indistinguishable. Here the *range* is the range of values assumed by the prototypes and ε is a certain threshold value less than 1.

Once this phase has been completed, A_is are represented in a concise manner as sequences of indexes $A_i = (i_1, i_2, \ldots, i_{ni})$. Note that each A_i is described in its own feature space. Denote by R_{ij} a set of features for which A_i and A_j overlap viz. the features being common to both information granules. The distance between these two granules expressed along the k-th feature is computed as the ratio

$$d(i, j, k) = \frac{|i_k - j_k|}{c_k} \tag{22}$$

where c_k denotes the number of prototypes (after eventual merging) projected on the k-th feature. Next the similarity between A_i and A_j is determined based on the above indexes determined for the individual features in the form

$$\text{sim}\left(A_i, A_j\right) = \frac{1}{P} \sum_{k \in R_{ij}} \left(1 - \frac{|i_k - j_k|}{c_k}\right) \tag{23}$$

where P is the overall number of features (attributes) present in the problem.

A. Linguistic Summarization of Numeric Prototypes
Consider symbolic representation of the prototypes $A_1, A_2, \ldots, A_c$, namely i_1, $i_2, \ldots, i_c$ where each i_k is a string of integer indexes where each index assumes values from 1 to c. The linguistic summarization of the prototype (or information granule) gives rise to the expressions such as *most* (attributes of granule are *high*), at *most* 50% (attributes are *low*), etc. Generally speaking, the summarization is of the format [22, 23]

$$\tau(\text{attributes of granule is } \mu) = \lambda \tag{24}$$

with λ standing for the quantification of the summarization. To realize the summarization, one needs to define in advance a collection of linguistic terms (descriptors) $\{\mu_1, \mu_2, \ldots, \mu_p\}$, say, *low, medium, high, very high*, etc. and a family of linguistic quantifiers $\{\tau_1, \tau_2, \ldots, \tau_r\}$, say at *least, about 50%, most*, etc. The semantics of the descriptors is quantified by means of the corresponding fuzzy sets. τ_i are defined over the space of integers $\{1, 2, \ldots, c\}$ whereas μ_j are defined over the space $\{1, 2, \ldots, c\}$. Given the symbolic representation of a certain information granule (a string of integers), $A = [a_1 a_2, \ldots, a_n]$ one computes the degrees of membership for μ_i as $\mu_i(a_1)$, $\mu_i(a_2)$, ..., $\mu_i(a_n)$. These calculations are carried out for all linguistic terms, $i = 1, 2, \ldots, p$. Considering the j-th linguistic quantifier τ_j one determines the value of the corresponding expression

$$\lambda_{ij} = \tau_j \left(\sum_{k=1}^{n} \mu_i(a_k)/n \right) \tag{25}$$

	low	*medium*	*high*
a few	0.88	0.57	0.67
about 50%	0.40	0.98	0.88
most	0.11	0.43	0.33

$j = 1, 2, \ldots, r$. The optimized result of the summarization results from the maximization

$$(i_0, j_0) = \arg\max_{ij} \lambda_{ij} \tag{26}$$

As an illustrative example, consider a prototype A expressed as a string of integers [4 1 5 6 3 9 9]; here $n = 7$ and $c = 9$. The linguistic terms and their membership functions expressed over the space of integers 1 2 …9 are

low [1.0 0.8 0.5 0.3 0.0 0.0 0.0 0.0 0.0]
medium [0.0 0.1 0.4. 0.8 1.0 0.8 0.4 0.1 0.0]
high [0.0. 0.0 0.0 0.0 0.0 0.3 0.5 0.8 1.0]
while the linguistic quantifiers defined over [0, 1] (proportion of entities) are defined as follows
about 50% $4\mu(1-\mu)$
most μ
a few $1-\mu$

The calculations carried out as outlined above lead to the following results

Thus the overall linguistic summarization returns the expression

about 50% attributes assume medium values (0.98) or
a few attributes assume low values (0.88)
about 50% attributes assume high values (0.88)

7 Granular Probes of Spatiotemporal Data

When coping with spatiotemporal data, (say time series of temperature recorded over in a given geographic region), a concept of spatiotemporal probes arises as an efficient vehicle to describe the data, capture their local nature, and articulate their local characteristics as well as elaborate on their abilities as modeling artifacts (building blocks). The experimental evidence is expressed as a collection of data $z_k = z(x_k, y_k, t_k)$ with the corresponding arguments describing the location (x, y), time (t) and the value of the temporal data z_k.

Here an information granule is positioned in a three-dimensional space:
(i) space of spatiotemporal variable z
(ii) spatial position defined by the positions (x, y)
(iii) temporal domain described by time coordinate
The information granule to be constructed is spanned over some position of the space of values z_0, spatial location (x_0, y_0) and temporal location t_0. With regard to the spatial location and the temporal location, we introduce some predefined level of specificity, which imposes a level of detail considered in advance. The coverage and specificity formed over the special and temporal domain are defined in the form

$$\mathrm{cov}(A) = \mathrm{card}\left\{z\left(x_k, y_k, t_k\right) \| z\left(x_k, y_k, t_k\right) - z_0 \| \leq \rho\right\} \tag{27}$$

$$\begin{aligned}
\mathrm{sp}(A) &= \max\left(0, 1 - \frac{|z - z_0|}{L_z}\right) \\
sp_x &= \max\left(0, 1 - \frac{|x - x_0|}{L_x}\right) \\
sp_y &= \max\left(0, 1 - \frac{|y - y_0|}{L_y}\right) \\
sp_t &= \max\left(0, 1 - \frac{|t - t_0|}{L_t}\right)
\end{aligned} \tag{28}$$

Where the above specificity measures are monotonically decreasing functions (linear functions in the case shown above). There are some cutoff ranges $(L_x, L_y, \ldots)$, which help impose a certain level of detail used in the construction of the information granule. Information granule A in the space of the spatiotemporal variable is carried out as before by maximizing the product of coverage of the data and the specificity, $cov(A)sp(A)$ however in this situation one considers the specificity that is associated with the two other facets of the problem. In essence, A is produced by maximizing the product $cov(A)sp(A)sp_x, sp_y, sp_t$.

8 Granular Models

8.1 The Concept

The paradigm shift implied by the engagement of information granules becomes manifested in several tangible ways including (i) a stronger dependence on data when building structure-free, user-oriented, and versatile models spanned over selected representatives of experimental data, (ii) emergence of models at various varying levels of abstraction (generality) being delivered by the specificity/generality of information granules, and (iii) building a collection of individual local models and supporting their efficient aggregation.

Here several main conceptually and algorithmically far-reaching avenues are emphasized. Notably, some of them have been studied to some extent in the past and several open up new directions worth investigating and pursuing. In what follows, we elaborate on them in more detail pointing at the relationships among them [52].

data → numeric models. This is a traditionally explored path being present in system modeling for decades. The original numeric data are used to build the model. There are a number of models, both linear and nonlinear exploiting various design technologies, estimation techniques and learning mechanisms associated with evaluation criteria where accuracy and interpretability are commonly exploited with the Occam razor principle assuming a central role. The precision of the model is an advantage however the realization of the model is impacted by the dimensionality of the data (making a realization of some models not feasible); questions of memorization and a lack of generalization abilities are also central to the design practices.

data → numeric prototypes. This path associates with the concise representation of data by means of a small number of representatives (prototypes). The tasks falling within this scope are preliminary to data analytics problems. Various clustering algorithms constitute generic development vehicles using which the prototypes are built as a direct product of the grouping method.

data → numeric prototypes → symbolic prototypes. This alternative branches off to symbolic prototypes where on purpose we ignore the numeric details of the prototypes with intent to deliver a qualitative view at the information granules. Along this line, concepts such as symbolic (qualitative) stability and qualitative resemblance of structure in data are established.

data → numeric prototypes → granular prototypes. This path augments the previous one by bringing the next phase in which the numeric prototypes are enriched by their granular counterparts. The granular prototypes are built in such a way so that they deliver a comprehensive description of the data. The principle of justifiable granularity helps quantify the quality of the granules as well as deliver a global view at the granular characterization of the data.

data → numeric prototypes → symbolic prototypes → qualitative modeling. The alternative envisioned here builds upon the one where symbolic prototypes are formed and subsequently used in the formation of qualitative models, viz. the models capturing qualitative dependencies among input and output variables. This coincides with the well-known subarea of AI known as qualitative modeling, see [13] with a number of applications [1, 6, 14, 16, 67, 68].

data → numeric prototypes → granular prototypes → granular models. This path constitutes a direct extension of the previous one when granular prototypes are sought as a collection of high-level abstract data based on which a model is being constructed. In virtue of the granular data, we refer to such models as granular models.

8.2 Construction of Granular Models

By granular model we mean a model whose result is inherently an information granule as opposed to numeric result produced by a numeric model. In terms of this naming, neural networks are (nonlinear) numeric models. In contrast, granular neural networks produce outputs that are information granules (say, intervals).

There are two fundamental ways of constructing granular models:

(i) stepwise development. One starts with a numeric model developed with the use of the existing methodology and algorithms and then elevate the numeric parameters of the models to their granular counterparts following the way outlined above. This design process dwells upon the existing models and in this way one take fully advantage of the existing modeling practices. By the same token, one can envision that the granular model delivers a substantial augmentation of the existing models. In this sense, we can talk about granular neural networks, granular fuzzy rule-based models. In essence, the design is concerned with the transformation $a \overset{\varepsilon}{\rightarrow} A = G(a)$ applied to the individual parameters where G is a certain formalism of information granulation. It is worth noticing that in the overall process, there are two performance indexes optimized: in the numeric model one usually considers the root mean squared error ($RMSE$) while in the granular augmentation of the model one invokes another performance index that takes into consideration the product of the coverage and specificity.

(ii) a single step design. One proceeds with the development of the granular model from scratch by designing granular parameters of the model. This process uses only a single performance index that is of interest to evaluate the quality of the granular result.

Proceeding with the first design process presented above, the example presented in Fig. 4 stresses a way in which granular models are formed.

One starts with a numeric (type-0) model $M(x; a)$ developed on a basis of input-output data $D = (x_k, target_k)$. Then one elevates it to type-1 by optimizing a level of information granularity allocated to the numeric parameters a thus making them granular.

The way of transforming (elevating) a numeric entity a to an information granule A is expressed formally as follows $A = G(a)$. Here G stands for a formal setting

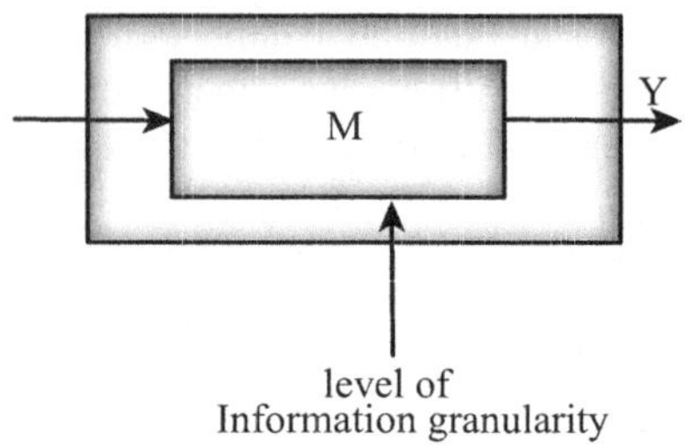

Fig. 4 From numeric to granular models by the development of granular parameters

of information granules. (say, intervals, fuzzy sets, etc.) and ε denotes a level of information granularity. One can envision two among a number of ways of elevating a certain numeric parameter a into its granular counterpart. If A is an interval, its bounds are determined as

$$A = [\min(a(1-\varepsilon), a(1+\varepsilon)), \max((a(1-\varepsilon), a(1+\varepsilon))], \varepsilon \in [0, 1] \quad (29)$$

Another option comes in the form

$$A = [\min(a/[+\varepsilon), a(1+\varepsilon)), \max(a/(1+\varepsilon), a(1+\varepsilon))], \varepsilon \geq 0 \quad (30)$$

If A is realized as a fuzzy set, one can regard the bounds of its support determined in the ways outlined above.

Obviously, the higher the value of ε, the broader the result and higher likelihood of satisfying the coverage requirement. Higher values of ε yield lower specificity of the results. Following the principle of justifiable granularity, we maximize the product of coverage and specificity by choosing a value of ε.

The performance of the granular model can be studied by analyzing the values of coverage and specificity for various values of the level of information granularity ε. Some plots of such relationships are presented in Fig. 5. In general, as noted earlier, increasing values of ε result in higher coverage and lower values of specificity. An important is a quantification of the changes encountered here. For instance,

By analyzing the pace of changes of the coverage versus the changes in the specificity, one can select a preferred value of ε as such beyond which the coverage does nor increase in a substantial way yet the specificity deteriorates significantly. One can refer to Fig. 5 where both the curves (a) and (b) help identify the suitable values of ε. One can develop a global descriptor of the quality of the granular model by computing the area under curve; the larger the area is, the better the overall quality of the model (quantified over all levels of information granularity) is. For instance, in Fig. 5, the granular model (a) exhibits better performance than (b).

The level of information granularity allocated to all parameters of the model is the same. Different levels of information granularity can be assigned to individual parameters; an allocation of these levels could be optimized in such a way that the values of the performance index V becomes maximized whereas a balance of information

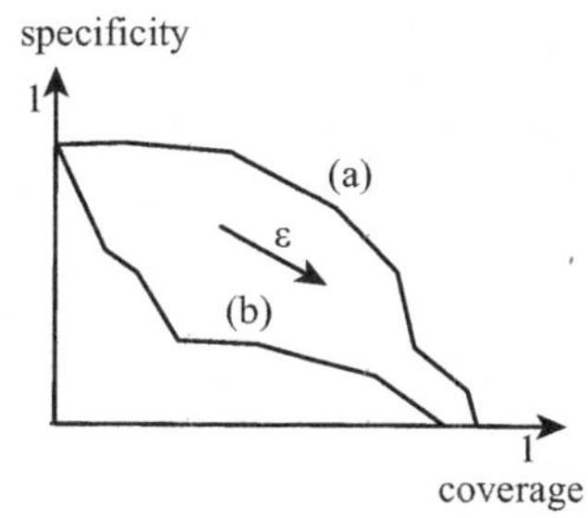

Fig. 5 Characteristics of granular models presented in the coverage-specificity coordinates

granularity is retained. In a formal way, we consider the following optimization task

$$\text{Min}_{\varepsilon}\, V \tag{31}$$

subject to constraints

$$\sum_{i=1}^{P} \varepsilon_i = P\varepsilon \quad \varepsilon_i \in [0, 1] \quad \text{for (27)}$$

$$\text{or} \sum_{i=1}^{P} \varepsilon_i = P\varepsilon \quad \varepsilon_i \geq 0 \quad \text{for (28)}$$

where the vector of levels of information granularity is expressed as $\varepsilon = [\varepsilon_1, \varepsilon_2, \ldots, \varepsilon_P]$ and P stands for the number of the parameters of the model. In virtue of the nature of this optimization problem, the use of evolutionary methods could be a viable option.

Granular Models of Higher Type

The design of granular models of higher type is associated by admitting the results that are information granules of higher type. For instance, in case of granular models of type-2, we envision the elevation of the type of information granularity

$$a \xrightarrow{\varepsilon} A(\text{type} -1) \text{ and } A \xrightarrow{\sigma} A \sim (\text{type} -2) \tag{32}$$

In the design process, as before one starts with the development of the numeric model. The granular model of type-1 is obtained by optimizing the granular results with the use of the principle of justifiable granularity. The type-1 granular model cover some data D; denote by D' the data not covered by this granular model. Using D', we construct granular model of type-2 again with the use of the principle of justifiable granularity, Fig. 6.

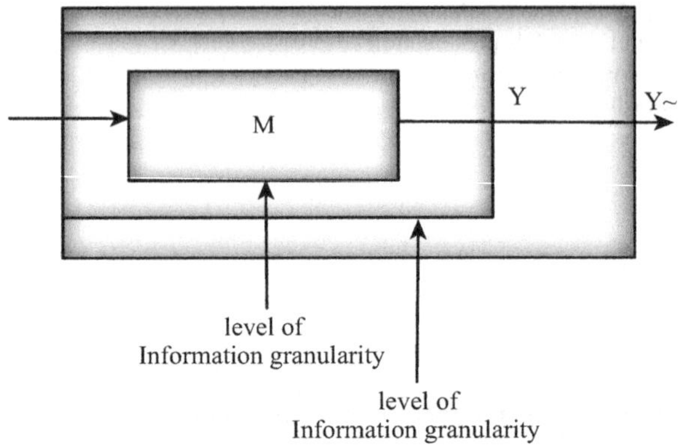

Fig. 6 Nesting granular models of higher type; from numeric to type-2 granular model

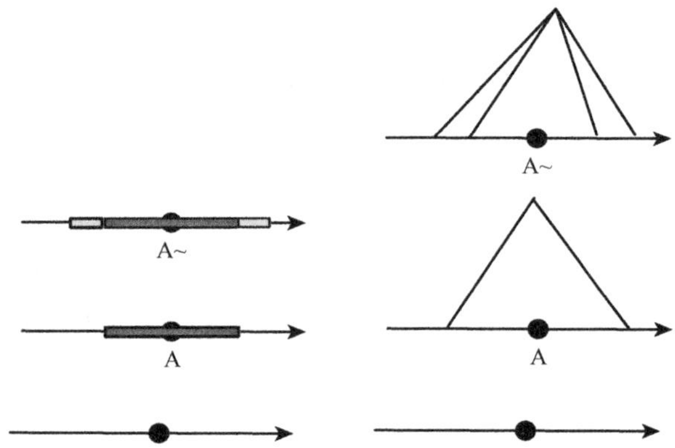

Fig. 7 From numeric to type-2 parameters of the models; shown are interval information granules and fuzzy sets

The formation of granular models of higher type is inherently associated with the parameters of the model which are also information granules of higher type, refer to Fig. 7; the types of granular parameters correspond with the types of the granular models.

9 Conclusions

The study has offered a focused overview of the fundamentals of Granular Computing positioned in the context of advanced system modeling. We identified a multifaceted role of information granules as meaningful conceptual entities being formed at the required level of abstraction. It has been emphasized that information granules are not only reflective of the nature of the data (the principle of justifiable granularity highlights the reliance of granules on available experimental evidence) but can efficiently capture some auxiliary domain knowledge conveyed by the user and in this way reflect the human-centricity aspects of the investigations and enhances the actionability aspects of the results. The interpretation of information granules at the qualitative (linguistic) level and their emerging characteristics such as e.g., stability enhancement of the interpretability capabilities of the framework of processing information granules is another important aspect of data analytics that directly aligns with the requirements expressed by the user. Several key avenues of system modeling based on the principles of Granular Computing were highlighted; while some of them were subject of intensive studies, some other require further investigations.

By no means, the study is complete; instead it can be regarded as a solid departure point identifying main directions of further far-reaching human-centric data analysis investigations. A number of promising avenues are open that are well aligned with the current challenges of data analytics including the reconciliation of results realized in the presence of various sources of knowledge (models, results of analysis), hierarchies of findings, quantification of tradeoffs between accuracy and interpretability (transparency).

References

1. Abou-Jaoud, W., Thieffry, D., Feret, J.: Formal derivation of qualitative dynamical models from biochemical networks. Biosystems **149**, 70–112 (2016)
2. Alefeld, G., Herzberger, J.: Introduction to Interval Computations. Academic, New York (1983)
3. Bargiela, A., Pedrycz, W.: Granular Computing: An Introduction. Kluwer Academic Publishers, Dordrecht (2003)
4. Bargiela, A., Pedrycz, W.: Toward a theory of granular computing for human-centered information processing. IEEE Trans. Fuzzy Syst. **16**(2), 320–330 (2008)
5. Bisi, C., Chiaselotti, G., Ciucci, D., Gentile, T., Infusino, F.G.: Micro and macro models of granular computing induced by the indiscernibility relation. Inf. Sci. **388–389**, 247–273 (2017)
6. Bolloju, N.: Formulation of qualitative models using fuzzy logic. Decis. Support Syst. **17**(4), 275–298 (1996)
7. Chiaselotti, G., Gentile, T., Infusino, F.: Granular computing on information tables: families of subsets and operators. Inf. Sci. **442–443**, 72–102 (2018)
8. Chiaselotti, G., Ciucci, D., Gentile, T.: Simple graphs in granular computing. Inf. Sci. **340–341**, 279–304 (2016)
9. Chiaselotti, G., Gentile, T., Infusino, F.: Knowledge pairing systems in granular computing. Knowl.-Based Syst. **124**, 144–163 (2017)
10. Dubois, D., Prade, H.: Outline of fuzzy set theory: an introduction. In: Gupta, M.M., Ragade, R.K., Yager, R.R. (eds.) Advances in Fuzzy Set Theory and Applications, pp. 27–39. North-Holland, Amsterdam (1979)
11. Dubois, D., Prade, H.: The three semantics of fuzzy sets. Fuzzy Sets Syst. **90**, 141–150 (1997)
12. Dubois, D., Prade, H.: An introduction to fuzzy sets. Clin. Chim. Acta **70**, 3–29 (1998)
13. Forbus, K.: Qualitative process theory. Artif. Intell. **24**, 85–168 (1984)
14. Guerrin, F.: Qualitative reasoning about an ecological process: interpretation in hydroecology. Ecol. Model. **59**, 165–201 (1991)
15. Goodfellow, I.J., Shlens, J., Szegedy, C.: Explaining and harnessing adversarial examples (2014). arXiv:1412.6572
16. Haider, W., Hu, J., Slay, J., Turnbull, B.P., Xie, Y.: Generating realistic intrusion detection system dataset based on fuzzy qualitative modeling. J. Netw. Comput. Appl. **87**, 185–192 (2017)
17. Han, Z., Zhao, J., Leung, H., Wang, W.: Construction of prediction intervals for gas flow systems in steel industry based on granular computing. Control Eng. Pract. **78**, 79–88 (2018)

18. Hirota, K.: Concepts of probabilistic sets. Fuzzy Sets Syst. **5**(1), 31–46 (1981)
19. Hoko, P.: Association discovery from relational data via granular computing. Inf. Sci. **234**, 136–149 (2013)
20. Hryniewicz, O., Kaczmarek, K.: Bayesian analysis of time series using granular computing approach. Appl. Soft Comput. **47**, 644–652 (2016)
21. Hu, H., Pang, L., Tian, D., Shi, Z.: Perception granular computing in visual haze-free task. Expert Syst. Appl. **41**, 2729–2741 (2014)
22. Kacprzyk, J., Zadrozny, S.: Computing With Words is an implementable paradigm: fuzzy queries, linguistic data summaries, and natural-language generation. IEEE Trans. Fuzzy Syst. **18**, 461–472 (2010)
23. Kacprzyk, J., Yager, R.R., Merigo, J.M.: Towards human-centric aggregation via ordered weighted aggregation operators and linguistic data summaries: A new perspective on Zadeh's inspirations. IEEE Comput. Intell. Mag. **14**, 15–30 (2019)
24. Klement, P., Mesiar, R., Pap, E.: Triangular Norms. Kluwer Academic Publishers, Dordrecht (2000)
25. Klir, G., Yuan, B.: Fuzzy Sets and Fuzzy Logic: Theory and Applications. Prentice-Hall, Upper Saddle River (1995)
26. Kurakin, A., Goodfellow, I., Bengio, S.: Adversarial examples in the physical world (2016). arXiv:1607.02533
27. Leng, J., Chen, Q., Mao, N., Jiang, P.: Combining granular computing technique with deep learning for service planning under social manufacturing contexts. Knowl.-Based Syst. **143**, 295–306 (2018)
28. Li, J., Mei, C., Xu, W., Qian, Y.: Concept learning via granular computing: a cognitive viewpoint. Inf. Sci. **298**, 447–467 (2015)
29. Liu, X., Pedrycz, W.: Axiomatic Fuzzy Set Theory and Its Applications. Springer, Berlin (2009)
30. Loia, V., Orciuoli, F., Pedrycz, W.: Towards a granular computing approach based on formal concept analysis for discovering periodicities in data. Knowl.-Based Syst. **146**, 1–11 (2018)
31. Liu, H., Xiong, S., Wu, C.-A.: Hyperspherical granular computing classification algorithm based on fuzzy lattices. Math. Comput. Model. **57**(3–4), 661–670 (2013)
32. Lu, W., Zhou, W., Shan, D., Zhang, L., Liu, X.: The linguistic modeling of interval-valued time series: a perspective of granular computing. Inf. Sci. **478**, 476–498 (2019)
33. Martłnez-Frutos, J., Martłnez-Castejn, P.J., Herrero-Prez, D.: Efficient topology optimization using GPU computing with multilevel granularity. Adv. Eng. Softw. **106**, 47–62 (2017)
34. Mendel, J.M., John, R.I., Liu, F.: Interval type-2 fuzzy logic systems made simple. IEEE Trans. Fuzzy Syst. **14**, 808–821 (2006)
35. Moore, R.: Interval Analysis. Prentice Hall, Englewood Cliffs (1966)
36. Moore, R., Kearfott, R.B., Cloud, M.J.: Introduction to Interval Analysis. SIAM, Philadelphia (2009)
37. Nguyen, H., Walker, E.: A First Course in Fuzzy Logic. Chapman Hall, CRC Press, Boca Raton (1999)
38. Pal, S.K., Chakraborty, D.B.: Granular flow graph, adaptive rule generation and tracking. IEEE Trans. Cybern. **47**(12), 4096–4107 (2017)
39. Pawlak, Z.: Rough sets. Int. J. Inf. Comput. Sci. **11**(15), 341–356 (1982)
40. Pawlak, Z.: Rough Sets. Theoretical Aspects of Reasoning About Data. Kluwer Academic Publishers, Dordrecht (1991)
41. Pawlak, Z.: Rough sets and fuzzy sets. Fuzzy Sets Syst. **17**(1), 99–102 (1985)
42. Pawlak, Z., Skowron, A.: Rough sets and Boolean reasoning. Inf. Sci. **177**, 41–73 (2007)
43. Pawlak, Z., Skowron, A.: Rudiments of rough sets. Inf. Sci. **177**, 3–27 (2007)
44. Pedrycz, A., Dong, F., Hirota, K.: Finite cut-based approximation of fuzzy sets and its evolutionary optimization. Fuzzy Sets Syst. **160**, 3550–3564 (2009)
45. Pedrycz, W., Bargiela, A.: Granular clustering: a granular signature of data. IEEE Trans. Syst. Man Cybern. **32**, 212–224 (2002)
46. Pedrycz, W.: Shadowed sets: representing and processing fuzzy sets. IEEE Trans. Syst. Man Cybern. Part B **28**, 103–109 (1998)

47. Pedrycz, W., Gacek, A.: Temporal granulation and its application to signal analysis. Inf. Sci. **143**(1–4), 47–71 (2002)
48. Pedrycz, W.: Interpretation of clusters in the framework of shadowed sets. Pattern Recognit. Lett. **26**(15), 2439–2449 (2005)
49. Pedrycz, W., Gomide, F.: Fuzzy Systems Engineering: Toward Human-Centric Computing. Wiley, Hoboken (2007)
50. Pedrycz, W., Homenda, W.: Building the fundamentals of granular computing: a principle of justifiable granularity. Appl. Soft Comput. **13**, 4209–4218 (2013)
51. Pedrycz, W.: Granular Computing. CRC Press, Boca Raton (2013)
52. Pedrycz, W.: Granular computing for data analytics: a manifesto of human-centric computing. IEEE/CAA J. Autom. Sin. **5**, 1025–1034 (2018)
53. Ray, S.S., Ganivada, A., Pal, S.K.: A granular self-organizing map for clustering and gene selection in microarray data. IEEE Trans. Neural Netw. Learn. Syst. **27**(9), 1890–1906 (2016)
54. Saberi, M., Mirtalaie, M.S., Hussain, F.K., Azadeh, A., Ashjari, B.: A granular computing-based approach to credit scoring modeling. Neurocomputing **122**, 100–115 (2013)
55. Salehi, S., Selamat, A., Fujita, H.: Systematic mapping study on granular computing. Knowl.-Based Syst. **80**, 78–97 (2015)
56. Savchenko, A.V.: Fast multi-class recognition of piecewise regular objects based on sequential three-way decisions and granular computing. Knowl.-Based Syst. **91**, 252–262 (2016)
57. Schweizer, B., Sklar, A.: Probabilistic Metric Spaces. North-Holland, New York (1983)
58. Singh, P., Dhiman, G.: A hybrid fuzzy time series forecasting model based on granular computing and bio-inspired optimization approaches. J. Comput. Sci. (2018) (In press)
59. Shen, Y., Pedrycz, W., Wang, X.: Clustering homogeneous granular data: formation and evaluation. IEEE Trans. Cybern. **49**, 1391–1402 (2019)
60. Shen, Y., Pedrycz, W., Wang, X.: Approximation of fuzzy sets by interval type-2 trapezoidal fuzzy sets. IEEE Trans. Cybern. (2019) (In press)
61. Tang, Y., Zhang, Y.Q., Huang, Z., Hu, X., Zhao, Y.: Recursive fuzzy granulation for gene subsets extraction and cancer classification. IEEE Trans. Inf. Technol. Biomed. **12**(6), 723–730 (2008)
62. Tang, Y., Hu, X., Pedrycz, W., Song, X.: Possibilistic fuzzy clustering with high-density viewpoint. Neurocomputing **329**, 407–423 (2019)
63. Wang, D., Pedrycz, W., Li, Z.: Granular data aggregation: an adaptive principle of justifiable granularity approach. IEEE Trans. Cybern. **49**, 417–426 (2019)
64. Wang, Q., Gong, Z.: An application of fuzzy hypergraphs and hypergraphs in granular computing. Inf. Sci. **429**, 296–314 (2018)
65. Wang, S., Pedrycz, W.: Data-driven adaptive probabilistic robust optimization using information granulation. IEEE Trans. Cybern. **48**, 450–462 (2018)
66. Wang, H., Yang, J., Wang, Z., Wang, Q.: A binary granular algorithm for spatiotemporal meteorological data mining. In: 2nd IEEE International Conference on Spatial Data Mining and Geographical Knowledge Services (ICSDM), pp. 5–11 (2015)
67. Wong, Y.H., Rad, A.B., Wong, Y.K.: Qualitative modeling and control of dynamic systems. Eng. Appl. Artif. Intell. **10**(5), 429–439 (1997)
68. Zabkar, J., Moina, M., Bratko, I., Demar, J.: Learning qualitative models from numerical data. Artif. Intell. **175**(9–10), 1604–1619 (2011)
69. Yao, Y., Wang, S., Deng, X.: Constructing shadowed sets and three-way approximations of fuzzy sets. Inf. Sci. **412–413**, 132–153 (2017)
70. Zadeh, L.A.: The concept of linguistic variables and its application to approximate reasoning I, II, III. Inf. Sci. **8**, 199–249, 301–357, 43–80 (1975)
71. Zadeh, L.A.: Towards a theory of fuzzy information granulation and its centrality in human reasoning and fuzzy logic. Fuzzy Sets Syst. **90**, 111–117 (1997)
72. Zadeh, L.A.: Toward a generalized theory of uncertainty (GTU) CC an outline. Inf. Sci. **172**, 1–40 (2005)
73. Zadeh, L.A.: From computing with numbers to computing with words-from manipulation of measurements to manipulation of perceptions. IEEE Trans. Circuits Syst. **45**, 105–119 (1999)

74. Zhongjie, Z., Jian, H.: Stabilizing the information granules formed by the principle of justifiable granularity. Inf. Sci. **503**, 183–199 (2019)
75. Zhou, J., Lai, Z., Miao, D., Gao, C., Yue, X.: Multigranulation rough-fuzzy clustering based on shadowed sets. Inf. Sci. (2018) (In press). Accessed 30 May 2018
76. Zhu, X., Pedrycz, W., Li, Z.: A development of granular input space in system modeling. IEEE Trans. Cybern. (2019) (In press)
77. Zhu, X., Pedrycz, W., Li, Z.: A design of granular Takagi-Sugeno fuzzy model through the synergy of fuzzy subspace clustering and optimal allocation of information granularity. IEEE Trans. Fuzzy Syst. **26**, 2499–2509 (2018)

Distributed Consensus Control for Nonlinear Multi-agent Systems

Xin Chen, Min Wu, Witold Pedrycz, Krzysztof Galkowski, and Wojciech Paszke

Abstract This chapter considers the distributed optimal consensus problem of discrete-time (DT) nonlinear multi-agent systems (MASs) with unknown dynamics. For this type of system, obtaining a coupled Hamilton–Jacobi–Bellman (HJB) equation is essential to solving the distributed optimal consensus problem. However, it is difficult to solve the coupled HJB equation of a system with unknown dynamics. In this chapter, a local value function is defined that takes into account local consensus errors, the behavior of agents, and the behavior of their neighbors. Based on adaptive dynamic programming (ADP) with the local value function, an action dependent heuristic dynamic programming based distributed consensus control method is put forward to realize the optimal consensus control (OCC). Furthermore, an ADP-based distributed model reference adaptive control method is also presented to achieve OCC for heterogeneous nonlinear MASs. Simulation examples are given to demonstrate the feasibility of the optimal consensus methods.

Keywords Adaptive dynamic programming · Optimal consensus · Multi-agent systems · Value function · Coupled Hamilton–Jacobi–Bellman equation

---

X. Chen (✉) · M. Wu
School of Automation, China University of Geosciences, Wuhan 430074, China
e-mail: chenxin@cug.edu.cn

Hubei Key Laboratory of Advance Control and Intelligent Automation for Complex Systems, Wuhan 430074, China

W. Pedrycz
Department of Electrical & Computer Engineering, University of Alberta, Edmonton, AB T6R 2V4, Canada

K. Galkowski · W. Paszke
Institute of Control and Computation Engineering, University of Zielona Gora, 65-516 Zielona Gora, Poland

M. Wu et al. (eds.), *Developments in Advanced Control and Intelligent Automation for Complex Systems*, Studies in Systems, Decision and Control 329,
https://doi.org/10.1007/978-3-030-62147-6_8

Abbreviations

MASs	Multi-agent systems
ADP	Adaptive dynamic programming
HJB	Hamilton–Jacobi–Bellman
RL	Reinforcement learning
CT	Continuous-time
DT	Discrete-time
OCC	Optimal consensus control
HDP	Heuristic dynamic programming
ADHDP	Action-dependent heuristic dynamic programming
NNs	Neural networks
MRAC	Model reference adaptive control
LQR	Linear quadratic regulator

1 Introduction

Over the past decade, interest in MASs has increased. Such systems provide an attractive means for modeling and designing complex distributed and concurrent systems. In a MASs, multiple agents cooperate or compete with each other to operate and share information in the same (virtual) environment.

1.1 Background and Related Work

In the field of control, distributed control of MASs has become a topic of widespread concern. It is widely used in distributed sensor networks [1], flocking [2], formation control [3], unmanned aerial vehicle [4], battery control [5], etc. [6, 7]. The basic problem in the distributed control field of MASs is to achieve synchronous consensus control for all agents [8–11]. A very challenging problem for MASs is optimal distributed consensus control. It requires all agents to be synchronized, and it needs to ensure that performance indexes are minimized.

The game theory framework, which is used to guarantee the optimality, can solve the multi-agent optimal decision control problem of dynamic interactive systems. Under this framework, each agent measures whether its control policy is optimal or not through corresponding performance indicator. However, since each subject is affected by the actions of its neighbors in addition to itself [12], this will produce a coupled HJB equation. In general, when a system model is unknown, it is difficult or even impossible to obtain an analytical solution of the coupled HJB equation [13].

ADP [14–17] has been adopted to handle the OCC problem to get around above challenge [18–28]. ADP has strong adaptability and self-learning ability, which combines adaptive control and RL [29].

ADP-based methods have been developed to handle CT homogeneous linear MASs [28] and heterogeneous linear MASs [21], DT linear MASs [19, 26, 27] and CT linear MASs [18, 20] with different control input matrices. We can see that many methods have been proposed to solve the OCC problem of linear MASs [18–21, 26–28]. However, in many cases, using a linear model to describe the actual agent dynamics is not enough. According to the actual needs, it is of great practical significance to study the OCC of nonlinear MASs.

There are known some ADP-based methods that have been proposed to solve OCC of nonlinear MASs. The compensation technique method has been used to enhance an extended system. This method is model-free OCC by increasing the state space dimension and using improved performance indicators [30]. In [25], with an additional compensator used, a nonlinear cooperative optimal control method using ADP was proposed for CT MASs, whose dynamics are completely unknown. Considered the situation of unknown nonlinearity and state delay in the system, a backstepping-based adaptive consensus control protocol has been introduced to solve a type of strict-feedback leader-following consensus problem [31]. Note that neural networks (NNs) have been used to identify uncertain non-linearities, which will increase the computational cost.

In addition, the dynamics of different agents in MASs may differ in practical applications. This type of system is called heterogeneous MASs. However, agents were considered homogeneous in the MASs studied above [25, 30, 31]. Therefore, research on OCC of heterogeneous MASs is necessary.

In recent years, the OCC problem of heterogeneous multi-agents has been a research hotspot, refer to such as [21, 32–37]. Modares et al. [32] proposes a model-free non-strategic RL algorithm in which a distributed adaptive observer is proposed to solve the optimal output synchronization problem. Yang et al. [33] further optimizes the control of active leaders. Through offline policy iteration, [35] proposes an optimal robust output suppression method for unknown heterogeneous linear MASs with variable system parameters. Two distributed compensators are designed to separately handle different agents and optimal performance index function [34]. However, these studies have only solved the consensus problem of linear heterogeneous MASs, while few studies have been done on the consensus problem of nonlinear heterogeneous MASs. In the optimal consensus problem of unknown nonlinear heterogeneous MASs, there are still many challenges to be solved.

Based on the above analysis, we can find two problems that need to be solved for in OCC. First, for unknown nonlinear MASs, the model identification of NNs incurs additional computational cost and introduces modeling errors. Second, it is very difficult to address the optimal consensus problems for unknown nonlinear heterogeneous MASs.

In this chapter, based on the above problems, we deal with a class of optimal distributed consensus control problems with DT nonlinear leader following MASs. As we all know, the ADP-based method can overcome the model-free problem in optimal control [38–40]. Therefore, in this chapter, a local-valued function is defined that takes into account local consensus errors, and the behavior of agents and their neighbors. Based on this, a distributed consensus control method based on ADHDP is proposed. This method combines local value functions and ADP to achieve OCC. In addition, we also propose a distributed model reference adaptive control method based on ADP to achieve OCC of DT heterogeneous nonlinear MAS with unknown dynamic. A simulation example is given to prove the effectiveness.

1.2 Preliminaries

First, some basic descriptions of graph theory for MASs are introduced as follows:

Let $\mathscr{G} = \{\mathscr{V}, \varepsilon\}$ be a directed graph, where $\mathscr{V} = \{v_1, \ldots, v_N\}$ is a nonempty finite set of N vertices and $\varepsilon \subseteq \mathscr{V} \times \mathscr{V}$ donates a set of edges. The associated adjacency matrix is represented by $\mathscr{A} = [a_{ij}] \in \mathbb{R}^{N \times N}$, with $a_{ij} > 0$ if $(v_j, v_i) \in \varepsilon$, which means agent j can transfer the information to agent i in G, otherwise $a_{ij} = 0$. For $i = 1, \ldots, N, a_{ii} = 0$. Denote the set of neighbors of node v_i as $\mathscr{N}_i = \{v_j : (v_j, v_i) \in \varepsilon\}$. $\mathscr{D} = diag\{d_i\}$ is the in-degree matrix whose diagonal elements are denoted by $d_i = \sum_{j \in \mathscr{N}_i} a_{ij}$. $\mathscr{L} = \mathscr{D} - \mathscr{A}$ denotes the graph Laplacian matrix, and the sum of its elements of in each row is equal to zero. If a directed path from the leader agent to any other agent is in the graph, we called the graph exists a spanning tree. Here, we assume that the leader agent is coupled by at least one agent and there are no duplicate paths.

2 ADHDP-Based Distributed Consensus Control for MASs

In this section, a distributed consensus control method based on ADHDP is introduced to solve the OCC problem of DT nonlinear MASs under unknown dynamics. The model identification is not used in this method. So, the identification error and the additional computational cost are avoided.

2.1 Problem Formulation

This subsection formulates the OCC problem of the DT nonlinear MASs. Consider a class of MASs with N followers and one leader, the leader can act as a signal generator, generating a set of specified trajectories for the followers. Here the follower does not know the process and result of the trajectory in advance.

The leader dynamics is described as

$$x_0(t+1) = f(x_0(t)), \tag{1}$$

where $x_0(t) \in \mathbb{R}^n$.

Each follower has the dynamics

$$x_i(t+1) = f(x_i(t)) + g(x_i(t))\,u_i(t), i = 1, \ldots, N \tag{2}$$

where the state of follower i is $x_i(t) \in \mathbb{R}^n$, and $u_i(t) \in \mathbb{R}^{m_i}$ represents the control input of follower i. In this chapter, $f(x_i(t)) \in \mathbb{R}^n$ and $g(x_i(t)) \in \mathbb{R}^{n \times m_i}$ for all i are local Lipschitz functions, and are considered to be unknown.

By using the state error between neighboring agents to measure the effect of synchronization, the distributed synchronization problem of MASs can be effectively solved. This error can also be called the local neighborhood consensus error, which is described as:

$$e_i(t) = \sum_{j \in \mathcal{N}_i} a_{ij}(x_i(t) - x_j(t)) + b_i(x_i(t) - x_0(t)), \tag{3}$$

where $e_i(t) \in \mathbb{R}^n$. Noted that the state of the connection between each follower and leader is described by the pinning gain b_i. When $b_i > 0$, it means that the follower i is connected to the leader and $b_i = 0$ otherwise.

The local neighborhood consensus error reflects the control error between individuals. In order to achieve a better synchronization effect, we also need to express the global consensus error to reflect the global control performance of the MASs. The global consensus error vector is

$$e(t) = ((\mathcal{L} + \mathcal{B}) \otimes I_n)x(t) - ((\mathcal{L} + \mathcal{B}) \otimes I_n)\underline{x}_0(t), \tag{4}$$

where $e(t) = [e_1^T(t), e_2^T(t), \ldots, e_N^T(t)]^T \in \mathbb{R}^{Nn}$, $\otimes$ represents the Kronecker product, $\underline{x}_0(t) = [x_0^T(t), x_0^T(t), \ldots, x_0^T(t)]^T \in \mathbb{R}^{Nn}$, $x(t) = [x_1^T(t), x_2^T(t), \ldots, x_N^T(t)]^T \in \mathbb{R}^{Nn}$, and $\mathcal{B} = \text{diag}\{b_i\} \in \mathbb{R}^{N \times N}$ denoted the diagonal matrix of the pinning gain.

The global disagreement vector is defined as

$$\xi(t) = x(t) - \underline{x}_0(t). \tag{5}$$

Then, the global consensus error vector is written as

$$e(t) = ((\mathcal{L} + \mathcal{B}) \otimes I_n)\xi(t). \tag{6}$$

From (6), if $(\mathcal{L} + \mathcal{B})$ is non-singular, we have $\|\xi(t)\| \to 0$ as $\|e(t)\| \to 0$. Therefore, the research goal of this chapter is to explore how the defined consensus error can converge to 0.

From (1)–(3), the local neighborhood consensus error dynamics is obtained as

$$\begin{aligned} e_i(t+1) = &\sum_{j\in\mathcal{N}_i} a_{ij}(f_{ei}(y_i(t)) - f_{ej}(y_j(t))) + b_i f_{ei}(y_i(t)) \\ &+ (d_i+b_i)g(x_i(t))u_i(t) - \sum_{j\in\mathcal{N}_i} a_{ij}g_j(x_j(t))u_j(t), \end{aligned} \tag{7}$$

where $f_{ei}(y_i(t)) = f(x_i(t)) - f(x_0(t))$, and $f_{ej}(y_j(t)) = f(x_j(t)) - f(x_0(t))$.

Equation (7) is non-autonomous, and the dynamics of the local neighborhood consensus error (7) are locally coupled due to the influence of the control policies of the agent i and its neighbors. When the system dynamics are unknown, it is difficult to design optimal controllers for such non-autonomous power systems. Therefore, we need to structure a model-free control method to solve the OCC problem faced here. This method should achieve the best control of the local error dynamic system (7) without knowing the dynamic characteristics of MASs (1) and (2).

In addition to achieving the consensus of the MAS, we also need to optimize the performance index. The local performance index to be optimized is defined as

$$J_i(e_i(t), u_i(t), u_{(j)}(t)) = \sum_{k=t}^{\infty} r_i(e_i(k), u_i(k), u_{(j)}(k)), \tag{8}$$

and the utility function is described as

$$r_i(e_i(k), u_i(k), u_{(j)}(k)) = e_i^T(t)V_{ii}e_i(t) + u_i^T(t)R_{ii}u_i(t) + \sum_{j\in\mathcal{N}_i} u_j^T(t)R_{ij}u_j(t), \tag{9}$$

where $u_{(j)}(t)$ denotes the control inputs $\{u_j \mid j \in \mathcal{N}_i\}$ of the neighbors of agent i, and symmetric time-invariant weighting matrices satisfy $V_{ii} > 0$, $R_{ii} > 0$, and $R_{ij} \geq 0$.

Therefore, the problem to be solved in this section focuses on designing optimal distributed control laws, which will minimize the energy of all agents while synchronizing all followers with the leader, in other words, minimize the local performance indicator function (8).

Definition 1 (*Admissible control*) the control policy u_i for each agent i is said to be admissible if $u_i(t)$ satisfies: (1) $u_i(t)$ is continuous on the consensus error state space E_i; (2) $u_i(t) = 0$ as $e_i(t) = 0$; (3) $u_i(t)$ stabilizes the system (7); (4) $u_i(t)$ guarantees the corresponding performance index function (8) finite.

Given an admissible control policy u_i for each agent i, the local value function $V_i(e_i(t))$ is defined as

$$V_i(e_i(t)) = \sum_{k=t}^{\infty} r_i(e_i(k), u_i(k), u_j(k)). \tag{10}$$

For the $u_i(t)$ and $u_j(t)$, the Hamiltonian function $H_i\big(e_i(t), \nabla V_i(e_i(t+1)), u_i(t), u_j(t)\big)$ satisfies the following coupled Hamiltonian–Jacobi equation:

$$
\begin{aligned}
0 &= H_i\left(e_i(t), \nabla V_i\left(e_i(t+1)\right), u_i(t), u_j(t)\right) \\
&= \nabla V_i^T\left(e_i(t+1)\right)\Bigg[\sum_{j\in\mathcal{N}_i} a_{ij}(f_{ei}(y_i(t)) - f_{ej}(y_j(t))) + b_i f_{ei}(y_i(t)) \\
&\quad + (d_i + b_i)g_i(x_i(t))u_i(t) - \sum_{j\in\mathcal{N}_i} a_{ij}g_j(x_j(t))u_j(t)\Bigg] + r_i(e_i(t), u_i(t), u_{(j)}(t)),
\end{aligned} \tag{11}
$$

where $\nabla V_i(e_i(t+1)) = \partial V_i(e_i(t+1))/\partial e_i(t+1)$, and $V_i(0) = 0$.

According to Bellman's optimal principle, we know that the optimal local value function $V_i^*(e_i(t+1))$ should satisfy the minimum value of the coupled HJB equation to be 0, so that we can get the optimal local control low $u_i^*(t)$. Therefore, the coupled HJB equation can be written as:

$$
\begin{aligned}
0 &= e_i^T(t)V_{ii}e_i(t) - \frac{1}{2}(d_i + b_i)^2 \nabla V_i{}^{*T}(e_i(t+1))g(x_i(t))R_{ii}^{-1}g^T(x_i(t))\nabla V_i{}^*(e_i(t+1)) \\
&\quad + \frac{1}{4}\sum_{j\in\mathcal{N}_i}[(d_j + b_j)^2 \nabla V_j{}^{*T}(e_j(t+1))g_j\left(x_j(t)\right)R_{jj}^{-1}R_{ij}R_{jj}^{-1}g_j^T\left(x_j(t)\right)\nabla V_j{}^*(e_j(t+1))] \\
&\quad + \nabla V_i{}^{*T}(e_i(t+1))[b_i f_{ei}(y_i(t)) + \sum_{j\in\mathcal{N}_i} a_{ij}(f_{ei}(y_i(t)) - f_{ej}(y_j(t)))] \\
&\quad + \frac{1}{4}(d_i + b_i)^2 \nabla V_i^{*T}(e_i(t+1))g(x_i(t))R_{ii}^{-1}g^T(x_i(t))\nabla V_i^*(e_i(t+1)) \\
&\quad + \frac{1}{2}\nabla V_i{}^{*T}(e_i(t+1))\sum_{j\in\mathcal{N}_i}[a_{ij}g_j\left(x_j(t)\right)(d_j + b_j)R_{jj}^{-1}g_j^T\left(x_j(t)\right)\nabla V_j{}^*(e_j(t+1))],
\end{aligned} \tag{12}
$$

where $\nabla V_i(e_i(t+1)) = \partial V_i(e_i(t+1))/\partial e_i(t+1)$, and $V_i(0) = 0$.

By solving the HJB equation (12), the OCC policy can be obtained. However, there are two difficulties. First, the HJB equation (12) depends on the agent i and its neighbors, so it is highly coupled. Usually, it is difficult to obtain an analytical solution of (12). Second, the solution of (12) requires knowledge of the system matrices $f(x_i(t))$ and $g(x_i(t))$ for each agent. In practical applications, it is very difficult or impossible to obtain system dynamics, which means that an accurate system matrix cannot be obtained. To overcome these difficulties, the next chapter will introduce a distributed consensus control method based on ADHDP to approximate the solution of the coupled HJB equation (12) without knowing the system dynamics.

2.2 ADHDP-Based Distributed Consensus Control Method

Accurate models of the MASs are the key to solving coupled HJB equations (12) and involve nonlinear deviation fractional equations. Therefore, it is impossible to solve (12) analytically. The following introduces ADHDP-based distributed consensus control method, which is an improved version of the ADP method.

Given any admissible control policy, there is a corresponding local value function for each agent i, which can be expressed as

$$V_i(e_i(t),u_i(t),u_{(j)}(t))=r_i(e_i(t),u_i(t),u_{(j)}(t))+V_i(e_i(t+1),\ u_i(t+1),\ u_{(j)}(t+1)). \tag{13}$$

The value function will perform an evaluation on all possible control inputs in each state, and we will continuously improve the policy based on the evaluation results to obtain optimal local control policy. Here we update the optimal policy by minimizing the local value function. Let the action space of the agent i be represented by $\mathbb{A}_i$, we have

$$u_i'(t) = \arg\min_{u_i(t)\in\mathbb{A}_i} V_i(e_i(t), u_i(t), u_{(j)}(t)). \tag{14}$$

Algorithm 1 Policy Iteration Based Distributed ADHDP Method

Initialization. Start with admissible initial policies u_i^0 for all i.

Step 1. Given the iterative admissible control policies u_i^l for all i, solve for the local value functions using the following iterative equation

$$\begin{aligned} V_i^l(e_i(t), u_i(t), u_{(j)}(t)) &= r_i(e_i(t), u_i(t), u_{(j)}(t)) \\ &+ V_i^l(e_i(t+1), u_i^l(t+1), u_{(j)}^l(t+1)) \end{aligned} \tag{15}$$

Step 2. Improve the control policies using the following equation

$$u_i^{l+1}(t) = \arg\min_{u_i(t)\in\mathbb{A}_i} V_i^l(e_i(t), u_i(t), u_{(j)}(t)) \tag{16}$$

If $\left\| V_i^{l+1}(e_i(t), u_i(t), u_{(j)}(t)) - V_i^l(e_i(t), u_i(t), u_{(j)}(t)) \right\| \le \varepsilon$ for all i, where ε represents a small constant, end. Else, let $l = l + 1$ and repeat Step 1, 2.

Algorithm 1 gives a policy iterative process based on the distributed ADHDP method, where l represents the iterative index.

This section introduces the ADHDP algorithm based on the distributed consensus control method. Next, we will introduce how to implement the algorithm in MASs.

2.3 *Implementation of the ADHDP-Based Distributed Consensus Control Method*

In this section, the ADHDP-based Distributed Consensus Control Method is implemented by a NNs-based actor-critic framework. The framework is divided into two parts. On the one hand, the critic network is designed to approximate the optimal

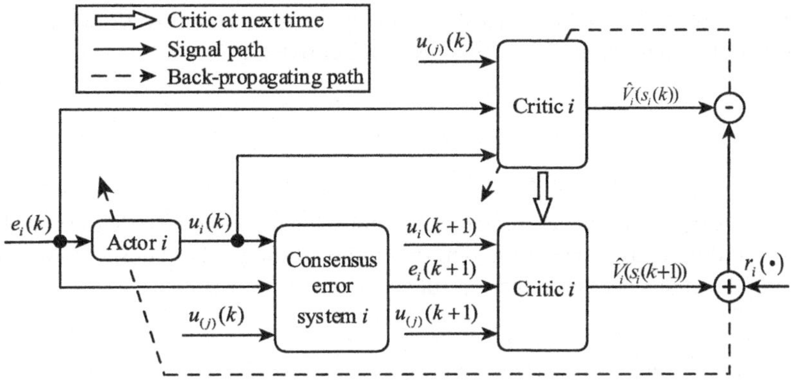

Fig. 1 Structure of control

local value function and serve the Bellman equation. On the other hand, the actor network is constructed to approximates the optimal control policy and minimize local value functions. Figure 1 shows the diagram of the introduced OCC structure for the agent i, where $s_i(t)$ is a row vector containing information from $e_i(t)$, $u_i(t)$, and $u_{(j)}(t)$.

A. NN-Based Critic Network

The local value function is approximated by the critic network that approximates (13) for all i,

$$V_i(s_i(t)) = w_{ci}^T \phi_i\left(h_i^T s_i(t)\right) + \varepsilon_{ci}, \tag{17}$$

where $\phi_i(\cdot)$ represents the activation function, the NN estimation error is denoted by ε_{ci}, and h_i and w_i represent the weights between the NN input layer and the NN hidden layer, the NN hidden layer and the NN output layer, respectively. Assume that the current weight estimates are $\hat{h}_i$ and $\hat{w}_{ci}$, and hence the critic network output is expressed as

$$\hat{V}_i(s_i(t)) = \hat{w}_{ci}^T \phi_i\left(\hat{z}_{ci}(t)\right), \tag{18}$$

where $\hat{z}_{ci}(t) = \hat{h}_i^T s_i(t)$.

Since there exist NN approximation errors in the critic network design, (18) is approximated as

$$e_{ci}(t) = r_i(s_i(t)) + \hat{V}_i(s_i(t+1)) - \hat{V}_i(s_i(t)). \tag{19}$$

The weights $\hat{w}_{ci}$ are optimized through minimizing the following objective function:

$$E_{ci}(t) = \frac{1}{2} e_{ci}^T(t) e_{ci}(t). \tag{20}$$

To minimize (20), gradient-based weight update rule is employed. Then the critic network weights updating rules for agent i are derived as

$$\hat{w}_{ci}^{l,t+1} = \hat{w}_{ci}^{l,t} - \alpha_{ci}\left[\frac{\partial E_{ci}^{l,t}}{\partial \hat{w}_{ci}^{l,t}}\right], \tag{21}$$

where $\alpha_{ci} > 0$ denotes the learning rate, l denotes the iteration index of the distributed policy iteration, and l denotes the iteration index of the gradient descent algorithm.

Using the chain-backpropagation rules, we derive

$$\hat{w}_{ci}^{l+1,t} = \hat{w}_{ci}^{l,t} - \alpha_{ci}\frac{\partial E_{ci}^{l,t}(t)}{\partial e_{ci}^{l,t}(t)}\frac{\partial e_{ci}^{l,t}(t)}{\partial \hat{w}_{ci}^{l,t}} = \hat{w}_{ci}^{l,t} - \alpha_{ci}e_{ci}^{l,t}(t)\phi_{ci}\left(\hat{z}_{ci}(t)\right). \tag{22}$$

B. NNs-Based Actor Network

The actor is utilized to approximate the optimal control policy (14) for each i,

$$\hat{u}_i(t) = \hat{w}_{ai}^T\varphi_i(\hat{z}_{ai}(t)), \tag{23}$$

where $\hat{z}_{ai}(t) = \hat{\kappa}_i^T e_i(t)$, $\varphi_i(\cdot)$ denotes the activation function, $\hat{\kappa}_i$ and $\hat{w}_{ai}$ denote the weights between the input layer and the hidden layer, the hidden layer and the output layer, respectively.

The role of the actor is to minimize the local value function. Thus, the error of the actor is defined as

$$e_{ai}(t) = \hat{V}_i(s_i(t)). \tag{24}$$

The objective function of the actor to be minimized is

$$E_{ai}(t) = \frac{1}{2}e_{ai}^T(t)e_{ai}(t). \tag{25}$$

Similarly, the update rule is represented as

$$\begin{aligned}
\hat{w}_{ai}^{l,t+1} &= \hat{w}_{ai}^{l,t} - \alpha_{ai}\left[\frac{\partial E_{ai}^{l}(t)}{\partial \hat{w}_{ai}^{l,t}(t)}\right] \\
&= \hat{w}_{ai}^{l,t} - \alpha_{ai}\frac{\partial E_{ai}^{l,t}(t)}{\partial e_{ai}^{l,t}(t)}\frac{\partial e_{ai}^{l,t}(t)}{\partial \hat{V}_i(s_i(t))}\frac{\partial \hat{V}_i(s_i(t))}{\partial \hat{u}_i(t)}\frac{\partial \hat{u}_i(t)}{\partial \hat{w}_{ai}^{l,t}(t)} \\
&= \hat{w}_{ai}^{l,t} - \alpha_{ai}\varphi_i(\hat{z}_{ai}(t))\hat{w}_{ci}^T\phi_i'\left(\hat{z}_{ci}(t)\right)\hat{\kappa}_i^T D_i\hat{w}_{ci}^T\phi_i\left(\hat{z}_{ci}(t)\right).
\end{aligned} \tag{26}$$

where $\alpha_{ai} > 0$ denotes the learning rate, $\phi_i'(\hat{z}_{ci}(t)) = \partial\phi_i(\hat{z}_{ci}(t))/\partial\hat{z}_{ci}(t)$, and $D_i = \partial s_i(t)/\partial\hat{u}_i(t)$ denotes a constant matrix.

Now, we are at the position to propose Algorithm 2 that shows the adjustment of a NN-based actor-critic network using only measurable data.

Algorithm 2: ADHDP-Based Distributed Consensus Control Algorithm

Initialization ($\forall i, i = 1, \ldots, N$).
$x_i(0)$, $x_0(0)$: initial states of the agents and leader.
$w_{ci}^{0,0}(0)$: the critic parameter
$w_{ai}^{0,0}(0)$: the actor parameter for an initial admissible control policy
ε: computation precision
$N_{c,\max}$, $N_{a,\max}$: maximum time steps for updating the critic and the actor
$E_{c,thr}$, $E_{a,thr}$: thresholds for the loss of the critic and the actor
α_{ci}, α_{ai}: learning rates
V_{ii}, R_{ii}, R_{ij}: positive definite matrices
End initialization
Let $k = 0, l = 0, t = 0$.
Calculate the local neighborhood consensus error $e_i(t) \leftarrow (3)$
Repeat
Calculate the control policy $\hat{u}_i(t) \leftarrow (23)$
Apply $\hat{u}_i(t)$ to the local error dynamic system and obtain $e_i(t+1)$
Calculate the control policy $\hat{u}_i(t+1) \leftarrow (23)$
Repeat
Calculate the prediction error of the critic $e_{ci}^t(t) \leftarrow (19)$
Calculate the objective function of the critic $E_{ci}^t(t) \leftarrow (20)$
Update the weights of the critic $\hat{w}_{ci}^{l,k+1} \leftarrow (22)$
Until ($E_{ci}^t \leq E_{c,thr}$ or $k \geq N_{c,\max}$; otherwise $k = k + 1$)
Set $w_{ci}^{l+1,0} = w_{ci}^{l,k+1}$, and $k = 0$
Repeat
Calculate the error of the actor $e_{ai}^t(t) \leftarrow (24)$
Calculate the objective function of the actor $E_{ai}^t \leftarrow (25)$
Update the weights of the actor $\hat{w}_{ai}^{l,k+1} \leftarrow (26)$
Until ($E_{ai}^t \leq E_{a,thr}$ or $k \geq N_{a,\max}$; otherwise $k = k + 1$)
Set $w_{ci}^{l+1,0} = w_{ci}^{l,k+1}$, and $k = 0$
Until ($\sum_{i=1}^{N} \left\| w_{ci}^{l+1,0} - w_{ci}^{l,0} \right\| / N \leq \varepsilon$; otherwise $l = l + 1, t = t + 1$)
Return $w_{ci}^{l+1,0}$, $w_{ai}^{l+1,0}$ for all i

2.4 Simulation Results

In this section, the effectiveness of the developed method is verified by a simulation. The communication topology of the DT nonlinear MASs is shown in Fig. 2, which consists of a leader and three followers.

The dynamics of these agents are described as:
Agent 1:

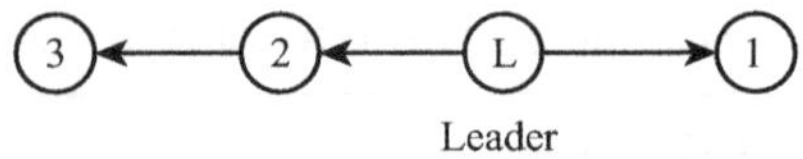

Fig. 2 Communication topology

Table 1 Parameters in Algorithm 2

$N_{c,\max}$	$N_{a,\max}$	$E_{c,thr}$	$E_{a,thr}$
200	200	10^{-4}	10^{-4}
l	α_{ci}	α_{ai}	Initial states
10^{-3}	0.001	0.001	Randomly in (0, 1)

$$x_{11}(t+1) = 0.04x_{12}(t) + x_{11}(t)$$

$$x_{12}(t+1) = -0.0001x_{11}(t) - 0.0335x_{11}^3(t) + x_{12}(t) + 0.05u_1(t)$$

Agent 2:

$$x_{21}(t+1) = 0.04x_{22}(t) + x_{21}(t)$$

$$x_{22}(t+1) = -0.0001x_{21}(t) - 0.0335x_{21}^3(t) + x_{22}(t) + 0.05u_1(t)$$

Agent 3:

$$x_{31}(t+1) = 0.04x_{32}(t) + x_{31}(t)$$

$$x_{32}(t+1) = -0.0001x_{31}(t) - 0.0335x_{31}^3(t) + x_{32}(t) + 0.05u_1(t)$$

The state trajectory of the leader is described as

$$x_{01}(t+1) = 0.04x_{02}(t) + x_{01}(t)$$

$$x_{02}(t+1) = -0.0001x_{01}(t) - 0.0335x_{01}^3(t) + x_{02}(t)$$

The pinning gain $b_1 = b_2 = 1, b_3 = 0$, and the edge weight $a_{32} = 1$. Let $V_{ii} = I_2,\ i = 1, 2, 3,\ R_{11} = R_{22} = R_{33} = R_{32} = 0.8$, and $R_{12} = R_{21} = R_{13} = R_{23} = R_{31} = 0$. Let the critic network activation function vector be $\phi_i(z_{ci}) = \left[e_{i1}^2, e_{i1}e_{i2}, e_{i1}u_i, e_{i2}^2, e_{i2}u_i, u_i^2\right]^T$ for $i = 1,2$, $\phi_3(z_{c3}) = \left[e_{31}^2, e_{31}e_{32}, e_{31}u_3, e_{31}u_2, e_{32}^2, e_{32}u_3, e_{32}u_2, u_3^2, u_3u_2, u_2^2\right]^T$, where $e_i = [e_{i1}, e_{i2}]^T$. Let the actor network activation function vector be $\varphi_i(z_{ai}) = \left[e_{i1}, e_{i2}, e_{i1}^2, e_{i1}e_{i2}, e_{i2}^2\right]^T$ for $i = 1, 2, 3$. Here, for simplicity, $u_i(t), e_i(t)$ is written as u_i, e_i, respectively.

The weights in the critic and actor network are tuned in the gradient descent circle, and the parameters of in Algorithm 2 are as presented at Table 1.

Figure 3 shows the consensus error of the MAS, from which we can see that the consensus errors will converge to zero eventually. Figures 4 and 5 shows the control inputs and the states of the MAS. Obviously, the input of each agent gradually comes

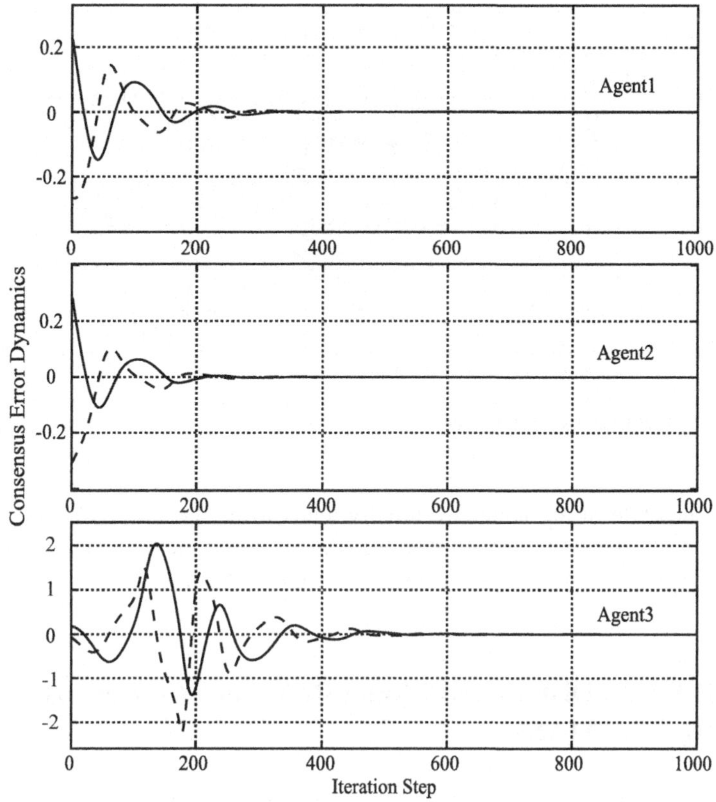

Fig. 3 Evolution of local consensus error

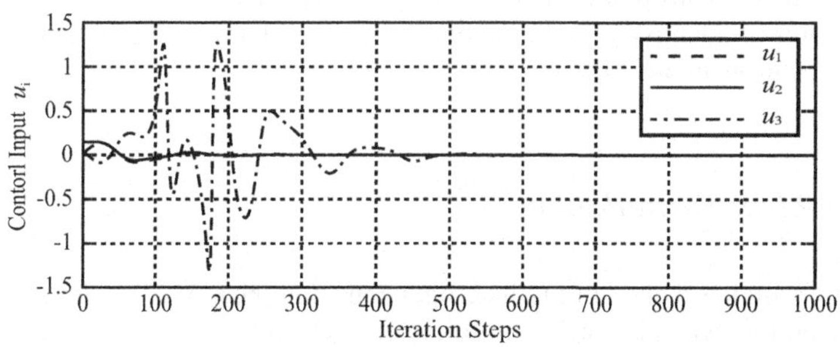

Fig. 4 Evolutions of the control input for three agents

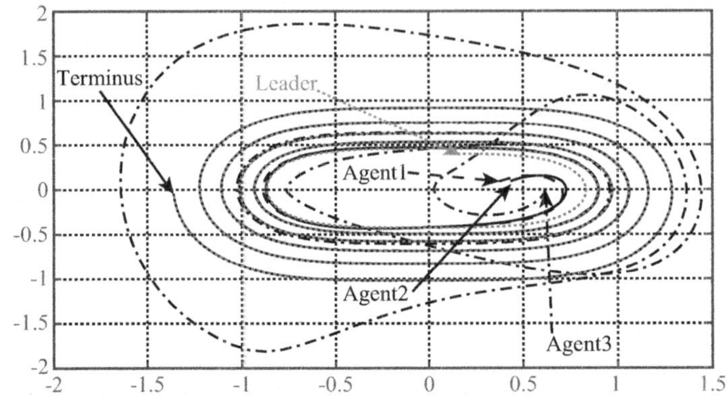

Fig. 5 The state trajectories of the agents

to 0, and the state of all agents eventually tends to be the same. Combining from Figs. 3, 4 and 5, we get the conclusion that the proposed method can not only solve the tracking control problem, but also guarantee the OCC.

3 ADP-Based Distributed Model Reference Consensus Control for MASs

In Sect. 2, homogenous MAS has been considered for unknown nonlinear systems, but in practical systems, the dynamic models of different agents are likely to be different, which means that the agents are heterogeneous. In this section, distributed model reference adaptive control method based on ADP is used to address the DT nonlinear heterogeneous MASs.

3.1 Problem Formulation

This section studies a class of DT heterogeneous nonlinear MASs with N followers and a leader, the dynamics of the ith follower is given by:

$$x_i(t+1) = \begin{bmatrix} 0_{m(n-1)\times m} & I_{m(n-1)\times m(n-1)} \\ 0_{m\times m} & 0_{m\times m(n-1)} \end{bmatrix} x_i(t) + \begin{bmatrix} 0_{m(n-1)\times m} \\ I_{m\times m} \end{bmatrix} (f_i(x_i(t)) + g_i(x_i(t))u_i(t)) \tag{27}$$

where $x_i(t) = \left[x_{i1}^T(t)\ x_{i2}^T(t)\ \ldots\ x_{in}^T(t)\right]^T \in R^{n\times m}$ represents the state vector, $x_{il}^T(t) \in R^m$, $l = 1, 2, \ldots, n$; $u_i(t) \in R^m$ denotes the control input vector; $f_i(x_i(t)) \in R^m$ and $g_i(x_i(t)) \in R^{m\times m}$ are the unknown nonlinear functions.

Assumption 1 For $\forall i \in 1, 2, \ldots, N$, the agent system is assumed to meet the following two conditions: the input dynamics terms $g_i(x_i(t))$ are nonsingular and bounded; the systems (27) are assumed to be controllable.

The dynamics of the leader which is represent by node v_0 is given by:

$$x_0(t+1) = Ax_0(t), \tag{28}$$

where $A = \begin{bmatrix} 0_{m(n-1)\times m} & I_{m(n-1)\times m(n-1)} \\ k_1 & k_2 \end{bmatrix}$, $k_1 \in R^{m\times m}$, $k_2 \in R^{m\times m(n-1)}$. Note that the eigenvalues of A should be outside or on the unit disk [26], and $x_0(t) \in R^{nm}$ represents the state vector.

Remark 1 One of the objectives is to design the consensus control laws $u_i(t)$ for each agent, which realizes that all followers and the leader reach consensus, i.e. $\lim_{t\to\infty} \|x_i(t) - x_0(t)\| = 0$, for $\forall i$.

Define the local neighborhood tracking error as:

$$e_i(t) = \sum_{j\in\mathcal{N}_i} a_{ij}(x_i(t) - x_j(t)) + b_i(x_i(t) - x_0(t)), \tag{29}$$

where $b_i \geq 0$ denotes the pinning gain [27].

Thus, overall tracking error can be calculated by

$$e(t) = ((\mathcal{L} + \mathcal{B}) \otimes I_{nm})(x(t) - \underline{x}_0(t)), \tag{30}$$

where $\mathcal{L} = [l_{ij}] \in R^{N\times N}$ represents the Laplacian matrix of the communication topology; $\mathcal{B} = [b_{ij}] \in R^{N\times N}$ is a diagonal matrix where the diagonal elements $b_{ii} = b_i$; $\otimes$ is the Kronecker product; $e(t) = \left[e_1^T(t)\ e_2^T(t)\ \cdots\ e_N^T(t)\right]^T \in R^{Nnm}$; $x(t) = \left[x_1^T(t)\ x_2^T(t)\ \cdots\ x_N^T(t)\right]^T \in R^{Nnm}$; $\underline{x}_0(t) = \left[x_0^T(t)\ x_0^T(t)\ \cdots\ x_0^T(t)\right]^T \in R^{Nnm}$; and $I_{nm} \in R^{nm\times nm}$ represents the identity matrix.

For agent i, the dynamics of the local neighborhood tracking error is denoted by:

$$\begin{aligned} e_i(t+1) = &\sum_{j\in\mathcal{N}_i} a_{ij}(F_i(t) - F_j(t) - G_j(t)u_j(t)) \\ &+ b_i \times (F_i(t) - Ax_0(t)) + (d_i + b_i)G_i(t)u_i(t), \end{aligned} \tag{31}$$

where $F_i(t) = \left[x_{i2}^T(t),\ x_{i3}^T(t),\ \cdots,\ x_{in}^T(t),\ f_i^T(x_i(t))\right]$, $G_i(t) = \left[0_{m\times m}\ 0_{m\times m}\ \ldots\ g_i^T(x_i(t))\right]^T$ and $i = 1, 2, \ldots, N$.

Accordingly, the global consensus error can be expressed as

$$\xi(t) = x(t) - \underline{x}_0(t). \tag{32}$$

Another objective is to minimize of the performance index function, which is defined as:

$$
\begin{aligned}
&J_i(e_i(t), u_i(t), u_{(j)}(t)) \\
&= \sum_{k=t}^{\infty} \alpha^{k-t} r_i(e_i(k), u_i(k), u_{(j)}(k)),
\end{aligned} \tag{33}
$$

where $r_i(e_i(t), u_i(t), u_{(j)}(t)) = e_i^T(t)V_{ii}e_i(t) + u_i^T(t)R_{ii}u_i(t) + \sum_{j \in \mathcal{N}_i} u_j^T(t)R_{ij}u_j(t)$; $u_{(j)}(t)$ is defined the same as (9); V_{ii}, R_{ii}, and R_{ij} are all positive defined matrix; and $0 < \alpha \le 1$ represents the discount factor. For simplicity, $r_i(t)$ is used to represent $r_i(e_i(t), u_i(t), u_{(j)}(t))$.

Assumed that u_i is an arbitrary admissible control law of agent i, (33) can be rewritten as [27]:

$$
J_i(e_i(t)) = r_i(t) + \alpha J_i(e_i(t+1)). \tag{34}
$$

Furthermore, the coupled HJB equation is defined by:

$$
J_i^*(e_i(t)) = \min_{u_i(t)}(r_i(t) + \alpha J_i^*(e_i(t+1))), \tag{35}
$$

where $J_i^*(e_i(t))$ is the optimal local performance index function. Thus, the optimal control law is formulated as

$$
u_i^*(t) = \arg\min_{u_i(t)}(r_i(t) + \alpha J_i^*(e_i(t+1))). \tag{36}
$$

The optimal response obtained by solving (35) and (36) satisfies the global Nash equilibrium, which also means that the optimal consensus is achieved. However, since the DT nonlinear MASs (27) is partially unknown, and the dynamics of each agent are different,it is very difficult to get the analytical solution of equation (35) for the coupled dynamic model (31). Fortunately, ADP technology provides a way to estimate the HJB equation of unknown nonlinear single agent system. However, due to the complexity of network structure of heterogeneous multi-agent systems, ADP technology is difficult to be directly used to solve their optimal consensus problem. Next, we will use a new optimal control strategy based on MRAC and ADP technology to address this problem.

3.2 *ADP-Based Distributed Model Reference Control Method*

In this subsection, for the MASs defined by (27) and (28), in order to achieve consensus and ensure the minimum performance index, we propose an ADP-based dis-

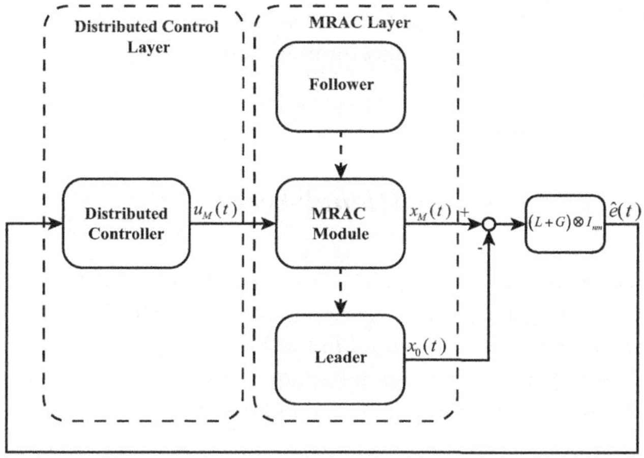

Fig. 6 Structure of the ADP-based distributed model reference adaptive control method

tributed model reference adaptive control strategy as shown in Fig. 6. It mainly consists of two layers: MRAC layer and distributed control layer.

In the MRAC layer, for each agent of the nonlinear heterogeneous MASs (27), a corresponding linear model, called MRAC model, is designed, so the MRAC layer consists of N MRAC models. In each MRAC modules, by using the MRAC scheme, the states of agent x_i will track the ones of their reference models x_{M_i}. In other words, we can control the states of nonlinear agents by controlling the corresponding MRAC model. That is to say, the OCC problem of system (27) is transformed into the states synchronization problem of the transformed linear homogeneous system, whose individual agent is the reference model and the leader.

For the transformed MAS whose topology is the same as the original systems, there are N distributed controllers. In order to keep the states of all reference models to be synchronized with that of the leader and ensure the optimality of the performance index, ADP technology is used in each distributed controller to solve the coupled HJB equation (35) online, thus we have $x_{M_i}(t) \rightarrow x_0(t)$. On the other hand, by using the MRAC scheme, the states of each agent will track the ones of its corresponding reference model, that is: $x_i(t) \rightarrow x_{M_i}(t)$. Finally we get $x_i(t) \rightarrow x_0(t)$. In other words, the states of all agents and the leader will reach a consensus indirectly.

In the existing MRAC structures in [42–44], model reference adaptive controller often integrates feedforward and feedback, and is approximated by multi-layer perceptrons. Because the MASs are nonlinear, the active function is also nonlinear. In other words, different from the tracking control [45–48], the input of the controller is a NNs function of the feedback tracking error and the reference input. Note that, its inverse function with regard to tracking error and the reference input need to be got for computation of the distributed control layer. In addition, the coupled HJB equation (35) of the transformed homogenous MAS is still a nonlinear equation rather than a constant Riccati matrix equation.

In the following subsections, adaptive feedforward and feedback control scheme and a distributed value iteration algorithm are studied respectively to address the above problems.

3.3 MRAC Scheme for Individual Agent

In this subsection, for realizing the states of agents tracking their corresponding reference ones, an adaptive feedforward and feedback control scheme based on the diagonal recurrent NN structure is used in the MRAC layer.

Firstly, for each system, a corresponding linear model, called reference model, is designed. The reference model of agent i is defined as:

$$x_{M_i}(t+1) = Ax_{M_i}(t) + \begin{bmatrix} 0_{m(n-1)\times m} \\ B_M \end{bmatrix} u_{M_i}(t), \tag{37}$$

where $x_{M_i}(t) = \left[x_{M_{i1}}^T(t)\, x_{M_{i2}}^T(t)\, \ldots\, x_{M_{in}}^T(t)\right]^T \in R^{nm}$ represents the reference state vector; $x_{M_{ij}}(t) \in R^m$, $j = 1, 2, \ldots, n$; $u_{M_i}(t) \in R^m$ is the reference input. Note that A is defined as in (28). The matrix $B_M \in R^{m\times m}$ is set to be nonsingular, and the reference model should be controllable. This is equivalent to building a bridge between the leader and followers. Therefore, the consensus problem between leader and followers is transformed into the problem of tracking control between each follower and its corresponding reference model and the consensus problem between reference models and the leader.

In order to realize the tracking control between each agent and its reference model, the model reference adaptive control based on adaptive feedforward and feedback control scheme is adopted. Once the states of each agent $x_i(t)$ track the ones of the corresponding reference model $x_{M_i}(t)$, the consensus problem of the original MASs can be achieved by realizing the consensus problem of linear MAS, in which the follower is the reference models.

By subtracting (37) from (27), the dynamics of the tracking error can be represented as:

$$\begin{aligned}\tilde{x}_i(t+1) = &\begin{bmatrix} 0_{m(n-1)\times m} & I_{m(n-1)\times m(n-1)} \\ 0_{m\times m} & 0_{m\times m(n-1)} \end{bmatrix} \tilde{x}_i(t) + \begin{bmatrix} 0_{m(n-1)\times m} \\ A_M \end{bmatrix} x_{M_i}(t) \\ &+ \begin{bmatrix} 0_{m(n-1)\times m} \\ I_{m\times m} \end{bmatrix} (f_i(x_i(t)) + g_i(x_i(t))u_i(t)) + \begin{bmatrix} 0_{m(n-1)\times m} \\ B_M \end{bmatrix} u_{M_i}(t),\end{aligned} \tag{38}$$

where $A_M = [k_1, k_2]$, k_1 and k_2 are defined in (28), and B_M is defined in (37).

According to the filtering method [49], the filtered tracking error of the model reference tracking error is calculated as

$$\bar{x}_i(t) = \Lambda_0 \tilde{x}_i(t), \tag{39}$$

where $\Lambda_0 = \left[\lambda_{n-1} I_m \ \lambda_{n-2} I_m \ \cdots \ I_m\right]$. By pre-multiplying Λ_0 both sides of Eq. (38), we have:

$$\bar{x}_i(t+1) = f_i(x_i(t)) + g_i(x_i(t))u_i(t) + \Lambda_0 \Lambda_1 \tilde{x}_i(t) - A_M x_{M_i}(t) - B_M u_{M_i}(t), \tag{40}$$

where $\Lambda_1 = \begin{bmatrix} 0_{m(n-1)\times m} & I_{m(n-1)\times m(n-1)} \\ 0_{m\times m} & 0_{m\times m(n-1)} \end{bmatrix}$.

The adaptive feedforward and feedback control law is designed as

$$u_i(t) = g_i^{-1}(x_i(t))(k_3 \bar{x}_i(t) - \Lambda_0 \Lambda_1 \tilde{x}_i(t) + A_M x_{M_i}(t) + B_M u_{M_i}(t)) + \bar{u}_i(t). \tag{41}$$

where the gain matrix $k_3 \in R^{m\times m}$ denotes a Hurwitz matrix and the virtual control $\bar{u}_i(t)$ will be discussed later. Note that, $g_i^{-1}(x_i(t))$ can be obtained by some model identification methods. Substituting (41) into (40) yields

$$\bar{x}_i(t+1) = k_3 \bar{x}_i(t) + f_i(x_i(t)) + g_i(x_i(t))\bar{u}_i(t). \tag{42}$$

NNs control method is used to realize MRAC, so the system (42) will be stable by designing the virtual control law $\bar{u}_i(t)$ as:

$$\bar{u}_i(t) = w_{ci}^T(t)\sigma_{ci}(t), \tag{43}$$

$$\sigma_{ci}(t) = \sigma(v_{ci}^T(t)x_i(t) + \mathrm{diag}(v_{cDi}(t))\sigma_{ci}(t-1)), \tag{44}$$

where w_{ci}, v_{ci} and v_{cDi} are the estimations of the ideal weight matrices w_{ci}^*, v_{ci}^* and v_{cDi}^*, respectively.

Remark 2 Different from the existing MRAC technologies [42–44], where the feedforward and feedback are integrated in the controller, the formulation of the feedforward and feedback control law used in our control strategy is the linear combination of the feedforward and feedback terms. Diagonal recurrent NNs is used to estimate the virtual control law, where the network input is $x_i(t)$. Accordingly, we use adaptive control structure to acquire the reference input $u_{M_i}(t)$ defined in Sect. 3.4.

The objective function is defined as:

$$E_m(t+1) = \frac{1}{2}\bar{x}_i^T(t+1)\bar{x}_i(t+1). \tag{45}$$

In order to minimize the objective function, the gradient-based adaption rule is used, and the weights are updated by:

$$w_{ci}(t+1) = w_{ci}(t) - \eta_{cwi}\sigma_{ci}(t)\bar{x}_i^T(t+1)g_i(x_i(t)), \tag{46a}$$

$$v_{ci}(t+1) = v_{ci}(t) - \eta_{cvi} x_i(t)\bar{x}_i^T(t+1) g_i(x_i(t)) \times w_{ci}^T(t) \text{diag}(\sigma_{ci}^{'}(t)), \tag{46b}$$

$$v_{cDi}(t+1) = v_{cDi}(t) - \eta_{cDi} \text{diag}(\sigma_{ci}(t-1)) \text{diag}(\sigma_{ci}^{'}(t)) \times w_{ci}(t) g_i^T(x_i(t)) \bar{x}_i(t+1), \tag{46c}$$

where η_{Iwi}, η_{Ivi} and η_{IDi} are the learning rates and $\sigma_{ci}^{'}(\cdot)$ is the partial derivative of $\sigma_{ci}(\cdot)$.

3.4 Distributed Value Iteration Algorithm

In the MRAC layer, all the individual agents in MASs (27) track their corresponding reference models on behavior by using the MRAC technologies. So the remaining task is the consensus problem between the leader and the reference models. Based on the adaptive dynamic programming, a distributed value iteration algorithm is developed to make the reference models and the leader achieve consensus, while ensure the optimality of the performance index. Thus, the states of the leader and individual agents will reach consensus indirectly. In the following parts, distributed value iteration algorithm and its implementation will be introduced, respectively.

A. Derivation of the Iteration Algorithm

In this part, based on results of the MRAC layer, a distributed value iteration algorithm is presented to make all the reference models achieve consensus with the leader.

Since the states of individual agent (27) track that of the reference model (37). The objective of the consensus control, $\lim_{t\to\infty} \|x_i(t) - x_0(t)\| = 0$ for $\forall i$, is formulated as $\lim_{t\to\infty} \left\|x_{M_i}(t) - x_0(t)\right\| = 0$ for $\forall i$.

Define

$$\xi_{M_i} = x_{M_i}(t) - x_0(t),$$

where $i = 1, 2, \ldots, N$.

Accordingly, local neighborhood tracking error (29) is reformulated as

$$\hat{e}_i(t) = \sum_{j\in\mathcal{N}_i} a_{ij}(x_{M_i}(t) - x_{M_j}(t)) + b_i(x_{M_i}(t) - x_0(t)), \tag{47}$$

where $b_i \geq 0$ is the pinning gain [27]. We obtain the overall tracking error vector

$$\hat{e}(t) = ((\mathcal{L} + \mathcal{B}) \otimes I_{nm})(x_M(t) - \underline{x}_0(t)), \tag{48}$$

where $\hat{e}(t) = \left[\hat{e}_1^T(t)\ \hat{e}_2^T(t)\ \cdots\ \hat{e}_N^T(t)\right]^T \in R^{Nnm}$, $x_M(t) = \left[x_{M_1}^T(t)\ x_{M_2}^T(t)\ \cdots x_{M_N}^T(t)\right]^T \in R^{Nnm}$, and $\underline{x}_0(t) = \left[x_0^T(t)\ x_0^T(t)\ \cdots\ x_0^T(t)\right]^T \in R^{Nnm}$. The dynamics of the local neighborhood tracking error in node v_i is given by

$$\hat{e}_i(t+1) = A\hat{e}_i(t) + (d_i + b_i)Bu_{M_i}(t) - \sum_{j\in\mathcal{N}_i} a_{ij}Bu_{M_j}(t), \tag{49}$$

where $B = \left[0_{m\times m}\ 0_{m\times m}\ \ldots\ B_M^T\right]^T$. Combined with the result of Sect. 3.3 that x_i is close to x_{mi}, we get:

$$r_i(t) \approx \hat{r}_i(t) \triangleq \hat{e}_i^T(t)V_{ii}\hat{e}_i(t) + u_i^T(t)R_{ii}u_i(t) + \sum_{j\in\mathcal{N}_i} u_j^T(t)R_{ij}u_j(t). \tag{50}$$

Let

$$u_{xi}(t) = g_i^{-1}(x_i(t))((k_3 - \Lambda_0\Lambda_1)\bar{x}_i(t) + A_M x_{M_i}(t)) + w_{ci}^T(t)\sigma_{ci}(t), \tag{51}$$

and

$$g_{Bi}(t) = g_i^{-1}(x_i(t))B_M. \tag{52}$$

Obviously, $u_{xi}(t)$ and $g_{Bi}(t)$ are only related to $x_i(t)$ and $x_{M_i}(t)$, so (41) can be rewritten as:

$$u_i(t) = u_{xi}(t) + g_{Bi}(t)u_{M_i}(t). \tag{53}$$

Remark 3 It follows from (51) and (52) that u_{xi} is a function with respect to $x_i(t)$ and $x_{M_i}(t)$, and $g_{Bi}(t)$ is a function with respect to x_i. Apparently, designing of the OCC law $u_i(t)$ in the original MAS (27) can be achieved by designing the reference input $u_{M_i}(t)$, which is

$$u_i^*(t) = u_{xi}(t) + g_{Bi}(t)u_{M_i}^*(t), \tag{54}$$

in which $u_{M_i}^*(t)$ is the optimal reference control input.

According to (53), one gets

$$\begin{aligned} u_j^T(t)R_{ij}u_j(t) = {} & u_{xj}^T(t)R_{ij}u_{xj}(t) + 2u_{xj}^T(t)R_{ij}g_{Bj}(t)u_{M_j}(t) \\ & + u_{M_j}^T(t)g_{Bj}^T(t)R_{ij}g_{Bj}(t)u_{M_j}(t). \end{aligned} \tag{55}$$

The reward function (50) can be approximately represented by

$$
\begin{aligned}
\hat{r}_i(t) =& \hat{e}_i^T(t)V_{ii}\hat{e}_i(t) + u_{xi}^T(t)R_{ii}u_{xi}(t) + 2u_{xi}^T(t)R_{ii} \\
&\times g_{Bi}(t)u_{M_i}(t) + u_{M_i}^T(t)g_{Bi}^T(t)R_{ii}g_{Bi}(t)u_{M_i}(t) \\
&+ \sum_{j\in\mathcal{N}_i}\big(u_{xj}^T(t)R_{ij}u_{xj}(t) + 2u_{xj}^T(t)R_{ij}g_{Bj}(t)u_{M_j}(t) \\
&+ u_{M_j}^T(t)g_{Bj}^T(t)R_{ij}g_{Bj}(t)u_{M_j}(t)\big).
\end{aligned} \tag{56}
$$

Given the admissible control laws $\forall i$, the local performance index functions can be rewritten as

$$
\bar{J}_i(\hat{e}_i(t)) = \hat{r}_i(t) + \alpha\bar{J}_i(\hat{e}_i(t+1)), \tag{57}
$$

where $\bar{J}_i(\hat{e}_i(t)) = \sum\limits_{k=t}^{\infty}\alpha^{k-t}\hat{r}_i(t)$.

Combined with (54), the optimal local performance index function can be expressed as

$$
\begin{aligned}
\bar{J}_i^*(\hat{e}_i(t)) &= \min_{u_i(t)}(\hat{r}_i(t)+\alpha\bar{J}_i^*(\hat{e}_i(t+1))) \\
&= \min_{u_{M_i}(t)}(\hat{r}_i(t)+\alpha\bar{J}_i^*(\hat{e}_i(t+1))).
\end{aligned} \tag{58}
$$

It follows from (49), (56), and (58) that the optimal reference control input is given as

$$
\begin{aligned}
u_{M_i}^*(t) =& \arg\min_{u_{M_i}(t)}(\hat{r}_i(t)+\alpha\bar{J}_i^*(\hat{e}_i(t+1))) \\
=& -(g_{Bi}^T(t)R_{ii}g_{Bi}(t))^{-1}\Bigg(g_{Bi}^T(t)R_{ii}u_{xi}(t) \\
&+ \frac{\alpha}{2}(d_i+b_i)B^T\frac{\partial\bar{J}_i^*(\hat{e}_i(t+1))}{\partial\hat{e}_i(t+1)}\Bigg).
\end{aligned} \tag{59}
$$

Note that, as $g_i^{-1}(x_i(t))$ and B_M are nonsingular matrices, it follows from (52) that $g_{Bi}^T(t)g_{Bi}(t)$ is the invertible matrix.

Note that, instead of constant Riccati matrix equation, (57) is a nonlinear equation. Therefore, LQR-based method [50] can not be applied here. Instead, we use distributed value iteration algorithm to solve the optimal consensus problem, which can be described as follows:

First, for each reference model i, initiate the local value function $\bar{J}_i^0(\hat{e}_i(t)), \forall\hat{e}_i(t)$, and solve the control laws by

$$
v_i^0(\hat{e}_i(t)) = \arg\min_{u_{M_i}(t)}\left\{\hat{r}_i(t) + \alpha\bar{J}_i^0(\hat{e}_i(t+1))\right\}. \tag{60}
$$

Then, update the value function by

$$\bar{J}_i^1(\hat{e}_i(t)) = \min_{u_{M_i}(t)} \left\{\hat{r}_i(t) + \alpha \bar{J}_i^0(\hat{e}_i(t+1))\right\} \hat{r}_i(t)\Big|_{u_{M_i}(t)=v_i^0(\hat{e}_i(t))} . \tag{61}$$

For the iteration index $l = 1, 2, \ldots$, the distributed value iteration algorithm iterates between

$$\begin{aligned} v_i^l(\hat{e}_i(t)) =& \arg\min_{u_{M_i}(t)} \left\{\hat{r}_i(t) + \alpha \bar{J}_i^l(\hat{e}_i(t+1))\right\} \\ =& -(g_{Bi}^T(t)R_{ii}g_{Bi}(t))^{-1}\Bigg(g_{Bi}^T(t)R_{ii}u_{xi}(t) \\ &+ \frac{\alpha}{2}(d_i+b_i)B^T\frac{\partial \bar{J}_i^l(\hat{e}_i(t+1))}{\partial \hat{e}_i(t+1)}\Bigg), \end{aligned} \tag{62}$$

and

$$\begin{aligned} \bar{J}_i^{l+l}(\hat{e}_i(t)) =& \min_{u_{M_i}(t)} \left\{\hat{r}_i(t) + \alpha \bar{J}_i^l(\hat{e}_i(t+1))\right\} \\ =& \hat{r}_i(t)\Big|_{u_{M_i}(t)=v_i^l(\hat{e}_i(t))} + \alpha \bar{J}_i^l(\hat{e}_i(t+1)). \end{aligned} \tag{63}$$

B. Implementation of the Iteration Algorithm

In this part, the detailed implementation of the distributed iteration algorithm will be introduced.

For acquiring the approximate optimal value function, an actor-critic framework is used in this chapter. Specifically, the critic network is designed by:

$$\hat{J}_i(\hat{e}_i(t)) = W_{ci}^T(t)\sigma(V_{ci}^T(t)\hat{e}_i(t)), \tag{64}$$

where $\hat{J}_i(\hat{e}_i(t))$ is the output of the network and represents the optimal value function. $W_{ci} \in R^{N_{chi}}$ and $V_{ci} \in R^{nm \times N_{chi}}$ represent weights in critic networks, where N_{chi} is the number of nodes in the hidden layer.

According to (57), there exists the following relationship:

$$\hat{r}_i(t) + \alpha \hat{J}_i(\hat{e}_i(t+1)) - \hat{J}_i(\hat{e}_i(t)) = 0. \tag{65}$$

However, due to the biased estimation of the neural network, (65) can be rewritten as:

$$e_{ci}(t) = \hat{r}_i(t) + \alpha \hat{J}_i(\hat{e}_i(t+1)) - \hat{J}_i(\hat{e}_i(t)). \tag{66}$$

where e_{ci} is the approximation error. So, the objective function can be defined as:

$$E_{ci} = \frac{1}{2}e_{ci}^2(t). \tag{67}$$

Gradient-descent-based method is used to minimize the objective function E_{ci}, so the updated weights of the network can be calculated by:

$$W_{ci}(t+1) = W_{ci}(t) + \eta_{ci} e_{ci}(t) \sigma(V_{ci}^T(t)\hat{e}_i(t)), \tag{68}$$

where η_{ci} is the learning rate.

Similarly, the action network can be expressed by:

$$u_{M_i}(t) = W_{ai}^T(t)\sigma(V_{ai}^T(t)z_i(t)). \tag{69}$$

where $u_{M_i}(t)$ is the actual reference input. $W_{ai} \in R^{N_{ahi}}$ and $V_{ai} \in R^{nm \times N_{ahi}}$ is the weights of action network, in which N_{ahi} is the number of nodes in the hidden layer. $z_i(t) = \left[\hat{e}_i^T(t),\ x_i^T(t),\ x_{M_i}^T(t)\right]^T$ is the inputs of the network.

It follows from (59) and (64) that the target value of $u_{M_i}(t)$ can be derived by

$$\begin{aligned} \bar{u}_{M_i}(t) = & -(g_{Bi}^T(t)R_{ii}g_{Bi}(t))^{-1}(g_{Bi}^T(t)R_{ii}u_{xi}(t) + \frac{\alpha}{2}(d_i + b_i) \\ & \times B^T \frac{\partial \hat{J}_i(\hat{e}_i(t+1))}{\partial \hat{e}_i(t+1)}). \end{aligned} \tag{70}$$

where $\sigma_c^{'}(t)$ is the derivative of $\sigma(V_{ci}^T(t)e_i(t))$ with respect to $V_{ci}^T(t)e_i(t)$.

Notice that, the target value of reference input $u_{M_i}(t)$ can be easily got by using the adaptive feedforward and feedback control scheme in the MRAC layer.

Define the actor NNs error as

$$e_{ai}(t) = \bar{u}_{M_i}(t) - u_{M_i}(t). \tag{71}$$

Accordingly, the objective function can be written as:

$$E_a = \frac{1}{2} e_{ai}^T(t) e_{ai}(t). \tag{72}$$

The weights of the action network is updated by minimizing the objective function E_a, which can be calculated by:

$$W_{ai}(t+1) = W_{ai}(t) - \eta_{ai}\sigma(V_{ai}^T(t)e_i(t))e_{ai}^T(t). \tag{73}$$

Obviously, by using the distributed value iteration algorithm, the states of all reference models will be consensus with those of the leader while guarantee the performance index is optimal. Combine with the results that each individual agent track the one of their corresponding reference model, we can get the conclusion that the states of all followers synchronize with the leader. Therefore, the DT heterogeneous nonlinear MASs with unknown dynamics achieve OCC by using the proposed control strategy.

Table 2 Definitions of the dynamics

g	l_i	D_i
Acceleration of gravity	Length of the manipulator	Viscous friction coefficient
M_i	G_i	$\theta_i(t)$
Mass of the payload	Moment of inertia	Angle position
$u_i(t)$	t	
Torque	Continuous time	

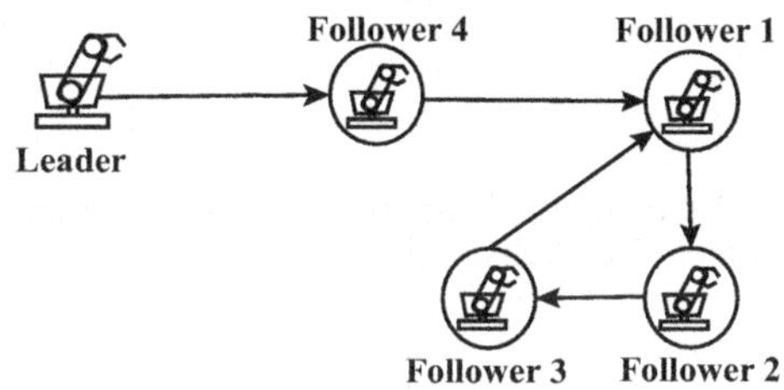

Fig. 7 Communication topology

To sum up, by using the proposed hierarchical and distributed control strategy, the OCC problem of nonlinear heterogeneous multi-agent systems is transformed into two problems: the tracking control problem of single agent system; the consensus control problem of corresponding linear homogeneous multi-agent system. Model reference adaptive algorithm and the distributed value iteration algorithm is used to address the above problems. Finally, we achieve the OCC of the original systems.

3.5 Simulation Studies

In this subsection, the proposed ADP-based distributed model reference adaptive control method has been applied in a heterogeneous DT nonlinear MAS. It is assumed that the dynamic models of the agents in the MAS are unknown.

We choose a multi-manipulator system to verify the effectiveness of our control strategy, where the system consists of multiple unknown single link manipulators. The dynamics of the manipulators are given by

$$\ddot{\theta}_i(t) = -\frac{M_i g l_i}{G_i}\sin(\theta_i(t)) - \frac{D_i}{G_i}\dot{\theta}_i(t) + \frac{1}{G_i}u_i(t), \tag{74}$$

where for the detailed definitions one can refer to Table 2.

The communication topology of the multi-manipulator system is depicted in Fig. 7. As can be seen, there are four followers and a leader in the multi-manipulator system.

Table 3 The parameter values

Para.	Agent			
	1	2	3	4
$\frac{2G_i - TD_i}{G_i}$	1.83	1.83	1.73	1.725
$\frac{G_i - TD_i}{G_i}$	0.83	0.83	0.73	0.725
$\frac{T^2 M_i g L_i}{G_i}$	$8.17 * 10^{-4}$	$1.23 * 10^{-3}$	$1.57 * 10^{-3}$	$3.31 * 10^{-3}$
$\frac{T^2}{G_i}$	$8.33 * 10^{-3}$	$8.33 * 10^{-3}$	$1.67 * 10^{-2}$	$1.25 * 10^{-2}$
Q_{ii}	$\begin{bmatrix} 1 & 0 \\ 0 & 1 \end{bmatrix}$	$\begin{bmatrix} 1 & 0 \\ 0 & 1 \end{bmatrix}$	$\begin{bmatrix} 1 & 0 \\ 0 & 1 \end{bmatrix}$	$\begin{bmatrix} 1 & 0 \\ 0 & 1 \end{bmatrix}$

By using the Euler methods with the sampling interval $T = 0.05$ s, the controlled object (74) is discretized to

$$\begin{cases} \theta_i(t+1) = \bar{\theta}_i(t) \\ \bar{\theta}_i(t+1) = \dfrac{2G_i - TD_i}{G_i}\bar{\theta}_i(t) - \dfrac{G_i - TD_i}{G_i}\theta_i(t) \\ \qquad - \dfrac{T^2 M_i g L_i}{G_i}\sin(\theta_i(t)) + \dfrac{T^2}{G_i}u_i(t), \end{cases} \tag{75}$$

where $\bar{\theta}_i(t) = \theta_i(t) + T\dot{\theta}_i(t)$.

The parameter values of the MAS are listed in Table 3.

Besides, according to Fig. 7, the communication typology parameter $R = [r_{ij}] = \begin{bmatrix} 1 & 1 & 0 & 1 \\ 0 & 1 & 1 & 0 \\ 1 & 0 & 1 & 0 \\ 0 & 0 & 0 & 1 \end{bmatrix}$ and $b = [b_1 \ \ b_2 \ \ b_3 \ \ b_4] = [0 \ \ 0 \ \ 0 \ \ 1]$. Set $a = [a_{ij}] = \begin{bmatrix} 0 & 0.8 & 0 & 0.7 \\ 0 & 0 & 0.6 & 0 \\ 0.8 & 0 & 0 & 0 \\ 0 & 0 & 0 & 0 \end{bmatrix}$, thus all of the parameter are determined.

The dynamic model of the leader mode is given by

$$\begin{cases} \theta_0(t+1) = \bar{\theta}_0(t) \\ \bar{\theta}_0(t+1) = -\theta_0(t) + 1.99\bar{\theta}_0(t), \end{cases} \tag{76}$$

According to the leader model (76) and the requirement of the controllable reference model, select the reference model as

$$\begin{cases} \theta_{M_i}(t+1) = \bar{\theta}_{M_i}(t), \\ \bar{\theta}_{M_i}(t+1) = -\theta_{M_i}(t) + 1.99\bar{\theta}_{M_i}(t) + 0.01u_{M_i}(t), \end{cases} \tag{77}$$

where $x_{M_{i1}}(t),\ x_{M_{i2}}(t) \in R$.

Table 4 Network parameter of the diagonal recurrent NNs structure

Number of neurons in hidden layer	Active function	Range of weights	c_w	c_v	cD
4	Log-sigmoid function	[−1, 1]	0.15	0.25	0.1

Table 5 Network parameter of the actor-critic NNs structure

Number of neurons in hidden layer	Active function	Range of weights	η_c	η_{ai}
4	Log-sigmoid function	[−1, 1]	0.1	0.1

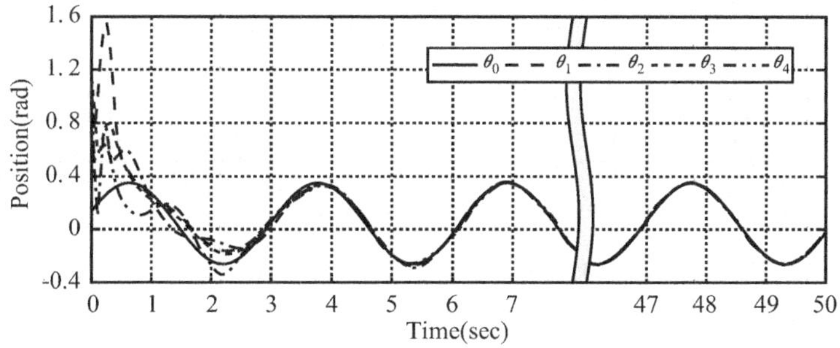

Fig. 8 Positions of the five agents

In the MRAC layer, by using the designed adaptive feedforward and feedback controller, the states of the multimanipulator (74) will track that of the reference model (77). The network parameters of the controller are shown in Table 4

For any agent, it should be noticed that the network parameters are the same.

Based on the results from MRAC layer, distributed value iteration algorithm are used to make the states of reference model synchronize with leader. A double NNs are used to approximate the critic network and action network, respectively. Accordingly, their network parameters are list in Table 5.

We select the parameters used in the ADP-based distributed model reference adaptive control method as $k_3 = -0.6, \alpha = 0.95, \Lambda_0 = [0.51], \Lambda_1 = \begin{bmatrix} 0 & 1 \\ 0 & 0 \end{bmatrix}$.

Figures 8 and 9 show the position and velocity of the MAS, respectively. Figure 10 is the evolution of local neighborhood tracking errors. For ease of observation, we plot the phase plane of the trajectories in Fig. 11. From Figs. 8, 9, 10 and 11, we can get the conclusion that all followers and the leader reach a consensus in a small bounded area, which proves the effectiveness of the strategy.

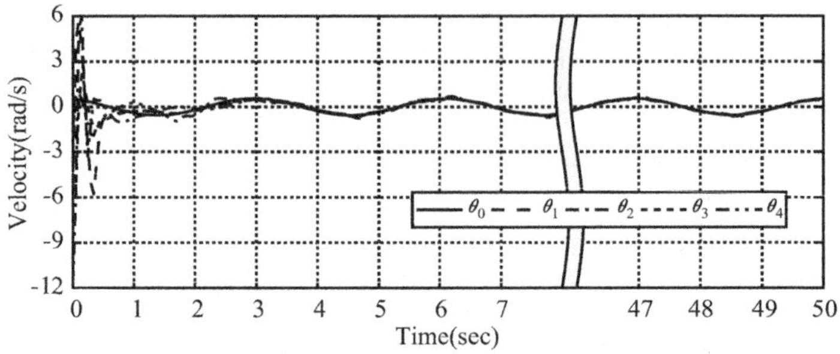

Fig. 9 Velocities of the five agents

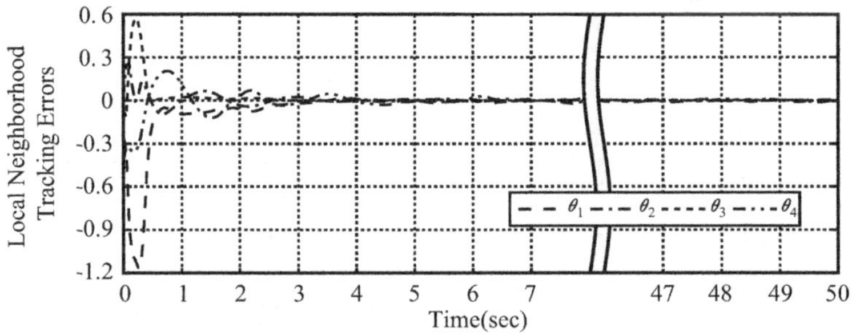

Fig. 10 Local neighborhood tracking errors of the four followers

Fig. 11 The evolution of the agent trajectories

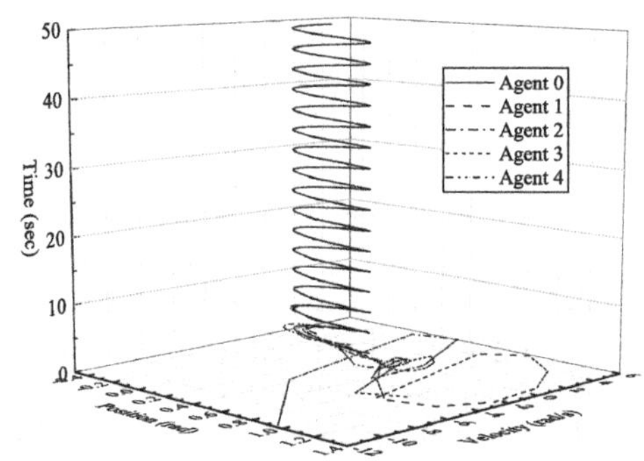

4 Conclusion

Optimal consensus control which not only guarantees synchronization but also minimizes the energy cost has been studied in the chapter. We have defined the local value function which not only consider the agent's local consensus error but also the control inputs from itself and its neighborhoods. Based on the local value function, a policy iteration method for multi-agent synchronization where the system dynamics does not need not to be known in priori have been proposed and used. The effectiveness of the used method is demonstrated by two simulations. We also developed the ADP-based distributed model reference adaptive control method to address the optimal consensus problem for DT heterogeneous nonlinear MASs.

References

1. Chen, J., Cao, X.H., Cheng, P., Xiao, Y., Sun, Y.X.: Distributed collaborative control for industrial automation with wireless sensor and actuator networks. IEEE Trans. Ind. Electron. **57**(12), 4219–4230 (2010)
2. Zhu, J.D., Lu, J.H., Yu, X.H.: Flocking of multi-agent non-holonomic systems with proximity graphs. IEEE Trans. Circuits Syst. I: Regul. Pap. **60**(1), 199–210 (2013)
3. Xiao, F., Wang, L., Chen, J., Gao, Y.P.: Finite-time formation control for multi-agent systems. Automatica **45**(11), 2605–2611 (2009)
4. Wang, X.H., Yadav, V., Balakrishnan, S.N.: Cooperative UAV formation flying with obstacle/collision avoidance. IEEE Trans. Control Syst. Technol. **15**(4), 672–679 (2007)
5. Wei, Q.L., Liu, D.R., Shi, G., Liu, Y.: Multibattery optimal coordination control for home energy management systems via distributed iterative adaptive dynamic programming. IEEE Trans. Ind. Electron. **62**(7), 4203–4214 (2015)
6. Fax, J.A., Murray, R.M.: Information flow and cooperative control of vehicle formations. IEEE Trans. Autom. Control **49**(9), 1465–1476 (2004)
7. Rehan, M., Jameel, A., Ahn, C.K.: Distributed consensus control of one-sided Lipschitz nonlinear multiagent systems. IEEE Trans. Syst. Man Cybern.: Syst. **48**(8), 1297–1308 (2018)
8. Wang, F., Chen, X., He, Y., Wu, M.: Finite-time consensus problem for second-order multi-agent systems under switching topologies. Asian J. Control **19**(5), 1756–1766 (2017)
9. Bu, X.H., Yu, Q.X., Hou, Z.S., Qian, W.: Model free adaptive iterative learning consensus tracking control for a class of nonlinear multiagent systems. IEEE Trans. Syst. Man Cybern.: Syst. **49**(4), 677–686 (2019)
10. Meng, W.C., Yang, Q.M., Sarangapani, J., Sun, Y.X.: Distributed control of nonlinear multiagent systems with asymptotic consensus. IEEE Trans. Syst. Man Cybern.: Syst. **47**(5), 749–757 (2017)
11. Liu, W., Huang, J.: Adaptive leader-following consensus for a class of higher-order nonlinear multi-agent systems with directed switching networks. Automatica **79**, 84–92 (2017)
12. Movric, K.H., Lewis, F.L.: Cooperative optimal control for multi-agent systems on directed graph topologies. IEEE Trans. Autom. Control **59**(3), 769–774 (2014)
13. Vamvoudakis, K.G., Lewis, F.L.: Multi-player non-zero-sum games: online adaptive learning solution of coupled Hamilton-Jacobi equations. Automatica **47**(8), 1556–1569 (2011)
14. Werbos, P.: Approximate dynamic programming for realtime control and neural modelling. Handbook of Intelligent Control: Neural, Fuzzy and Adaptive Approaches, pp. 493–525 (1992)
15. Wang, D., Liu, D.R., Li, H.L., Luo, B., Ma, H.W.: An approximate optimal control approach for robust stabilization of a class of discrete-time nonlinear systems with uncertainties. IEEE Trans. Syst. Man Cybern.: Syst. **46**(5), 713–717 (2016)
16. Wei, Q.L., Lewis, F.L., Liu, D.R., Song, R.Z., Lin, H.Q.: Discrete-time local value iteration adaptive dynamic programming: convergence analysis. IEEE Trans. Syst. Man Cybern.: Syst. **48**(6), 875–891 (2018)
17. Wang, Z., Liu, X.P., Liu, K.F., Li, S., Wang, H.Q.: Backstepping-based Lyapunov function construction using approximate dynamic programming and sum of square techniques. IEEE Trans. Cybern. **47**(10), 3393–3403 (2017)
18. Vamvoudakis, K.G., Lewis, F.L., Hudas, G.R.: Multi-agent differential graphical games: online adaptive learning solution for synchronization with optimality. Automatica **48**(8), 1598–1611 (2012)

19. Zhong, X.N., He, H.B.: GrHDP solution for optimal consensus control of multiagent discrete-time systems. IEEE Trans. Syst. Man Cybern.: Syst. (2018). https://doi.org/10.1109/TSMC.2018.2814018
20. Vamvoudakis, K.G.: Q-learning for continuous-time graphical games on large networks with completely unknown linear system dynamics. Int. J. Robust Nonlinear Control **27**(16), 2900–2920 (2017)
21. Wei, Q.L., Liu, D.R., Lewis, F.L.: Optimal distributed synchronization control for continuous-time heterogeneous multi-agent differential graphical games. Inf. Sci. **317**, 96–113 (2015)
22. Zhang, H.G., Zhang, J.L., Yang, G.H., Luo, Y.H.: Leader-based optimal coordination control for the consensus problem of multiagent differential games via fuzzy adaptive dynamic programming. IEEE Trans. Fuzzy Syst. **23**(1), 152–163 (2015)
23. Tatari, F., Naghibi-Sistani, M.B., Vamvoudakis, K.G.: Distributed learning algorithm for non-linear differential graphical games. Trans. Inst. Meas. Control **39**(2), 173–182 (2017)
24. Kamalapurkar, R.K., Dinh, H.Y., Walters, P., Dixon, W.: Approximate optimal cooperative decentralized control for consensus in a topological network of agents with uncertain nonlinear dynamics. In: Proceedings of 2013 American Control Conference, pp. 1320–1325 (2013)
25. Zhang, J.L., Zhang, H.G., Feng, T.: Distributed optimal consensus control for nonlinear multiagent system with unknown dynamic. IEEE Trans. Neural Netw. Learn. Syst. **29**(8), 3339–3348 (2018)
26. Zhang, H.G., Jiang, H., Luo, Y.H., Xiao, G.Y.: Data-driven optimal consensus control for discrete-time multi-agent systems with unknown dynamics using reinforcement learning method. IEEE Trans. Ind. Electron. **64**(5), 4091–4100 (2017)
27. Abouheaf, M.I., Lewis, F.L., Vamvoudakis, K.G., Haesaert, S., Babuska, R.: Multi-agent discrete-time graphical games and reinforcement learning solutions. Automatica **50**(12), 3038–3053 (2014)
28. Li, J.N., Modares, H., Chai, T.Y., Lewis, F.L., Xie, L.H.: Off-policy reinforcement learning for synchronization in multiagent graphical games. IEEE Trans. Neural Netw. Learn. Syst. **28**(10), 2434–2445 (2017)
29. Sutton, R.S., Barto, A.G.: Reinforcement Learning: An Introduction. MIT Press, Cambridge (1998)
30. Murray, J.J., Cox, C.J., Lendaris, G.G., Saeks, R.: Adaptive dynamic programming. IEEE Trans. Syst. Man Cybern. Part C **32**(2), 140–153 (2002)
31. Chen, K.R., Wang, J.W., Zhang, Y., Liu, Z.: Leader-following consensus for a class of nonlinear strick-feedback multiagent systems with state time-delays. IEEE Trans. Syst. Man Cybern.: Syst. (2018). https://doi.org/10.1109/TSMC.2018.2813399
32. Modares, H., Nageshrao, S.P., Lopes, G.A.D., Babuška, R., Lewis, F.L.: Optimal model-free output synchronization of heterogeneous systems using off-policy reinforcement learning. Automatica **71**, 334–341 (2016)
33. Yang, Y.L., Modares, H., Wunsch, D.C., Yin, Y.X.: Leader-follower output synchronization of linear heterogeneous systems with active leader using reinforcement learning. IEEE Trans. Neural Netw. Learn. Syst. **29**(6), 2139–2153 (2018)
34. Zhang, H.G., Liang, H.J., Wang, Z.S., Feng, T.: Optimal output regulation for heterogeneous multiagent systems via adaptive dynamic programming. IEEE Trans. Neural Netw. Learn. Syst. **28**(1), 18–29 (2017)
35. Zuo, S., Song, Y.D., Lewis, F.L., Davoudi, A.: Optimal robust output containment of unknown heterogeneous multiagent system using off-policy reinforcement learning. IEEE Trans. Cybern. **48**(11), 3197–3207 (2018)
36. Modares, H., Lewis, F.L., Kang, W., Davoudi, A.: Optimal synchronization of heterogeneous nonlinear systems with unknown dynamics. IEEE Trans. Autom. Control **63**(1), 117–131 (2018)
37. Kiumarsi, B., Lewis, F.L.: Output synchronization of heterogeneous discrete-time systems: a model-free optimal approach. Automatica **84**, 86–94 (2017)
38. Luo, B., Liu, D.R., Wu, H.N., Wang, D., Lewis, F.L.: Policy gradient adaptive dynamic programming for data-based optimal control. IEEE Trans. Cybern. **47**(10), 3341–3354 (2017)

39. Chen, X., Xie, P.H., Xiong, Y.H., He, Y., Wu, M.: Two-phase iteration for value function approximation and hyperparameter optimization in Gaussian-kernel-based adaptive critic design. Math. Probl. Eng. (2015)
40. Wang, W., Chen, X.: Model-free optimal containment control of multi-agent systems based on actor-critic framework. Neurocomputing **314**, 242–250 (2018)
41. Zhao, D.B., Xia, Z.P., Wang, D.: Model-free optimal control for affine nonlinear systems with convergence analysis. IEEE Trans. Autom. Sci. Eng. **12**(4), 1461–1468 (2015)
42. Ari, E.O., Kocaoglan, E.: An SRWNN-based approach on developing a self-learning and self-evolving adaptive control system for motion platforms. Int. J. Control **89**(2), 380–396 (2016)
43. Kumar, R., Srivastava, S., Gupta, J.R.P.: Diagonal recurrent neural network based adaptive control of nonlinear dynamical systems using Lyapunov stability criterion. ISA Trans. **67**, 407–427 (2017)
44. Khanesar, M.A., Oniz, Y., Kaynak, O., Gao, H.J.: Direct model reference adaptive fuzzy control of networked SISO nonlinear systems. IEEE/ASME Trans. Mechatron. **21**(1), 205–213 (2016)
45. Wang, N., Sun, Z., Yin, J.C., Zou, Z.J., Su, S.F.: Fuzzy unknown observer-based robust adaptive path following control of underactuated surface vehicles subject to multiple unknowns. Ocean Eng. **176**, 57–64 (2019)
46. Wang, N., Deng, Q., Xie, G.M., Pan, X.X.: Hybrid finite-time trajectory tracking control of a quadrotor. ISA Trans. **90**, 278–286 (2019)
47. Wang, N., Xie, G.M., Pan, X.X., Su, S.F.: Full-state regulation control of asymmetric underactuated surface vehicles. IEEE Trans. Ind. Electron. **66**(11), 8741–8750 (2019)
48. Wang, N., Su, S.F., Pan, X.X., Yu, X., Xie, G.M.: Yaw-guided trajectory tracking control of an asymmetric underactuated surface vehicle. IEEE Trans. Ind. Inform. **15**(6), 3502–3513 (2019)
49. Fu, H., Chen, X., Wang, W.: A model reference adaptive control with ADP-to-SMC strategy for unknown nonlinear systems. In: Proceedings of 2017 11th Asian Control Conference, pp. 1537–1542 (2018)
50. Zhang, H.G., Feng, T., Liang, H.J., Luo, Y.H.: LQR-based optimal distributed cooperative design for linear discrete-time multiagent systems. IEEE Trans. Neural Netw. Learn. Syst. **28**(3), 599–611 (2017)

Stochastic Consensus Control of Multi-agent Systems under General Noises and Delays

Xiaofeng Zong, Ji-Feng Zhang, and George Yin

Abstract This chapter is devoted to studying the consensus control of continuous-time multi-agents systems with general noises and delays. By using the stochastic analysis, matrix theory, and algebraic graph theory, conditions on the stochastic approximation-type protocol are obtained for mean square and almost sure weak and strong consensus under the general martingale noises, which is a distinctive feature of this work. For the delay-free case, the necessary and sufficient conditions are given for the mean square weak consensus, and sufficient conditions are given for the almost sure weak consensus. For the delay case, the precondition on the delay, control gain function, and the topology graph is presented. Under this precondition, the sufficient conditions and necessary conditions for stochastic weak and strong consensus are obtained.

Keywords Multi-agent system · Noise · Delay · Stochastic consensus

---

X. Zong (✉)
School of Automation, China University of Geosciences, Wuhan 430074, China

Hubei Key Laboratory of Advanced Control and Intelligent Automation
for Complex Systems, Wuhan 430074, China
e-mail: zongxf@cug.edu.cn

J.-F. Zhang
The Key Laboratory of Systems and Control, Institute of Systems Science, Academy of Mathematics and Systems Science, Chinese Academy of Sciences, Beijing 100190, China

School of Mathematical Sciences, University of Chinese Academy of Sciences,
Beijing 100149, China

G. Yin
Department of Mathematics, University of Connecticut, Storrs, CT 06269-1009, USA

M. Wu et al. (eds.), *Developments in Advanced Control and Intelligent Automation for Complex Systems*, Studies in Systems, Decision and Control 329,
https://doi.org/10.1007/978-3-030-62147-6_9

Basic Notation

$\mathbb{R}^n$	n-dimensional Euclidean space
$\mathbb{C}^n$	n-dimensional complex vector space
$\lvert\lambda\rvert$	modulus of the complex number λ (or absolute value if λ is real)
$\Re(\lambda)$	real part of the complex number λ
$\Im(\lambda)$	imaginary part of the complex number λ
A^T	transpose of the matrix or vector A
A^*	conjugate transpose of the matrix or vector A
$\lVert A\rVert$	Euclidean norm of the matrix or vector A
$\mathbf{1}_n$	n-dimensional column vector with all ones
$\eta_{N,i}$	N-dimensional column vector with the ith element being 1 and others being zero
I_N	N-dimensional identity matrix
$(\Omega, \mathscr{F}, \mathbb{P})$	a complete probability space with a filtration $\{\mathscr{F}_t\}_{t\geq 0}$ satisfying the usual conditions, namely, it is right continuous and increasing while $\mathscr{F}_0$ contains all $\mathbb{P}$-null sets
$\mathbb{E}X$	mathematical expectation of the random variable or vector X
$\langle M\rangle(t)$	quadratic variation of the continuous martingale $M(t)$
$C([-\tau, 0], \mathbb{R}^n)$	the space of all continuous $\mathbb{R}^n$-valued functions φ defined on $[-\tau, 0]$ with a norm $\lVert\varphi\rVert_C = \sup_{t\in[-\tau,0]} \lVert\varphi(t)\rVert$

1 Introduction

Due to the demands in distributed computing [1], sensor networks [2, 3], and statistical physics [4], distributed coordination control of multi-agent systems, involving a collection of decision-making components with limited processing capabilities, locally sensed information, and limited inter-component communications, has attracted more and more attention in recent years. One of the most fundamental problems in the distributed coordination control of multi-agent systems is consensus, that is, all agents seek to achieve a collective objective. The related reviews and progress reports were given in [5–9], to name just a few.

Delay is an important phenomenon and may lead to the limitations in information processing and transmission. Therefore, control of multi-agent systems with delays has become an attractive subject. For the continuous-time models, Olfati-Saber and Murray [10] gave a direct connection between the robustness margin to delays and the maximum eigenvalue of the network topology under the undirected networks. Cepeda-Gomez and Olgac [11] and Cepeda-Gomez [12] studied the exact delay bound for the high-order linear multi-agent systems. Zhu and Jiang [13] investigated the event-based control for the linear leader-following multi-agent systems with input delays between controller and actuator, and presented a necessary condition and two sufficient conditions for leader-following consensus. Abdessameud et al. [14] proposed a distributed control algorithm to study the coordinated control problem of under-actuated thrust-propelled vehicles with the varying communication delays.

There are also abundant results for the discrete-time models; see [15–18], and the references therein.

Note that most of the works above concentrated on deterministic models, which often fail to capture the essence of the system uncertainty existing in the complex environment. So multi-agent systems subject to stochasticity have received much attention. Stochastic approximation approach has been examined in many works [19–24] for the consensus problem of the discrete-time models. For continuous-time models with balanced graphs, the necessary and sufficient conditions were developed in Li and Zhang [25] for the mean square strong consensus. For the case of general digraphs (directed graphs), the mean square strong consensus conditions were studied in [26], and the almost sure strong consensus conditions were investigated in [27]. Our recent work [28] gave the necessary and sufficient conditions on the almost sure strong consensus under the general digraphs. Besides these, we also developed some necessary conditions and sufficient conditions for the mean square weak consensus and the almost sure weak consensus under the general digraphs. However, the noises are assumed to be the Gaussian white noises. This might be restrictive if the communication channels are subject to more complicated non-Gaussian environments. For example, for the case with general martingale noises, stochastic consensus (including mean square and almost sure weak and strong consensus) remains to be unclear.

In general, delays and communication noises are commonly encountered in many physical systems, engineering, and control, and they often produce the difficulty in analyzing the consensus problem. To date, only a few works have concentrated on the stochastic consensus of continuous-time multi-agent systems with additive noises and delays. Liu et al. [29] revealed some sufficient conditions with a delay bound for the mean square average-consensus under the balanced graphs. Our recent work [31] gave the necessary and sufficient conditions for the stochastic weak and strong consensus under the general digraphs. Zhang et al. [32] contributed the stochastic leader-following consensus under the measurement noises and delays. However, for the multi-agent systems with general noises and delays, the necessary and sufficient conditions for stochastic weak and strong consensus have not been established. This chapter is to fill in this gap.

Based on [28, 31], this chapter addresses the necessary conditions and sufficient conditions on the control gain function for stochastic weak and strong consensus under the general noises and delays. By the algebraic graph theory and the matrix theory, the consensus problem is converted to the asymptotic stability analysis of stochastic differential (or delay) equations driven by continuous martingale. The martingale describes more general random processes in the communication channels. Note also that the topology graph is assumed to be the general digraph, and the eigenvalues of the corresponding Laplacian matrix may not be real. These lead to the difficulty in examining the stability condition of the corresponding stochastic differential (or delay) equations. In this work, the variation of constants formula, the Law of the Iterated Logarithm for martingales, and the backward inference method are applied to overcome the difficulty.

For the delay-free case, by the semi-decoupled and backstepping methods, we first show that the sufficient conditions on the control gain function $\alpha(t)$ for the mean square weak consensus are $\int_0^\infty \alpha(t)dt = \infty$ and $\lim_{t\to\infty} \int_0^t \exp(-2\underline{\lambda} \int_s^t \alpha(u)du)$ $\alpha^2(s)\,\bar{\gamma}(s)ds = 0$, where $\underline{\lambda}$ is the minimum real part of the eigenvalues of the Laplacian matrix, and $\bar{\gamma}(t)$ is the maximum derivative of the martingale quadratic variation induced by the general noises. The necessary conditions are $\int_0^\infty \alpha(t)dt = \infty$ and $\lim_{t\to\infty} \int_0^t \exp(-2\bar{\lambda} \int_s^t \alpha(u)du)\ \alpha^2(s)\underline{\gamma}(s)ds = 0$, where $\bar{\lambda}$ is the maximum real parts of the eigenvalues of the Laplacian matrix, and $\underline{\gamma}(t)$ is the minimum derivative of the martingale quadratic variation induced by the general noises. Then, from the Law of the Iterated Logarithm for Continuous Martingales, it is proved that conditions $\int_0^\infty \alpha(t)dt = \infty$ and $\lim_{t\to\infty} \bar{\gamma}(t)\alpha(t)\log\int_0^t \alpha(s)ds = 0$ are sufficient and conditions $\int_0^\infty \alpha(t)dt = \infty$ under the general digraphs, and $\liminf_{t\to\infty} \underline{\gamma}(t)\alpha(t)\log\int_0^t \alpha(s)ds = 0$ are necessary under the undirected graphs for the almost sure weak consensus. Finally, by semi-martingale and martingale convergence theorems, we get that $\int_0^\infty \alpha(t)dt = \infty$ and $\int_0^\infty \bar{\gamma}(s)\alpha^2(s)ds < \infty$ are sufficient and $\int_0^\infty \alpha(t)dt = \infty$ and $\int_0^\infty \underline{\gamma}(s)\alpha^2(s)ds < \infty$ are necessary for the almost sure and mean square strong consensus. Especially, the almost sure strong consensus and the mean square strong consensus are equivalent if the martingale noises have the same quadratic variation, and under the undirected graphs, the almost sure weak consensus implies the mean square weak consensus.

For the case with the delay τ, we use the precondition that there exists a constant $t_0 > 0$ such that $\tau\alpha(t)\max_{2\le j\le N} \frac{|\lambda_j|^2}{Re(\lambda_j)} < 1$ for all $t > t_0$, which aims to guarantee the exponent decaying estimation of a complex differential resolvent function. Under the pre-condition and $\int_0^\infty \alpha(t)dt = \infty$, we use the comparison methods for the delay and delay-free cases and show that condition $\lim_{t\to\infty} \int_0^t e^{-\kappa_0 \int_s^t \alpha(u)du}\alpha^2(s)\bar{\gamma}(s)ds = 0$ is sufficient, and condition $\lim_{t\to\infty} \int_0^t e^{-2\bar{\lambda} \int_s^t \alpha(u)du}\alpha^2(s)\underline{\gamma}(s)ds = 0$ is necessary for the mean square weak consensus, where κ_0 is a constant depending on the eigenvalues of the Laplacian matrix; Then, it is shown that $\lim_{t\to\infty} \alpha(t)\bar{\gamma}(t)\log\int_0^t \alpha(s)ds = 0$ is sufficient, and $\lim_{t\to\infty} \alpha(t)\underline{\gamma}(t)\log\int_0^t \alpha(s)ds = 0$ is necessary under the undirected graphs for almost sure weak consensus. Finally, we get similar necessary and sufficient conditions for mean square and almost sure strong consensus as that of delay-free case. These results are based on the delay-free case.

The rest of this chapter is organized as follows. Section 2 serves as an introduction to the networked systems and consensus problems, and introduces an auxiliary lemma for later use. Section 3 gives the necessary and sufficient conditions on the control gain function for the mean square weak consensus, and the sufficient conditions for almost sure weak consensus for multi-agent systems with the additive noises. Section 4 furthers our investigation on the stochastic consensus of multi-agent systems with additive noises and delays. Section 5 gives the simulations to confirm the theoretical results. Section 6 concludes the work and discusses possible future research topics.

2 Problem Formulation and Preliminary

In this chapter, we consider a general digraph $\mathscr{G} = \{\mathscr{V}, \mathscr{E}, \mathscr{A}\}$, where $\mathscr{V} = \{1, 2, ..., N\}$ denotes the set of nodes with i representing the ith agent, $\mathscr{E}$ denotes the set of edges, and $\mathscr{A}=[a_{ij}]\in\mathbb{R}^{N\times N}$ is the adjacency matrix of $\mathscr{G}$ with element $a_{ij} = 1$ or 0 indicating whether or not there is an information flow from agent j to agent i directly. The Laplacian matrix of $\mathscr{G}$ is defined as $\mathscr{L} = \mathscr{D} - \mathscr{A}$, where $\mathscr{D} = \text{diag}(\text{deg}_1, ..., \text{deg}_N)$, and $\text{deg}_i = \sum_{j=1}^{N} a_{ij}$ is the degree of i. Then, $\mathscr{L}$ admits a zero eigenvalue, denoted by $\lambda_1(\mathscr{L})$. We also use N_i to denote the set of i's neighbors, that is, for $j \in N_i$, $a_{ij} = 1$.

Let $x_i(t) \in \mathbb{R}^n$ denote agent i's state at time t with the dynamics

$$\dot{x}_i(t) = u_i(t) \tag{1}$$

with $u_i(t) \in \mathbb{R}^n$ denoting the control input of the ith agent. Taking measurement noises and delays into account, we assume that the relative state measurement from agent i's neighbor j has the following form

$$z_{ji}(t) = x_j(t - \tau_{ji}) - x_i(t - \tau_{ji}) + \mathbf{1}_n\sigma_{ji}\xi_{ji}(t), \tag{2}$$

where $\tau_{ji} \geq 0$ is the delay, $\xi_{ji}(t) \in \mathbb{R}$ denotes the measurement noise, and $\sigma_{ji} \geq 0$ is the noise intensity function. To see the consensus conditions clearly, the channel dependent delays τ_{ji} are assumed to be equal, that is, $\tau_{ji} = \tau$. We also assume $\sigma_{ji} > 0$ for some j, i, and the measurement noises are independent martingale noises defined below.

Assumption 1 $\xi_{ji}(t) \in \mathbb{R}$ satisfies $\int_0^t \xi_{ji}(s)ds = W_{ji}(t), t \geq 0, j, i = 1, 2, \ldots, N$, where $\{W_{ji}(t), i, j = 1, 2, ..., N\}$ are independent continuous martingales with $W_{ji}(0) = 0$ on $(\Omega, \mathscr{F}, \mathbb{P})$. The quadratic variation of $W_{ji}(t)$ is $\langle W_{ji}\rangle(t) = \int_0^t \gamma_{ji}(s)ds$ with deterministic function $\gamma_{ji}(s) \geq 0$.

In the following, we let $\bar{\gamma}(t) = \max_{ji} \gamma_{ji}(t)$ and $\underline{\gamma}(t) = \min_{ji} \gamma_{ji}(t)$. Based on the measurement (2), we hope to find the following type control

$$u_i(t) = \alpha(t) \sum_{j=1}^{N} a_{ij} z_{ji}(t), \ i = 1, 2, ..., N, \tag{3}$$

such that the consensus of multi-agent system (1) can be solved in the sense of probability to be defined below. Here, $\alpha(t) \in C((0, \infty); [0, \infty))$ is the control gain function to be designed. So the control issues of multi-agent systems with martingale noises are actually to find the conditions on the control gain function $\alpha(t)$ such that the consensus can be solved.

Note that substituting (3) into (1) may produce N closed-loop delay systems, which are stochastic differential delay systems driven by the continuous martingale3 (see (5) below). Hence, we need to give the initial data $x(t) = \varphi(t), t \in [-\tau, 0]$, where we assume that $\varphi \in C([-\tau, 0]; \mathbb{R}^{nN})$. To proceed, we give some definitions on the stochastic consensus, including the mean square and the almost sure weak and strong consensus.

Definition 1 We say that the control (3) solves the mean square weak consensus if system (1) has the property that for any initial data $\varphi \in C([-\tau, 0], \mathbb{R}^{Nn})$ and all distinct $i, j \in \mathscr{V}$, $\lim_{t\to\infty} \mathbb{E}\|x_i(t) - x_j(t)\|^2 = 0$. If, in addition, there is a random vector $x^* \in \mathbb{R}^n$, such that $\mathbb{E}\|x^*\|^2 < \infty$ and $\lim_{t\to\infty} \mathbb{E}\|x_i(t) - x^*\|^2 = 0$, $i = 1, 2, \ldots, N$, then we say that the control (3) solves the mean square strong consensus.

Definition 2 We say that the control (3) solves the almost sure weak consensus if the system (1) has the property that for any initial data $\varphi \in C([-\tau, 0], \mathbb{R}^{Nn})$ and all distinct $i, j \in \mathscr{V}$, $\lim_{t\to\infty} \|x_i(t) - x_j(t)\| = 0$, a.s. If, in addition, there is a random vector $x^* \in \mathbb{R}^n$, such that $\mathbb{P}\{\|x^*\| < \infty\} = 1$ and $\lim_{t\to\infty} \|x_i(t) - x^*\| = 0$, a.s. $i = 1, 2, \ldots, N$, then we say that the control (3) solves the almost sure strong consensus.

Note that there is no direct relationship between the mean square convergence and almost sure convergence. So in this chapter we have to study the mean square consensus and almost sure consensus separately.

The stochastic strong consensus (mean square strong consensus and almost sure strong consensus) of the continuous-time multi-agent systems with additive noises was addressed under the following conditions (see [25, 29] and the references therein),

(A1) $\int_0^\infty \alpha(s)ds = \infty$;
(A2) $\int_0^\infty \alpha^2(s)ds < \infty$;
(A3) $\lim_{t\to\infty} \alpha(t) = 0$.

Call $\alpha(t)$ the stepsize. Then the above conditions are essentially the stepsize conditions in stochastic approximation; see [30] for the discrete-time counterpart. Recently, we also proposed some new conditions for stochastic weak consensus (mean square weak consensus and almost sure weak consensus) for the continuous-time systems in [28, 31]. The corresponding conditions are all based on the Gaussian white noises. Then, a nature question is: What are the consensus conditions if the Gaussian white noises become more general martingale noises in Assumption 1. In fact, different noises may yield different consensus control design. This will be discussed in detail in this work. We need the following auxiliary lemma, which can be founded in [28].

Lemma 1 *For the Laplacian matrix $\mathscr{L}$, we have the following assertions:*

1. *There exists a probability measure π such that $\pi^T \mathscr{L} = 0$.*
2. *There exists a matrix $\widetilde{Q} \in \mathbb{R}^{N\times(N-1)}$ such that the matrix $Q = (\frac{1}{\sqrt{N}}\mathbf{1}_N, \widetilde{Q}) \in \mathbb{R}^{N\times N}$ is nonsingular and*

$$Q^{-1} = \begin{pmatrix} \nu^T \\ \widehat{Q} \end{pmatrix}, \; Q^{-1}\mathscr{L}Q = \begin{pmatrix} 0 & 0 \\ 0 & \widehat{\mathscr{L}} \end{pmatrix}, \tag{4}$$

where $\overline{Q} \in \mathbb{R}^{(N-1)\times N}$, $\widetilde{\mathscr{L}} \in \mathbb{R}^{(N-1)\times(N-1)}$ *and* ν *is a left eigenvector of* $\mathscr{L}$ *such that* $\nu^T \mathscr{L} = 0$ *and* $\frac{1}{\sqrt{N}}\nu^T \mathbf{1}_N = 1$.

3. *The digraph* $\mathscr{G}$ *contains a spanning tree if and only if all the eigenvalues of* $\widetilde{\mathscr{L}}$ *have positive real parts. Moreover, if the digraph* $\mathscr{G}$ *contains a spanning tree, then the probability measure* π *is unique and* $\nu = \sqrt{N}\pi$.

3 Networks with Additive Noises

In this section, we assume that the delay vanishes ($\tau = 0$). The results of this section are important for us to examine the case with delays in the next section. We first study the mean square consensus for the delay-free case ($\tau = 0$).

3.1 Mean Square Weak Consensus

We need the following conditions:

(A4) $\lim_{t\to\infty} \int_0^t e^{-2\underline{\lambda}\int_s^t \alpha(u)du}\alpha^2(s)\bar{\gamma}(s)ds = 0$, where $\underline{\lambda} = \min_{2\le i\le N} Re(\lambda_i(\mathscr{L}))$.

(A4') $\lim_{t\to\infty} \int_0^t e^{-2\bar{\lambda}\int_s^t \alpha(u)du}\alpha^2(s)\underline{\gamma}(s)ds = 0$, where $\bar{\lambda} = \max_{2\le i\le N} Re(\lambda_i(\mathscr{L}))$.

Theorem 2 *Suppose Assumption 1 holds and* $\tau = 0$. *Then, the control* (3) *solves the mean square weak consensus if* $\mathscr{G}$ *contains a spanning tree and Conditions (A1) and (A4) hold, and only if* $\mathscr{G}$ *contains a spanning tree and Conditions (A1) and (A4') hold.*

Proof The proof is divided into the following three steps.

Step 1: Transform stochastic consensus problem into stochastic stability problem. Substituting (3) into (1) and using Assumption 1, we have

$$dx(t) = -\alpha(t)(\mathscr{L} \otimes I_n)x(t)dt + \alpha(t)\sum_{i,j=1}^{N} a_{ij}\sigma_{ji}(\eta_{N,i} \otimes \mathbf{1}_n)dW_{ji}(t). \quad (5)$$

Let $J_N = \frac{1}{\sqrt{N}}\mathbf{1}_N\nu^T$, where ν is the left eigenvector of Laplacian matrix $\mathscr{L}$ defined in Lemma 1. Bearing in mind that $\mathscr{L}\mathbf{1}_N = 0$ and $\nu^T\mathscr{L} = 0$, one can get that $(I_N - J_N)\mathscr{L} = \mathscr{L} = \mathscr{L}(I_N - J_N)$. Defining $\Delta(t) = [(I_N - J_N) \otimes I_n]x(t)$, we have

$$d\Delta(t) = -\alpha(t)(\mathscr{L} \otimes I_n)\Delta(t)dt + \alpha(t)\sum_{i,j=1}^{N} a_{ij}\sigma_{ji}((I_N - J_N)\eta_{N,i} \otimes \mathbf{1}_n)dW_{ji}(t). \quad (6)$$

Let $\widetilde{\Delta}(t) = (Q^{-1} \otimes I_n)\Delta(t) = [\widetilde{\Delta}_1^T(t), \ldots, \widetilde{\Delta}_N(t)^T]^T$ and $\overline{\Delta}(t) = [\widetilde{\Delta}_2^T(t), \ldots, \widetilde{\Delta}_N(t)^T]^T$, where Q^{-1} is given in Lemma 1. Then, we can easily obtain from the definition of Q^{-1} that $\widetilde{\Delta}_1(t) = (\nu^T \otimes I_n)\Delta(t) = (\nu^T(I_N - J_N) \otimes I_n)x(t) = 0$ and

$$d\overline{\Delta}(t) = -\alpha(t)(\widetilde{\mathscr{L}} \otimes I_n)\overline{\Delta}(t)dt + dM(t), \tag{7}$$

where $M(t) = \sum_{i,j=1}^N a_{ij}\sigma_{ji}(\bar{q}_i \otimes \mathbf{1}_n)\int_0^t \alpha(s)dW_{ji}(s)$, $\bar{q}_i = \overline{Q}(I_N - J_N)\eta_{N,i}$, and $\overline{Q}$ is defined in Lemma 1. The matrix theorem tells us that there exists a complex invertible matrix R such that $R\widetilde{\mathscr{L}}R^{-1} = J$, Here, J is the Jordan normal form of $\widetilde{\mathscr{L}}$, i.e.,

$$J = diag(J_{\lambda_1,n_1}, J_{\lambda_2,n_2}, \ldots, J_{\lambda_l,n_l}), \ \sum_{k=1}^{l} n_k = N - 1,$$

where $\lambda_1, \lambda_2, \ldots, \lambda_l$ are all eigenvalues of $\widetilde{\mathscr{L}}$ and

$$J_{\lambda_k,n_k} = \begin{pmatrix} \lambda_k & 1 & \cdots & 0 & 0 \\ 0 & \lambda_k & \cdots & 0 & 0 \\ \vdots & \vdots & \ddots & \vdots & \vdots \\ 0 & 0 & \cdots & \lambda_k & 1 \\ 0 & 0 & \cdots & 0 & \lambda_k \end{pmatrix} \tag{8}$$

is the corresponding Jordan block of size n_k with eigenvalue λ_k. Letting $X(t) = (R \otimes I_n)\overline{\Delta}(t) = [X_1(t), \ldots, X_N(t)]^T$ with $X_j(t) \in \mathbb{R}^n$, we have from (7) that

$$dX(t) = -\alpha(t)(J \otimes I_n)X(t)dt + (R \otimes I_n)dM(t). \tag{9}$$

Considering the kth Jordan block and its corresponding component $\chi_k = [\chi_{k,1}, \ldots, \chi_{k,n_k}]^T$ and $R(k) = [R_{k,1}, \ldots, R_{k,n_k}]^T$, where $\chi_{k,j} = X_{k_j}$ and $R_{k,j} = R_{k_j}$ is k_jth row of R with $k_j = \sum_{l=1}^{k-1} n_l + j$, we have

$$d\chi_k(t) = -\alpha(t)(J_{\lambda_k,n_k})\chi_k(t)dt + (R(k) \otimes I_n)dM(t).$$

This implies that

$$d\chi_{k,n_k}(t) = -\alpha(t)\lambda_k\chi_{k,n_k}(t)dt + \mathbf{1}_n dM_{k,n_k}(t) \tag{10}$$

and for $j = 1, \ldots, n_k - 1$,

$$d\chi_{k,j}(t) = -\alpha(t)\lambda_k\chi_{k,j}(t)dt - \alpha(t)\chi_{k,j+1}(t)dt + \mathbf{1}_n dM_{k,j}(t), \tag{11}$$

where $M_{k,j}(t) = \sum_{i=1}^N r_{k_j,i}\sum_{l=1}^N a_{il}\sigma_{li}\int_0^t \alpha(s)dW_{li}(s)$, $r_{k_j,i} = R_{k_j}\bar{q}_i \in \mathbb{C}$, $j = 1, \ldots, n_k$. Note that $\Delta_i(t) = x_i - \sum_{k=1}^N \pi_k x_k(t) = \sum_{i=1}^N \pi_j(x_i(t) - x_j(t))$, $i = 1, \ldots, N$. Then, for $i \neq j$, $\lim_{t\to\infty}\mathbb{E}\|x_j(t) - x_i(t)\|^2 \leq 2\lim_{t\to\infty}\mathbb{E}\|x_j(t) -$

$\sum_{k=1}^{N}\pi_k x_k(t)\|^2 + 2\lim_{t\to\infty}\mathbb{E}\,\|x_i(t)-\sum_{k=1}^{N}\pi_k x_k(t)\|^2 = 2\lim_{t\to\infty}\mathbb{E}\|\Delta_j(t)\|^2 + 2\lim_{t\to\infty}\mathbb{E}\|\Delta_i(t)\|^2$. These tell us that the mean square weak consensus equals that $\lim_{t\to\infty}\mathbb{E}\|\chi_{k,j}(t)\|^2 = 0, k = 1,\ldots,l, j = 1,2,\ldots,n_k$ for any initial value $x(0)$.

Step 2: Prove the Sufficiency of Conditions (A1) and (A4). Resorting to a variation of constants formula, we can obtain from (10)

$$\chi_{k,n_k}(t) = e^{-\lambda_k\int_0^t \alpha(u)du}\chi_{k,n_k}(0) + \mathbf{1}_n\Xi_{k,n_k}(t), \tag{12}$$

where $\Xi_{k,n_k}(t) = \int_0^t e^{-\lambda_k\int_s^t \alpha(u)du}dM_{k,n_k}(s)$. Consider the quadratic variation process of the stochastic integral $\int_0^t \alpha(s)dW_{ji}(s)$, we have

$$\mathbb{E}|\int_0^t \alpha(s)dW_{ji}(s)|^2 = \mathbb{E}\int_0^t \alpha^2(s)d\langle W_{ji}\rangle(s) = \int_0^t \alpha^2(s)\gamma_{ji}(s)ds.$$

This together with (12) yields

$$\begin{aligned}\mathbb{E}\|\chi_{k,n_k}(t)\|^2 &= e^{-2Re(\lambda_k)\int_0^t \alpha(u)du}\|\chi_{k,n_k}(0)\|^2 \\ &\quad + \int_0^t e^{-2Re(\lambda_k)\int_s^t \alpha(u)du}\alpha^2(s)C_1(s)ds,\end{aligned} \tag{13}$$

where $C_1(s) = n\sum_{i=1}^{N}|r_{k_{n_k},i}|^2\sum_{j=1}^{N}a_{ij}\sigma_{ji}^2\gamma_{ji}(s)$. Hence, one can easily obtain that under (A1) and (A4),

$$\lim_{t\to\infty}\mathbb{E}\|\chi_{k,n_k}(t)\|^2 = 0.$$

Now, we use backward inference method to prove $\lim_{t\to\infty}\mathbb{E}\|\chi_{k,j}(t)\|^2 = 0, k = 1,\ldots,l, j = 1,2,\ldots,n_k$. Supposing that $\lim_{t\to\infty}\mathbb{E}\|\chi_{k,j+1}(t)\|^2 = 0$ for some fixed $j < n_k$, we will prove $\lim_{t\to\infty}\mathbb{E}\|\chi_{k,j}(t)\|^2 = 0$. Using the variation of constants formula for (11), we obtain

$$\begin{aligned}\chi_{k,j}(t) &= e^{-\lambda_k\int_0^t \alpha(u)du}\chi_{k,j}(0) + \mathbf{1}_n\Xi_{k,j}(t) \\ &\quad - \int_0^t e^{-\lambda_k\int_s^t \alpha(u)du}\alpha(s)\chi_{k,j+1}(s)ds.\end{aligned} \tag{14}$$

Then, we have

$$\begin{aligned}\mathbb{E}\|\chi_{k,j}(t)\|^2 &\le 2e^{-2Re(\lambda_k)\int_0^t \alpha(u)du}\|\chi_{k,j}(0)\|^2 + \int_0^t e^{-2Re(\lambda_k)\int_s^t \alpha(u)du}\alpha^2(s)C_1(s)ds \\ &\quad +2\mathbb{E}\|\int_0^t e^{-\lambda_k\int_s^t \alpha(u)du}\alpha(s)\chi_{k,j+1}(s)ds\|^2.\end{aligned}$$

From Conditions (A1) and (A4), one can see that the first two terms tend to zero. So the remaining is to prove that the last term tends to zero. Fix k, j and write $\chi_{k,j+1}(s) = [y_1(s),\ldots,y_n(s)]^T \in \mathbb{C}^n$. Then, we have $\lim_{t\to\infty}\mathbb{E}|y_m(s)|^2 = 0, m = 1,\ldots,n$, and

$$
\begin{aligned}
&\mathbb{E}\|\int_0^t e^{-\lambda_k \int_s^t \alpha(u)du}\alpha(s)\chi_{k,j+1}(s)ds\|^2 \\
&= \mathbb{E}\sum_{m=1}^{n}|\int_0^t e^{-\lambda_k \int_s^t \alpha(u)du}\alpha(s)y_m(s)ds|^2 \\
&\leq \sum_{m=1}^{n}\mathbb{E}\Big(\int_0^t e^{-Re(\lambda_k)\int_s^t \alpha(u)du}\alpha(s)|y_m(s)|ds\Big)^2.
\end{aligned} \tag{15}
$$

Defining $\widetilde{X}_m(t) = \int_0^t e^{-Re(\lambda_k)\int_s^t \alpha(u)du}\alpha(s)|y_m(s)|ds$ and using integral Minkowski's inequality, we have

$$
\sqrt{\mathbb{E}(\widetilde{X}_m(t))^2} \leq \int_0^t e^{-Re(\lambda_k)\int_s^t \alpha(u)du}\alpha(s)\sqrt{\mathbb{E}|y_m(s)|^2}ds.
$$

If $\int_0^\infty e^{Re(\lambda_k)\int_0^s \alpha(u)du}\alpha(s)\sqrt{\mathbb{E}|y_m(s)|^2}ds < \infty$, then we obtain from Condition (A1) that

$$
\lim_{t\to\infty}\sqrt{\mathbb{E}(\widetilde{X}_m(t))^2} \leq \lim_{t\to\infty} e^{-Re(\lambda_k)\int_0^t \alpha(u)du}\int_0^t e^{Re(\lambda_k)\int_0^s \alpha(u)du}\alpha(s)\sqrt{\mathbb{E}|y_m(s)|^2}ds = 0.
$$

Bearing in mind that $\lim_{t\to\infty}\mathbb{E}|y_m(s)|^2 = 0$, if

$$
\int_0^\infty e^{-Re(\lambda_k)\int_0^s \alpha(u)du}\alpha(s)\sqrt{\mathbb{E}|y_m(s)|^2}ds = \infty,
$$

we still obtain from L'Hôpital's rule that

$$
\lim_{t\to\infty}\sqrt{\mathbb{E}(\widetilde{X}_m(t))^2} \leq \lim_{t\to\infty}\frac{\int_0^t e^{Re(\lambda_k)\int_0^s \alpha(u)du}\alpha(s)\sqrt{\mathbb{E}|y_m(s)|^2}ds}{e^{Re(\lambda_k)\int_0^t \alpha(u)du}} = \lim_{t\to\infty}\frac{\sqrt{\mathbb{E}|y_m(t)|^2}}{Re(\lambda_k)} = 0.
$$

Hence, we must have $\lim_{t\to\infty}\mathbb{E}|\widetilde{X}_m(t)|^2 = 0$. Now, we have showed that $\lim_{t\to\infty}\mathbb{E}\|\chi_{k,j}(t)\|^2 = 0$ for the fixed $j < n_k$. Repeating the above arguments yields

$$
\lim_{t\to\infty}\mathbb{E}\|\chi_{k,j}(t)\|^2 = 0
$$

for all $j = 1, \ldots, n_k$. Therefore, $\lim_{t\to\infty}\mathbb{E}\|\chi_{k,j}(t)\|^2 = 0$ for all $k = 1, \ldots, l$ and $j = 1, \ldots, n_k$. That is, the control (3) solves the mean square weak consensus and the proof of sufficiency is completed.

Step 3: Prove the Necessity of Conditions (A1) and (A4'). From (13), we have

$$
\begin{aligned}
\mathbb{E}\|\chi_{k,n_k}(t)\|^2 \geq\ & e^{-2Re(\lambda_k)\int_0^t \alpha(u)du}\|\chi_{k,n_k}(0)\|^2 \\
&+C_2\int_0^t e^{-2Re(\lambda_k)\int_s^t \alpha(u)du}\alpha^2(s)\underline{\gamma}(s)ds,
\end{aligned} \tag{16}
$$

where $C_2 = n\sum_{i=1}^{N}|r_{k_{n_k},i}|^2\sum_{j=1}^{N}a_{ij}\sigma_{ji}^2$. Note that the mean square weak consensus implies $\lim_{t\to\infty}\mathbb{E}\|\chi_{k,n_k}(t)\|^2 = 0, k = 1,\ldots,N$. Then, by (16), we obtain that Conditions (A1) and (A4') hold and $Re(\lambda_k) > 0, k = 1,\ldots,N$. By Lemma 1, the digraph $\mathscr{G}$ contains a spanning tree, and the proof is completed. □

Note that Conditions (A4) and (A4') may be difficult to verify. In fact, the two conditions can be guaranteed by the following conditions, respectively.

(A3') $\lim_{t\to\infty}\alpha(t)\bar{\gamma}(t) = 0$;
(A3") $\lim_{t\to\infty}\alpha(t)\underline{\gamma}(t) = 0$;

Corollary 1 *Suppose that Assumption 1 holds and $\tau = 0$. Then, the control* (3) *solves mean square weak consensus if $\mathscr{G}$ contains a spanning tree, and Conditions (A1) and (A3') hold. Especially, if $\alpha(t)\underline{\gamma}(t)$ is decreasing, then the control* (3) *solves mean square weak consensus if and only if $\mathscr{G}$ contains a spanning tree, and Conditions (A1) and (A3') hold.*

Proof By Conditions (A1), (A3'), and L'Hôpital's rule, we have

$$\begin{aligned}
&\lim_{t\to\infty}\int_0^t \exp\Big\{-\lambda\int_s^t\alpha(u)du\Big\}\alpha^2(s)\bar{\gamma}(s)ds\\
&=\lim_{t\to\infty}\frac{\int_0^t\exp\Big\{\lambda\int_0^s\alpha(u)du\Big\}\alpha^2(s)\bar{\gamma}(s)ds}{\exp\Big\{\lambda\int_0^t\alpha(u)du\Big\}}\\
&=\frac{1}{\lambda}\lim_{t\to\infty}\alpha(t)\bar{\gamma}(t)=0,\ \forall\,\lambda>0.
\end{aligned}$$

Hence, Condition (A3') under Condition (A1) implies Condition (A4). By Theorem 2, the desired mean square weak consensus follows.

Assume that $\alpha(t)\underline{\gamma}(t)$ is a decreasing function. Then, the necessity of the existence of spanning tree of digraph $\mathscr{G}$ and Condition (A1) can be obtained in Theorem 2. Note that $\alpha(t)\underline{\gamma}(t)$ is decreasing. Then, for any $\lambda > 0$,

$$\begin{aligned}
\int_0^t e^{-2\lambda\int_s^t\alpha(u)du}\alpha^2(s)\underline{\gamma}(s)ds &\geq \alpha(t)\underline{\gamma}(t)e^{-2\lambda\int_0^t\alpha(u)du}\int_0^t e^{2\lambda\int_0^s\alpha(u)du}\alpha(s)ds\\
&=\frac{\alpha(t)\underline{\gamma}(t)}{2\lambda}(1-e^{-2\lambda\int_0^t\alpha(u)du}).
\end{aligned}$$

This implies $\alpha(t)\underline{\gamma}(t) \leq 2\lambda\int_0^t e^{-2\lambda\int_s^t\alpha(u)du}\alpha^2(s)ds(1-e^{-2\lambda\int_0^t\alpha(u)du})^{-1}$, which tends to zero as $t\to\infty$. Hence, Condition (A4') under Condition (A1) also guarantees Condition (A3"), and thus, the necessity follows. □

Remark 1 If there is a leader in this multi-agent systems, then the mean square weak consensus implies the mean square leader-following consensus. Hence, Theorem 2 also produces the necessary conditions and sufficient conditions for the mean square leader-following consensus. Wang et al. [34] considered the containment control for the multi-leaders case and showed that Conditions (A1) and (A3) can guarantee containment control. When the multi-leaders case becomes the one-leader case, Theorem 2 gives the weaker conditions for the mean square leader-following consensus than that in [34].

3.2 Almost Sure Weak Consensus

For the almost sure weak consensus, we introduce the following conditions.

(A5) $\lim_{t\to\infty} \alpha(t)\bar{\gamma}(t) \log \int_0^t \alpha(s)ds = 0$;
(A5') $\liminf_{t\to\infty} \alpha(t)\underline{\gamma}(t) \log \int_0^t \alpha(s)ds = 0$.

Theorem 3 *Suppose that Assumption 1 holds and $\tau = 0$. Then, the control* (3) *solves the almost sure weak consensus if $\mathcal{G}$ contains a spanning tree and Conditions (A1) and (A5) hold, and only if $\mathcal{G}$ contains a spanning tree and Condition (A1).*

Proof From (10) and (11), we know that the almost sure weak consensus is equivalent to $\lim_{t\to\infty} \|\chi_{k,j}(t)\| = 0$, a.s., for all $k = 1, \ldots, l$, $j = 1, 2, \ldots, n_k$ and any initial value $x(0)$. So in the following, we examine the necessary conditions and sufficient conditions for almost sure weak consensus from the conditions for the almost sure asymptotic stability of (10) and (11).

First, letting k be fixed, we show $\lim_{t\to\infty} \|\chi_{k,n_k}(t)\| = 0$ a.s. under Conditions (A1), (A5), and $Re(\lambda_k) > 0$. One can easily see that $\lim_{t\to\infty} e^{-\lambda_1 \int_0^t \alpha(s)ds} = 0$ under (A1) and $Re(\lambda_k) > 0$. Then, in order to obtain $\lim_{t\to\infty} \|\chi_{k,n_k}(t)\| = 0$, a.s., we need to show $\lim_{t\to\infty} \Xi_{k,n_k}(t) = 0$, a.s. Let

$$M_k^1(t) = \sum_{m=1}^{N} Re(r_{k_{n_k},m}) \sum_{j=1}^{N} a_{mj}\sigma_{jm} \int_0^t \alpha(s)dW_{jm}(s)$$

and

$$M_k^2(t) = \sum_{m=1}^{N} Im(r_{k_{n_k},m}) \sum_{j=1}^{N} a_{mj}\sigma_{jm} \int_0^t \alpha(s)dW_{jm}(s).$$

Then, $M_{k,n_k}(t) = M_k^1(t) + iM_k^2(t)$, where $i^2 = -1$. Defining

$$\bar{M}_k(t) = \int_0^t e^{\lambda_k \int_0^s \alpha(u)du} dM_{k,n_k}(s),$$

we have

$$\bar{M}_k(t) = \int_0^t e^{-Re(\lambda_k) \int_0^s \alpha(u)du} \cos(p_k(s))dM_k^1(s)$$

$$
\begin{aligned}
&-\int_0^t e^{-Re(\lambda_k)\int_0^s \alpha(u)du}\sin(p_k(s))dM_k^2(s)) \\
&+i\int_0^t e^{-Re(\lambda_k)\int_0^s \alpha(u)du}\sin(p_k(s))dM_k^1(s) \\
&+i\int_0^t e^{-Re(\lambda_k)\int_0^s \alpha(u)du}\cos(p_k(s))dM_k^2(s) \\
&=: \bar{m}_{k1}(t) - \bar{m}_{k2}(t) + i\bar{m}_{k3}(t) + i\bar{m}_{k4}(t),
\end{aligned} \tag{17}
$$

where $p_k(s) = -Im(\lambda_k)\int_0^s \alpha(u)du$. Note that $M_k^1(t)$ and $M_k^2(t)$ are real-valued continuous martingales. Hence, $\{\bar{m}_{kj}(t)\}_{j=1}^4$ are real-valued continuous martingales. By the definition of $\Xi_{k,n_k}(t)$ defined in (12), we need to prove that

$$
z_{kj}(t) = e^{-Re(\lambda_k)\int_0^t \alpha(u)du}\cos(p_k(t))\bar{m}_{kj}(t)
$$

and

$$
\bar{z}_{kj}(t) = e^{-Re(\lambda_k)\int_0^t \alpha(u)du}\sin(p_k(t))\bar{m}_{kj}(t), \;\; j = 1, 2, 3, 4,
$$

tend to zero almost surely as $t \to \infty$. We only show $\lim_{t\to\infty} z_{k1}(t) = 0$ a.s. and the others can be proved similarly. For the real-valued continuous martingale $\bar{m}_{11}(t)$, we either have that

$$
\lim_{t\to\infty}\langle \bar{m}_{k1}\rangle(t) = \int_0^\infty e^{2Re(\lambda_k)\int_0^s \alpha(u)du}\cos^2(p_k(s))\alpha^2(s)C_3(s)ds < \infty, \tag{18}
$$

or

$$
\lim_{t\to\infty}\langle \bar{m}_{k1}\rangle(t) = \infty, \tag{19}
$$

where $C_3 = \sum_{m=1}^N Re^2(r_{k_{n_k},m})\sum_{j=0}^N a_{mj}\sigma_{jm}^2\gamma_{ji}(s)$. In case (18), the martingale convergence theorem [33, Proposition 1.8, p. 183] tells us that $M_1(t)$ must converge almost surely to a finite limit. Noting Condition (A1) holds and

$$
z_{k1}(t) = e^{-Re(\lambda_k)\int_0^t \alpha(u)du}\cos(p_k(t))\bar{m}_{1l}(t),
$$

we can immediately obtain $\lim_{t\to\infty} z_{k1}(t) = 0$, a.s. In case (19), resorting to the Law of the Iterated Logarithm for Martingales (see [33, p. 186]), we have

$$
\limsup_{t\to\infty}\frac{|\bar{m}_{k1}(t)|}{\sqrt{2\langle \bar{m}_{k1}\rangle(t)\log\log(\langle \bar{m}_{k1}\rangle(t))}} = 1, a.s. \tag{20}
$$

This implies that for almost all $\omega \in \Omega$, there is a finite $T(\omega) > 0$ such that for all $t > T(\omega)$, $\log\log(\langle M_1\rangle(t)) > 0$ and

$$
|z_{kj}(t)| \le 2\sqrt{e^{-2Re(\lambda_k)\int_0^t \alpha(s)ds}\langle \bar{m}_{k1}\rangle(t)\log\log(\langle \bar{m}_{k1}\rangle(t))}, a.s. \tag{21}
$$

Let $\bar{C}(t) = \int_0^t e^{-2Re(\lambda_k)\int_s^t \alpha(u)du} \cos^2(p_k(s))\alpha^2(s)\bar{\gamma}(s)ds \log \int_0^t \alpha(s)ds$. Then, applying L'Hôpital's rule and Condition (A5) yields

$$\begin{aligned}
\lim_{t\to\infty} \bar{C}(t) &= \lim_{t\to\infty} \frac{\int_0^t e^{2Re(\lambda_k)\int_0^s \alpha(u)du} \cos^2(p_k(s))\alpha^2(s)\bar{\gamma}(s)ds}{e^{2Re(\lambda_k)\int_0^t \alpha(u)du} \log^{-1} \int_0^t \alpha(s)ds} \\
&= \lim_{t\to\infty} \frac{\alpha(t)\bar{\gamma}(t) \log \int_0^t \alpha(u)du \cos^2(p_k(t))}{2Re(\lambda_k) - \frac{1}{\int_0^t \alpha(u)du \log \int_0^t \alpha(s)ds}} = 0. \qquad (22)
\end{aligned}$$

Hence, for any $\epsilon \in (0, 1)$ satisfying $p_1\epsilon < 1$, there exists $T(\epsilon) > 0$ such that $\int_0^t \alpha(s)ds > e$ and $\int_0^t e^{2\lambda_k \int_0^s \alpha(u)du}\alpha^2(s)\bar{\gamma}(s)ds \le \epsilon e^{2\lambda_k \int_0^t \alpha(u)du}$ for all $t > T(\epsilon)$. These also gives

$$\log\log(p_1 \int_0^t e^{2\lambda_k \int_0^s \alpha(u)du}\alpha^2(s)\bar{\gamma}(s)ds) \le \log(2\lambda_k) + \log \int_0^t \alpha(u)du.$$

This together with (21) produces that for $t > \max\{T(\omega), T(\epsilon)\}$,

$$|z_{k1}(t)|^2 \le 4p_1 \int_0^t e^{-2\lambda_k \int_s^t \alpha(u)du}\alpha^2(s)\bar{\gamma}(s)ds[\log(2\lambda_k) + \log \int_0^t \alpha(s)ds].$$

Thus, from (22), we have $\lim_{t\to\infty} \|z_{k1}(t)\| = 0$, a.s. Similarly, $\lim_{t\to\infty} \|z_{kj}(t)\| = 0$, a.s. and $\lim_{t\to\infty} \|\bar{z}_{kj}(t)\| = 0$, a.s., $j = 1, 2, 3, 4$. In view of the definition of $\Xi_{k,n_k}(t)$, we obtain $\lim_{t\to\infty} \Xi_{k,n_k}(t) = 0$ a.s.. Therefore, $\lim_{t\to\infty} \|\chi_{k,n_k}(t)\| = 0$, a.s.

Now, we use backward inference method to prove $\lim_{t\to\infty} \|\chi_{k,j}(t)\| = 0$, a.s. for all $j = 1, \ldots, n_k$. Assume we have obtained that $\lim_{t\to\infty} \|\chi_{k,j+1}(t)\| = 0$, a.s. for certain $j < n_k$. We prove that $\lim_{t\to\infty} \|\chi_{k,j}(t)\| = 0$, a.s., where $\chi_{k,j}(t)$ is defined in (11). By the same estimation as $\chi_{k,n_k}(t)$ above, we can get $\lim_{t\to\infty} e^{-\lambda_k \int_0^t \alpha(u)du}$ $\chi_{k,j}(0) = 0$ and $\lim_{t\to\infty} \|\Xi_{k,j}(t)\| = 0$, a.s. Hence, the remaining is to show

$$\lim_{t\to\infty} \int_0^t e^{-\lambda_k \int_s^t \alpha(u)du}\alpha(s)\chi_{k,j+1}(s)ds = 0 \; a.s. \qquad (23)$$

Let k, j be fixed and still write $\chi_{k,j+1}(s) = [y_1(s), \ldots, y_n(s)]^T \in \mathbb{C}^n$. Then, we know that $\lim_{t\to\infty} |y_m(s)| = 0$, m=$1, \ldots, n$. Let $G(t)$=$\int_0^t e^{-Re(\lambda_k)\int_0^s \alpha(u)du}\alpha(s)|y_m(s)|ds$. Then,

$$\begin{aligned}
A_m(t) &:= |\int_0^t e^{-\lambda_k \int_s^t \alpha(u)du}\alpha(s)y_m(s)ds| \\
&\le \int_0^t e^{-Re(\lambda_k)\int_s^t \alpha(u)du}\alpha(s)|y_m(s)|ds \\
&= e^{-Re(\lambda_k)\int_0^t \alpha(u)du}G(t).
\end{aligned}$$

Note that we have either

$$\int_0^\infty e^{-Re(\lambda_k)\int_0^s \alpha(u)du}\alpha(s)|y_m(s)|ds < \infty, \tag{24}$$

or

$$\int_0^\infty e^{-Re(\lambda_k)\int_0^s \alpha(u)du}\alpha(s)|y_m(s)|ds = \infty, \tag{25}$$

since $G(t)$ is nondecreasing. In case (24), we get from Condition (A1)

$$\lim_{t\to\infty} A_m(t) \le \lim_{t\to\infty} e^{-Re(\lambda_k)\int_0^t \alpha(u)du}\int_0^\infty e^{Re(\lambda_k)\int_0^s \alpha(u)du}\alpha(s)|y_m(s)|ds = 0.$$

In case (25) by L'Hôpital's rule, we still have

$$\lim_{t\to\infty} A_m(t) = \lim_{t\to\infty} \frac{\int_0^t e^{-Re(\lambda_k)\int_0^s \alpha(u)du}\alpha(s)|y_m(s)|ds}{e^{Re(\lambda_m)\int_0^t \alpha(u)du}} = \frac{1}{Re(\lambda_m)}\lim_{t\to\infty}|y_m(t)| = 0, \ a.s.$$

Hence, $\lim_{t\to\infty}\|\chi_{k,j}(t)\| = 0$, a.s. By repeating the processes above for $j = 1, \ldots, n_k$, we can obtain $\lim_{t\to\infty}\|\chi_{k,j}(t)\| = 0$, a.s. for all $j = 1, \ldots, n_k$. Therefore, for $k = 1, 2, \ldots, l$, we have $\lim_{t\to\infty}\|\Delta(t)\| = 0$, a.s. Thus, the almost sure weak consensus is solved.

Note that the first term on the right side of (12) is deterministic and convergent for each $\chi_{k,n_k}(0)$. Then, $\lim_{t\to\infty}\|\chi_{k,n_k}(t)\| = 0$, a.s. gives $\lim_{t\to\infty} e^{-\lambda_k\int_0^t c_1(u)du} = 0$, which also implies $Re(\lambda_k) > 0$ and $\int_0^\infty \alpha(s)ds = \infty$. The proof is complete. □

Theorem 3 gives the necessity of the existence of spanning tree and Condition (A1) for almost sure weak consensus, but it does not tell us the necessity of Condition (A5). In fact, we have no idea on how to prove it under the general digraph. However, if all the eigenvalues of the Laplacian matrix $\mathscr{L}$ are real, we can have a necessary Condition (A5'). So we make the following assumption.

Assumption 4 All the eigenvalues of the Laplacian matrix $\mathscr{L}$ are real.

Theorem 5 *Suppose that Assumptions 1 and 4 hold and $\tau = 0$. Then, the control* (3) *solves the almost sure weak consensus only if $\mathscr{G}$ contains a spanning tree and Conditions (A1) and (A5') hold.*

Proof Note that the necessity of the existence of spanning tree and Condition (A1) is proved above. Hence, the remaining is to prove necessity of Condition (A5') for almost sure stability under (A1) and $\lambda_k > 0$, $k = 1, \ldots, l$.

Since the Laplacian matrix $\mathscr{L}$ admits the real eigenvalues, the Jordan matrix decomposition in producing (9) can be obtained by a real invertible matrix R such that $R^{-1}DR = J_D$. In view of this, the coefficients in (10) and (11) are real. Note that the almost sure weak consensus gives that for any initial value $\chi(0)$, $\lim_{t\to\infty}\|\chi_{k,n_k}(t)\| = 0$, a.s., $k = 1, 2, \ldots, l$. This together with (12) implies that

$$\lim_{t\to\infty}\|\Xi_{k,n_k}(t)\| = 0, a.s. \tag{26}$$

Let $\hat{M}_{k,j}(t) = \int_0^t e^{-\lambda_k \int_0^s \alpha(u)du} dM_{k,j}(s)$, $j = 1, \ldots, n_k$. Then, $\Xi_{k,j}(t) = e^{-\lambda_k \int_0^t \alpha(u)du} \hat{M}_{k,j}$, and

$$\lim_{t\to\infty} \langle \hat{M}_{k,j}\rangle(t) = \int_0^t e^{2\lambda_k \int_0^s c_1(u)du} \alpha^2(s) C_{k,j}(s)ds, \; j = 1, 2, \ldots, n_k, \tag{27}$$

where $\bar{C}_{k,j}(s) = n\sum_{i=1}^N |r_{k_j,i}|^2 \sum_{j=1}^N a_{ij}\sigma_{ji}^2 \gamma_{ji}(s)$. Let $C_{k,j} = n\sum_{i=1}^N |r_{k_j,i}|^2 \sum_{j=1}^N a_{ij}\sigma_{ji}^2$. Note that $\sigma_{ji} > 0$ for some j, i, and R is invertible. Hence, there must exist k, j such that $C_{k,j} > 0, 0 < j \le n_k$. Let k be fixed. Without loss of generality, we assume that $C_{k,n_k} > 0$. We use the proof by contradiction to see that (26) implies

$$\liminf_{t\to\infty} \alpha(t)\underline{\gamma}(t) \log \int_0^t \alpha(s)ds = 0.$$

Otherwise, one must have

$$\liminf_{t\to\infty} \alpha(t)\underline{\gamma}(t) \log \int_0^t \alpha(s)ds \ne 0,$$

which implies that there is $\underline{c} > 0$ such that

$$\liminf_{t\to\infty} \alpha(t)\underline{\gamma}(t) \log \int_0^t \alpha(s)ds > \underline{c}, \tag{28}$$

Thus, for some $T > 0$,

$$\alpha(t)\underline{\gamma}(t) \log \int_0^t \alpha(s)ds > \frac{\underline{c}}{2}, \; t > T, \tag{29}$$

and for $t > T$,

$$\int_0^t e^{2\lambda_k \int_0^s \alpha(u)du} \alpha^2(s)\underline{\gamma}(s)ds \ge \frac{\underline{c}}{2} \frac{\int_T^t e^{2\lambda_k \int_0^s \alpha(u)du}\alpha(s)ds}{\log \int_0^t \alpha(u)du}. \tag{30}$$

From (27), we obtain $\lim_{t\to\infty}\langle \hat{M}_{k,n_k}\rangle(t) = \infty$, which facilitates us to use the Law of the Iterated Logarithm for Martingales. Hence, we have $\limsup_{t\to\infty} \frac{\|\Xi_{k,n_k}(t)\|^2}{\Xi_1(t)} = 1$ a.s., where $\Xi_1(t) = 2e^{-2\lambda_k \int_0^t \alpha(s)ds}\langle \hat{M}_{k,n_k}\rangle(t) \log\log \langle \hat{M}_{k,n_k}\rangle(t)$. This implies that if $\lim_{t\to\infty} \|\Xi_{k,n_k}(t)\| = 0$ a.s., then $\liminf_{t\to\infty} \Xi_1(t) = 0$. Note that (30) implies that

$$\liminf_{t\to\infty} \frac{\log\log \int_0^t e^{2\lambda_k \int_0^s \alpha(u)du}\alpha^2(s)\underline{\gamma}(s)ds}{\log \int_0^t \alpha(u)du} \ge 1 \tag{31}$$

and

$$e^{-2\lambda_k \int_0^t \alpha(s)ds} \int_0^t e^{2\lambda_k \int_0^s \alpha(u)du} \alpha^2(s)\underline{\gamma}(s)ds \log \int_0^t c_(u)du$$
$$\geq \frac{\underline{c}}{2} \frac{\int_T^t e^{2\lambda_k \int_0^s \alpha(u)du} \alpha(s)ds}{e^{2\lambda_k \int_0^t \alpha(s)ds}}, t > T. \tag{32}$$

Then, by L'Hôpital's rule and Condition (A1), we get

$$\lim_{t\to\infty} \frac{\int_T^t e^{2\lambda_k \int_0^s \alpha(u)du} \alpha(s)ds}{e^{2\lambda_k \int_0^t \alpha(s)ds}} = \frac{1}{2\lambda_k}. \tag{33}$$

Combining the definition of $\Xi_1(t)$, (31), (32), and (33) yields

$$\liminf_{t\to\infty} \Xi_1(t) \geq C_{k,n_k}\underline{c} \liminf_{t\to\infty} \frac{\log\log \int_0^t e^{2\lambda_k \int_0^s \alpha(u)du} \alpha^2(s)\underline{\gamma}(s)ds}{\log \int_0^t \alpha(u)du}$$
$$\times \frac{\int_T^t e^{2\lambda \int_0^s \alpha(u)du} \alpha(s)ds}{e^{2\lambda \int_0^t \alpha(s)ds}} > \frac{C_{k,n_k}\underline{c}}{2\lambda_k} > 0,$$

which is a contradiction, and thus, Condition (A5') holds. If $C_{k,n_k} = 0$, we define $l_0 = \max\{n_k > j > 1, C_{k,j} > 0\}$. Then, we have from (14) that

$$\|\Xi_{k,l_0}(t)\| \leq \|\chi_{k,l_0}(t)\| + e^{-2\lambda_k \int_0^t c_1(u)du} \|\chi_{k,l_0}(0)\| + \|S_{k,l_0}(t)\|. \tag{34}$$

Note that $\lim_{t\to\infty} \|\chi_{k,l_0}(t)\| = 0$, $\lambda_k > 0$ and $\lim_{t\to\infty} \|S_{k,l_0}(t)\| = 0$. Hence, $\lim_{t\to\infty} \|\Xi_{k,l_0}(t)\|^2 = 0$. By the same methods as the case $C_{k,n_k} > 0$ above, we can obtain the necessity of Condition (A5') under $C_{k,l_0} > 0$. Therefore, the proof is complete. □

Remark 2 Here, one still requires the eigenvalues of the Laplacian matrix to be real such that we can obtain the necessary Conditions (A1) and (A5') for the almost sure weak consensus. It is of great interest to study the case with one non-zero real eigenvalue for the Laplacian matrix.

3.3 Mean Square and Almost Sure Strong Consensus

For the case with Gaussian white noises, previous works concluded that Conditions (A1) and (A2) are necessary and sufficient for almost sure and mean square strong consensus, and then the mean square strong consensus and almost sure strong consensus are equivalent [27, 28]. But for the case with general noises, Condition (A2) is changed to the following forms,

(A6) $\int_0^\infty \alpha(s)^2\bar{\gamma}(s)ds < \infty$;
(A6') $\int_0^\infty \alpha(s)^2\underline{\gamma}(s)ds < \infty$.

Theorem 6 *Suppose Assumption 1 holds and $\tau = 0$. Then, the control* (3) *solves the almost sure and mean square strong consensus if $\mathscr{G}$ contains a spanning tree and Conditions* (A1) *and* (A6) *hold, and only if $\mathscr{G}$ contains a spanning tree and Conditions* (A1) *and (A6') hold.*

Proof It is easy to see that Condition (A6) implies Condition (A4). That is, if the digraph $\mathscr{G}$ contains a spanning tree and Conditions *(A1)* and *(A5)* hold, then the mean square weak consensus follows. Note that $\widetilde{\mathscr{L}}$ is Hurwitz. Then, there exists a unique positive definite matrix P such that

$$-P\widetilde{\mathscr{L}} - \widetilde{\mathscr{L}}^T P = -I_{N-1}.$$

Let $V(\overline{\Delta}(t)) = \overline{\Delta}(t)^T (P \otimes I_n)\overline{\Delta}(t)$. Then, using Itô's formula to (7) yields

$$\begin{aligned} V(\overline{\Delta}(t)) &= V(\overline{\Delta}(0)) - \int_0^t \alpha(s)|\overline{\Delta}(s)|^2 ds \\ &\quad + \int_0^t \alpha^2(s)C_2(s)ds + 2\int_0^t \overline{\Delta}(s)^T (P \otimes I_n)dM(s), \end{aligned} \tag{35}$$

where $C_2(s) = n\sum_{i,j=1}^N a_{ij}\sigma_{ji}^2[\bar{q}_i^T P\bar{q}_i]\gamma_{ji}(s)$. Note that $\int_0^t \alpha(s)|\overline{\Delta}(s)|^2 ds$ is increasing and Condition (A6). By the semi-martingale convergence theorem, we have $\lim_{t\to\infty} V(\overline{\Delta}(t)) < \infty$ almost surely. Bear in mind that we obtained the mean square weak consensus and the mean square weak consensus implies the existence of a convergent subsequence $\{V(\overline{\Delta}(t_i))\}_{i=1}^\infty$. This together with the uniqueness of the limit implies $\lim_{t\to\infty} V(\overline{\Delta}(t)) = 0$ almost surely. That is, the almost sure weak consensus follows.

By the property of the matrix $\mathscr{L}$ and (5), we have

$$(\pi^T \otimes I_n)x(t) = (\pi^T \otimes I_n)x(0) + \bar{M}(t), \tag{36}$$

where where $\bar{M}(t) = \sum_{i,j=1}^N a_{ij}\sigma_{ji}\int_0^t \alpha(s)[\pi^T\eta_{N,i} \otimes \mathbf{1}_n]dW_{ji}(s)$. Then, Condition (A6) guarantees that

$$\mathbb{E}|\bar{M}(t)|^2 = n\sum_{i,j=1}^N a_{ij}^2\sigma_{ji}^2\pi_i^2\int_0^t \alpha^2(s)\gamma_{ji}(s)ds < \infty,\ \forall\, t > 0. \tag{37}$$

This together with the martingale convergence theorem yields that $(\pi^T \otimes I_n)x(t)$ converges to a random variable in the mean square and almost sure senses, denoted by x^*. By definition of $\Delta(t)$, we have $\Delta_i(t) = x_i - \sum_{k=1}^N \pi_k x_k(t), i = 1, \ldots, N$. Hence, for each i, $\lim_{t\to\infty}\mathbb{E}\|x_i(t) - x^*\|^2 \le 2\lim_{t\to\infty}\mathbb{E}\|x_i(t) - \sum_{k=1}^N \pi_k x_k(t)\|^2 + 2\lim_{t\to\infty}\mathbb{E}\|\sum_{k=1}^N \pi_k x_k(t) - x^*\|^2 = 0$, and $\lim_{t\to\infty}\|x_i(t) - x^*\| \le \lim_{t\to\infty}\|x_i(t) - \sum_{k=1}^N \pi_k x_k(t)\| + \lim_{t\to\infty}\|\sum_{k=1}^N \pi_k x_k(t) - x^*\| = 0$, almost surely.

Note that strong consensus implies weak consensus. So, both the mean square strong consensus and the almost sure strong consensus imply the necessity of the existence of spanning tree and Condition (A1). It can be seen that the mean square and almost sure strong consensus imply that $(\pi^T \otimes I_n)x(t)$ converges to a random variable in the mean square and almost sure sense, respectively. Note that $(\pi^T \otimes I_n)x(t)$ converges in the two senses if and only if the limit of the continuous local martingale $\bar{M}(t)$ exists, denoted by $\bar{M}(\infty)$. But this is also equivalent to $\lim_{t\to\infty}\langle \bar{M}\rangle(t) < \infty$ a.s.; see [33]. Note also that $\langle \bar{M}\rangle(t) = n\sum_{i,j=1}^{N} a_{ij}^2\sigma_{ji}^2\pi_i^2 \int_0^t \alpha^2(s)\gamma_{ji}(s)ds$ (see [33]). Hence, $(\pi^T \otimes I_n)x(t)$ converges almost surely if and only if $\langle \bar{M}\rangle(t) < \infty$, which implies (A6'). □

If the quadratic variations of martingales $W_{ji}(t)$ are the same, then we can obtain the necessary and sufficient conditions for the mean square and almost sure strong consensus directly from Theorem 6. We conclude this result as in the following corollary. In view of this, we can see that the equivalence between the mean square strong consensus and almost sure strong consensus still holds for the case with martingale noises.

Corollary 2 *Suppose that Assumption 1 holds, $\tau = 0$ and the martingales $\{W_{ji}(t)\}$ have the same quadratic variations ($\bar{\gamma}(s) = \underline{\gamma}(s)$). Then, the control* (3) *solves the almost sure and mean square strong consensus if and only if $\mathscr{G}$ contains a spanning tree and Conditions (A1) and and (A6) hold.*

In this section, we investigate the consensus conditions of multi-agent systems with martingale noises. It can be observed that the consensus conditions or the control gain designs depend not only on the topology $\mathscr{G}$, but also on the quadratic variations of martingales. Moreover, the decaying quadratic variations of martingales may weak the consensus conditions on the control gain function, which is also revealed in Sect. 5. The next section will take the delay into account and develop new conditions for the control gain design.

4 Networks with Additive Noises and Delays

Because of the delays, the corresponding closed-loop stochastic systems are stochastic differential delay equations driven by additive noises. To find more accurate consensus conditions, we need to examine the following linear scalar equation

$$\dot{\Upsilon}(t) = -\lambda\alpha(t)\Upsilon(t-\tau), t > 0, \tag{38}$$

$\Upsilon(t) = \varpi(t)$ for $t \in [-\tau, 0]$, where $\lambda > 0$, $\tau \geq 0$ and $\varpi \in C([-\tau, 0], \mathbb{C})$. The solution to (38) can have the form (see [35, p. 295])

$$\Upsilon(t) = \Pi(t, s)\Upsilon(s),\ \forall t \geq s \geq 0, \tag{39}$$

where $\Pi(t, s)$ is the differential resolvent function, satisfying $\Pi(t, t) = 1$ for $t > 0$, $\Pi(t, s) = 0$ for $t < s$ and

$$\frac{\partial}{\partial t}\Pi(t,s) = -\lambda\alpha(t)\Pi(t-\tau,s). \tag{40}$$

The following lemma [31] is to reveal the estimate of the differential resolvent function $\Pi(t,s)$ for $0 < s < t$.

Lemma 2 *If there is a constant $t_0 > 0$ such that $|\lambda|^2\tau\alpha(t) < Re(\lambda)$, $\forall\, t > t_0$, then the solution to* (40) *satisfies*

$$|\Pi(t,s)|^2 \le b(\lambda)e^{\kappa(\lambda)\tau\bar{c}_{t_0}}e^{-\kappa(\lambda)\int_s^t \alpha(u)du},\ t > s > t_0, \tag{41}$$

where $b(\lambda)$, $\bar{c}_{t_0} = \sup_{t\ge t_0}\alpha(t)$ and $\kappa(\lambda) < Re(\lambda) - |\lambda|^2\tau\bar{c}_{t_0}$ are two positive constants depending on λ.

4.1 Mean Square Weak Consensus

We first examine the mean square consensus. Let $\kappa(\lambda)$ be defined in Lemma 2 and introduce another conditions on the control gain:

(A3''') There exists a constant $t_0 > 0$ such that $\tau\alpha(t)\max_{2\le j\le N}\frac{|\lambda_j|^2}{Re(\lambda_j)} < 1, t > t_0$;

(A4'') $\lim_{t\to\infty}\int_0^t e^{-\kappa_0\int_s^t\alpha(u)du}\alpha^2(s)\bar{\gamma}(s)ds = 0$, where $\kappa_0 = \min_{1\le j\le N}\kappa(\lambda_j)$.

Theorem 7 *Suppose that Assumption 1 and Condition (A3''') holds. Then, the control* (3) *solves the mean square weak consensus if $\mathscr{G}$ contains a spanning tree and Conditions (A1) and (A4'') hold, and only if $\mathscr{G}$ contains a spanning tree and Condition (A4) holds under (A1).*

Proof Similar to (6), we have

$$d\Delta(t) = -\alpha(t)(\widetilde{\mathscr{L}}\otimes I_n)\Delta(t-\tau)dt + \alpha(t)\sum_{i,j=1}^{N}a_{ij}\sigma_{ji}((I_N - J_N)\eta_{N,i}\otimes\mathbf{1}_n)dW_{ji}(t),$$

Let us still define $\widetilde{\Delta}(t) = (Q^{-1}\otimes I_n)\Delta(t) = [\widetilde{\Delta}_1^T(t),\ldots,\widetilde{\Delta}_N(t)^T]^T$, and $\overline{\Delta}(t) = [\widetilde{\Delta}_2^T(t),\ldots,\widetilde{\Delta}_N(t)^T]^T$. Similarly,

$$d\overline{\Delta}(t) = -\alpha(t)(\widetilde{\mathscr{L}}\otimes I_n)\overline{\Delta}(t-\tau)dt + dM(t), \tag{42}$$

where $M(t)$ is defined in (7). Letting $Y(t) = (R\otimes I_n)\overline{\Delta}(t) = [Y_1(t),\ldots,Y_N(t)]^T$ with $Y_j(t)\in\mathbb{R}^n$, we have from (42) that

$$dY(t) = -\alpha(t)(J\otimes I_n)Y(t-\tau)dt + (R\otimes I_n)dM(t). \tag{43}$$

Let $\zeta_k = [\zeta_{k,1},\ldots,\zeta_{k,n_k}]^T$ and $R(k) = [R_{k,1},\ldots,R_{k,n_k}]^T$, where $\zeta_{k,j} = Y_{k_j}$, $R_{k,j} = R_{k_j}$ is k_jth row of R with $k_j = \sum_{l=1}^{k-1}n_l + j$. Considering the kth Jordan block of

(43), we have

$$d\zeta_k(t) = -\alpha(t)(J_{\lambda_k,n_k})\zeta_k(t-\tau)dt + (R(k)\otimes I_n)dM(t).$$

This together with the definition of J_{λ_k,n_k} in (8) implies that

$$d\zeta_{k,n_k}(t) = -\alpha(t)\lambda_k\zeta_{k,n_k}(t-\tau)dt + \mathbf{1}_n dM_{k,n_k}(t) \tag{44}$$

and for $j = 1, \ldots, n_k - 1$,

$$d\zeta_{k,j}(t) = -\alpha(t)\lambda_k\zeta_{k,j}(t-\tau)dt - \alpha(t)\zeta_{k,j+1}(t-\tau)dt + \mathbf{1}_n dM_{k,j}(t), \tag{45}$$

where $M_{k,j}(t)$ is defined in (11).

Then the mean square weak consensus equals that $\lim_{t\to\infty}\mathbb{E}\|\zeta_{k,j}(t)\|^2 = 0$, $k = 1, \ldots, l$, $j = 1, 2, \ldots, n_k$. We use $\Pi_k(t,s)$ to denote the differential resolvent function defined by (40), where λ is replaced with λ_k. Resorting to the variation of constants formula for (44), we obtain

$$\zeta_{k,n_k}(t) = \Pi_k(t,t_0)\zeta_{k,n_k}(t_0) + \mathbf{1}_n \Xi_{k,n_k}(t,t_0), \tag{46}$$

where $\Xi_{k,n_k}(t,t_0) = \int_{t_0}^t \Pi_k(t,s)dM_{k,n_k}(s)$. Hence,

$$\mathbb{E}\|\zeta_{k,n_k}(t)\|^2 = |\Pi_k(t,t_0)|^2\|\zeta_{k,n_k}(t_0)\|^2 + \int_{t_0}^t |\Pi_k(t,s)|^2\alpha^2(s)C_4(s)ds, \tag{47}$$

where $C_4 = n\sum_{i=1}^N |r_{k_{n_k},i}|^2 \sum_{j=0}^N a_{ij}\sigma_{ji}^2\gamma_{ji}(s)$. This together with Lemma 2 produces

$$\begin{aligned}\mathbb{E}\|\zeta_{k,n_k}(t)\|^2 \le\, & b(\lambda_k)e^{\kappa(\lambda_k)\tau\bar{c}}e^{-\kappa(\lambda_k)\int_0^t \alpha(u)du}\|\zeta_{k,n_k}(t_0)\|^2\\ & + C_1 b(\lambda_k)e^{\kappa(\lambda_k)\tau\bar{c}}\int_{t_0}^t e^{-\kappa(\lambda_k)\int_s^t \alpha(u)du}\alpha^2(s)ds,\end{aligned} \tag{48}$$

Then, from Conditions (A1) and (A4"), we have $\lim_{t\to\infty}\mathbb{E}\|\zeta_{k,n_k}(t)\|^2 = 0$. Now, we assume that $\lim_{t\to\infty}\mathbb{E}\|\zeta_{k,j+1}(t)\|^2 = 0$ for some fixed $j < n_k$, and aim to show $\lim_{t\to\infty}\mathbb{E}\|\zeta_{k,j}(t)\|^2 = 0$. Applying the variation of constants formula for (45), we obtain

$$\zeta_{k,j}(t) = \Pi_k(t,t_0)\zeta_{k,j}(t_0) + \mathbf{1}_n\Xi_{k,j}(t,t_0) - \int_{t_0}^t \Pi_k(t,s)\alpha(s)\zeta_{k,j+1}(s)ds. \tag{49}$$

Hence, taking mean square expectations, we have

$$\mathbb{E}\|\zeta_{k,j}(t)\|^2 \le 2|\Pi_k(t,t_0)|^2\mathbb{E}\|\zeta_{k,j}(t_0)\|^2 + C_1\int_{t_0}^t |\Pi_k(t,s)|^2\alpha^2(s)ds$$

$$+2\mathbb{E}\|\int_{t_0}^{t}\Pi_k(t,s)\alpha(s)\zeta_{k,j+1}(s)ds\|^2.$$

Conditions (A1) and (A4”) imply that the first two terms converge to zero. Then, we need to prove that the last term tends to zero. Similar to the delay-free case, we fix k, j and write $\zeta_{k,j+1}(s) = [y_1(s), \ldots, y_n(s)]^T \in \mathbb{C}^n$. Then, $\lim_{t\to\infty}\mathbb{E}|y_m(s)|^2 = 0$, $m = 1, \ldots, n$, and

$$\begin{aligned}
&\mathbb{E}\|\int_{t_0}^{t}\Pi_k(t,s)\alpha(s)\zeta_{k,j+1}(s)ds\|^2\\
&=\mathbb{E}\sum_{m=1}^{n}|\int_{t_0}^{t}\Pi_k(t,s)\alpha(s)y_m(s)ds|^2\\
&\le b(\lambda_k)e^{\kappa(\lambda_k)\tau\bar{c}}\sum_{m=1}^{n}\mathbb{E}\Big(\int_{t_0}^{t}e^{-0.5\kappa(\lambda_k)\int_s^t\alpha(u)du}\alpha(s)|y_m(s)|ds\Big)^2
\end{aligned}$$

Let $Y_m(t) = \int_0^t e^{-0.5\kappa(\lambda_k)\int_s^t\alpha(u)du}\alpha(s)|y_m(s)|ds$. By the similar skills used in the proof of Theorem 2, we have $\lim_{t\to\infty}\mathbb{E}|Y_m(t)|^2 = 0$, and then $\lim_{t\to\infty}\mathbb{E}\|\zeta_{k,j}(t)\|^2 = 0$ for the fixed $j < n_k$. By a similar argument, $\lim_{t\to\infty}\mathbb{E}\|\zeta_{k,j}(t)\|^2 = 0$ for all $j = 1, \ldots, n_k$, and therefore $\lim_{t\to\infty}\mathbb{E}\|\zeta_{k,j}(t)\|^2 = 0$ for all $k = 1, \ldots, l$ and $j = 1, \ldots, n_k$. That is, the mean square weak consensus is solved by the protocol (3) under Conditions (A1) and (A4”) and the “if" part is proved.

Now, we show the “only if" part. Note that $\mathscr{L}$ at least has two zero eigenvalues if $\mathscr{G}$ does not contain a spanning tree. This together with Lemma 1 implies that $\widetilde{\mathscr{L}}$ admits a zero eigenvalue, denoted by λ_1. Then, we obtain from (12)

$$\zeta_{1,n_1}(t) = \zeta_{1,n_1}(0) + \mathbf{1}_n M_{1,n_1}(t), \tag{50}$$

This implies

$$\mathbb{E}\|\zeta_{1,n_1}(t)\|^2 = \|\zeta_{1,n_1}(0)\|^2 + n\mathbb{E}|M_{1,n_1}(t)|^2 > 0,$$

which conflicts with the definition of mean square weak consensus. That is, $\mathscr{G}$ must contain a spanning tree. The remaining is to prove that Condition (A4’) is necessary for the mean square weak consensus. Let $G_k(t) = \zeta_{k,n_k}(t) - \zeta_{k,n_k}(t-\tau)$. Note that the mean square weak consensus yields $\lim_{t\to\infty}\mathbb{E}\|\zeta_{k,n_k}(t)\|^2 = 0$. Then, we have $\lim_{t\to\infty}\mathbb{E}\|G_k(t)\|^2 = 0$. Note that

$$d\zeta_{k,n_k}(t) = -\alpha(t)\lambda_k\zeta_{k,n_k}(t)dt + \alpha(t)\lambda_k G_k(t)dt + \mathbf{1}_n dM_{k,n_k}(t).$$

This together with the variation of constants formula implies

$$\zeta_{k,n_k}(t) = e^{-\lambda_k\int_0^t\alpha(u)du}\zeta_{k,n_k}(0) + \int_0^t e^{-\lambda_k\int_s^t\alpha(u)du}\alpha(t)\lambda_k G_k(t)dt + \mathbf{1}_n\Xi_{k,n_k}(t)$$

$$= \chi_{k,n_k}(t) + \int_0^t e^{-\lambda_k \int_s^t \alpha(u)du} \alpha(t)\lambda_k G_k(t)dt, \tag{51}$$

where χ_{k,n_k} is defined by (11). Thus,

$$\mathbb{E}\|\chi_{k,n_k}(t)\|^2 \leq 2\mathbb{E}\|\zeta_{k,n_k}(t)\|^2 + 2\mathbb{E}\| \int_0^t e^{-\lambda_k \int_s^t \alpha(u)du} \alpha(t)\lambda_k G_k(t)dt\|^2.$$

Similar to (15), we get

$$\lim_{t\to\infty} \mathbb{E}\| \int_0^t e^{-\lambda_k \int_s^t \alpha(u)du} \alpha(t)\lambda_k G_k(t)dt\|^2 = 0,$$

and then $\lim_{t\to\infty} \mathbb{E}\|\chi_{k,n_k}(t)\|^2 = 0$. This together with (13) produces (A4). The proof is complete. □

Similar to Corollary 1, the following simple conditions can be used to guarantee the mean square weak consensus, which is concluded in the following corollary. The proof is omitted.

Corollary 3 *Suppose that Assumption 1 and Condition (A3"') hold. Then, the control* (3) *solves the mean square weak consensus if $\mathcal{G}$ contains a spanning tree and Conditions (A1) and (A3') hold. Especially, if $\alpha(t)\underline{\gamma}(t)$ is decreasing, then the control* (3) *solves the mean square weak consensus if and only if $\mathcal{G}$ contains a spanning tree, and Conditions (A1) and (A3") hold.*

4.2 Almost Sure Weak Consensus

Theorem 8 *Suppose that Assumption 1, Conditions (A1), and (A3"') hold. Then, the control* (3) *solves the almost sure weak consensus if $\mathcal{G}$ contains a spanning tree and Condition (A5) holds. Moreover, if $\mathcal{G}$ is undirected, then the control* (3) *solves the almost sure weak consensus only if $\mathcal{G}$ is connected and Condition (A5') holds.*

Proof Let $\delta_{k,n_k}(t) = \chi_{k,n_k}(t) - \zeta_{k,n_k}(t)$. Then, we have from (10) and (44)

$$\dot{\delta}_{k,n_k}(t) = -\alpha(t)\lambda_k\delta_{k,n_k}(t-\tau) + \alpha(t)g_{k,n_k}(t), \tag{52}$$

where $g_{k,n_k}(t) = \lambda_k(\chi_{k,n_k}(t-\tau) - \chi_{k,n_k}(t))$ is continuous. By Theorem 3, we have that $\lim_{t\to\infty} \|g_{k,n_k}(t)\| = 0$, a.s. Applying the variation of constants formula for Eq. (52), we obtain

$$\delta_{k,n_k}(t) = \Pi_k(t, t_0)\delta_{k,n_k}(t_0) + \int_{t_0}^t \Pi_k(t, s)\alpha(s)g_{k,n_k}(s)ds,$$

where $\Pi_k(t, s)$ is the differential resolvent function. By (41), we get

$$\|\delta_{k,n_k}(t)\| \leq |\Pi_k(t,t_0)|\|\delta_{k,n_k}(t_0)\| + \int_{t_0}^{t} \|\Pi_k(t,s)\|\alpha(s)\|g_{k,n_k}(s)\|ds$$
$$\leq \sqrt{b_0}e^{-0.5\kappa_0\int_{t_0}^{t}\alpha(u)du}\|\delta_{k,n_k}(t_0)\| + \sqrt{b_0}\int_{t_0}^{t} e^{-0.5\kappa_0\int_s^t\alpha(u)du}\alpha(s)\|g_{k,n_k}(s)\|ds.$$

Let $B(t) = \int_0^t e^{0.5\kappa_0\int_0^s\alpha(u)du}\alpha(s)\|g_{k,n_k}(s)\|ds$ and $Y(t) = e^{-0.5\kappa_0\int_0^t\alpha(u)du}B(t)$. Note that $B(t)$ is increasing. Then, we have either $\lim_{t\to\infty} B(t) < \infty$ or $\lim_{t\to\infty} B(t) = \infty$. One can easily see from (A1) that $\lim_{t\to\infty}\|Y(t)\| = 0$ if $\lim_{t\to\infty} B(t) < \infty$. But if $\lim_{t\to\infty} B(t) = \infty$, L'Hôpital's rule tells us that

$$\lim_{t\to\infty}\|Y(t)\| = \lim_{t\to\infty}\frac{\int_0^t e^{0.5\kappa_0\int_0^s\alpha(u)du}\alpha(s)\|g_{k,n_k}(s)\|ds}{e^{0.5\kappa_0\int_0^t\alpha(u)du}}$$
$$= \frac{2}{\kappa_0}\lim_{t\to\infty}\|g_{k,n_k}(t)\| = 0, a.s. \tag{53}$$

Hence, $\lim_{t\to\infty}\|\delta_{k,n_k}(t)\| = 0$, a.s. Note that $\delta_{k,n_k}(t)=\chi_{k,n_k}(t) - \zeta_{k,n_k}(t)$ and $\lim_{t\to\infty}\|\chi_{k,n_k}(t)\| = 0$, a.s.. Therefore, we obtain $\lim_{t\to\infty}\|\zeta_{k,n_k}(t)\| = 0$, a.s.

We now prove $\lim_{t\to\infty}\|\zeta_{k,j+1}(t)\| = 0$, a.s. for all $j < n_k$. Assuming that $\lim_{t\to\infty}\|\zeta_{k,j+1}(t)\| = 0$, a.s. for certain $j < n_k$, we will prove that $\lim_{t\to\infty}\|\zeta_{k,j}(t)\| = 0$, a.s. Let $\tilde{g}_{k,j+1} = \chi_{k,j+1}(t) - \chi_{k,j+1}(t-\tau)$. Then, we get from Theorem 3 that $\lim_{t\to\infty}\|\tilde{g}_{k,j+1}\| = 0$ a.s. and

$$d\delta_{k,j}(t) = -\alpha(t)\lambda_k\delta_{k,j}(t-\tau) + \alpha(t)g_{k,j}(t) - \alpha(t)\tilde{g}_{k,j+1}(t)dt. \tag{54}$$

Applying the variation of constants formula again yields

$$\delta_{k,j}(t) = \int_0^t \Pi_k(t,s)\alpha(s)g_{k,j}(s)ds - \int_0^t \Pi_k(t,s)\alpha(s)\alpha(s)\tilde{g}_{k,j+1}(s)ds.$$

Repeating the same argument used in estimating $\|\delta_{k,n_k}(t)\|$, one can obtain $\lim_{t\to\infty}\|\delta_{k,j}(t)\| = 0$, a.s. Noting that $\lim_{t\to\infty}\|\chi_{k,j}(t)\| = 0$ a.s., we get $\lim_{t\to\infty}\|\zeta_{k,j}(t)\| = 0$, a.s. That is, the almost sure weak consensus is solved and the first assertion is proved.

Next, we prove the second assertion. If $\mathscr{G}$ does not contain a spanning tree and the almost sure weak consensus is solved, then we have from (50) that in order for $\lim_{t\to\infty}\zeta_{1,n_1}(t) = 0$, a.s., the martingale $\mathbf{1}_n M_{1,n_1}(t)$ must converge to $-\zeta_{1,n_1}(0)$ for any initial data ψ, which is impossible since $\mathbf{1}_n M_{1,n_1}(t)$is independent of the initial data. Note that all corresponding component of $Y(t)$ under undirected graph $\mathscr{G}$ has the form (44) with $\lambda_k > 0$, $k = 2, \ldots, N$, $n_k = 1$. To show Condition (A5'), the remaining is to prove $\lim_{t\to\infty}\chi_{k,n_k}(t) = 0$, a.s., which implies (A5')(see Theorem 3). Bearing in mind that $\lim_{t\to\infty}\|G_k(t)\| = 0$ a.s. and $\lim_{t\to\infty}\zeta_{k,n_k}(t) = 0$ a.s., we obtain from (51) that

$$\|\chi_{k,n_k}(t)\| \leq \|\zeta_{k,n_k}(t)\| + \int_0^t e^{-\lambda\int_s^t\alpha(u)du}\|G_k(s)\|\alpha(s)ds.$$

Then, applying the same skills in obtaining (23), we have $\lim_{t\to\infty} \|\chi_{k,n_k}(t)\| = 0$. Therefore, Condition (A5′) holds, and the proof is complete. □

4.3 Mean Square and Almost Sure Strong Consensus

Theorem 9 *Suppose that Assumption 1 and Condition (A3”) hold. Then, the control* (3) *solves the almost sure and mean square strong consensus if $\mathscr{G}$ contains a spanning tree and Conditions (A1) and (A6) hold, and only if $\mathscr{G}$ contains a spanning tree and Condition (A6’) holds under Condition (A1).*

Proof If $\mathscr{G}$ contains a spanning tree and Conditions (A1) and (A6) hold, we can use similar method used in Theorem 6 and obtain mean square weak consensus of the delay-free case. Note that the existence of a spanning tree and Conditions (A1) and (A6) can guarantee almost sure weak consensus, which is proved in Theorem 6. Hence, $\lim_{t\to\infty} \chi_{k,j}(t) = 0$ a.s., $k = 2, \ldots, l$, $j = 1, 2, \ldots, n_k$. By means of the skills in proving the first assertion in Theorem 8, one can easily get almost sure weak consensus for multi-agent system with noises and time delays.

Next, we show under Conditions (A1) and (A6), the almost sure strong consensus is solved. By Lemma (1), and

$$dx(t) = -\alpha(t)(\mathscr{L} \otimes I_n)x(t-\tau)dt + \alpha(t)\sum_{i,j=1}^{N} a_{ij}(\eta_{N,i} \otimes \sigma_{ji})dW_{ji}(t). \quad (55)$$

we have

$$(\pi^T \otimes I_n)x(t) = (\pi^T \otimes I_n)x(0) + \bar{M}(t),$$

where $\bar{M}(t) = \sum_{i,j=1}^{N} a_{ij}\sigma_{ji} \int_0^t \alpha(s)[\pi^T \eta_{N,i} \otimes \mathbf{1}_n]dW_{ji}(s)$. Hence, each agent converges to a common random variable if and only if the martingale $\bar{M}(t)$ is convergent. Note that

$$\langle \bar{M} \rangle(t) = \mathbb{E}|\bar{M}(t)|^2 = n\sum_{i,j=1}^{N} a_{ij}^2\sigma_{ji}^2\pi_i^2 \int_0^t \alpha^2(s)\gamma_{ji}(s)ds$$

Hence, $\bar{M}(t)$ is convergent if Condition (A6) holds, and only if (A6’) holds [33]. These prove the “if" part and the necessity of Condition (A6’) for mean square and almost sure strong consensus.

Note that mean square strong consensus implies mean square weak consensus, which together with Theorem 7 also implies that graph $\mathscr{G}$ contains a spanning tree and Condition (A1) holds. Similarly, we can obtain the necessity of the existence of a spanning tree for the almost sure strong consensus. Therefore, the proof is complete. □

In this section, we can see the estimate of the differential resolvent function (Lemma 2) plays an important role in obtaining the sufficient conditions of stochastic consensus. In the proof of sufficient conditions for the mean square consensus, Lemma 2 is applied to get mean square estimation of the semi-decoupled solutions (44) and (45). But this can not be used in getting the almost sure weak consensus since we can not obtain the almost sure asymptotic behavior of $\Xi_{k,j}(t, t_0)$. So we resort to the asymptotic behavior of the delay-free system to obtain the sufficient conditions. For the necessary conditions for stochastic consensus, the consensus conditions for the delay-free case is also very important. By comparison between the delay and delay-free cases, the necessary conditions for mean square and almost sure consensus are obtained. That is, the study of delay-free case is helpful for us to examine consensus of the delay case.

5 Simulations

This section is to conduct some simulations for a four-agent systems with the topology graph $\mathcal{G} = \{\mathcal{V}, \mathcal{E}, \mathcal{A}\}$, where $\mathcal{V} = \{1, 2, 3, 4\}$, $\mathcal{E} = \{(1, 2), (1, 3), (2, 3), (3, 4), (4, 1)\}$ and $\mathcal{A} = [a_{ij}]_{4\times4}$ with $a_{21} = a_{31} = a_{32} = a_{43} = a_{14} = 1$ and other being zero. One can easily see that the digraph $\mathcal{G}$ contains a spanning tree. The eigenvalues of the corresponding Laplacian matrix are $\lambda_1 = 0$, $\lambda_2 = 1.5 + 0.866i$, $\lambda_3 = 1.5 - 0.866i$, and $\lambda_4 = 3$. Then, we have $\lambda_0 := \max_{2\le j\le 4} \frac{|\lambda_j|^2}{Re(\lambda_j)} = 3$. Assume that $\tau = 0.2$ and the initial state $x(t) = [-9, 6, 1, -6]^T$ for $t \in [-\tau, 0]$. Let $W_{ji}(t) = \int_0^t \sqrt{\gamma(s)}dw_{ji}(s)$, where $\{w_{ji}(s)\}$ are standard Brownian motions. We will take $\gamma(s) = \frac{1}{1+s}$ and $\gamma(s) = \frac{1}{(1+s)^2}$, separately.

For the case $\gamma(s) = \frac{1}{1+s}$, we first choose the fixed control gain $\alpha(t) = 1$. Then, $\tau \max_{2\le j\le N} \frac{|\lambda_j|^2}{Re(\lambda_j)} \approx 0.4 < 1$. By Theorem 8, the control (3) solves the almost sure weak consensus of the system with four agents. This is depicted in Fig. 1. Then, choosing $\alpha(t) = \frac{1}{\sqrt{1+t}}$, by Theorem 9, we have that the control (3) solves the almost sure strong consensus, which is shown in Fig. 2.

For the case $\gamma(s) = \frac{1}{(1+s)^2}$, we still choose the fixed control gain $\alpha(t) = 1$. Then, Conditions (A3''') and (A5) hold. By Theorem 8, the control (3) solves the almost sure weak consensus. This is depicted in Fig. 3. Then, we choose $\alpha(t) = \sqrt{1+t}$, which implies Condition (A5). Hence, from Theorem 8, the control (3) solves the almost sure weak consensus, which is shown in Fig. 4.

Based on the simulations above, we can see that the conditions on control gain function can be relaxed if its quadratic variation of the martingale noise is convergent. In fact, in order to get the high convergence rate, it is recommended to use large control gain as much as possible. So taking the decay of the noise into consideration may be helpful for us to design the control gain. However, the control gain can not be too large and the balance is up to the necessary conditions developed in this work.

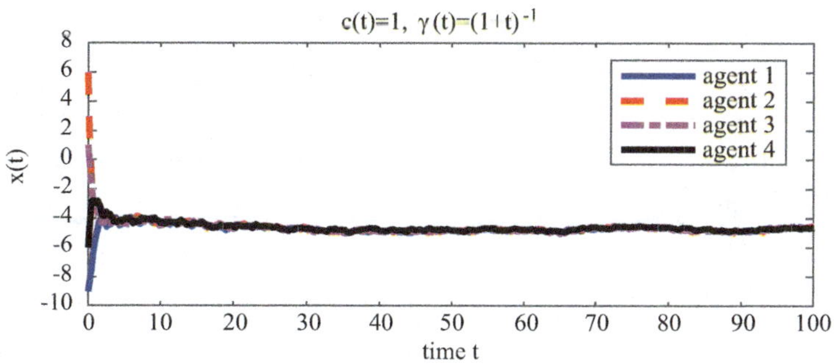

Fig. 1 States of the four agents with martingale noises and delays: $\alpha(t) = 1$ and $\gamma(s) = \frac{1}{1+s}$

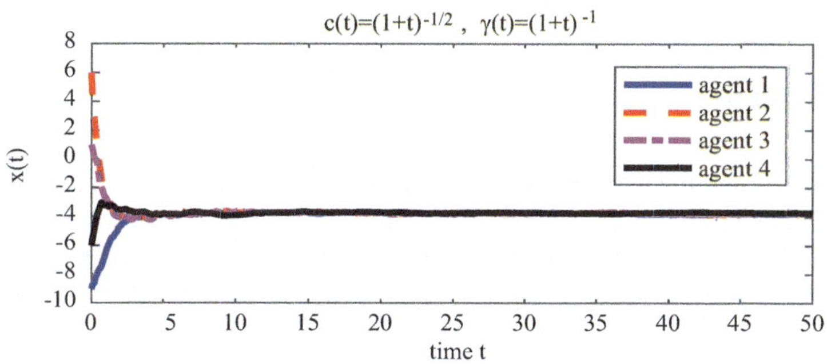

Fig. 2 States of the four agents with martingale noises and delays: $\alpha(t) = \frac{1}{\sqrt{1+t}}$ and $\gamma(s) = \frac{1}{1+s}$

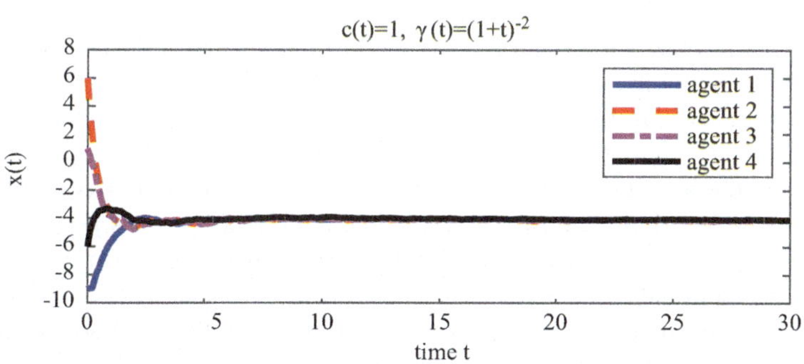

Fig. 3 States of the four agents with martingale noises and delays: $\alpha(t) = 1$ and $\gamma(s) = \frac{1}{(1+s)^2}$

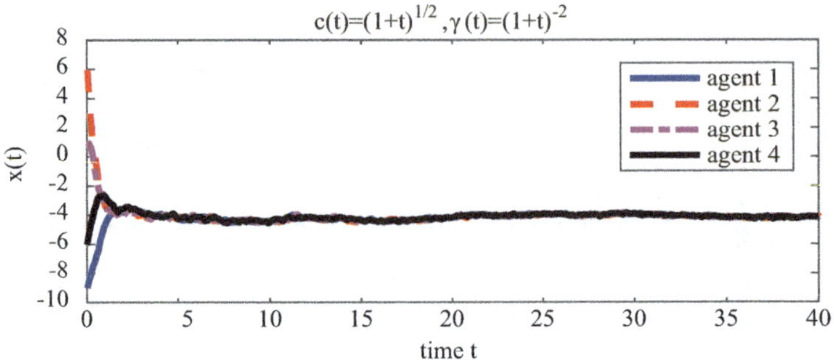

Fig. 4 States of the four agents with martingale noises and delays: $\alpha(t) = \sqrt{1+t}$ and $\gamma(s) = \frac{1}{(1+s)^2}$

6 Conclusion

This work studied the consensus conditions of the continuous-time multi-agent systems with general noises and delays. The general noise assumption allows more models in complex environments than the Gaussian white noise. Generally, the analysis of almost sure consensus is much more difficult than that of mean square consensus. Moreover, delays bring a difficulty to the stability analysis. By converting the consensus problem into the stability analysis of non-autonomous stochastic differential (or delay) equations driven by the continuous martingale and examining the stochastic stability conditions, the stochastic weak and strong consensus conditions are obtained under the general digraphs.

The ideas in this chapter can be extended to other control design problems. (a) Using similar methods, we can consider the leader-following tracking and containment control issues. In this case, it is unnecessary to consider the strong consensus since the target is just the leader or the convex hull spanned by the leaders. (b) Considering the martingale noise, the uncertainty can be modeled as multiplicative noise. The corresponding consensus problems with delays are also of significance. One needs to resort to other methods to obtain the consensus conditions. Lyapunov function or functionals used in [36, 37] may work well to deal with; (c) We can study stochastic consensus under other communication noises, like Poisson jump, or the general Lévy processes. In this case, the mean square weak and strong consensus can be obtained similarly. But for the almost sure consensus, its condition may be different from the case with continuous martingale since general Lévy process may be discontinuous and the Law of the Iterated Logarithm may fail.

There are still many interesting topics worthy of investigation in this direction. (1) Consensus under switching topology. This work is conducted under the fixed topology. In the future, the time-varying topologies, including the deterministic and the Markovian switching topologies [38, 39] need to be considered for stochastic consensus seeking under a weak connectivity condition. (2) Consensus under non-uninform delays. This work assumes that the delays in all channels are equal. It is still difficult to extend current results to the case with non-uninform delays. (3) Asynchronous algorithms can be considered [40]. (4) High order multi-agent consensus. The obtained results focus on the first order system. It is of interest to study the second order multi-agent systems (or general liner multi-agent systems) with general noises and delays.

References

1. Lynch, N.A.: Distributed Algorithms. Morgan Kaufmann Publishers Inc., San Francisco (1996)
2. Akyildiz, I.F., Su, W., Sankarasubramaniam, Y., Cayirci, E.: A survey on sensor networks. IEEE Commun. Mag. **40**(8), 102–114 (2002)
3. Ogren, P., Fiorelli, E., Leonard, N.E.: Cooperative control of mobile sensor networks: adaptive gradient climbing in a distributed environment. IEEE Trans. Autom. Control **49**(8), 1292–1302 (2004)
4. Vicsek, T., Czirók, A., Ben-Jacob, E., Cohen, T., Shochet, O.: Novel type of phase transition in a system of self-driven particles. Phys. Rev. Lett. **75**(6), 1226 (1995)
5. Jadbabaie, A., Lin, J., Morse, A.S.: Coordination of groups of mobile autonomous agents using nearest neighbor rules. IEEE Trans. Autom. Control **48**(6), 988–1001 (2003)
6. Olfati-Saber, R., Fax, J.A., Murray, R.M.: Consensus and cooperation in networked multi-agent systems. Proc. IEEE **95**(1), 215–233 (2007)
7. Cao, Y., Yu, W., Ren, W., Chen, G.: An overview of recent progress in the study of distributed multi-agent coordination. IEEE Trans. Ind. Inf. **9**(1), 427–438 (2013)
8. Radenkovic, M., Bose, T.: On multi-agent self-tuning consensus. Automatica **55**, 46–54 (2015)
9. Fanti, M.P., Mangini, A.M., Mazzia, F., Ukovich, W.: A new class of consensus protocols for agent networks with discrete time dynamics. Automatica **54**, 1–7 (2015)
10. Olfati-Saber, R., Murray, R.M.: Consensus problems in networks of agents with switching topology and time-delays. IEEE Trans. Autom. Control **49**(9), 1520–1533 (2004)
11. Cepeda-Gomez, R., Olgac, N.: An exact method for the stability analysis of linear consensus protocols with time delay. IEEE Trans. Autom. Control **56**(7), 1734–1740 (2011)
12. Cepeda-Gomez, R.: Finding the exact delay bound for consensus of linear multi-agent systems. Int. J. Syst. Sci. **47**(11), 2598–2606 (2016)
13. Zhu, W., Jiang, Z.P.: Event-based leader-following consensus of multi-agent systems with input time delay. IEEE Trans. Autom. Control **60**(5), 1362–1367 (2015)
14. Abdessameud, A., Polushin, I., Tayebi, A.: Motion coordination of thrust-propelled underactuated vehicles with intermittent and delayed communications. Syst. Control Lett. **79**, 15–22 (2015)
15. Liu, S., Li, T., Xie, L.: Distributed consensus for multiagent systems with communication delays and limited data rate. SIAM J. Control Optim. **49**(6), 2239–2262 (2011)
16. Hadjicostis, C., Charalambous, T.: Average consensus in the presence of delays in directed graph topologies. IEEE Trans. Autom. Control **59**(3), 763–768 (2014)
17. Park, M.J., Kwon, O.M., Choi, S.G., Cha, E.J.: Consensus protocol design for discrete-time networks of multiagent with time-varying delay via logarithmic quantizer. Complexity **21**(1), 163–176 (2015)
18. Charalambous, T., Yuan, Y., Yang, T., Pan, W., Hadjicostis, C., Johansson, M.: Distributed finite-time average consensus in digraphs in the presence of time-delays. IEEE Trans. Control Netw. Syst. **2**(4), 370–381 (2015)
19. Li, T., Zhang, J.-F.: Consensus conditions of multi-agent systems with time-varying topologies and stochastic communication noises. IEEE Trans. Autom. Control **55**(9), 2043–2057 (2010)
20. Huang, M., Manton, J.H.: Coordination and consensus of networked agents with noisy measurements: stochastic algorithms and asymptotic behavior. SIAM J. Control Optim. **48**(1), 134–161 (2009)

21. Huang, M., Manton, J.H.: Stochastic consensus seeking with noisy and directed inter-agent communication: fixed and randomly varying topologies. IEEE Trans. Autom. Control **55**(1), 235–241 (2010)
22. Huang, M., Dey, S., Nair, G.N., Manton, J.H.: Stochastic consensus over noisy networks with markovian and arbitrary switches. Automatica **46**(10), 1571–1583 (2010)
23. Aysal, T.C., Barner, K.E.: Convergence of consensus models with stochastic disturbances. IEEE Trans. Inf. Theory **56**(8), 4101–4113 (2010)
24. Xu, J., Zhang, H., Xie, L.: Stochastic approximation approach for consensus and convergence rate analysis of multiagent systems. IEEE Trans. Autom. Control **57**(12), 3163–3168 (2012)
25. Li, T., Zhang, J.-F.: Mean square average-consensus under measurement noises and fixed topologies: necessary and sufficient conditions. Automatica **45**(8), 1929–1936 (2009)
26. Cheng, L., Hou, Z.G., Tan, M.: A mean square consensus protocol for linear multi-agent systems with communication noises and fixed topologies. IEEE Trans. Autom. Control **59**(1), 261–267 (2014)
27. Wang, B., Zhang, J.-F.: Consensus conditions of multi-agent systems with unbalanced topology and stochastic disturbances. J. Syst. Sci. Math. Sci. **29**(10), 1353–1365 (2009)
28. Zong, X., Li, T., Zhang, J.-F.: Consensus conditions for continuous-time multi-agent systems with additive and multiplicative measurement noises. SIAM J. Control Optim. **56**(1), 19–52 (2018)
29. Liu, J., Liu, X., Xie, W.C., Zhang, H.: Stochastic consensus seeking with communication delays. Automatica **47**(12), 2689–2696 (2011)
30. Kushner, H.J., Yin, G.: Stochastic Approximation and Recursive Algorithms and Applications. Springer, New York (2003)
31. Zong, X., Li, T., Zhang, J.-F.: Consensus conditions for continuous-time multi-agent systems with time-delays and measurement noises. Automatica **99**, 412–419 (2019)
32. Zhang, Y., Li, R., Zhao, W., Huo, X.: Stochastic leader-following consensus of multi-agent systems with measurement noises and communication time-delays. Neurocomputing **282**(22), 136–145 (2018)
33. Revuz, D., Yor, M.: Continuous Martingales and Brownian Motion. Springer, New York (1999)
34. Wang, Y., Cheng, L., Hou, Z.G., Tan, M., Wang, M.: Containment control of multi-agent systems in a noisy communication environment. Automatica **50**(7), 1922–1928 (2014)
35. Gripenberg, G., Londen, S.O., Staffans, O.: Volterra Integral and Functional Equations. Cambridge University Press, Cambridge (1990)
36. Li, T., Wu, F., Zhang, J.-F.: Multi-agent consensus with relative-state-dependent measurement noises. IEEE Trans. Autom. Control **59**(9), 2463–2468 (2014)
37. Zong, X., Li, T., Yin, G., Wang, L.Y., Zhang, J.F.: Stochastic consentability of linear systems with time delays and multiplicative noises. IEEE Trans. Autom. Control **63**(4), 1059–1074 (2018)
38. Yin, G., Sun, Y., Wang, L.Y.: Asymptotic properties of consensus-type algorithms for networked systems with regime-switching topologies. Automatica **47**, 1366–1378 (2011)
39. Yin, G., Wang, L.Y., Sun, Y.: Stochastic recursive algorithms for networked systems with delay and random switching: multiscale formulations and asymptotic properties. SIAM J.: Multiscale Model. Simul. **9**, 1087–1112 (2011)
40. Yin, G., Yuan, Q., Wang, L.Y.: Asynchronous stochastic approximation algorithms for networked systems: regime-switching topologies and multi-scale structure. SIAM J.: Multiscale Model. Simul. **11**, 813–839 (2013)

Multimodal Emotion Recognition and Intention Understanding in Human-Robot Interaction

Luefeng Chen, Zhentao Liu, Min Wu, Kaoru Hirota, and Witold Pedrycz

Abstract Emotion recognition and intention understanding are important components of human-robot interaction. In multimodal emotion recognition and intent understanding, feature extraction and selection of recognition methods are related to the calculation of affective computing and the diversity of human-robot interaction. Therefore, by studying multimodal emotion recognition and intention understanding we to create an emotional and human-friendly human-robot interaction environment. This chapter introduces the characteristics of multimodal emotion recognition and intention understanding, presents different modalities emotion feature extraction methods and emotion recognition methods, proposes the intention understanding method, and finally applies them in practice to achieve human-robot interaction.

Keywords Multimodal emotion recognition · Intention understanding · Human-robot interaction

---

L. Chen (✉) · Z. Liu · M. Wu
School of Automation, China University of Geosciences, Wuhan 430074, China

Hubei Key Laboratory of Advanced Control and Intelligent Automation for Complex Systems, Wuhan 430074, China
e-mail: chenluefeng@cug.edu.cn

K. Hirota
Tokyo Institute of Technology, Yokohama 226-8502, Japan

W. Pedrycz
Department of Electrical and Computer Engineering, University of Alberta, Edmonton, AB TR6 2G7, Canada

M. Wu et al. (eds.), *Developments in Advanced Control and Intelligent Automation for Complex Systems*, Studies in Systems, Decision and Control 329,
https://doi.org/10.1007/978-3-030-62147-6_10

Abbreviations

PCA	Principal component analysis
LDA	Linear discriminant analysis
SIFT	Scale-invariant feature transform
SURF	Speed-up robost features
ELM	Extreme learning machine
BPNN	Back propagation neural networks
CNN	Convolution Neural Network
DBN	Deep Belief Network
RNN	Recurrent Neural Network
SVM	Support vector machine
TLWFSVR	Three-layer weighted fuzzy support vector regression
FSVR	Fuzzy support vector regressions
ROI	Regions of Interest
SC	Sparse coding
FFT	Fast Fourier Transformation
SR	Softmax Regression
DSAN	Deep Sparse Autoencoder Network
BPA	Basic Probability Assignment
RF	Random Forest
FCM	Fuzzy c-means

1 Introduction

The emotion plays an important role in our daily life. A large deal of emotional information is transferred during people's communication, which promotes harmonious and natural relationship. At present, cognitive scientists compare emotion with classical cognitive processes such as perception, learning, memory, and speech. Studies on emotion itself and the interaction with other cognitive processes have become hotspots in cognitive science. Affective computing is to establish a harmonious human-robot environment by enabling the robot to recognize, understand, and express human emotions, and make robots more intelligent.

The two key techniques in the affective computing are how to enable robots to recognize human emotions, and how robots can understand human emotions and express their own emotions based on human emotional states. This chapter systematically introduces the multimodal emotion recognition and human-robot interaction from three aspects: multimodal emotion recognition, emotional intention understanding and emotional human-robot interaction system.

The concepts of multimodal emotion recognition, emotional intention understanding and emotional human-robot interaction system are introduced in this section, which explains the complete process of emotional human-robot interaction from the theory.

1.1 Multimodal Emotion Recognition

In our daily communication, the human being is able to capture the emotional change of the other person by observing the facial expression and the body gesture, listening to the voice, which is because the human brain can perceive and understand the man's emotional state information, at the sound of his voice and visual signals (such as the special speech word, the change of the intonation and facial expression, etc.). The multimodal emotion recognition is the simulation of the human emotion perception and understanding process by robots, and its task is to extract emotional features from multimodal signals and find the mapping relationship between human emotions and emotional features.

Facial expression, speech signal and body gesture often appear in the process of human-robot interaction at the same time, which are used to analyze and inference each other's emotional states and intention in real time, and then guide them to make different reactions [1]. It is conceivable that if robots acquire the same visual, auditory, and cognitive abilities as humans, they could be able to make personalized and adaptive responses to each other's states, just like humans. In terms of emotion recognition, most of research is based on the different modes emotional data. The multimodal emotion recognition by constructing a corresponding feature set based on different modal of emotional information can effectively reflect the emotional state.

At present, research of affective computing mainly focuses on facial expression recognition, speech emotion recognition and body gesture recognition. In the emotional human-robot interaction, multimodal emotion recognition is mainly consists of two steps: multimodal emotion feature extraction and emotion recognition, among which the multimodal emotion feature extraction is a very important link. The ability of feature sets to represent emotional information directly affects the results of emotion recognition.

Facial expression is the most visible way to express emotions. At present, there are two main types of methods for extracting facial features: deformation-based facial feature extraction and motion-based facial feature extraction. The main facial feature extraction methods based on deformation mainly include principal component analysis (PCA), linear discriminant analysis (LDA), geometric features and models, Gabor wavelet transform etc. The methods of extracting facial features based on motion are to take the facial expression as a sports field, and analyze and recognize facial expressions through the information of facial motion changes. The main core is to use motion changes as recognition features. At present, the motion feature extraction methods mainly include optical flow method and feature point tracking method.

For speech emotion recognition, speech emotion features can be divided into acoustic and language features [2]. The two types of feature extraction methods and their contributions to speech emotion recognition are different depending on the selected speech database. If the speech database is a text-based database, the language features can be ignored and if the database is close to current real life, language features will play a very important role. The acoustic features used for speech emotion recognition can be summarized into prosodic features, spectral-based features, and voice quality features. These features are often extracted from the frames, and participate in emotion recognition in the form of global feature statistics. Some studies have shown that various gestures and movements expressed by the body movements in interactive behaviors can express or assist in expressing people's thoughts, emotions, intentions, etc. So that body gesture is critical to understanding emotional communication. Feature extraction of body gesture can be divided into global feature extraction and local feature extraction. The global features include color, texture, motion energy image, and motion history image, etc. The local features include gradient histogram, scale-invariant feature transform (SIFT), speed-up robost features (SURF), spatio-temporal points of interest, etc.

Most of the current emotion databases describe emotion as an adjective label form (such as anger, happiness, sadness, fear, surprise and neutrality), which is a discrete emotion description model. Therefore, it is usually reduced to the classification of standard pattern of emotion recognition based on discrete emotion description model and corresponding emotion database. When the training data and test data come from different people or different databases, in addition to requiring good emotional representation ability for emotional features, the design of the emotion recognition classifier also faced with higher requirements. The follows mainly introduces the methods of discrete emotion recognition [3].

Research on unimodal emotion recognition algorithm is nearly mature [4–6]. Although, multimodal emotion recognition has received widespread attention, there are few related studies. Most of them are based on bi-modal information, such as speech-facial emotion recognition [7], posture-facial emotion recognition [8], physiological signal-facial emotion recognition [9], etc. Multimodal information fusion method is of great significance for making full use of multimodal emotion information and improve the performance of emotion recognition. Information fusion, as the theoretical basis of multimodal emotion recognition, involves a wide range of fields. When conducting multimodal emotional interactions, multi-channel sensors are used to obtain the signals of different emotional state of interviewers, and then data fusion and decision-making are performed, the key of which is the multimodal emotion recognition algorithm. By integrating the emotional feature data of each channel is fused, and the decision is made according to certain rules, so as to determine the emotion category attributes corresponding to the multimodal information. Multimodal emotional information fusion can be classified as feature-level fusion and decision-level fusion.

Feature-level fusion first extracts features from the original information obtained by the sensors, and then analyzes and processes the information comprehensively. Generally speaking, the extracted feature information should be a sufficient statistics of the pixel information, and the multi-sensor data should be classified, aggregated, and integrated according to the feature information. The fusion system first preprocesses the data to complete the data calibration, and then implements parameter correlation and state vector estimation. In multimodal information fusion, the feature-level fusion strategy is to firstly extracts the emotional feature data in each mode separately, and then cascade the feature data of all modalities into a feature vector for emotion recognition. Only one classifier is designed for the emotional feature data, and the output of the classifier is the predicted result of emotional type.

For decision-level fusion, before proceeding with the fusion, the corresponding processing components of each local sensor have independently completed the decision-making or classification task. The essence is to coordinate according to certain criteria and the credibility of each sensor to make a global optimal decision. Decision level fusion is a joint decision result, which is theoretically more precise and reliable than any single sensor decision. At the same time, it is also a high-level fusion. The result can provide a basis for the final decision. Therefore, decision-level fusion must fully account of the needs of specific decision-making issues, and make full use of the various types of feature information of the measurement object, then use appropriate fusion technology. Decision-level fusion is directly targeted at specific decision-making goals, and the result of which directly affects the level of decision-making. In the emotion recognition, the decision-level fusion strategy first designs a corresponding emotion classifier for the emotional feature data, and then makes a decision on the output of each classifier to synthesize the finally obtains the final result of emotion recognition according to the comprehensive analysis of certain decision rules. The methods used for decision-level fusion include Bayesian reasoning, Dempster-Shafer evidence theory, and fuzzy reasoning, etc.

The current emotion recognition methods mainly include neural network, support vector machine, and extreme learning machine (ELM). Neural network algorithms are widely used in the field of emotion recognition. Back propagation neural networks (BPNN) are characterized by forward signal transfer and error back propagation, which also make the calculation process of BPNN into two major steps. When the signal is transmitted forward, the signal enters the hidden layer from the input layer, and the result is in the output layer. The processing result of the previous layer will directly affect the processing effect of the subsequent layer. If the output layer fails to achieve the desired result, it will switch to back propagation. Adjust the network weights and thresholds according to the prediction error until the output result is close to the desired output result. Deep learning algorithms have also been applied to the field of emotion recognition for learning and recognition of emotion features. Deep learning combines low-level features to form more abstract high-level representation or features to discover distributed feature representations of data. At present, typical deep learning models include: Convolution Neural Network (CNN), Deep Belief Network (DBN), Autoencoder, and Recurrent Neural Network (RNN).

Support vector machine (SVM) has great advantages in dealing with small samples and non-linear problems. The essence of SVM is to find the optimal linear classification hyperplane, which includes two cases: sample linearly separable and sample nonlinearly separable. For the first case, SVM tries to find the best among the classification boundaries that completely separate the samples. For the second case, SVM uses the kernel function to solve the linear discrimination in the high-dimensional feature space, and also solves the problem of large amount of computation in high-dimensional feature space. Because the feature parameters of emotion are not completely linearly separable in the input space, the non-linearly separable cases are used for emotion recognition.

The extreme learning machine (ELM) has only a single hidden layer. Different from the traditional learning theory, which needs to adjust all parameters of the forward transfer neural network, it randomly assigns input weights and thresholds in the neuron. Then the weights are calculated and output through the regularization principle, so that the neural network can still approach any continuous system. It is proved that the random acquisition of the hidden layer node parameters of the single hidden layer neural network will not affect the network's convergence ability, so the network training speed of the over-limit learning machine is much faster than the traditional BP neural network and SVM.

1.2 Emotional Intention Understanding

Emotional intention understanding is based on the analysis of emotional intent correlation, and further studies with more detailed theories, methods and technologies of intention understanding from emotional surface information and deep cognitive information, so that according to users' personal emotional intentions based on their emotions, surface communication information, and specific scenarios, and eventually achieve natural harmony of human-robot interaction.

Humans are inherently capable of expressing, predicting, and understanding intentions, regardless of whether the intentions are expressed in an explicit or implicit manner [10]. Intention is an important part of interpersonal communication and the human cognitive system. Although not directly obtained, it can be inferred from the behavior, physiological indicators, and atmosphere. Research on intent understanding mainly focused on the research of physical and psychological behavior. With the continuous development of artificial intelligence and computer technology, many scholars have begun to introduce intention understanding into human-robot interaction so as to machines have the same intention understanding ability as humans, thereby promoting harmonious human-machine interaction.

In our previous researches [11], in order to deeply understand human internal thinking, an intention understanding model based on two-layer fuzzy support vector regression is proposed in human-Crobot interaction. [12] proposed a Two-layer Fuzzy SVR-TS Model for emotion understanding in human-robot interaction. [13] proposed a three-layer weighted fuzzy support vector regression (TLWFSVR) model for understanding human intention, and it is based on the emotion-identification information in human-robot interaction.

1.3 Emotional Human-Robot Interaction System

With the integration and rapid development of robotics and artificial intelligence, more and more robots have begun to enter people's daily life. The personification of robots will definitely be an important direction for future development. Researchers have tried to give robots a human-like shape to make them more acceptable to us, so there have been many humanoid robots that can execute human commands, which performs excellently in tour guide welcome, education and teaching, disaster relief, and rehabilitation [14, 15].

However, the expectations for humanoid robots don't stop there. Robots can't satisfy human needs by doing some mechanical repetitive commands. They expect robots to have human-like intelligence, so that robots are endowed with the ability of emotion perception and expression. It is an important direction for future development to build a natural and autonomous cognitive robot interaction system that meets the psychological needs. The construction of the emotional robot system is generally based on the ordinary humanoid robot, constructing emotional information acquisition frame, emotional understanding frame, and emotional robot interaction frame, which finally form an emotional robot system that resembles a humanoid appearance and has human emotional cognitive capabilities. Existing research on emotional robot systems mainly focuses on two aspects, one is to acquire emotional information acquisition by recognizing facial expressions, speech signals, body gestures, etc, and another is to guide the robot to make behavioral responses based on the acquired emotions. For example, [16] proposed a companion robot system with emotional interaction capabilities. The robot detects the user's basic emotions through visual and audio sensors. Depending on the user's emotional state, the robot will play appropriate music, and generate a scent that soothes the user's mood. The robotic system can also automatically navigate to accompany the user. A robot-based emotion recognition and rehabilitation system based on a browser interface is developed [17] , which combined physiological signals and facial expressions to infer the user's emotional state, and then combined emotional expression technology to alleviate the negative emotions of the user or strengthen the positive emotions. Another research explores the robot's multimodal emotional expression. By adjusting the facial expressions, speech mechanisms, and body gestures, robots are endowed with rich emotional expression capabilities like humans. Klug and Zell [18] developed a humanoid robot ROBIN, which can express almost any facial expression of human beings, and generate expressive gestures. It is also equipped with a speech synthesizer to achieve emotional transformation, selection, expression, and evaluate, and some auxiliary functions. Boccanfuso et al. [19] developed an emotional robot system to accompany children with autism. By communicating with children with autistic children, the emotional robot can recognize the emotions expressed by the children and correctly classify them.

There are many shortcomings in current emotional robot systems. For example, the current research does not consider environmental and scene information, which will affect the emotional expression of the operator. In addition, it fails to consider the deep intention information behind the emotional expression, and lacks a cognitive analysis of user behavior. It is necessary to comprehensively consider multimodal emotional information and environmental information on the basis of existing emotional robot systems, introduce an emotional intention understanding framework, and build an emotional robot system with emotional intention interaction capabilities under the network architecture.

1.4 The Structure of the Chapter

This chapter first introduces the basic knowledge of multimodal emotion recognition, emotional intention understanding and emotional human-robot interaction system in the introduction, and theoretically explains the complete process of emotional human-robot interaction. Then combined with the proposed algorithms, we briefly describe some of our existing research results, including algorithm theory of multimodal emotion feature extraction, multimodal emotion recognition, and emotion intention understanding. Finally, some simulation experiments and application results of our emotional human-robot interaction system are shown.

2 Multimodal Emotion Feature Extraction

Multimodal emotion feature extraction is an indispensable part of multimodal emotion recognition. In this section, three methods are proposed for three modes‘ features, namely, regions of interest based feature extraction in facial expression, sparse coding-SURF based feature extraction in body gesture and FCM based feature extraction in speech.

2.1 Regions of Interest based Feature Extraction in Facial Expression

Before the feature extraction, we need to perform facial expression image segmentation, size adjustment, and gray balance operation on the regions of interest (ROI).

The eyebrows, eyes, and mouth have obvious texture and shape in the facial expression. The changes of these three key parts have a greater impact on the expression, so we use these areas for feature extraction.

For the JAFFE database, the four dimensional coordinate information of the eyebrows, mouth, and eyes in the ROI of the expression image is manually obtained, as shown in Table 1. Each column in the coordinates represents the horizontal and

Table 1 Comparison of emotion recognition experimental results

Key parts	Corners coordinates	Clipping region
	$\begin{bmatrix} Xleftup\ Xrightup\ Xleftdown\ Xrightdown \\ Yleftup\ Yrightup\ Yleftdown\ Yrightdown \end{bmatrix}$	$[Length, Width]$
Eyebrows	$\begin{bmatrix} 74.21\ 182.26\ 182.26\ 74.21 \\ 100.14\ 100.24\ 120.05\ 120.05 \end{bmatrix}$	$\begin{bmatrix} 108.05\ 19.91 \end{bmatrix}$
Eyes	$\begin{bmatrix} 74.22\ 182.25\ 182.25\ 74.22 \\ 120.85\ 120.85\ 140.12\ 140.12 \end{bmatrix}$	$\begin{bmatrix} 108.03\ 19.27 \end{bmatrix}$
Mouth	$\begin{bmatrix} 95.55\ 162.01\ 162.01\ 95.55 \\ 180.03\ 180.03\ 210.05\ 210.05 \end{bmatrix}$	$\begin{bmatrix} 66.46\ 30.02 \end{bmatrix}$

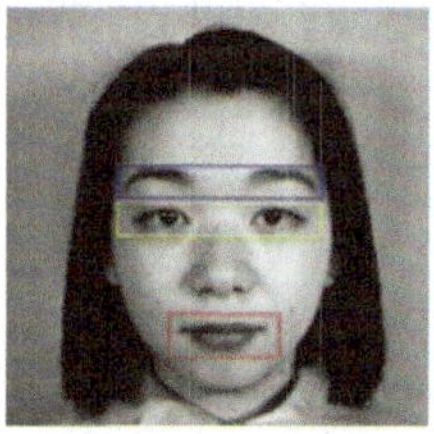

Fig. 1 Specific ROI crop areas image

vertical coordinates (x and y) of the four points along the clockwise direction; the matrix of the rectangular cropped areas represents the height and width of the areas.

By trimming the ROI region, while reducing the interference of irrelevant redundant images on facial expression features, it also reduces the amount of data, which improves the calculation speed. The extracted ROI image is shown in Fig. 1.

2.2 Sparse Coding-SURF based Feature Extraction in Body Gesture

For gesture feature extraction, deep feature representations of gesture images is more benefit to gesture recognition. The mapping between general image features such as SURF and gesture categories are high-dimensional and non-linear relationships [20–22], and it is difficult to directly construct a training model and achieve higher recognition effects. It is necessary to construct deep feature representations of gesture images.

This section mainly introduces two image feature extraction algorithms, namely, including Speeded-up robust features (SURF) and Sparse coding (SC) algorithm. The SURF algorithm is mainly used to extract the position coordinates and feature descriptors of the boundary corner points of the target images. The sparse coding algorithm is to further extract high-dimensional features than the SURF algorithm, and remove redundant information and speed up the identification process.

Integral images [23] enable the SURF algorithm to detect features with higher speed and lower computational overhead. It uses a Hessian matrix to detect potential feature points. The Hessian matrix discriminant for each pixel in the gesture image is

$$Det(H)=L_{xx} * L_{yy} - (L_{xy})^2 \tag{1}$$

where $L_{xx}(x, \delta) = \frac{\partial^2 g(\delta)}{\partial x^2} * I(x, y)$, which is a quadratic Gaussian convolution of the integrated image, and representing the Gaussian scale space of the image. δ denotes the scale parameter of the point (x, y).

Box filters are used instead of quadratic Gaussian derivatives to speed up the computation process [23], so that the filtering of the image is transformed into the addition and subtraction of the sum of pixels in different areas of the image under a given size sliding window. Such calculations can be easily performed by searching the integral image.

$$Det(H)=D_{xx}D_{yy} - (wD_{xy})^2 \tag{2}$$

where $D_{(\cdot)}$ is the approximate solution corresponding to $L_{(\cdot)}$, and the value of w is set to 0.9 according to [23], the role of which is to balance the error caused by the use of the box filter. Different scale of images can be obtained by using different size box filter templates, which could help in search for the extreme point of the speckle response.

For candidate feature points, use a non-maximum suppression algorithm to identify the final feature point. For any pixel, compare its value with 8 pixels in the surrounding neighborhood and 18 pixels in the adjacent scale(a total of 26 pixels). If the point is a local maximum point and the (Hessian) discriminant is greater than a given threshold, the point is a feature point of the area, and its coordinates and scales are recorded. Next, linear interpolation is used to obtain sub-pixel-accurate feature points, while unstable candidate points are eliminated.

We calculate the Haar wavelet features of 16 sub-areas in a square area around each feature point, and then obtain the following 4 values in each sub-area. Therefore, the feature dimension of each descriptor is 16 × 4 = 64.

$$\left[\sum dx, \sum dy, \sum |dx|, \sum |dy|\right] \tag{3}$$

As an unsupervised learning method, sparse coding (SC) algorithm can implement feature selection and remove redundant information to obtain deeper image representations, [24] which is very useful for classification and recognition of images. The algorithm tries to find an incomplete set of base vectors, and uses linear combinations of these vectors to express most or all of the original feature vectors.

Suppose X is the feature descriptor extracted by the SURF algorithm, ϕ_m is an over-complete base vector, then X can be represented by a linear combination of ϕ_m:

$$X=\sum_{m=1}^{M}\alpha_m\phi_m \tag{4}$$

where α is the sparsity coefficient, which is determined by the input vector X. The objective function optimized by the sparse coding algorithm is:

$$a\hat{} = \min_{\alpha,\phi}\sum_{j=1}^{m}\left\|x^{(j)}-\sum_{i=1}^{k}\alpha_i^{(j)}\phi_i\right\|^2+\lambda\sum_{i=1}^{k}S(\alpha_i^{(j)}) \tag{5}$$

$$subject\ to \qquad \|\phi\|^2\le C, \forall i = 1, ..., k \tag{6}$$

The first term in the above formula is the reconstruction term, and the second term is the scattered term. λ is a regularization parameter that balances the influence of the reconstruction term and the sparse term, and S is the cost function of the sparse term. In practice, the general choice of the $S(.)$ is the $L1$ norm cost function $S(\alpha_i) = |\alpha_i|_1$ and the logarithmic cost function $S(\alpha_i) = \log(1+\alpha_i^2)$.

The SC process includes training phase and encoding phase. In the training phase we find an over-complete base vectors. Firstly, the dictionary ϕ is randomly initialized and fixed, α_i are adjusted to minimize the objective function, and then fixed α to minimize the objective function. This stepwise iteration training until the objective function is minimized, so that we have an over-complete dictionary. In the encoding stage, the trained over-complete dictionary ϕ is used for feature vector encoding, and a linear combination of eigenvector X is found, that is, the sparse representation of the feature vector.

2.3 FCM based Feature Extraction in the Speech Emotion

We adopt the non-personalized speech emotion features to supplement personalized emotion speech features, and achieve universal and negotiable speech emotion feature extraction. Firstly, the openSMILE [25] toolkit (version 2.3) was used to calculate the speech emotion features, and 16 basic features and their 1st derivatives are extracted, which including F0, zero crossing rate (ZCR), RMS energy, and MFCC 1-12. Derivative features are less affected by different persons, which are considered non-personalized features, and 12 statistical values of these basic features are calculated.

The ZCR mentioned above refers to the fact that the current speech signal is recorded when passes through the zero level. ZCR of the speech signal $\chi(m)$ is

$$Z=\frac{1}{2}Z=\frac{1}{2}\sum_{m=0}^{N-1}|sgn[x(m)]-sgn[x(m-1)]| \tag{7}$$

The constructed personalized and non-personalized speech emotion feature sets are mainly composed of the following 32 emotional low-level descriptors: pitch frequency and its 1st derivative, root mean square frame energy and its 1st derivative, ZCR and its 1st derivatives, harmonic-to-noise ratio and its 1st derivatives, and 1–12 order MFCC coefficient and its 1st derivative. Since the dimensions of the feature matrix of the speech emotion bottom descriptor will change with the length of the speech signal, the classifier generally cannot directly process feature matrices of different lengths. It is necessary to use a feature statistical function to convert the feature matrix into a feature vector. We select the 12 statistical functions of average, standard deviation, kurtosis, skewness, minimum value, maximum value, relative position, range of change, two linear regression coefficients and their mean square error to describe the above 32 underlying emotional descriptors. The constructed feature matrix is transformed to form a 384-dimensional personalized and non-personalized speech emotion feature set.

The Fuzzy C-Means (FCM) algorithm is used to cluster speech emotion features. It is a fuzzy clustering algorithm based on the objective function, which is very sensitive to the initial number of clusters. The initial number of clusters C needs to be set in advance, which contains l samples, and each sample has n evaluation indexes. The sample data set is divided into C categories according to different evaluation indicators. The FCM algorithm calculates the minimum value of the objective function to obtain the membership degree of the ith sample element to different clusters, and divides the sample element i according to the membership degree. The training data set is divided into multiple subsets using the FCM algorithm, which is based on the minimization of the following objective function:

$$\begin{aligned} &\min J_m(U, V) = \sum_{k=1}^{L}\sum_{o=1}^{N}(\mu_{io})^m D_{ko}^2 \\ &D_{ko}^2 = \|y_o - c_k\|^2 \\ &\text{s.t.} \begin{cases} \sum_{k=1}^{L} \mu_{ko} = 1 \\ 0 < \mu_{ko} < 1 \end{cases} k = 1, \cdots, L, o = 1, \cdots, N \end{aligned} \tag{8}$$

where μ_{ko} is the membership value of the oth sample in the kth cluster, U is the related fuzzy partition matrix consisting of μ_{ko}, $V = (c_1, c_2, \ldots, c_L)$ is the cluster center matrix, L is the number of clusters, m is the fuzzification exponential which has an important regulatory effect on the fuzziness degree of clusters, and usually $m = 2$. D_{ko} is the Euclidean distance between oth sample y_o and kth cluster center c_k.

3 Multimodal Emotion Recognition

After multimodal feature extraction, it is necessary to emotion recognize for implementing a human-robot interaction system with certain emotion recognition.In this section, four methods are proposed for emotion recognition, namely, softmax regression based deep sparse autoencoder network, multi-SVM based dempster-shafer theory using sparse coding feature, two-layer fuzzy multiple random forest and two-stage fuzzy fusion based convolution neural network for dynamic facial expression and speech emotion recognition.

3.1 Softmax Regression based Deep Sparse Autoencoder Network for Facial Emotion Recognition

The coordinates of ROI region in the facial expression is extracted as facial emotional features, and the model is trained using a deep sparse autoencoder network. Firstly, the network is pre-trained using the greedy algorithm to obtain the initial weight, and then the sparse parameters are optimized to obtain the optimal network model, including the number of hidden layers and its nodes. Finally, Softmax regression is used for feature classification, and gradient descent algorithm is also used to speed up the training process. In addition, the BP algorithm is also used to fine-tune the model weights obtained in the previous step to enhance the robustness of the deep learning network and improve the effect of facial expression emotion recognition.

Assuming that the numbers of nodes in the input layer and the hidden layer are v and h respectively, the joint probability distribution functions of v and h using input data and pre-trained parameters are as follows

$$p(h_j = 1|v) = \sigma(c_j + \sum_i w_{ij} v_i) \tag{9}$$

where the sigmoid function is as follows

$$\sigma = \frac{1}{1 + \exp(x)} \tag{10}$$

The model is trained using a deep sparse autoencoder network. x is the network input data, the network output data is defined as $h_{w,b}(x)$. $w_{ij} (i = 1, 2, \ldots, n)$ is the initial weighted matrix. The reconstructed signal in the decoding stage is

$$h_{w,b}(x) = g({w_i}^T u + b_{i+1}) \tag{11}$$

We consider the Softmax classifier for feature classification, whose output is the probability of each classifier of samples. We firstly calculate the probability of each category and then compare the result with a given threshold ϕ to convert the threshold into a binary classification problem as follows:

$$h_\theta(x) = g\left(\theta^T x\right)\frac{1}{1+e^{-\theta^T x}} = p(y = 1|x;\theta) \tag{12}$$

where $h_\theta(x)$ is the probability of 1, and θ is the model parameter.

The training set of SR is: $\left\{\left(x^{(1)}, y^{(1)}\right), \ldots, \left(x^{(m)}, y^{(m)}\right)\right\}$, where $y^{(i)} \in \{1, 2, \ldots, k\}$, and $k = 7$. That is 7 kinds of emotion (neutral, happiness, anger, sadness, surprise, disgust, and fear).

We add an attenuation in the SR's cost function to penalize too large parameter, and the SR's cost function is rewritten as follows,

$$J(\theta) = -\frac{1}{m}\left[\sum_{i=1}^{m}\sum_{j=1}^{k} 1\left\{y^{(i)} = j\right\}\log\frac{e^{\theta_j^T x^{(i)}}}{\sum_{l=1}^{k} e^{\theta_j^T x^{(i)}}}\right] + \frac{\lambda}{2}\sum_{i=1}^{k}\sum_{j=0}^{n}\theta_{ij}^2 \tag{13}$$

Gradient descent algorithm is also used to speed up the training process, which is expressed as:

$$\theta_j = \theta_j - \alpha\nabla_{\theta_j} J(\theta)(j = 1, 2\ldots, k) \tag{14}$$

3.2 *Multi-SVM based Dempster-Shafer Theory for Gesture Recognition Using Sparse Coding Feature*

We constructed a multi-class support vector classification model to recognize body pose images. The decision function of SVM is:

$$f(z) = \sum_{i=1}^{n}\alpha_i y_j k(z_i, z_j) + b = \omega^T z + b \tag{15}$$

In the above (15), $k(z_i, z_j)$ is the kernel function, and y_j is the category label. In order to reduce the training time, we choose a linear kernel function, that is $k(z_i, z_j)=z_i^T z_j$.

We construct the following pooling function, assuming that α is the result of sparse encoding of feature descriptor X:

$$\begin{gathered} z = F(\alpha) \\ z_j = \max\left\{\left|\alpha_{1j}\right|, \left|\alpha_{2j}\right|, \ldots, \left|\alpha_{Mj}\right|\right\} \end{gathered} \tag{16}$$

We Define a pooling function $F(\cdot)$ in each column of α as the maximum value of the it. Each column in α corresponds to the response of dictionary ϕ to a certain feature descriptor. We use the pooling function to construct the linear kernel function as follows:

$$k(z_i, z_j) = {z_i}^T z_j = \sum_{l=0}^{2} \sum_{s=1}^{2^l} \sum_{t=1}^{2^l} < z_i^l(s,t), z_j^l(s,t) > \tag{17}$$

The "one-to-all" strategy is used to construct a multi-class linear SVM. During the training process, samples of a certain category are classified into one category, and the remaining samples that do not belong to this category are classified into another category. Then, m category samples can be constructed with the m classifiers. During the test process, input test samples into the m models for prediction, and then get m prediction labels. These prediction labels are recorded, and the largest number is the final discrimination label.

The collected gesture images include RGB and depth images. Because of the heterogeneity of the data,we need to perform feature extraction and classification recognition for the two types of images, and introduce a certain criterion to get the final results. Here we introduced Dempster-Shafer Evidence Theory to make decision for multimodal gesture image recognition [26].

In the D-S evidence theory, assuming $\varDelta$ is a hypothesis in the recognition framework $\varTheta$, the combination rules for two Basic Probability Assignment (BPA) $m(i)$ and $m(j)$ are as follows:

$$m_{i,j}(\varDelta) = \frac{1}{1-K} \sum_{\varDelta_1 \cap \varDelta_2 = \varDelta} m_i(\varDelta_1) m_j(\varDelta_2) \; K = \sum_{\varDelta_1 \cap \varDelta_2 = \emptyset} m_i(\varDelta_1) m_j(\varDelta_2) \tag{18}$$

where K is the normalization factor, while $1 - K$ reflects the degree of evidence conflict.

3.3 Two-Layer Fuzzy Multiple Random Forest for Speech Emotion Recognition

Random forest is a flexible machine learning algorithm that will bring good results in most cases even without hyper-parameter tuning. In addition, it can effectively solve the problem of overfitting by averaging decision trees. A random forest consists of a set of decision trees $h(x, \theta_i)$, where θ_i is the number of decision trees, and it is an random vector with independent and uniformly distributed. x is the input data, each decision tree vote to determine the optimal classification result. The steps for generating a random forest are as follows:

Step 1: The bootstrap method is used to randomly extract k samples from the input speech emotion features and use them as training sets for a decision tree in the forest.

Step 2: Let the feature dimension of each sample be M, and select m ($m \ll M$) subsets from the M features to establish the nodes of the decision tree. Each time the tree is split, the best feature is selected to build nodes from the m features based on the Gini coefficient. The value of m remains constant during the growth of the decision tree.

Step 3: Do not perform any cutting with each tree grows to a maximum.

Step 4: Repeat Step1 to Step3 until the entire forest is completed.

The output of the random forest is determined by voting, that is, the output of each tree decision in the forest is counted, and the category with the most votes is the final output of the random forest. The decision process of random forest is as follows

$$H(x) = \arg\max \sum_{i=1}^{M} I\left(h_i(x) = Y\right) \tag{19}$$

where H_x is the output of ensemble classifier, $h_i(x)$ represents single decision tree model, $I(.)$ refers to indicatory function, Y refers to target tag, which is the types of emotions here.

For a given input sample D, there are N categories, and the number of nth categories is C_n, then the Gini coefficient of sample D is:

$$Gini(D) = \sum_{i=1}^{K} p_i\left(1 - p_i\right) = 1 - \sum_{i=1}^{K} P_i^2 \tag{20}$$

where p_i refers to probability of the of class i. In the case of two categories, it is relatively easier, assuming that the first sample probability is p, then the Gini coefficient formula is as follow

Some specific emotions are difficult to identify to a certain extent. In response to this problem, multiple random forests are established to individually identify certain specific emotions. Assume that there are N types of emotions to be identified. Each time a random forest is used to identify them, the number of random forests required is

$$M_{RF} = 2\left|\frac{N}{2}\right| - 1, N = 1, 2, \ldots, n \tag{21}$$

3.4 Two-stage Fuzzy Fusion based Convolution Neural Network for Dynamic Facial Expression and Speech Emotion Recognition

Visual information and speech information are both important in judging emotional states, and they have a certain degree of complementarity. In this section, a two-stage fuzzy fusion strategy (TSFFS) is used to fuse the facial and speech emotion modes to obtain the final emotion recognition result. And the steps for implementation of TSFFS are given as follows:

Step 1: The convolutional neural network is used to extract the high-level emotional semantic features of facial expression and speech, which are labeled as F_h

and B_h, and then the principal component analysis method is used to reduce the feature dimensions. The reduced-dimensional emotional semantic features of facial expression and speech are F_p and B_p.

Step 2: Perform canonical correlation analysis (CCA) on F_p and B_p to find the relevant facial and speech features, which is F_c and B_c. The fusion feature $FB = [F_c, B_c]$ is the concatenation of them.

Step 3: The emotion feature FB is input to the Softmax classifier for recognition, and the confidence data obtained is recorded as C_{FB}. At the same time, the confidence data obtained from the facial expression and speech are recorded as C_F, C_B. C_{FB}, C_F and C_B are connected in series to form a decision vector, so $C = [C_F, C_B, C_{FB}]$.

Step 4: Produce recognition results by using Fuzzy Broad Learning System (FBLS) for final decision from confidence data C.

The FBLS algorithm is a combination of Takagi-Sugeno fuzzy model and Broad Learning. The algorithm has the high calculation speed as broad learning, and also has the ability to map continuous functions like T-S fuzzy model. In addition, the model also shows the advantages of ensemble learning, whose structure is shown in Fig. 2.

As can be seen from Fig. 2, the first layer of FBLS is the fuzzification. The confidence data $C \in \Re^{U \times V}$ input at this layer is fuzzified by a Gaussian membership function,

$$\xi_{iv}^{m}(c) = e^{-\left(\frac{c-\kappa_{iv}^{m}}{\sigma_{iv}^{m}}\right)^2} \tag{22}$$

where κ_{iv}^m and σ_{iv}^m are the center and width parameters, respectively

The second layer is the fuzzy inference layer. Each node represents a fuzzy rule, which is the product of the membership of the previous layer and the trigger intensity of the rule. The formula for the trigger strength is

$$\omega_{ui}^{m} = \prod_{v=1}^{V} \xi_{iv}^{m}(c_{uv}) \tag{23}$$

We normalize the trigger intensity of the second layer to the third layer, so the nodes in the third layer correspond to the nodes in the second layer, the trigger intensity is given as

$$\bar{\omega}_{ui}^{m} = \frac{\omega_{ui}^{m}}{\sum_{i=1}^{I_m} \omega_{ui}^{m}} \tag{24}$$

In the fourth layer, the trigger strength of the third layer is weighted as

$$G_{ui}^{m} = \bar{\omega}_{ui}^{m} g_{ui}^{m} \tag{25}$$

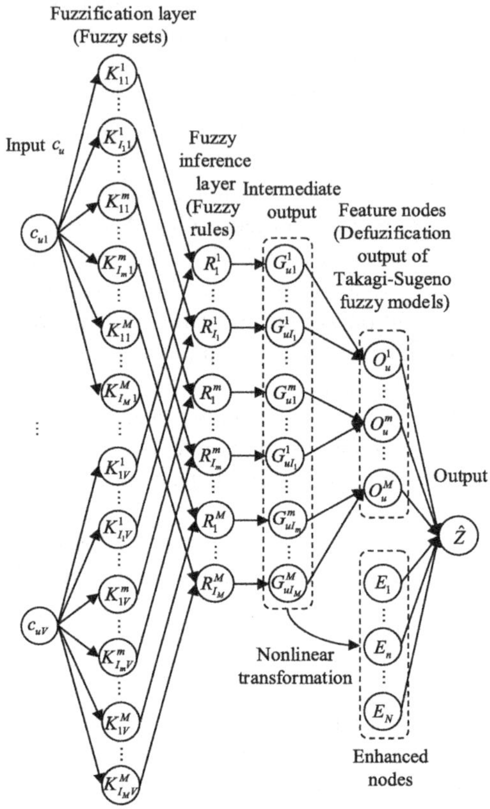

Fig. 2 Structure of FBLS

4 Emotion Intention Understanding

This section mainly introduces two models of emotion intention understanding, the first is three-layer weighted fuzzy support vector regression (TLWFSVR) model, the other is Two-layer Fuzzy SVR-TS Model (TLFSVR-TS). In TLWFSVR, nationality, gender, and age are used as additional criteria for understanding emotion intention. Adjusted weighted kernel FCM (AWKFCM) for data clustering constitute a three-layer network with fuzzy support vector regressions (FSVR) for information understanding and weighted fusion for intention understanding. In two-layer FSVR (TLFSVR), the local learning layer and the global learning layer form a two-layer network.

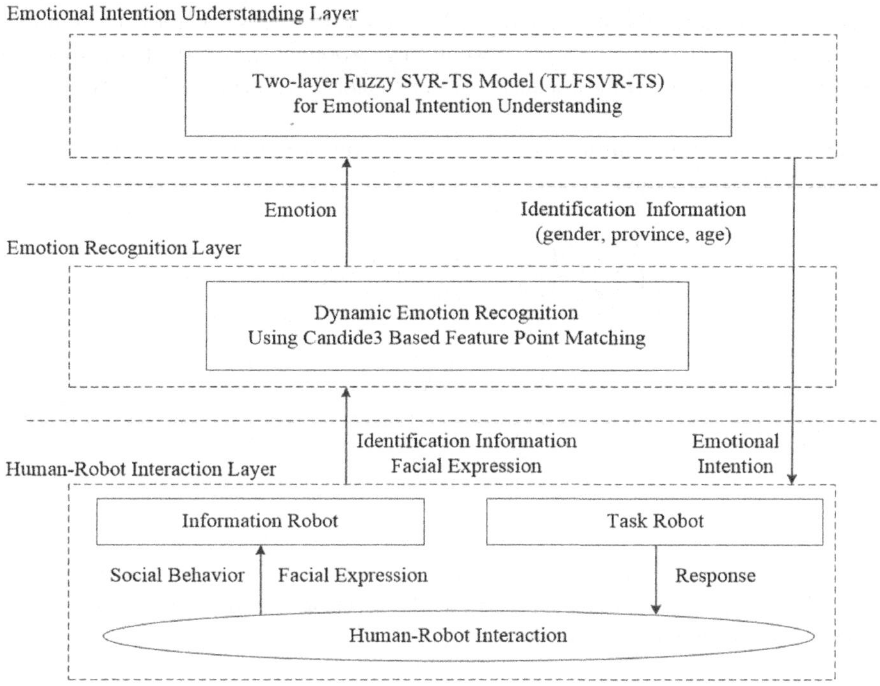

Fig. 3 Structure of dynamic emotion recognition and understanding

4.1 Three-Layer Weighted Fuzzy Support Vector Regression for Emotion Intention Understanding

The structure of the TLFSVR model is shown in Fig. 3. This model mainly realizes emotional understanding in the process of human-robot interaction. Firstly, the facial image is collected in real time through the Candide3 model and feature points are matched [27, 28], and then human emotional understanding is realized on this basis. We introduce human identity information (including gender, age, and nationality) [29], and divide the robots used into two categories, and information robot and task robot. Information robot mainly collect the identity information of the interactors, including age, gender and nationality, and the interaction of Types of drinks that will be selected under different emotional states. The task robot executes the corresponding feedback behavior according to the selection of the information robot.

First of all, by using the fuzzy c-means (FCM) algorithm [30, 31], the training data are classified into several subsets, and it aims to minimize the objective function as follows,

$$f(U, V) = \sum_{i=1}^{l} \sum_{k=1}^{C} (\mu_{ik})^m \, \|X_i - V_k\|^2 \tag{26}$$

where $l \in N$ denotes the number of training data, $C \in [2, l)$ denotes the number of clusters, $m \in [1, +\infty)$ denotes a weighting exponent, $\mu_{ik} \in [0, 1]$ denotes the membership grade of training data X_i in the cluster k, V_k denotes the center of the cluster k.

Approximate optimization of (26) is via iteration, with the updating of membership and the cluster center V_k by

$$\mu_{ik} = \left(\sum_{k=1}^{C} \left(\frac{d_{ij}}{d_{ik}} \right)^{\frac{2}{(m-1)}} \right)^{-1}, 1 \leqslant i \leqslant l, 1 \leqslant j \leqslant C \tag{27}$$

$$V_k = \frac{\sum_{i=1}^{l} (\mu_{ik})^m X_i}{\sum_{i=1}^{l} (\mu_{ik})^m}, 1 \leqslant j \leqslant C \tag{28}$$

where $d_{ik} = \|X_i - V_k\|$.

When termination criterion is satisfied, i.e., $\|X_i - V_k\| \leqslant \epsilon$, the iteration will stop, where $V = (v_1, v_2, \cdots, v_C)$, and ϵ is the given sensitivity threshold (Table 2).

Then the spread width is obtained by

$$\delta_k^o = \sqrt{\frac{\sum_{i=1}^{l} (\mu_{ik})^m \left\| x_i^o - v_i^o \right\|^2}{\sum_{i=1}^{l} (\mu_{ik})^m}}, 1 \leqslant k \leqslant C, 1 \leqslant o \leqslant N. \tag{29}$$

Based on the cluster centers and spread width, the training data is separated into subsets as (30),

$$D_k = \left\{ (X_i, y_i) \| v_k^o - \eta\delta_k^o \leqslant x_i^o \leqslant v_k^o + \eta\delta_k^o \right\}, 1 \leqslant i \leqslant l, 1 \leqslant o \leqslant N, 1 \leqslant k \leqslant C \tag{30}$$

where η denotes a constant, which is to control the overlap region of the training subsets, and it will increase as the size of training subsets increase.

Second, regression function of each cluster is calculated as shown below,

$$SVR_k = \sum_{i=1}^{l_k} \left(\alpha_{i,k}^* - \alpha_{i,k} \right) k(X_i, X) + b_k, X \in D_k \tag{31}$$

where l_k is the number of training data in the kth subset, the $\alpha_{i,k}^*$, $\alpha_{i,k}$ and b_k are obtained via the training of SVR.

Table 2 Two-layer fuzzy support vector regression algorithm.

Algorithm 2.
1. Initialization.
Input: *X, N, C, l, V*.
Output: y.
where X is the training data, $l\epsilon$ N is the number of training data, C is the number of clusters, V is
the center of cluster and $V = (v_1, v_2, ...v_C)$
2. Termination Check.
(a) If $\parallel X_i - V_k \parallel \geq \epsilon$,then the iteration will stop.
(b) Else go to the 3.
3. For i=1,...l Do
(a) Update the membership and the cluster centers, according to (26), and the output $y(X_i)$.
(b) Calculate the spread width using (26), then divide training data into several subsets.
(c) Take age, gender, province into account, respectively, and construct k_{th} the regression function,
according to (28).
(d) Obtain the fuzzy weighted of the k_{th} SVR, according to (29).
(e) Get the output $y(X_i)$ of the proposed TLFSVR.
4. End For.
5. Update $y(X_i)$.
6. Goto 2.

Finally, by using the fuzzy weighted average algorithm [32] fusion model, output of global learning is constructed based on the triangle membership functions and the SVRs. The membership function is defined as,

$$A_k(X_i) = \max\left(\min\left(\frac{x_i^o-(v_k^o-\eta\delta_i^o)}{v_k^o-(v_k^o-\eta\delta_k^o)}, \frac{(v_k^o+\eta\delta_k^o)-x_i^o}{(v_k^o+\eta\delta_k^o)-v_k^o}\right), 0\right) \tag{32}$$

where $A_k(X_i)$ is the fuzzy weight of the kth SVR.

Then, the global output of the proposed TLFSVR is calculated as follow,

$$I_i(X_i) = \frac{\sum_{k=1}^{C} A_k(X_i)SVR_k(X_i)}{\sum_{k=1}^{C} A_k(X_i)}, 1 \leqslant i \leqslant l, 1 \leqslant k \leqslant C. \tag{33}$$

4.2 Dynamic Emotion Understanding in Human-Robot Interaction Based on Two-layer Fuzzy SVR-TS Model

The TS fuzzy model can be used to describe a complicated nonlinear system, where input sets are decomposed into several subsets and each one can be represented by a simple linear regression model. The typical TS fuzzy rule is described as follow,

R^i: If $z_1(t)$ is F_i^1 and if $z_2(t)$ is F_i^2 and... and if $z_n(t)$ is F_i^n, then $\dot{x}(t) = Ax(t)$.

And the nonlinear model is given as follows,

$$\dot{x}(t) = f(x(t)) \tag{34}$$

where $f(\cdot)$ is a nonlinear function.

Deriving from TS fuzzy model, a model is exited as follow,

$$\dot{x}(t) = \sum_{i=1}^{R} h_i(z(t))A_i x(t) = A_z x(t) \tag{35}$$

where $x \epsilon \mathbb{R}^n$ denotes the state vector, $h_i(z)$ denotes the membership functions, $z(t)$ denotes the premise variable with bounded and smooth in a compact set of the state space.

In (35), the membership functions are calculated as

$$h_i(z) = \frac{\omega_i(z)}{\sum_{i=1}^{R} \omega_i(z)} \tag{36}$$

$$\overline{y} = \frac{\sum_{i=1}^{R_n} h_i y_i}{\sum_{i=1}^{R_n} h_i} \tag{37}$$

$$\sigma = \frac{\sum_{i=1}^{R_n} (y_i - \overline{y})\, h_i}{\sum_{i=1}^{R_n} h_i} \tag{38}$$

$$\mu(y_i, \overline{y}, \sigma) = \begin{cases} \frac{\sigma - |y_i - \overline{y}|}{\sigma}, \; y \in [\overline{y}_i - \sigma, \overline{y}_i + \sigma] \\ 0, \quad y \in R - [\overline{y} - \sigma, \overline{y} + \sigma] \end{cases} \tag{39}$$

where $\omega_i(z)$ is the intensity of intention to drink in different emotions. Note that $h_i(\cdot)$ satisfies $0 \le h_i(\cdot) \le 1$. If the input vector is $X = [x_1, x_2, \cdots x_N]$, then the output will be obtained, shown as follow,

$$y_i = f[\mu(y_i, \overline{y}, \sigma)] \tag{40}$$

At the same time, thirty volunteers' intentions to drink in the home environment have been collected. From the data, we can find that women's and men's drink-

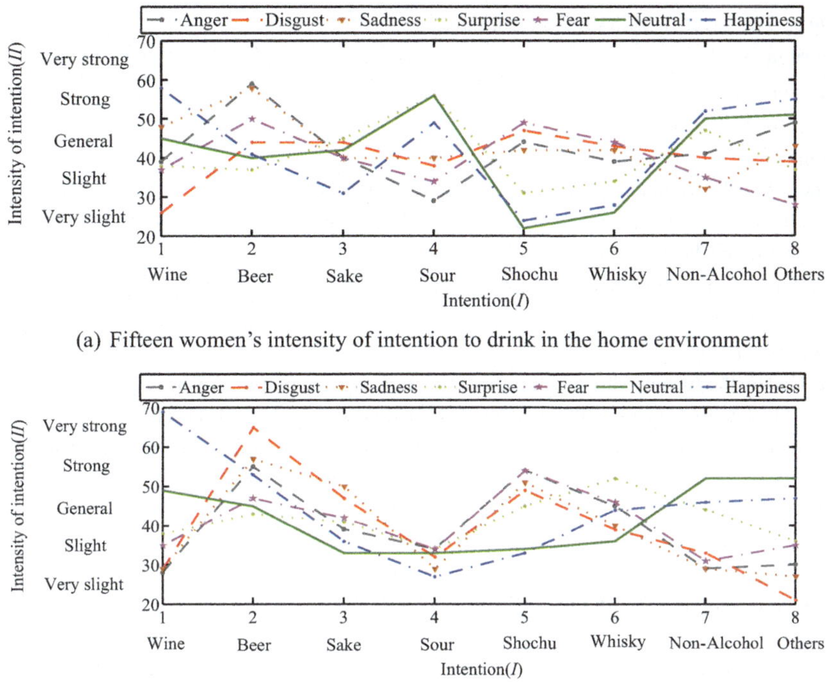

(a) Fifteen women's intensity of intention to drink in the home environment

(b) Fifteen men's intensity of intention to drink in the home environment

Fig. 4 Human's intensity of intention to drink

ing intention is different in some expression. The women's and men's intensity of intention to drink are shown in Fig. 4a and b, respectively.

As drinking preferences are not the same, the TLFSVR-TS is proposed to solve the above problems. And then, different membership functions are used to revise the outputs, as follows,

$$\hat{y} = \begin{cases} \max\left[\mu\left(y_i, \overline{y}, \sigma\right)\right] \\ mean\left[\mu\left(y_i, \overline{y}, \sigma\right)\right] \\ median\left[\mu\left(y_i, \overline{y}, \sigma\right)\right] \end{cases} \tag{41}$$

R^i: If $E = E_i$, $G = G_i$, $I = I_i$, $II = II_i$, then

$$y_i = \begin{cases} I_{TLFSVR} \quad \& \quad h_{TLFSVR} = II_{TLFSVR} / \sum_{j=1}^{R_n} II_j, \quad I_i = I_{TLFSVR} \\ I_i \quad \& \quad h_i = II_i / \sum_{j=1}^{R_n} II_j, \quad I_i \neq I_{TLFSVR} \end{cases} \tag{42}$$

Table 3 Two-layer fuzzy SVR-TS algorithm.

Algorithm 3.
1. Initialization.
Input: X, N, C, l, V.
Output: y.
where X is the training data, $l \in N$ is the number of training data,
C is the number of clusters, V is the center of cluster and $V = (v_1, v_2, ...v_C)$.
2. Termination Check.
(a) If $\| X_i - V_k \| \geq \epsilon$, then the iteration will stop.
(b) Else go to 3.
3. For i=1,...l Do
(a) Update the membership and the cluster centers, and the output $y(X_i)$.
(b) Calculate the spread width, then divide training data into several subsets.
(c) Take gender, province, age into account, respectively, and construct $k_{th}$the regression function.
(d) Obtain the fuzzy weighted of the k_{th} SVR.
(e) If the output is not matched with the real intention, then use the TS model, according to (41), else go to (f).
(f) Get the output $y(X_i)$ of the proposed TLFSVR-TS.
4. End For.
5. Update $y(X_i)$.
6. Goto 2.

where i is the number of fuzzy rules, from 1 to 56 (= 7 emotions $\times$ 8 intentions), and each emotion corresponds to 8 intentions, thus $R_n = 8$.

Then the output of the proposed TLFSVR-TS can be culculated as (Table 3)

$$\hat{y} = \frac{\sum_{i=1}^{R_n} h_i y_i}{\sum_{i=1}^{R_n} h_i} \tag{43}$$

5 Experiments and Applications of Emotional Human-Robot Interaction System

In order to realize a human-robot interaction system with certain emotion recognition and intention understanding, and to establish a natural and harmonious human-robot interaction process, this section proposes a human-robot interaction system scheme based on affective computing. First, we presented the overall architectural design of the system, and then presented the application experience based on the human-robot emotional interaction system based on application scenarios.

5.1 Multimodal Emotional Human-Robot Interaction System

The ability of robots to correctly recognize emotional states, understand emotional intentions, and give emotional feedback is an important content and challenge in the study of emotional robot systems. Therefore, we have constructed a natural interactive emotional robot system under the Internet architecture, enabling users to conduct natural emotional interactions with emotional robots through multiple interaction channels.

Based on the facial expression, speech emotion, body gesture, and personalized information, as well as environmental information during human-robot interaction, we analyze the correlation between human-robot interaction process data and emotional intention in specific environmental scenarios, and establish multi-dimensional emotional intention understanding model, to build an emotional robot system under the Internet, and realize real-time remote information interaction of multiple emotional robots in specific environmental scenarios.

The system structure can be divided into three layers: Web client layer, network service layer and robot local service layer. The web client layer mainly provides users with an human-robot interactive control interface, and monitors the task execution and the operation of the robot equipment in real time through the interface; network service layer mainly receive instructions issued by the web client layer, forward specific instructions to the robot local service layer, or directly call the corresponding service. At the same time, this layer also needs to accept the emotional intent-related information obtained by robot local service layer and complete the corresponding computing tasks (feature extraction and recognition of emotional intention-related information, multimodal emotional intention understanding) and feed back the calculation results (behavioral decision and coordination of affective robots) to the local control layer. The robot local service layer mainly obtain information about emotional intention and send it to the network service layer, and perform calculation results fed back from the network service layer.

The emotional robot in the system has functions of facial expression recognition, speech recognition, body gesture recognition, emotion expression, emotional intention understanding, and natural human-robot interaction. We use Kinect and wearable device to obtain facial expression, speech, body gesture. In the Windows/Linux environment, by using languages such as Matlab, C++/C, or Python, the proposed algorithms for multimodal emotion recognition, emotional intention understanding, and natural human-robot interaction technology are programmed and compare with existing algorithms to verify the feasibility and superiority, and then to verify the correctness and effectiveness of the theoretical research.

The system takes the school teaching scene and the laboratory discussion scene as the background, constructs the human-robot interaction process, and explores the application of the natural interactive emotional robot system under the Internet. In the school teaching scene, the emotional robot identifies the teachers and students' emotion and intention information to obtain the students' degree of knowledge understanding during the lecture, provide reference for the optimization of teaching mode, and make the teaching more intelligent. At the same time, the emotional robot also plays an active atmosphere in the classroom, helping the teaching process to progress smoothly. In the laboratory discussion scene, the emotional robot recognizes the speech content of the participants and record the content of the meeting. It also uses the facial expression, speech, body gesture of the participants to analyze their emotional intention and generate corresponding emotional expression to adjust the scene atmosphere and provide certain services. In the acquisition of communication information, the emotional robots obtain communication information through sensing devices, including facial expression, speech, body gesture, and personalized information. Each emotional robot transmits the communication information to the emotional intention understanding via Ethernet / wireless network. The high-performance computer in the emotional intention understanding perform complex computing on the multidimensional emotional intention model based on the analysis of emotional intention-related information. The goal of behavior decision-making and coordination is mainly to determine the robot's behavior based on the current emotional intention. Based on the calculation of the reward and punishment function, the high-performance computer performs robot behavior decision-making and coordination, and sends the emotion to the execution mechanism of each emotional robot in real time.

5.2 *The Application Experiment of Emotional Human-Robot Interaction System*

In this section, we apply the proposed emotional recognition algorithm and emotional intention understanding algorithm to our emotional robot system, and mainly introduce the research results of softmax regression based deep sparse autoencoder network for facial emotion recognition and TLWFSVR model for emotional intention understanding in emotional robot system.

System workflow uses kinect that on the top of wheeled robot to track facial expression images firstly, then invokes facal emotion recognition algorithm to feature extraction and emotion recognition, which is designed by MATLAB, and relies on the affective computing workstations as shown in Fig. 5.

Moreover, JAFFE facial expression database from Japan is used as training sample, which contains 213 facial expression images, including ten subjects, seven types of basic expressions. The sample images of JAFFE are shown in Fig. 6.

Visualization of underlying characteristics for weights that learned by sparse autoencoder network is designed, and the numbers of neurons node in the hidden layer are set to 140 to obtain an initial image of characteristics' visualization.

Figures 7 and 8 are characteristic matrix of weights visualization images before and after fine-tune overall weights, it can be seen that the features self-learned by overall network look more sophisticated after fine-tune the weights to ensure high

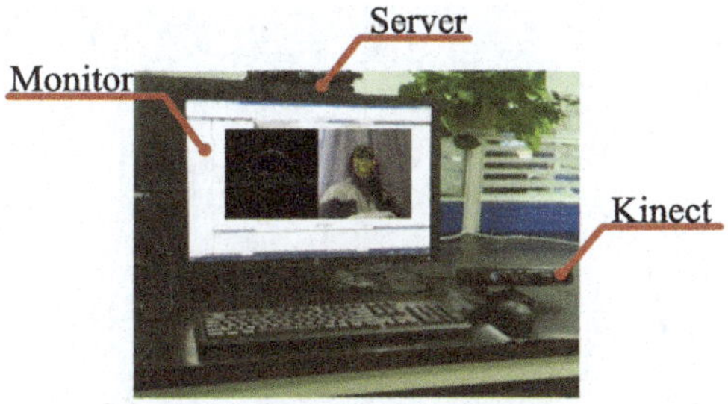

Fig. 5 Emotional computing workstation

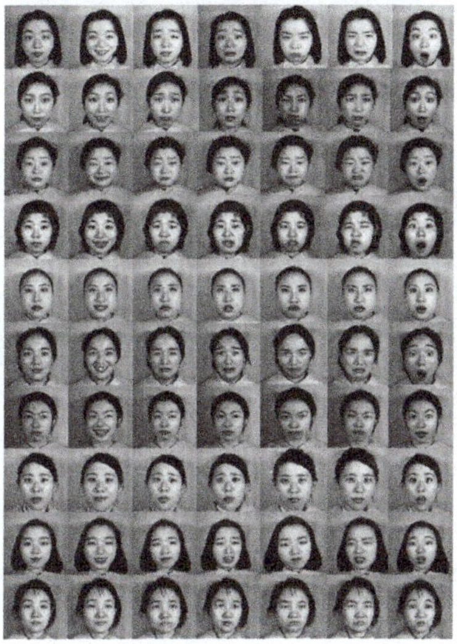

Fig. 6 JAFFE facial expression library samples

recognition accuracy. The relation curve between the seven kinds of facial emotion recognition rate and the sparsity parameter is shown in Fig. 9.

According to the comparison of the Fig. 10 with Fig. 11, it indicates that fine-tune can make the overall cost function converge faster, and in the actual test of 182 times when they have converged and stopped training.

It can be seen from Figs. 12 and 13 that increase the numbers of hidden layer nodes increase the recognition rate of emotion, however, too large numbers of hidden layer nodes are helpless to improve the recognition rate of emotion. It will cause network over fitting and not conducive to expression feature extraction. After fine-tune the weight matrix, the recognition rate will be improved to a certain extent, and offset the impact of the hidden layer nodes numbers change, shown in Fig. 14.

To verify the accuracy of emotion recognition, we performed simulation experiments using MATLAB and the results are shown in Table 4.

Fig. 7 The weights visualization of the underlying characteristics

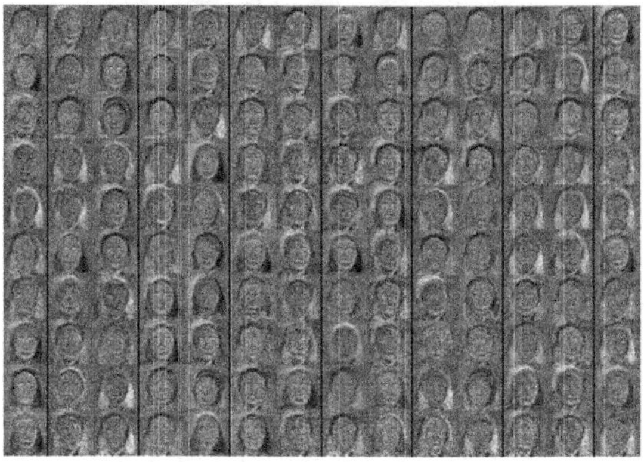

Fig. 8 The fine-tuning the weights visualization of the underlying characteristics

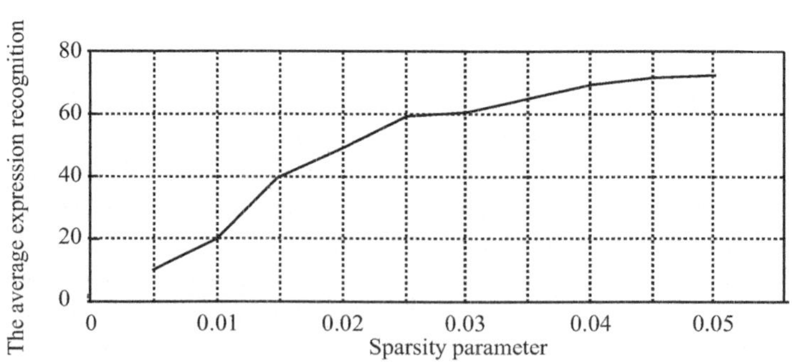

Fig. 9 The influence of sparse parameter and fine-tuning the weights on the rate of facial emotion recognition

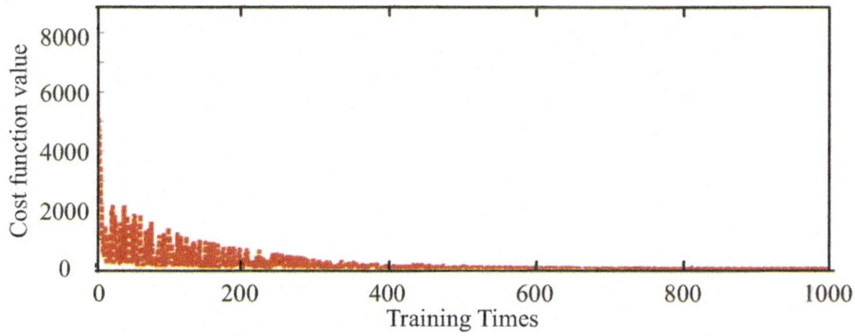

Fig. 10 The convergence situation of overall cost function before fine-tune

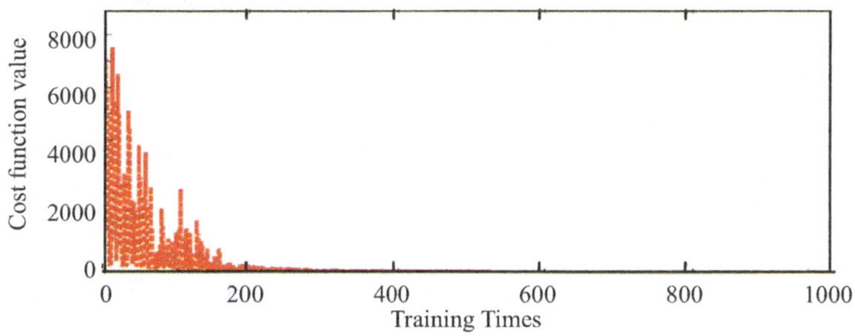

Fig. 11 The convergence situation of overall cost function after fine-tune

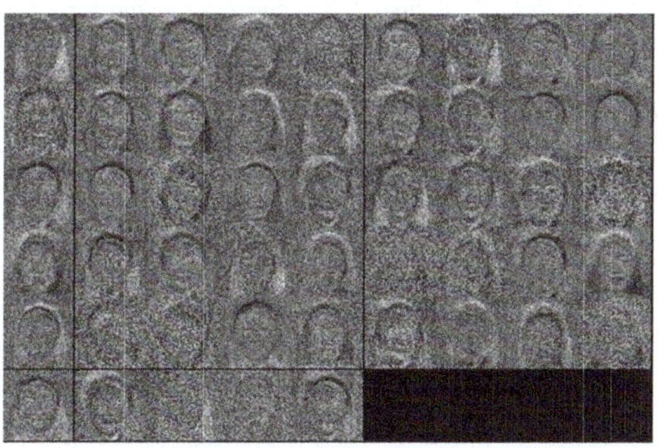

Fig. 12 Weights visualizations for the hidden layer node number is 50

According to Table 4, the average recognition accuracy of Softmax regression is 73.333% in the final test. Nevertheless, if we use unlabeled training data to train deep autoencoder network firstly, and then train the Softmax regression model, and we can find that the times of iteration convergence are only 181, and the accuracy is 94.761%. By fusing the Softmax retrogression in deep learning, it can be seen that the features self-learned by overall network looks more sophisticated after fine-tune, and fine-tune makes the overall cost function converge faster, which overcomes the local

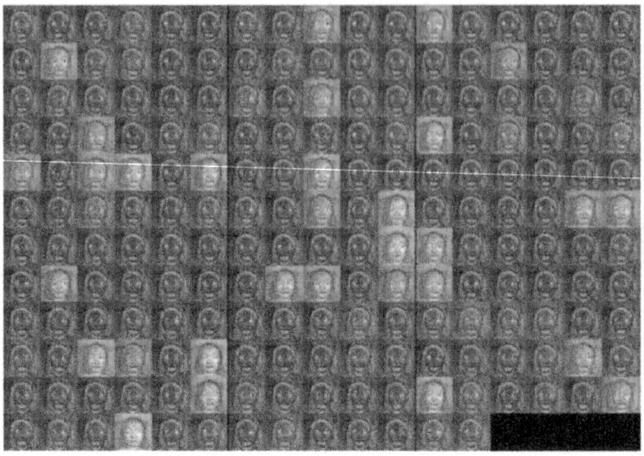

Fig. 13 Weights visualizations for the hidden layer node number is 200

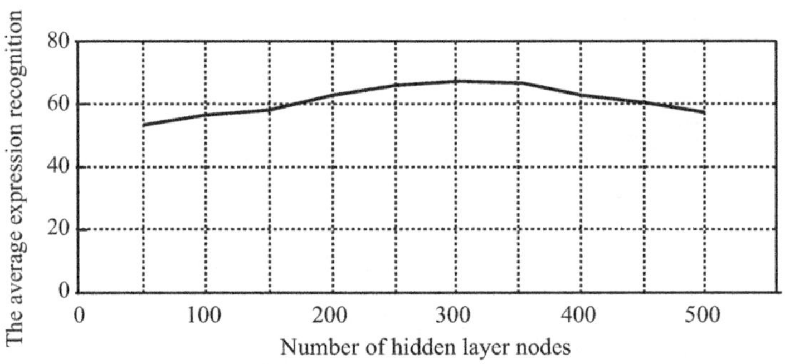

Fig. 14 The influence of the number of hidden layer nodes on the expression recognition rate

Table 4 Comparison of emotion recognition experimental results

Emotion	Softmax regression (%)	SRDSAN (%)
Natural	86.667	100.00
Happy	80.000	93.333
Sad	63.333	100.00
Fear	63.333	76.667
Angry	83.333	100.00
Disgust	66.667	93.333
Surprise	70.000	100.00
Average	73.333	94.761

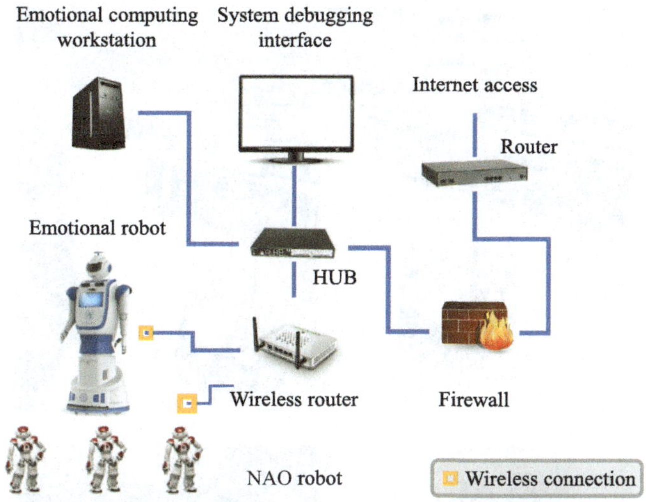

Fig. 15 Emotional social robot system

extrema and gradient diffusion problems. Moreover, this shows that the characteristics learned by our self-learning sparse autoencoder network are more representative than the characteristics in initial input data, which is the typical difference between conventional training methods and deep learning training methods. In addition, texture and shape feature changes in the three key parts (region of interest, ROI) such as eyebrows, eyes, mouth may reflect changes in facial expression features exactly.

The proposed Softmax-regression-based deep sparse autoencoder network will be applied to the preliminary application experiments by using the developing ESRS, as shown in Fig. 16. It is an important way to autonomous robots for emotion recognition. First, the Kinect is used for image acquisition, then, the image was conducted on ROI clipping. With that, the image information is applied to the proposed method Softmax-regression-based deep sparse autoencoder network. In the end, facial expression results will be output as shown in Fig. 16.

6 Conclusion

Research on multimodal emotion feature extraction can promote emotion recognition, regions of interest based feature extraction is designed for facial expression. In gesture feature extraction, high-level gesture image representation is completed, which first performs foreground segmentation and SURF feature extraction, then sparse coding is used to extract deep features. With regard to speech emotional feature extraction, we adopt the non-personalization speech emotional feature based on

Fig. 16 The preliminary application experiments

derivative, and the FCM is used for data clustering. The ability of feature sets to represent emotional information directly affects the results of emotion recognition.

The task of multimodal emotion recognition is to find out the mapping (relationship) between the emotional features and human emotions. Softmax regression-based deep sparse autoencoder network is proposed to recognize facial emotion in human-robot interaction. It aims to handle large data in the output of deep learning by using SR. In gesture recognition, multi-SVM based Dempster-Shafer theory using sparse coding feature is proposed. And in the aspect of speech emotional, we introduced a two-layer fuzzy multiple random forest for speech emotion recognition. In addition, the two-stage fuzzy fusion based convolution neural network is proposed for dynamic facial expression and speech emotion recognition.

Emotional intention understanding is to discover more detailed intention from emotional surface information and deep cognitive information. Two-layer fuzzy SVR-TS model is proposed for emotion understanding in human-robot interaction, where the realtime dynamic emotion recognition is realized by using Candide3 based feature point matching method, and emotional intention understanding is obtained mainly based on human emotions and identification information. In addition, we proposed a three-layer weighted fuzzy support vector regression (TLWFSVR) model for understanding human intention, and it is based on the emotion-identification information in human-robot interaction.

Finally, in order to realize a human-robot interaction system with certain emotion recognition and intention understanding, and to establish a natural and harmonious human-robot interaction process, we proposed a human-robot interaction system scheme based on affective computing. Combining specific scenarios, the proposed multimodal emotion recognition and intention understanding algorithm are applied to human-robot interaction systems, which achieved good results.

References

1. Mehrabian, A.: Communication without words. Psychol. Today **2**(4), 53–56 (1968)
2. Schuller, B., Batliner, A., Steidl, S., Seppi, D.: Recognising realistic emotions and affect in speech: state of the art and lessons learnt from the first challenge. Speech Commun. **53**(9), 1062–1087 (2011)
3. Chen, L., Zheng, S.K.: Speech emotion recognition: Features and classification models. Digital Signal Process. **22**(6), 1154–1160 (2012)
4. Chen, L.F., Zhou, M.T., Su, W.J., Wu, M., She, J.H., Hirota, K.: Softmax regression based deep sparse autoencoder network for facial emotion recognition in human-robot interaction. Inf. Sci. **428**, 49–61 (2018)
5. Chen, L.F., Feng, Y., Maram, M.A., Wang, Y.W., Wu, M., Hirota, K., Pedrycz, W.: Multi-SVM based dempster-shafer theory for gesture intention understanding using sparse coding feature. Appl. Soft Comput. (2019). https://doi.org/10.1016/j.asoc.2019.105787
6. Chen, L.F., Su, W.J., Feng, Y., Wu, M., She, J.H., Hirota, K.: Two-layer fuzzy multiple random forest for speech emotion recognition in human-robot interaction. Inf. Sci. **509**, 150–163 (2020)
7. Kim, Y., Provost, E.M.: ISLA: temporal segmentation and labeling for audio-visual emotion recognition. IEEE Trans. Affect. Comput. **2702653**, 2017 (2017). https://doi.org/10.1109/TAFFC
8. Barros, P., Jirak, D., Weber, C.: Multimodal emotional state recognition using sequence-dependent deep hierarchical features. Neural Netw. **72**, 140–151 (2015)
9. Soleymani, M., Asghariesfeden, S., Fu, Y.: Analysis of EEG signals and facial expressions for continuous emotion detection. IEEE Trans. Affect. Comput. **7**(1), 17–28 (2016)
10. Wegner, D.M.: The Illusion of Conscious Will. MIT Press, Cambridge (2002)
11. Emotion-Age-Gender-Nationality based intention understanding in human-robot interaction using two-layer fuzzy support vector regression. Int. J. Soc. Robot. **7**(5), 709–729 (2015)
12. Chen, L.F., Wu, M., Zhou, M.T.: Dynamic emotion understanding in human-robot interaction based on two-layer fuzzy SVR-TS model. IEEE Trans. Syst. Man Cybern. Syst. **50**(2), 490–501 (2017)
13. Chen, L.F., Zhou, M.T., Wu, M.: Three-layer weighted fuzzy support vector regression for emotional intention understanding in human-robot interaction. IEEE Trans. Fuzzy Syst. **26**(5), 2524–2538 (2018)
14. Lu, X., Bao, W., Wang, S.: Three-dimensional interfacial stress decoupling method for rehabilitation therapy robot. IEEE Trans. Ind. Electron. **64**(5), 3970–3977 (2017)
15. Turner, J., Meng, Q., Schaefer, G.: Distributed task rescheduling with time constraints for the optimization of total task allocations in a multi-robot system. IEEE Trans. Cybern. **48**(9), 2583–2597 (2018)

16. Lui, J.H., Samani, H., Tien, K.Y.: An affective mood booster robot based on emotional processing unit. In: Proceedings of 2017 International Automatic Control Conference, Pingtung, China, pp. 1–6 (2018)
17. Somchanok, T., Michiko, O.: Emotion recognition using ECG signals with local pattern description methods. Int. J. Affect. Eng. **15**(2), 51–61 (2016)
18. Klug, M., Zell, A.: Emotion-based human-robot interaction. In: Proceedings of the 9th IEEE International Conference on Computational Cybernetics, Tihany, Hungary, pp. 365–368 (2013)
19. Boccanfuso, L., Barney, E., Foster, C.: Emotional robot to examine differences in play patterns and affective response of children with and without ASD. In: Proceedings of the 11th ACM/IEEE International Conference on Human-Robot Interaction, Christchurch, New Zealand, pp. 19–26 (2016)
20. Phaisangittisagul, E., Thainimit, S., Chen, W.: Predictive high-level feature representation based on dictionary learning. Expert Syst. Appl. **69**(2017), 101–109 (2017)
21. Li, B., Zhao, F., Su, Z.: Example-based image colorization using locality consistent sparse representation. IEEE Trans. Image Process **26**(11), 5188–5202 (2017)
22. Stefania, B., Alfonso, C., Peelen, M.V.: View-invariant representation of hand postures in the human lateral occipitotemporal cortex. NeuroImage **181**(2018), 446–452 (2018)
23. Bay, H., Ess, A., Tuytelaars, T., Gool, L.V.: SURF: speeded up robust features. In: Proceedings of Computer Vision and Image Understanding, pp. 346–359 (2008)
24. Zhu, X., Li, X., Zhang, S.: Robust joint graph sparse coding for unsupervised spectral feature selection. IEEE Trans. Neural Netw. Learn. Syst. **28**(6), 1263–1275 (2017)
25. Eyben, F., Wollmer, M., Graves, A.: Online emotion recognition in a 3-D activation-valence-time continuum using acoustic and linguistic cues. J. Multimodal User Interfaces **3**, 7–19 (2010)
26. Chen, C., Jafari, R., Kehtarnavaz, N.: Improving human action recognition using fusion of depth camera and inertial sensors. IEEE Trans. Human-Mach. Syst. **45**(1), 51–61 (2015)
27. Wang, J.N., Xiong, R., Chu, J.: Facial feature points detecting based on Gaussian mixture models. Pattern Recognit. Lett. **53**, 62–68 (2015)
28. Asteriadis, S., Nikolaidis, N., Pitas, I.: Facial feature detection using distance vector fields. Pattern Recognit. **42**(7), 1388–1398 (2009)
29. Chen, L.F., Liu, Z.T., Wu, M., Ding, M., Dong, F.Y., Hirota, K.: Emotion-age-gender-nationality based intention understanding in human-robot interaction using two-layer fuzzy support vector regression. Int. J. Soc. Robot. **7**(5), 709–729 (2015)
30. Lin, K.P.: A novel evolutionary kernel intuitionistic fuzzy c-means clustering algorithm. IEEE Trans. Fuzzy Syst. **22**(5), 1074–1087 (2014)
31. Nguyen, D.D., Ngo, L.T.: Multiple kernel interval type-2 fuzzy c-means clustering. In: IEEE International Conference on Fuzzy Systems, Hyderabad, India, pp. 1–8 (2013)
32. Dong, W.M., Wong, F.S.: Fuzzy weighted averages and implementation of the extension principle. Fuzzy Sets Syst. **21**(2), 183–199 (1987)

Dynamic Multi-objective Optimization for Multi-objective Vehicle Routing Problem with Real-time Traffic Conditions

Changhe Li, Shengxiang Yang, and Sanyou Zeng

Abstract Dynamic multi-objective optimization plays an important role in planning and decision-making. This chapter provides an example of the application of dynamic multi-objective optimization to a practical dynamic multi-objective optimization problem which called multi-objective vehicle routing problem with real-time traffic conditions and then introduces an offline optimization first and then online optimization mechanism. In the offline optimization phase, we combined the adaptive local search algorithm based on Pareto control with the dynamic constrained multi-objective optimization algorithm to obtain a high-quality offline solution to reduce the pressure of online optimization. Experimental results show that the dynamic optimization mechanism can obtain an excellent solution that satisfies the constraints of the multi-objective vehicle routing problem with real-time traffic conditions.

Keywords Vehicle routing problem · Local search · Dynamic multi-objective optimization · Constrained optimization

C. Li (✉)
School of Automation, China University of Geosciences, Wuhan 430074, China

Hubei Key Laboratory of Advanced Control and Intelligent Automation for Complex Systems, Wuhan 430074, China
e-mail: lichanghe@cug.edu.cn

S. Yang
School of Computer Science and Informatics, De Montfort University, Leicester LE1 9BH, United Kingdom

S. Zeng
School of Mechanical Engineering and Electronic Information, China University of Geosciences, Wuhan 430074, China

M. Wu et al. (eds.), *Developments in Advanced Control and Intelligent Automation for Complex Systems*, Studies in Systems, Decision and Control 329,
https://doi.org/10.1007/978-3-030-62147-6_11

Abbreviations

DMOP	Dynamic multi-objective optimization problem
VRP	Vehicle routing problem
VRPTW	VRP with time windows
MOVRPRTC	Multi-objective VRP with real-time traffic conditions
DVRP	Dynamic vehicle routing problem
ALNS	Adaptive large neighborhood search
EA	Evolutionary algorithm
DMOA	Dynamic multi-objective optimization algorithm
PS	Pareto optimal set
PF	Pareto optimal front

1 Introduction

There are many optimization problems in real life. Including general daily consumptions and some industry manufactures. Optimization plays an important role when facing a choice or making a decision such as keep a balance between expense and life qualities, the trade-off between the cost of production and profits, and so on. T.T. Nguyen has taken an investigation about real dynamic optimization applications [1]. The results show that over 68% of statistical models have more than one objective. In a real application, optimization indicators or objectives may change with time which makes it be a dynamic multi-objective optimization problem (DMOP). The multi-objective vehicle routing problem with real-time traffic conditions is one of them.

In recent years, logistics go through a thriving time with the development of internet shopping. Transporting, storing, packing and dispatching goods has been a complicated task in the whole process of logistics. In 2018, China's total social logistics expenditure accounted for 14.8% of GDP. Every 1% diminution in the total logistics expenses in GDP will reduce hundreds of billions of costs. In the same year, China's total social logistics expense was 13.3 trillion yuan, among which transportation cost was 6.9 trillion yuan. It can be seen that transportation is a significant problem in logistics. The premise of transportation is route planning. Therefore, it makes sense to optimize the vehicle routing problem (VRP).

VRP is a problem that consists of a group of fleets departing from a depot and a group of customers to be delivered by that fleet. The task of this problem is to get a set of lowest cost routes that meet the needs of the customer. The cost can be traveling distance, traveling time, number of vehicles, and so on. The demonstration of VRP is shown in Fig. 1. In this figure, node 0 is the depot, and nodes 1 to 7 are the customers. Since the question was proposed, many variants have appeared, such as the VRP with time windows (VRPTW) [2], the VRP with pickup and delivery [3], the VRP with stochastic demands [4], etc. But, few models take real-time traffic conditions into account. In the actual distribution process, the dynamic change of traffic conditions is one of the main challenges for planning vehicle routes since traffic conditions directly affect the driving time of vehicles on the road network. So we propose a multi-objective VRP with real-time traffic conditions (MOVRPRTC) which based on real road networks.

For decades, various models and methods have been proposed for different dynamic vehicle routing problems (DVRPs) [4]. But heuristic algorithms are the only feasible ways to solve practical DVRPs since exact algorithms could not solve complex DVRPs in an acceptable time [5]. Mohamed [6] proposes a technology that integrates information about future customer needs to improve decision-making on dynamic vehicle routing problem. Chen et al. [7] used an adaptive large neighborhood search (ALNS) algorithm which combined with a metaheuristic procedure to solve the DVRP with a limited number of vehicles and hard-time windows. Sabar et al. [8] proposed an evolutionary algorithm (EA) with a self-adaptive mechanism for vehicle routing problems with traffic congestion. The proposed EA will evolve a set of configurations that encoded into the DVRP solution. In the search process, the different configurations were used to handle dynamic changes. Jia et al. [9] designed a set-based particle swarm optimization algorithm to solve a vehicle routing problem with new orders.

Previous research on DVRP shows that how to deal with dynamics is still a problem to be solved. There are three weaknesses in these methods. The first is that the algorithm does not use the historical information of the problem, so these algorithms cannot quickly adapt to the dynamic environment. The second weakness is that these algorithms do not apply some excellent dynamic multi-objective optimization strategies such as diversity introduction [10, 11], diversity maintenance [12–14], memory-based methods [15, 16], prediction-based methods [17, 18] and other strategies [19, 20]. The third disadvantage is that none of these algorithms use constraint optimization techniques to deal with constraints. To solve the above three weaknesses, we use historical traffic condition information to predict the traffic conditions on a working day, and use prediction-based dynamic multi-objective optimization strategies to establish an offline optimization and then online optimization mechanism, so that the algorithm can calmly cope with the dynamic changes of the environment. In the offline optimization phase, we first apply the dynamic constrained multi-objective optimization algorithm to the VRP, which provides a new idea of constraint treatment in the combinatorial optimization problem.

The rest of this chapter is divided into five parts. Section 2 introduces the relevant content of the dynamic multi-objective optimization algorithm. Section 3 describes the MOVRPRTC. Then we introduce our proposed optimization algorithm for MOVRPRTC in Sect. 4. The experimental results are given in Sect. 5. Section 6 makes a conclusion.

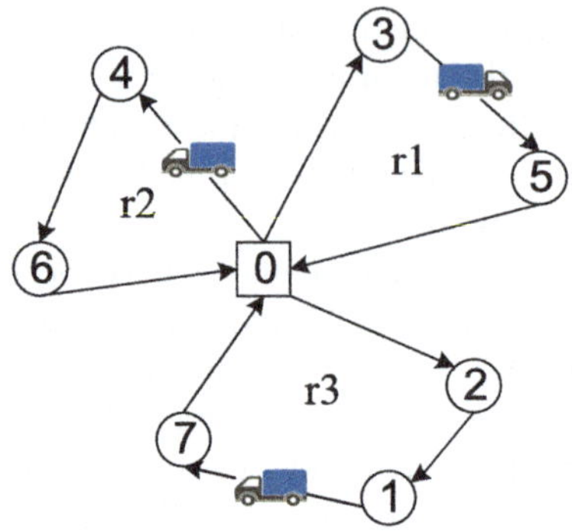

Fig. 1 Demonstration of VRP

2 Background

A dynamic multi-objective optimization problem can be defined as below:

$$\begin{cases} \min\limits_{\mathbf{x}\in R^n} \mathbf{f}(\mathbf{x}(t),t) = \{f_1(\mathbf{x}(t),t), f_2(\mathbf{x}(t),t), ..., f_m(\mathbf{x}(t),t)\} \\ s.t.\ \mathbf{g}(\mathbf{x}(t),t) \le 0 \\ \quad \mathbf{h}(\mathbf{x}(t),t) = 0 \end{cases} \tag{1}$$

In (1), $\mathbf{f}(\mathbf{x}(t), t)$ is an objective vector that includes m objectives, and all sub-objective may change with time. $\mathbf{g}(\mathbf{x}(t), t)$ and $\mathbf{h}(\mathbf{x}(t), t)$ are inequality and equality constraints respectively. And they can both be influenced by dynamic environments.

2.1 *Basic Definitions*

Definition 1 (*Feasible Solution*) For a solution $\mathbf{x} \in R^n$, if $\mathbf{x}$ satisfies the constraints $\mathbf{g}(\mathbf{x}(t), t) \le 0,\ \mathbf{h}(\mathbf{x}(t), t) = 0$ in (1), then $\mathbf{x}$ is called a feasible solution.

Definition 2 (*Feasible Solution Set*) Set of all feasible solutions, recorded as $\mathbf{X}_f$, and $\mathbf{X}_f \subseteq R^n$.

Definition 3 (*Pareto Dominance*) Suppose $\mathbf{x_1}$, $\mathbf{x_2}$ are two feasible solutions of a dynamic multi-objective optimization problem. $\mathbf{x_1}$ dominates $\mathbf{x_2}$, if and only if, $\forall i = 1, 2, \ldots, m,\ f_i(\mathbf{x_1}) \le f_i(\mathbf{x_2}) \wedge \exists j = 1, 2, \ldots, m,\ f_j(\mathbf{x_1}) < f_j(\mathbf{x_2})$, recorded as $\mathbf{x}_1 \prec \mathbf{x}_2$. As shown in Fig.2, $b \prec a$, $c \prec a$, b and c do not dominate each other.

Definition 4 (*Pareto Optimal Solution*) A feasible solution $\mathbf{x}^* \in \mathbf{X}_f$ is called Pareto optimal solution if and only if the following conditions are satisfied $\nexists \mathbf{x} \in \mathbf{X}_f : \mathbf{x} \prec \mathbf{x}^*$.

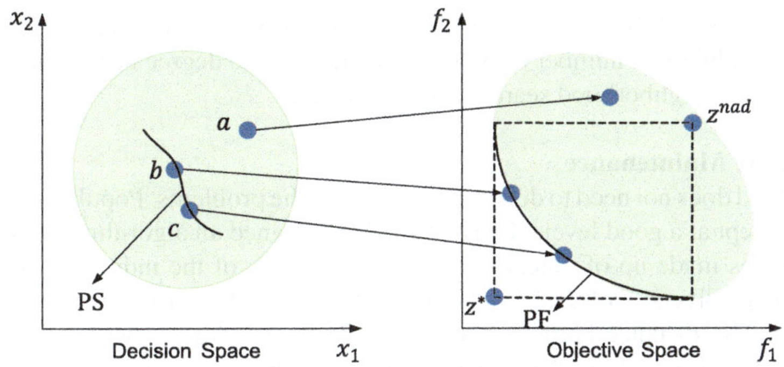

Fig. 2 Correspondence between decision space and objective space

Definition 5 (*Pareto Optimal Set*) The Pareto optimal set is the set of all Pareto optimal solutions, as defined below, $\mathbf{PS}^* \triangleq \left\{\mathbf{x}^* | \nexists x \in \mathbf{X}_f : \mathbf{x} \prec \mathbf{x}^*\right\}$.

Definition 6 (*Pareto Optimal Front*) The surface composed of the objective vectors corresponding to all Pareto optimal solutions in Pareto optimal set is called Pareto optimal front,
that is, $\mathbf{PF}^* \triangleq \left\{\mathbf{F}(\mathbf{x}^*) = (f_1(\mathbf{x}^*), f_2(\mathbf{x}^*), \ldots, f_m(\mathbf{x}^*))^{\mathrm{T}} | x \in \mathbf{PS}^*\right\}$.

2.2 Dynamic Multi-objective Optimization Algorithms

For DMOPs, not only the balance between sub-objectives be considered but also need to respond to dynamic changes for a dynamic multi-objective optimization algorithm (DMOA). The most challenging point is that when the population approximately converges to a PF, genetic diversity will lose heavily. The population will be hard to search for new PS or PF in a limited time once environments changed. Therefore, besides good diversity should be kept, algorithms should also have quick convergence when facing with dynamic changes. There have been several strategies to deal with changing environments.

A. Diversity Introduction

This strategy first detects the changes in the environments and then responds to it. The easiest method is re-initialize once a change is detected in any objective values. But it seems too simple to improve the search efficiency. Cobb [21] investigated the dynamic optimization problems by hyper-mutation. Algorithms detect the state of the environments. The rate of mutation is low when the environments are stable; otherwise, it gets bigger. Deb et al. [10] designed two schemes. One considers replacing a proportion of worst individuals with hyper-mutation individuals in the population. The other one replaces them with random individuals. A method named variable relocation was also proposed [11]. After detecting the changes, each dimension of the individuals will be relocated by the variation of the objectives. Other methods include introducing a number of individuals through the degree of the changes [22] and variables neighborhood search [23] and so on.

B. Diversity Maintenance

This method does not need to detect the changes in the problems. Population diversity is always kept at a good level. Ghosh et al. [24] designed an algorithm whose fitness function was made up of objective values and the age of the individual which can help keep the diversity of the population. Andrea et al. [12] took diversity as another objective. The distance between individuals and sort value is two new objectives that need to be optimized. The same strategy can be found in [25, 26]. Charged swarms are a new scheme inspired by physic. Blackwell [13] designed a charged swarm in dynamic environments and "electron" in the population will keep exploring the search space, and "neutron" will converge to optima or PS. Another effective scheme to maintain diversity is the multi-population algorithm because it has a natural attribute that sub-populations distribute in different places. Sub-population search independently and communicate with each other. There are some strategies to manage sub-populations, concrete designing can refer to [14, 27–29].

C. Memory-based

The memory-based strategy is designed for problems whose optima or PS may change periodically. Yang [15] adopted an archive to store the best individual in every iteration, and retrieve the most similar one as a seed to initialize new population when the environments change. Hendrik [30] divided the whole search space into many subspaces and statistic historical optima located in each subspace. Finally, initializing a new population by the proportion of each subspace. Other memory-based algorithms can refer to [16, 31, 32]

D. Prediction-based

The prediction-based strategy is similar to memory-based that they all want to find a correct place where the PS will be located next time to initialize the new population. For prediction-based methods, they are suitable for dynamic problems to have specific change mode. Iason et al. [33] designed a forward-looking approach to predict the new location of the PS. K.W.Tat et al. [34] predicted the change direction and amplitude of the environments through archive individuals' statistical features. Zhou et al. [17] predicted the new PS by historical PS's centroid and manifold. And Arrchana et al. [18] used the Kalman filter to predict optima.

According to the review of strategies to deal with changing environments, we can use the prediction-based strategy to solve MOVRPRTC. Because the structure of the road network in a specific area remains almost unchanged in the short term, and most people's daily activities are regular. So, the dynamic characteristic of MOVRPRTC has a specific change mode. Based on historical traffic conditions, we can predict the traffic conditions of the working day. Based on the predicted traffic conditions, we can simulate the delivery of orders that need to be delivered on a working day, i.e., offline optimization. After the vehicles depart from the depot, i.e., the working day, the traffic conditions will be different from the predicted traffic conditions, and there may be some new orders that need to serve. For these environmental changes, we can use real-time information to adjust the offline solution and use historical information to predict whether it can meet the needs of new orders, i.e., online optimization.

3 Multi-objective Vehicle Routing Problem with Real-time Traffic Conditions

MOVRPRTC is a dynamic optimization problem based on road networks and real-time traffic conditions. MOVRPRTC is introduced in the following two parts: (1) road network topology; (2) formulation of MOVRPRTC.

3.1 *Road Network Topology*

There are two differences between the road network-based VRP and complete undirected graph-based VRP: (1) the travel time between a vertex in the road network will change according to real-time traffic conditions, where the travel time between vertex is fixed in a complete undirected graph-based VRP; (2) the travel time between any two vertexes in the road network is asymmetric. The road network is a directed graph.

In this chapter, the road network topology in Wuhan Guanggu is taken as an example. The road network is shown in Fig. 3. The Fig. 4 can better explain the role of the road network in MOVRPRTC. In Fig. 4, vehicles can travel from node 1 to node 2 via route 1 or route 2. Vehicles should be able to choose a route with the shortest travel time $T_{12}(t)$ based on real-time traffic conditions. We assume that Route 1 is selected. According to the real-time traffic condition trend in Fig. 5, the $T_{12}(t)$ will be different at different time.

Based on this road network topology, we can establish a multi-objective vehicle routing problem with real-time traffic conditions.

3.2 *Formulation of MOVRPRTC*

Because of the MOVRPRTC is based on a real road network, so the connection relationship between customers can only exist in the route of the road network. In general, the MOVRPRTC is a problem that includes real-time traffic conditions and considers soft time window constraints.

MOVRPRTC is described as follows: in road network topology $G = (V, E)$, $V = \{v_0, \ldots, v_N\}$ represents all the N nodes in the road network, $E = \{\langle i, j\rangle | z_{ij} = 1\}$

Fig. 3 Road network in Wuhan Guanggu Area

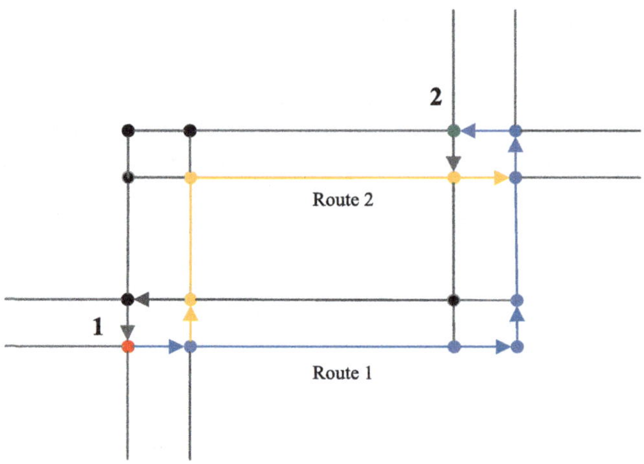

Fig. 4 Example of route selection

denote all the edges formed by the connection in the road network, where z_{ij} are elements of the matrix $Z_{N\times N}$. If $z_{ij} = 1$, the link from v_i to v_j is feasible. The fleet starts from the depot v_0. They need to serve n customers $c_1, c_2, \ldots, c_n$ selected from V and then return to depot v_0. The d_{ij} represents the length of the path from the node v_i to v_j, and the travel time $T_{ij}(t)$ of the path between the node v_i and v_j changes with the time t. The load Q_k of the vehicle k cannot exceed its maximum capacity C. The time window for each customer c_i to be served is $[b_i, e_i]$, and if the arriving time $a_i(t)$ of the vehicle arriving at the customer c_i exceeds e_i, the vehicle is allowed to unload, but there will be a delayed penalty to describe customer dissatisfaction. The waiting time of the vehicle will be accumulated if the arriving time $a_i(t)$ of the vehicle arriving at the customer c_i is earlier than b_i. The time for the vehicle to serve

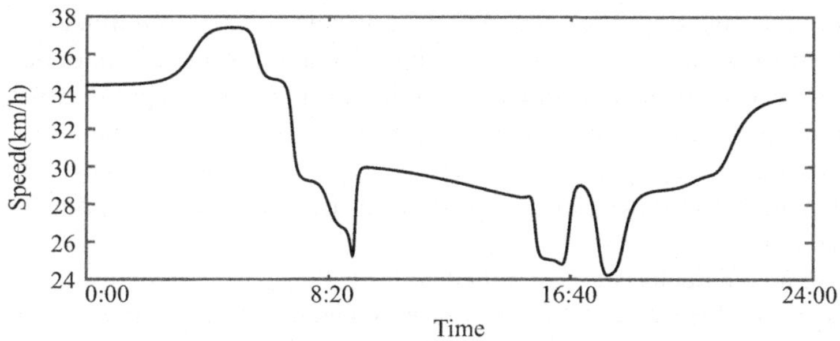

Fig. 5 Traffic Trend within 24 h of a Road

the customer c_i is s_i. d_i and p_i respectively denote the delivery demand and pick-up demand of the customer c_i. After the vehicle leaves the depot, there will be some new orders. These vehicles need to serve these new orders as much as possible. The position of new orders is selected from $V = \{v_0, \ldots, v_N\}$ randomly. And the time of new orders' generation should be 60 min before its earliest time window b_i.

According to the above description, the formulation of the MOVRPRTC can be expressed as

$$f_1 = \sum_{k=1}^{K} \sum_{i=0}^{n} \sum_{j=0, j=i}^{n} d_{ij} x_j^k z_{ij} \tag{2}$$

$$f_2(t) = \sum_{k=1}^{K} \sum_{i=1}^{n} \max\{a_i(t) \cdot m_{ik} - e_i, 0\} \tag{3}$$

$$f_3(t) = \sum_{k=1}^{K} \sum_{i=1}^{n} \max\{b_i - a_i(t) \cdot m_{ik}, 0\} \tag{4}$$

st:

$$g_1 = \sum_{k=1}^{K} m_{ik} = 1, i = 1, 2, 3, \ldots, n \tag{5}$$

$$g_2 = \sum_{i=1}^{n} (d_i \cdot m_{ik}) - \sum_{i=1}^{n} (d_i \cdot m_{ik}) + \sum_{i=1}^{n} (p_i \cdot m_{ik}) \leq C, k = 1, 2, 3, \ldots, K \tag{6}$$

$$g_3(t) = a_0(t) \leq e_0 \tag{7}$$

where K is the number of vehicles used. x_{ij}^k is the number of times vehicle k accesses v_j from the v_i. If the customer c_i has been served by vehicle k, $m_{ik} = 1$, otherwise $m_{ik} = 0$. f_1 represents the total length of these routes. $f_2(t)$ is the sum of the delay time of all vehicles. $f_3(t)$ is the sum of the waiting time of all vehicles, which is used to measure the delivery efficiency. All objectives need to be minimized. g_1 means that each customer is served once and g_2 means that the vehicle cannot be overload at any time. $g_3(t)$ means that vehicles cannot return after the depot is closed.

In this problem, the number of customers (no more than 802), the depot opening time, the maximum capacity of the vehicle are free to set. Demands and time windows of the customers are determined according to the real online shopping data after desensitization.

Its dynamic characteristics are reflected in two aspects: (1) the traffic conditions will change in real-time. And the vehicle needs to adjust the traveling route according to the real-time traffic conditions; (2) new orders appear after the vehicle departs from the depot. The vehicle should serve these new orders as much as possible.

We can solve this dynamic optimization problem by the mechanism of offline optimization first and then online optimization.

4 Offline Optimization and Online Optimization for MOVRPRTC

In order to reduce the pressure of online optimization, longer-term offline optimization can be used to obtain a solution with the highest quality possible. In the offline optimization phase, the search operator, multi-objective optimization algorithm, and constraint optimization techniques are used to simulate the delivery of orders that need to be delivered on the second working day. In this phase, we use the traffic prediction model to get the vehicle speed. So we can assume that the speed of the vehicle is known at any time. The algorithm for offline optimization is called adaptive local search based on a dynamic constrained multi-objective genetic algorithm framework (ALSDCMOEA). After vehicles depart from the depot, the online optimization phase begins. In this phase, because of the real-time traffic conditions and new order data, the vehicles need to adjust the distribution route in real-time.

4.1 Framework of ALSDCMOEA

DCMOEA is a framework of constraint processing. It has two characteristics [35] for MOVRPRTC: (1) a constraint optimization problem equivalent converted into a dynamic constraint multi-objective optimization problem with an original objective and a constraint-violation objective. (2) the purpose of reducing constraint boundaries is to deal with constraint difficulty.

Inspired by DCMOEA, f_2 and f_3 in MOVRPRTC can be converted into constraints. Then the problem can be converted into VRPRTC:

$$\min f_1 = \sum_{k=1}^{K}\sum_{i=0}^{n}\sum_{j=0,j\neq i}^{n} d_{ij}x_{ij}^{k}z_{ij} \tag{8}$$

$$st : G = \{f_2, f_3, g_1, g_2, g_3\} \tag{9}$$

Then, according to the method in [35], VRPRTC can be converted to DCMOVRP-TRC:

$$DCMOVRPRTC^{(s)} \begin{cases} \min(f(x), cv(x)) \\ st : \vec{g}(x) \leq \vec{\varepsilon}^{(s)} \end{cases} \tag{10}$$

$$cv(x) = \frac{1}{4}\sum_{i=1}^{4}\frac{G_i(x)}{\max_{x\in \mathbf{P}(0)}\{G_i(x)\}} \tag{11}$$

$$G_i(x) = \max\{g_i(x), 0\}, i = 1, 2, \ldots, 4 \tag{12}$$

where x in(10) denotes a solution of DCMOVRPRTC, which consists of a set of routes. S is the maximum number of environmental changes, s is the environmental state, $s = 0, 1, \ldots, S$. $\vec{\varepsilon}^{(s)}$ is the dynamic constraint boundary. The number 4 in (11) is the number of constraints in (9) except for g_1. Because g_1 indicates that each customer is served only once. It is a binary value.

One benefit of converting MOVRPTC to DCMOVRPRTC is to increase individual selection pressure. Since the probability of individuals not dominating each other increases with the number of objectives, so many individuals in the same rank. When selecting offspring, there is a high probability that individuals in the first few ranks will be selected, which causes the problem of poor diversity.

Algorithm 1: Framework of ALSDCMOEA

step1: Get Initial population and set the iteration counter $t = 0$;
step2: Set $\vec{\varepsilon} = \vec{\varepsilon}^{(0)}$, , $s = 0$ and the local counter $t_s = 0$;
step3:
if *the population is εfeasible* **then**
  Reduce constraint boundary;
  Reset population $\vec{\varepsilon}$-feasibility and other components;
else
  $t_s = t_s + 1$;
step4: ALS();
step5: Select the next parent population;
step6: $t = t + 1$;
step7:
if *s reaches the final state S or t reaches AbortingT* **then**
  Go to **step8** ;
else
  Go back to **step3** ;
step8: Output results;

The framework of ALSDCMOEA for DCMOVRPRTC is shown in Algorithm 1. In **step3**, the method of changing environment is shown as follow

$$\varepsilon_i^{(s)} = A_i e^{-\left(\frac{s}{B_i}\right)^{cp}} - \delta, i = 1, 2, \ldots, m \tag{13}$$

where δ is a positive number which close to zero ($\delta = 1e - 8$ in this chapter) and cp can control the decreasing level. According to [35], we have the initial constraint boundary $\varepsilon_i^{(0)} = \max_{\mathbf{x} \in \mathbf{P}(0)} \{G_i(x)\}, i = 1, 2, \ldots, m$ at $s = 0$ (see (Eq.12)), and the final constraint boundary is $\overrightarrow{\varepsilon}^{(S)} = \overrightarrow{0}$ at $s = S$. From (13), we can obtain two equations

$$\begin{cases} \varepsilon_i^{(0)} = A_i - \delta & s = 0 \\ 0 = A_i e^{-\left(\frac{S}{B_i}\right)^{cp}} - \delta & s = S \end{cases} \tag{14}$$

where $i = 1, 2, \ldots, m$. According (14), we get

$$\begin{cases} A_i = \varepsilon_i^{(0)} + \delta \\ B_i = \dfrac{S}{\sqrt[cp]{\ln\left(\frac{\varepsilon_i^{(0)} + \delta}{\delta}\right)}} \end{cases} \tag{15}$$

The multi-objective optimization algorithms in **step5** is NSGA-II [36] in this chapter. Different from the common CMOEA, the Pareto domination is $\overrightarrow{\varepsilon}$ constrained Pareto domination (Definition 7).

Definition 7 (*$\overrightarrow{\varepsilon}$ constrained Pareto domination*) $\forall \mathbf{a}, \mathbf{b} \in \mathbf{R}^{\mathbf{n}}$:

- if both are $\overrightarrow{\varepsilon}$ feasible, then the Pareto domination is used;
- if one is $\overrightarrow{\varepsilon}$ feasible and the other $\overrightarrow{\varepsilon}$ infeasible, then the $\overrightarrow{\varepsilon}$ feasible one wins;
- if both are $\overrightarrow{\varepsilon}$ infeasible, then the one with a smaller constraint-violation cv wins.

Next, we will introduce the critical steps in the algorithm. Such as the solution initialization in **step1**, and the adaptive local search (ALS) in **step4**.

A. Solution Initialization

Before introducing the initialization method, we need to explain the encoding of the solution first. Since MOVRPRTC contains the road network, a solution should include not only the serving sequence of customers but also the detailed traveling sequence of coordinate points of each vehicle. Figure 6 shows an example, where sequence A indicates the serving sequence of the customers, sequence B indicates the traveling sequence of the coordinate points, gray dots between the depot and the customer 1 indicates the point in the road network that the vehicle needs to travel from the depot to the customer 1. The optimization algorithm only operates sequence A to reallocate customers for vehicles and change their serving sequences. Sequence B is only used for evaluation.

To get the initialization solution, we can simply use the random initialization method. In the random initialization method, we randomly select one of the customers

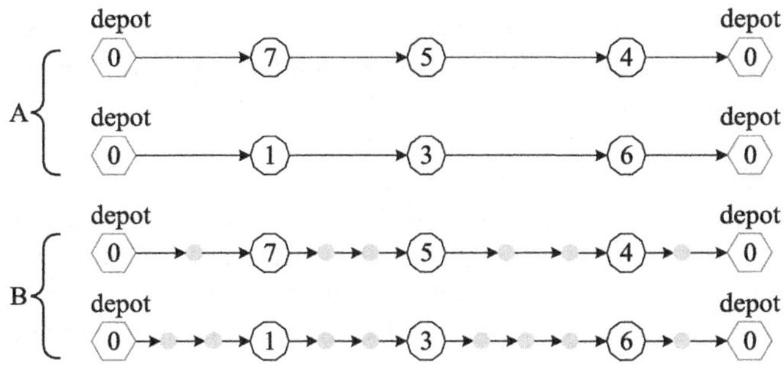

Fig. 6 Solution representation

and place it in the first vehicle. If the maximum capacity of the first vehicle is exceeded, then add it to the second vehicle, and repeat the above steps until all customers are assigned. Then, we can use ALS and NSGA-II to optimize the initial population.

B. Adaptive Local Search for MOVRPRTC

The fundamental tasks of solving MOVRPRTC are the allocation of customers for vehicles and the sequencing of customer servings. In order to address these two issues, we design eight local search operators and use an adaptive framework to improve search efficiency. The eight operators have shown as below:

1. **LS1**: Randomly choose two different customers from a random vehicle, and swap their serving orders;
2. **LS2**: Randomly choose two different customers from a random vehicle, and reverse the serving sequence between these two customers (the length of the sequence shall not exceed 3);
3. **LS3**: Randomly choose one vehicle and randomly intercept a sequence from it, then insert the sequence into another randomly selected vehicle. Then the entire serving sequence of the inserted vehicle should be re-sorted by using Nearest Neighbor Search;
4. **LS4**: Find two customers with the maximum wait time and with the maximum delay time, respectively. Insert the customer with the maximum delay time at the position right before the customer with the maximum wait time;
5. **LS5**: Randomly select a customer from a random vehicle, and insert it at a random location in another randomly selected vehicle;
6. **LS6**: Repeat **LS5** several times (no more than 4 times);
7. **LS7**: Find the customer with the maximum traveling time, and insert it at the position right before the customer who has the maximum waiting time;
8. **LS8**: Find the customer with the maximum traveling time and insert it at a random location.

LS1 and **LS2** are designed for changing the serving sequence, and the other operators are designed for changing the allocation of customers. **LS3**, **LS4**, **LS7**, and **LS8**

both utilize prior knowledge of the problem. After the local search is operated on a solution, the solution path will be reconstructed by a specific function. The main content of this function is the A* algorithm [37], which can get the shortest route for every vehicle according to the traffic conditions.

The subtlety part is the adaptive mechanism. At the beginning of the algorithm, each operator has the same weight, and then it will be adjusted according to the performance of the offspring generated by the operator in each iteration. If a better solution is generated, the corresponding operator score will be incremented by 1. In each iteration, the operators will be selected by the roulette. If the average best solution remains unchanged for several iterations, the algorithm will be considered as being converged. At this time, the scores of all operators are reset to 1, and then execute the next iteration until the termination condition is met. The pseudo-code of ALS is shown in Algorithm 2.

Algorithm 2: Adaptive local search

```
Input: solution x;
success_flag = false;
cnt = 0; //cnt represent the times of x not change
for i = 0 → popSize do
    x^t = LS_w(x); //w = 1, ...8, roulette wheel selection from LS1 to LS8;
    if x^t dominating x then
        score[w] + +; //score for operators
        x = x^t;
        success_flag = true;
    else
        x^t = x;
        cnt + +;
if cnt ≥ 100 then
    score[8] = [1, ..., 1];
    if success_flag == true then
        success_flag = false;
        cnt = 0;
```

The adaptive local search has two advantages: (1) the probability of each operator's selection depends on the dominance of parent and offspring, which can effectively reduce useless searches; (2) the weight distribution mechanism after convergence can effectively prevent the extremes of operator weights.

The optimization time of offline optimization can be long to find a good solution. After the offline optimization is complete, we need to choose an optimal solution. This best plan will be implemented in the online phase. In this chapter, the definition of the best solution is the solution with the minimum delay time (f_2).

4.2 Online Optimization

Based on the higher quality solution obtained during the offline phase, the online optimization should be able to achieve the goal of quickly processing new orders. The task of online optimization is to serve new orders as much as possible and adjust the vehicle's traveling route based on real-time traffic conditions. There are three rules for inserting new orders:

- **Rule 1:** If the vehicle arrives at the current delivery point later than the time when the new order appears, judge **Rule 2**; otherwise, reject it;
- **Rule 2:** After inserting a new order, use historical traffic information to estimate the impact of the inserted new order on subsequent orders (adaptive local search to adjust customer orders). If inserting a new order result in an average delay of more than 45 min per vehicle in the entire plan, the new order is rejected; otherwise, execute **Rule 3**.
- **Rule 3:** If the maximum capacity of the vehicle is exceeded after inserting a new order, the new order will be rejected.

Different from Algorithm 2, the adaptive local search in **Rule 2** only used three operators: **LS1**, **LS2**, and **LS4**. These three operators can only search among unserved customers in a single route during the online optimization phase.

5 Experiment

All algorithms are implemented in C++. And the experiments are run on a PC equipped with Core-i7 3.4 GHz and 16 GB of RAM.

Problem parameter settings: The opening time of the depot is from 8:00 to 24:00. The vehicle type is the same one, with a maximum capacity of 3 tons. The number of customers known before the working day is 100. The number of new orders is 15.

Algorithm parameter setting: The population size is 100. The maximum number of iterations is 10000.

To prove the effectiveness of ALSDCMOEA, we can compare the initial and final solutions generated by ALSDCMOEA in the objectives space. From Fig. 7, we can know the initial solution set is entirely dominated by the final solution set. This shows that in the offline optimization phase, ALSDCMOEA can solve MOVRPRTC.

Aiming to illustrate the full effect of ALSDCMOEA more intuitively, we can compare the best one of the initial solutions with the best one of the final solutions. The number of vehicles in the final best solution is 11. Their objectives are shown in Table 1. The average traveling distance is 45.86 km, the average waiting time is 99.21 min, and the average delay is 1.87 min. Customer waiting time characterizes customer satisfaction. The shorter the delay time, the higher the customer satisfaction. From the above results, compared to the initial solution, executing the distribution task according to the final best solution can improve customer satisfaction. At the

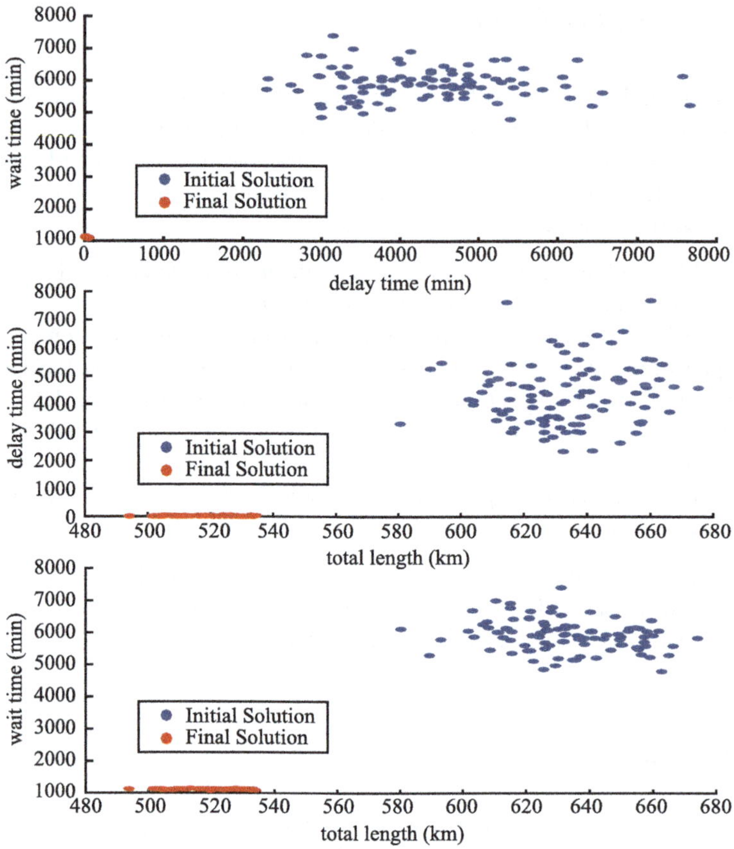

Fig. 7 Dominance of initial solutions and final solutions of offline phase

Table 1 The objectives of the initial best solution and the best final solution of the offline phase

	Total length	Wait time	Delay time
Initial solution	608.629 km	4884.91 min	2941.29 min
Final solution	**504.505** km	**1091.35** min	**20.5629** min
Reduce ratio	17.11%	77.66%	99.3%

same time, the traveling distance and waiting time of vehicles decreased by 17.11% and 77.66%, respectively.

The results of online optimization are shown in Table 2. The average traveling distance is 53.53 km, the average waiting time is 78.87 min, and the average delay is 65.79 min. Compared to the final solution of offline optimization, the distance traveled by the vehicle and the waiting time of the customer is increased, and the waiting time of the vehicle is reduced. This is because new orders were inserted into

Table 2 The result of online optimization

	Total length	Wait time	Delay time	Order remaining
Online begin	504.505 km	1091.35 min	20.5629 min	15
Online end	588.813 km	868.637 min	723.73 min	0

the offline solution, which caused the vehicle's driving distance to increase, and the vehicle's arrival time was later than originally planned, which led to a reduction in waiting time and an increase in delay time. The average delay is 65.79 min. It exceeds the prescribed 45 min in **Rule 2**. This is because of 45 min in **Rule 2** are estimated by historical traffic conditions. There is a certain deviation between historical traffic conditions and real-time traffic conditions.

6 Conclusion

A multi-objective vehicle routing problem with real-time conditions and a dynamic optimization mechanism for this problem was introduced in this chapter. From the experimental results, we can know the dynamic optimization mechanism can effectively solve MOVRPRTC. But some other dynamic optimization methods can apply to this practical dynamic problem to get better solution, such as the multi-population method. Besides, the NSGA-II can be replaced by some other multi-objective algorithms such as MOEA/D.

References

1. Nguyen, T.T.: Continuous dynamic optimisation using evolutionary algorithms. PhD thesis, University of Birmingham (2011)
2. Iqbal, S., Kaykobad, M., Rahman, M.S.: Solving the multi-objective vehicle routing problem with soft time windows with the help of bees. Swarm Evolut. Comput. **24**, 50–64 (2015)
3. Parragh, S.N., Doerner, K.F., Hartl, R.F.: A survey on pickup and delivery problems. J. für Betriebswirtschaft **58**(1), 21–51 (2008)
4. Marinakis, Y., Iordanidou, G.R., Marinaki, M.: Particle swarm optimization for the vehicle routing problem with stochastic demands. Appl. Soft Comput. **13**(4), 1693–1704 (2013)
5. Zhou, Y., Wang, J.: A local search-based multiobjective optimization algorithm for multiobjective vehicle routing problem with time windows. IEEE Syst. J. **9**(3), 1100–1113 (2015)
6. Barkaoui, M.: A co-evolutionary approach using information about future requests for dynamic vehicle routing problem with soft time windows. Memetic Comput. **10**(3), 307–319 (2018)
7. Chen, S., Chen, R., Wang, G.G., Gao, J., Sangaiah, A.K.: An adaptive large neighborhood search heuristic for dynamic vehicle routing problems. Comput. Electrical Eng. **67**, 596–607 (2018)
8. Sabar, N.R., Bhaskar, A., Chung, E., Turky, A., Song, A.: A self-adaptive evolutionary algorithm for dynamic vehicle routing problems with traffic congestion. Swarm Evolut. Comput. **44**, 1018–1027 (2019)

9. Jia, Y.H., Chen, W.N., Gu, T., Zhang, H., Yuan, H., Lin, Y., Yu, W.J., Zhang, J.: A dynamic logistic dispatching system with set-based particle swarm optimization. IEEE Trans. Syst., Man, Cybern.: Syst. **48**(9), 1607–1621 (2017)
10. Deb, K., Karthik, S.: Dynamic multi-objective optimization and decision-making using modified NSGA-II: a case study on hydro-thermal power scheduling. In: International Conference on Evolutionary Multi-criterion Optimization, pp. 803–817 (2007)
11. Woldesenbet, Y.G., Yen, G.G.: Dynamic evolutionary algorithm with variable relocation. IEEE Trans. Evolut. Comput. **13**(3), 500–513 (2009)
12. Toffolo, A., Benini, E.: Genetic diversity as an objective in multi-objective evolutionary algorithms. Evolut. Comput. **11**(2), 151–167 (2003)
13. Blackwell, T.M., Bentley, P.J.: Dynamic search with charged swarms. In: Proceedings of the 4th Annual Conference on Genetic and Evolutionary Computation, pp. 19–26 (2002)
14. Li, C., Yang, S.: A general framework of multipopulation methods with clustering in undetectable dynamic environments. IEEE Trans. Evolut. Comput. **16**(4), 556–577 (2012)
15. Yang, S.: Memory-based immigrants for genetic algorithms in dynamic environments. In: Proceedings of the 7th Annual Conference on Genetic and Evolutionary Computation, pp. 1115–1122 (2005)
16. Sahmoud, S., Topcuoglu, H.R.: A memory-based NSGA-II algorithm for dynamic multi-objective optimization problems. In: European Conference on the Applications of Evolutionary Computation, pp. 296–310 (2016)
17. Zhou, A., Jin, Y., Zhang, Q.: A population prediction strategy for evolutionary dynamic multiobjective optimization. IEEE Trans. Cybern. **44**(1), 40–53 (2013)
18. Muruganantham, A., Tan, K.C., Vadakkepat, P.: Evolutionary dynamic multiobjective optimization via Kalman filter prediction. IEEE Trans. Cybern. **46**(12), 2862–2873 (2015)
19. Goh, C.K., Tan, K.C.: A competitive-cooperative coevolutionary paradigm for dynamic multiobjective optimization. IEEE Trans. Evolut. Comput. **13**(1), 103–127 (2008)
20. Jiang, M., Huang, Z., Qiu, L., Huang, W., Yen, G.G.: Transfer learning-based dynamic multiobjective optimization algorithms. IEEE Trans. Evolut. Comput. **22**(4), 501–514 (2017)
21. Cobb, H.G.: An investigation into the use of hypermutation as an adaptive operator in genetic algorithms having continuous, time-dependent nonstationary environments. Naval Research Lab Washington DC (1990)
22. Liu, M., Zheng, J., Wang, J., Liu, Y., Jiang, L.: An adaptive diversity introduction method for dynamic evolutionary multiobjective optimization. IEEE Congress Evolut. Comput. **2014**, 3160–3167 (2014)
23. Vavak, F., Jukes, K., Fogarty, T.C.: Learning the local search range for genetic optimisation in nonstationary environments. In: Proceedings of 1997 IEEE International Conference on Evolutionary Computation, pp. 355–360 (1997)
24. Ghosh, A., Tsutsui, S., Tanaka, H.: Function optimization in nonstationary environment using steady state genetic algorithms with aging of individuals. In: Proceedings of 1998 IEEE International Conference on Evolutionary Computation, pp. 666–671 (1998)
25. Abbass, H.A., Deb, K.: Searching under multi-evolutionary pressures. International Conference on Evolutionary Multi-criterion Optimization 391–404 (2003)
26. Bui, L.T., Abbass, H.A., Branke, J.: Multiobjective optimization for dynamic environments. 2005 IEEE Congress Evolut. Comput. **3**, 2349–2356 (2005)
27. Li, C., Yang, S., Yang, M.: An adaptive multi-swarm optimizer for dynamic optimization problems. Evolut. Comput. **22**(4), 559–594 (2014)
28. Li, C., Nguyen, T.T., Yang, M., Yang, S., Zeng, S.: Multi-population methods in unconstrained continuous dynamic environments: the challenges. Inf. Sci. **296**, 95–118 (2015)
29. Li, C., Nguyen, T.T., Yang, M., Mavrovouniotis, M., Yang, S.: An adaptive multipopulation framework for locating and tracking multiple optima. IEEE Trans. Evolut. Comput. **20**(4), 590–605 (2015)
30. Richter, H., Yang, S.: Learning behavior in abstract memory schemes for dynamic optimization problems. Soft Comput. **13**(12), 1163–1173 (2009)

31. Bravo, Y., Luque, G., Alba, E.: Global memory schemes for dynamic optimization. Nat. Comput. **15**(2), 319–333 (2016)
32. Xu, X., Tan, Y., Zheng, W., Li, S.: Memory-enhanced dynamic multi-objective evolutionary algorithm based on Lp decomposition. Appl. Sci. **8**(9), 1673 (2018)
33. Hatzakis, I., Wallace, D.: Dynamic multi-objective optimization with evolutionary algorithms: a forward-looking approach. In: Proceedings of the 8th Annual Conference on Genetic and Evolutionary Computation, pp. 1201–1208 (2006)
34. Koo, W.T., Goh, C.K., Tan, K.C.: A predictive gradient strategy for multiobjective evolutionary algorithms in a fast changing environment. Memetic Comput. **2**(2), 87–110 (2010)
35. Zeng, S., Jiao, R., Li, C., Li, X., Alkasassbeh, J.S.: A general framework of dynamic constrained multiobjective evolutionary algorithms for constrained optimization. IEEE Trans. Cybern. **47**(9), 2678–2688 (2017)
36. Deb, K., Pratap, A., Agarwal, S., Meyarivan, T.: A fast and elitist multiobjective genetic algorithm: NSGA-II. IEEE Trans. Evolut. Comput. **6**(2), 182–197 (2002)
37. Hart, P.E., Nilsson, N.J., Raphael, B.: A formal basis for the heuristic determination of minimum cost paths. IEEE Trans. Syst. Sci. Cybern. **4**(2), 100–107 (1968)

Intelligent Robot System Design and Control

Dielectric Elastomer Intelligent Devices for Soft Robots

Yawu Wang, Jundong Wu, Wenjun Ye, Peng Huang, Kouhei Ohnishi, and Chunyi Su

Abstract With desirable performances of impressive actuation strain, high degree of electromechanical coupling and high mechanical compliance, dielectric elastomer (DE) materials have been widely used as intelligent actuators and intelligent sensors in the fields of soft robots. However, there are many challenges to characterize the DE materials due to their inherent nonlinearity, complicated electromechanical coupling and time-varying viscoelastic feature. Acting as actuators, most studies focused on the planar DE intelligent actuators (DEIAs). However, the studies on some other functional shapes are scarce. By investigating a conical DEIA with the material of polydimethylsiloxane, we propose a dynamic model based on the principle of the nonequilibrium thermodynamics. When implemented as sensors, most works of the DE intelligent sensors (DEISs) devote to the sensor structure design. However, the studies on the modelling of the DEIS are insufficient. A mathematical model of the DEIS is built to depict its sensing property. Because the DEIS is used to measure the displacement and force synchronously, it can serve as a soft force and displacement intelligent sensor (SFDIS). Then, we identify the parameters of the proposed models by dint of the differential evolution algorithm. The results of the model validations demonstrate the effectiveness of the proposed models.

Keywords Soft robot · Dielectric elastomer intelligent actuator · Dielectric elastomer intelligent sensor · Soft force and displacement intelligent sensor · Differential evolution algorithm

---

Y. Wang · P. Huang
School of Automation, China University of Geosciences, Wuhan 430074, China

Hubei Key Laboratory of Advanced Control and Intelligent Automation for Complex Systems, Wuhan 430074, China

J. Wu · W. Ye · C. Su (✉)
Department of Mechanical, Industrial and Aerospace Engineering, Concordia University, Montreal, QC H3G 1M8, Canada
e-mail: cysu@alcor.concordia.ca

K. Ohnishi
Department of System Design Engineering, Keio University, Yokohama 223-8522, Japan

M. Wu et al. (eds.), *Developments in Advanced Control and Intelligent Automation for Complex Systems*, Studies in Systems, Decision and Control 329,
https://doi.org/10.1007/978-3-030-62147-6_12

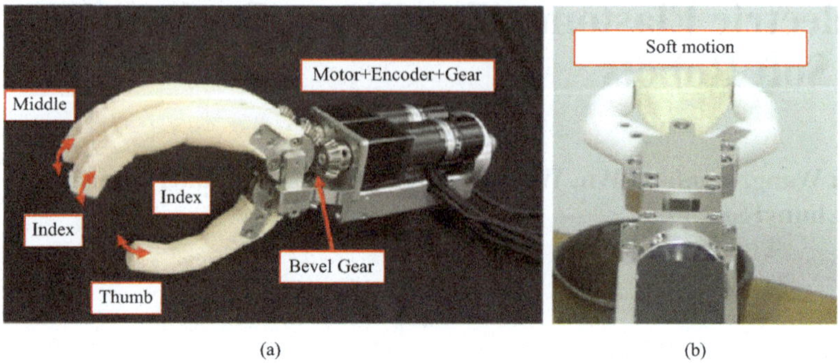

Fig. 1 Preliminary works of soft robots. **a** Haptic prosthetic hand, and **b** Robotic gripper

Abbreviations

DE	Dielectric elastomer
PDMS	Polydimethylsiloxane
DEIA	Dielectric elastomer intelligent actuator
DEIS	Dielectric elastomer intelligent sensor
SFDIS	Soft force and displacement intelligent sensor

1 Dynamic Modelling of Dielectric Elastomer Intelligent Actuator (DEIA)

1.1 Introduction

Soft robots have attracted great interests among engineers and scientists in recent years [1]. Albeit traditional rigid robots have made significant developments in the fields of the manufacturing, soft robots are more flexible and have broader application prospects [2]. Because soft robots are commonly manufactured by soft materials, they are highly deformable and can adapt to complex environments [3]. In general, soft robots use soft actuators made of soft materials [4]. Soft actuators can produce large deformations while meeting functional requirements [5].

One of the coauthors has done some preliminary works on soft robots shown in Fig. 1. In [6], a new haptic prosthetic hand was developed (see Fig. 1a), which achieves an intuitive grasping and flexible adaptation to the shape of the target object with a simple sensor-free structure. In [7], a robotic gripper system was designed (see Fig. 1b), which implements real haptics to achieve recording and playback of the motion.

Dielectric elastomer (DE) materials are new intelligent materials, which have the superiorities of large deformation, high energy density and fast response [8]. A DE intelligent actuator (DEIA) is composed of two compliant electrodes and a DE membrane sandwiched between the electrodes [9]. Once a high voltage is exerted to the electrodes, the DE membrane will expand in area and decrease in thickness [8, 10]. Because of the merit of large electrical deformation, DEIAs have been used to actuate soft robots, for instance, crawling robot [11, 12], gripping robot [13] and fish robot [14].

The dynamic model is the foundation of deeply understanding the inherent nonlinearity, complicated electromechanical coupling and time-varying viscoelastic feature of the DEIA. The early researches mainly focused on the modelling of the planar DEIA. In [15], a model was built to depict the nonlinear time-varying electromechanical response of the planar DEIA. As for other shapes, there were studies on the cylindrical shape. In [16], a model was developed to describe the transient characteristic of the cylindrically stacked DEIA. The conical DEIA has a more complex shape. In [17], a static model was built to analyze the static large deformation of the conical DEIA. However, the static model fails to illustrate the time-varying viscoelastic feature of the DEIA. In [18], a dynamic model was established to depict the motion behavior of a conical DEIA that is loaded by a mass and a linear spring, and the Ogden model is employed to depict the material properties. But, the influence of the inertia was neglected.

The DE materials used in the above researches are mostly very high bond tapes (VHB for short), which is a kind of polyacrylate materials manufactured by Minnesota Mining and Manufacturing company. However, the VHB has a defect of the high viscoelasticity. With the recent development of the DE materials, the polydimethylsiloxane (PDMS) material has been commercially available. A feasible solution for the defect of the high viscoelasticity is employing the PDMS to replace the VHB. However, there are few studies on exploring the dynamic behaviors of the DEIA based on the PDMS.

Combining new material and complex shape, we select the PDMS to manufacture a conical DEIA. According to the deformation mechanism of the PDMS material and the principle of the nonequilibrium thermodynamics, a dynamic model is established to depict the complex motion features of the conical DEIA. Based on the experimental data, the undetermined parameters in the established dynamic model are identified by using the differential evolution algorithm. The model validation verifies that the dynamic model can depict the viscoelastic characteristic and the electromechanical response of the conical DEIA.

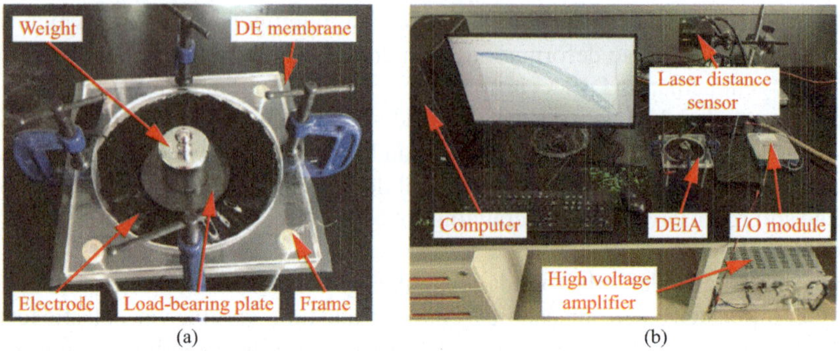

Fig. 2 System set up: **a** Conical DEIA; **b** Experiment platform

1.2 DEIA Manufacture and Experiment Platform Description

The manufacture of the conical DEIA is introduced briefly. Moreover, the experiment platform is also introduced in this subsection.

A. DEIA Manufacture

As shown in Fig. 2a, a conical DEIA is manufactured. It mainly consists of five parts: (1) DE membrane (Starting material: PDMS; Maker: Wacker Chemie, Germany; Initial thickness: $d_0 = 200\,\mu\text{m}$). (2) Frame (Starting material: Polymethyl methacrylate; Inner circle radius: $R = 6\,\text{cm}$). (3) Load-bearing plate (Starting material: Polymethyl methacrylate; Radius: $R_0 = 3\,\text{cm}$). (4) Electrode (Type: DD-10; Maker: Saidi Technology, China). (5) Weight (Mass: $m = 0.2\,\text{kg}$).

B. Experiment Platform

The experiment platform (see Fig. 2b) mainly consists of four parts: (1) High voltage amplifier (Type: 10/40A-HS-H-CE; Maker: TREK, America); (2) Laser distance sensor (Type: LK-H152; Maker: Keyence, Japan); (3) I/O module (Type: PCIe-6361; Maker: NI, America); (4) Computer (CPU: i7-8700; ARM: 16G).

The I/O module is employed to produce the primary voltage (From 0V to 10V) for the high voltage amplifier. Meanwhile, the I/O module captures the displacement data of the DEIA from the laser distance sensor. The high voltage amplifier is employed to amplify the primary voltage by one thousand times. And then, the amplified voltage is applied to the electrodes of the DEIA. The signal transfer diagram of the whole system is shown in Fig. 3.

1.3 DEIA Modelling

The dynamic model of the DEIA with conical shape is built in this subsection. To facilitate the understanding, three different statuses of the DEIA are declared. These

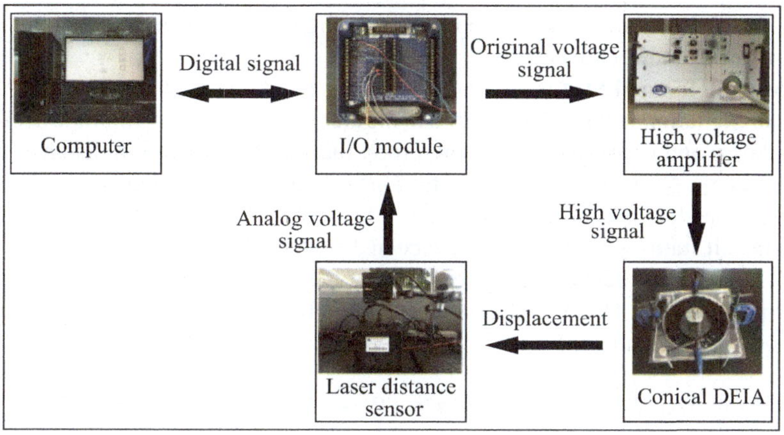

Fig. 3 Signal transfer diagram of experiment platform

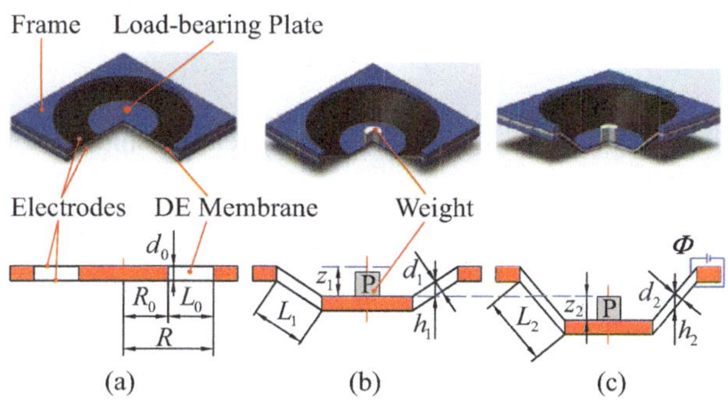

Fig. 4 Statuses of DEIA. **a** Undeformed status, **b** Prestretched status, and **c** Electrodeformed status

are the undeformed status, the prestretched status and the electrodeformed status, shown in Fig. 4a–c, respectively.

A. Undeformed Status

The DE membrane is clamped by the frame. The initial thickness of the DE membrane is defined as d_0, and the inner circle radius of the frame is defined as R. A load-bearing plate is stuck to the center of the DE membrane, whose radius is defined as R_0. Two sides of the DE membrane (two annular regions) are coated with the compliant electrodes. Therefore, the radial length of the DEIA can be defined to be $L_0 = R - R_0$.

B. Prestretched Status

The weight with the mass m is attached to the center of the load-bearing plate. Subjecting to the gravity P, the weight will drop a distance that is defined as z_1 to reach the equilibrium position. As thus, the DE membrane is prestretched to be a conical shape. As we defined in Fig. 4b, L_1, d_1 and h_1 are the sizes of the DEIA in regard to the prestretched status, in which L_1 represents the generatrix length, d_1

represents the thickness, and h_1 represents the height difference between the upper surface and the lower surface.

C. Electrodeformed Status

Once an excitation voltage Φ is exerted to the electrodes, the DE membrane decreases in thickness and expands in area. Therefore, the weight will drop a distance that is defined as z_2. As we defined in Fig. 4c, L_2, d_2 and h_2 are the sizes of the DEIA in regard to the electrodeformed status.

So far, all statuses of the DEIA have been declared. Next, we develop the dynamic model of the DEIA according to the theory of the nonequilibrium thermodynamics. The volumes of the DEIA in the undeformed status, the prestretched status and the electrodeformed status are

$$V_0 = \pi d_0 \left(R^2 - R_0^2\right), \quad V_1 = \pi h_1 \left(R^2 - R_0^2\right), \quad V_2 = \pi h_2 \left(R^2 - R_0^2\right) \tag{1}$$

Since the DEIA is incompressible [19], the volume of the DEIA is constant. Hence, V_0 is equal to V_1 also equal to V_2. According to (1), one can obtain

$$d_0 = h_1 = h_2 \tag{2}$$

Based on (2), the mathematical relationships among z_1, z_2, d_1 and d_2 can be derived as

$$d_1 = h_1 \frac{L_0}{L_1} = d_0 \frac{L_0}{\sqrt{z_1^2 + L_0^2}}, \quad d_2 = h_2 \frac{L_0}{L_2} = d_0 \frac{L_0}{\sqrt{(z_1 + z_2)^2 + L_0^2}} \tag{3}$$

To make the description easier, the generatrix, the thickness and the circumferential stretches are used to depict the status of the conical DEIA. In the prestretched status, the prestretches of the DEIA are defined as $\lambda_{pre,L}$, $\lambda_{pre,d}$ and $\lambda_{pre,C}$, respectively. In the electrodeformed status, the stretches of the DEIA are defined as λ_1, λ_2 and λ_3, respectively. From Fig. 4, the following equations hold:

$$\lambda_{pre,L} = L_1/L_0, \ \lambda_{pre,d} = d_1/d_0, \ \lambda_{pre,C} = 2\pi/2\pi = 1 \tag{4}$$

$$\lambda_1 = L_2 L_0, \ \lambda_2 = d_2/d_0, \ \lambda_3 = 2\pi/2\pi = 1 \tag{5}$$

Based on (2)—(5), the following equation is established:

$$\lambda_1 \lambda_2 \lambda_3 = \lambda_{pre,L} \lambda_{pre,d} \lambda_{pre,C} = 1 \tag{6}$$

The mathematical relationship between the charge Q and the voltage Φ is

$$Q = \Phi C = \Phi \varepsilon \pi L_2 (R + R_0)/d_2 = \varepsilon \Phi \pi \left(R^2 - R_0^2\right) \lambda_1^2/d_0 \tag{7}$$

Fig. 5 Element displacement

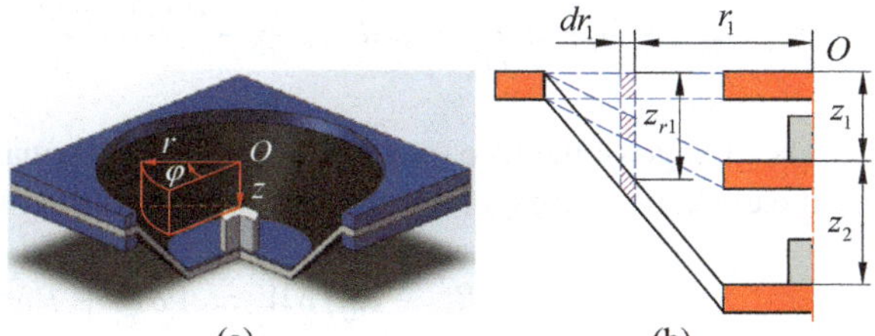

where ε denotes the dielectric constant of the DE material, and C denotes the capacitance.

Based on (3)–(6), the mathematical relationship between $\delta\lambda_1$ and δz_2 is

$$\frac{\delta z_2}{\delta\lambda_1} = \frac{L_2 L_0}{\sqrt{L_2^2 - L_0^2}} \tag{8}$$

From (6) and (7), the change of the charge on the electrode is

$$\delta Q = \frac{\varepsilon\pi\left(R^2 - R_0^2\right)}{d_0}\left(\lambda_1^2 \delta\Phi + 2\Phi\delta\lambda_1\right) \tag{9}$$

In order to calculate the work of the inertial forces, we employed a cylindrical coordinate displayed in Fig. 5, in which O is defined as the coordinate origin of the cylindrical coordinate, r is defined as the radial distance, φ is defined as the azimuth angle, and z is defined as the height of the cylindrical coordinate.

As we displayed in Fig. 5b, an infinitesimal element is constructed, whose inner radius is r_1 and outer radius is $r_1 + dr_1$. In the electrodeformed status, the displacement of the element along the r-direction, the φ-direction and the z-direction are 0, 0 and z_{r1}, respectively. Hence, the mathematical relationship between z_{r1} and z_2 can be derived as

$$z_{r1} = (z_1 + z_2)\frac{R - r_1}{R - R_0} \tag{10}$$

The inertial forces in each element along the r-direction, the φ-direction and the z-direction are 0, 0 and dF_{r1}, respectively. According to D'Alembert's principle, one can obtain

$$dF_{r1} = -\rho \cdot 2\pi d_0 r_1 dr_1 \cdot \frac{d^2 z_{r1}}{dt^2} \tag{11}$$

where ρ denotes the density of the DE material.

As that, the change of works done by the inertial forces are 0, 0 and $\delta H_{I,z}$, respectively. Based on (10) and (11), the work done by the inertial force dF_{r1} is

$$\delta H_{I,z} = \int_{R_0}^{R} \delta z_{r1} dF_{r1} = -\frac{\rho\pi d_0 L_0 (R + 3R_0)}{6}\frac{d^2 z_2}{dt^2}\delta z_2 \tag{12}$$

The change of the free energy of the DEIA is equal to the sum of the works done by the excitation voltage, the gravity and the inertial forces. Namely,

$$\pi d_0 \left(R^2 - R_0^2\right) \delta W = \Phi \delta Q + P \delta z_2 + \delta H_{I,z} \tag{13}$$

where W represents the free energy density of the DEIA, and δW represents the change of the free energy.

Submitting (9) and (12) into (13), one can get

$$\begin{aligned}\delta W = & \frac{\varepsilon\Phi\left(\lambda_1^2\delta\Phi + 2\Phi\lambda_1\delta\lambda_1\right)}{d_0^2} + \frac{P\delta z_2}{\pi d_0\left(R^2 - R_0^2\right)} \\ & - \frac{\rho(R + 3R_0)}{6(R + R_0)}\frac{d^2 z_2}{dt^2}\delta z_2\end{aligned} \tag{14}$$

Submitting (8) into (14), one has

$$\frac{\partial W}{\partial \lambda_1} = \frac{2\varepsilon\Phi^2\lambda_1}{d_0^2} + \frac{PL_2}{\pi d_0 (R + R_0)\sqrt{L_2^2 - L_0^2}} - \frac{\rho L_2 L_0 (R + 3R_0)}{6(R + R_0)\sqrt{L_2^2 - L_0^2}}\frac{d^2 z_2}{dt^2} \tag{15}$$

where

$$\frac{d^2 z_2}{dt^2} = \frac{-L_0^4}{\left(L_2^2 - L_0^2\right)^{\frac{3}{2}}}\left(\frac{d\lambda_1}{dt}\right)^2 + \frac{L_2 L_0}{\sqrt{L_2^2 - L_0^2}}\frac{d^2\lambda_1}{dt^2} \tag{16}$$

The expression of $\partial W/\partial\lambda_1$ is obtained by analysing the electromechanical coupling relationship of the DEIA. On the other hand, W depends on the elastic energy density and the electric displacement of the DEIA [2]. Thus, one can deduce the other expression of $\partial W/\partial\lambda_1$. And then, the dynamic model of the DEIA is established from these two expressions.

In deed, the PDMS material has viscoelasticity. There are some mathematical models that can describe the viscoelasticity of the materials, such as Kelvin-Voigt model, standard linear solid model, generalized Maxwell model and so on. Since the generalized Maxwell model is flexible and comprehensive, it is adopted in this chapter. The generalized Maxwell model (see Fig. 6) has two parallel units [20]. The part A is only composed of a spring α_0, as well as the part B is composed of n parallel formations, and each formation is composed of a spring α_i ($i = 1, 2, \ldots, n$) with a series-wound dashpot. In this chapter, we set each dashpot as a Newtonian fluid with viscosity η_i. Letting ξ_{ij} ($j = 1, 2$) represent the stretches with regard to the dashpots, the stretches of the spring α_i can be calculated according to the multiplication rules. That is, $\lambda_{i1}^e = \lambda_1/\xi_{i1}$ and $\lambda_{i2}^e = \lambda_2/\xi_{i2} = \lambda_1^{-1}\xi_{i2}^{-1}$.

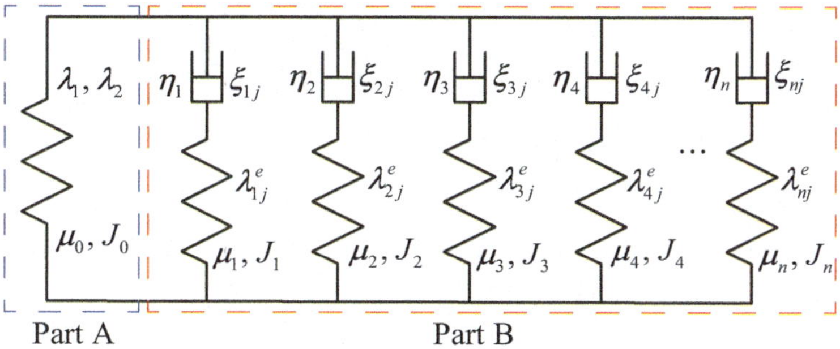

Fig. 6 Generalized Maxwell model

According to [19], W can be described as

$$W = W_s + D^2/2\varepsilon \tag{17}$$

where W_s is the elastic energy density of the DEIA, and D is the electric displacement. From [15], D is equal to

$$D = \frac{Q}{\pi L_2 (R + R_0)} \tag{18}$$

We select the Gent model [15, 21] to depict the elastic energy density of the DEIA. For each spring in the generalized Maxwell model shown in Fig. 6, the elastic energy density is

$$\begin{cases} W_{ela}^{\alpha_0} = -\frac{\mu_0 J_0}{2} \ln\left(1 - \frac{\lambda_1^2+\lambda_2^2+\lambda_1^{-2}\lambda_2^{-2}-3}{J_0}\right) \\ W_{ela}^{\alpha_i} = -\frac{\mu_i J_i}{2} \ln\left(1 - \frac{(\lambda_{i1}^e)^2+(\lambda_{i2}^e)^2+(\lambda_{i1}^e\lambda_{i2}^e)^{-2}-3}{J_i}\right) \quad (i = 1,\ 2,\ \ldots,\ n) \end{cases} \tag{19}$$

where μ_i and J_i $(i = 0,\ 1,\ \ldots,\ n)$ are the shear modulus and the deformation limits of the spring α_i, respectively.

Therefore, W_s is equal to the sum of the elastic energy density of each spring $W_{ela}^{\alpha_i}$ $(i = 0,\ 1,\ \ldots,\ n)$, that is

$$W_s = \sum_{i=0}^{n} W_{ela}^{\alpha_i} = -\frac{\mu_0 J_0}{2} \ln\left(1 - \frac{\lambda_1^2 + \lambda_2^2 + \lambda_1^{-2}\lambda_2^{-2} - 3}{J_0}\right) - \sum_{i=1}^{n} \frac{\mu_i J_i}{2} \ln\left(1 - \frac{\lambda_1^2\xi_{i1}^{-2} + \lambda_2^2\xi_{i2}^{-2} + \lambda_1^{-2}\lambda_2^{-2}\xi_{i1}^2\xi_{i2}^2 - 3}{J_i}\right) \tag{20}$$

According to (5)–(7) and (17)–(20), one can get

$$
\begin{aligned}
W = {} & \frac{\varepsilon \Phi^2 \lambda_1^2}{2 d_0^2} - \frac{\mu_0 J_0}{2} \ln \left(1 - \frac{\lambda_1^2 + \lambda_1^{-2} - 2}{J_0}\right) \\
& - \sum_{i=1}^{n} \frac{\mu_i J_i}{2} \ln \left(1 - \frac{\lambda_1^2 \xi_{i1}^{-2} + \lambda_1^{-2} \xi_{i2}^{-2} + \xi_{i1}^2 \xi_{i2}^2 - 3}{J_i}\right)
\end{aligned} \tag{21}
$$

On the basis of Newton's third law of motion, the stresses of the spring α_i are equal to the stresses of the dashpot η_i $(i = 1, 2, \ldots, n)$. So,

$$
-\xi_{ij} \frac{\partial W_{ela}^{\alpha_i}}{\partial \xi_{ij}} = \eta_i \frac{d\xi_{ij}}{dt} \quad (i = 1, 2, \ldots, n;\ j = 1, 2) \tag{22}
$$

From (20) and (22), the strain rates of the dashpots are formulated to be

$$
\begin{cases}
\dfrac{d\xi_{i1}}{dt} = -\dfrac{\mu_i}{\eta_i} \dfrac{-\lambda_1^2 \xi_{i1}^{-2} + \xi_{i1}^2 \xi_{i2}^2}{1 - \dfrac{\lambda_1^2 \xi_{i1}^{-2} + \lambda_1^{-2} \xi_{i2}^{-2} + \xi_{i1}^2 \xi_{i2}^2 - 3}{J_i}} \\
\dfrac{d\xi_{i2}}{dt} = -\dfrac{\mu_i}{\eta_i} \dfrac{-\lambda_1^{-2} \xi_{i2}^{-2} + \xi_{i1}^2 \xi_{i2}^2}{1 - \dfrac{\lambda_1^2 \xi_{i1}^{-2} + \lambda_1^{-2} \xi_{i2}^{-2} + \xi_{i1}^2 \xi_{i2}^2 - 3}{J_i}}
\end{cases}
(i = 1,\ 2,\ \ldots,\ n) \tag{23}
$$

where $\eta_i / \mu_i = \tau_i(t)$ denotes the relaxation time of the DEIA.

Submitting (21) into (15), and combining with (23), the dynamic model of the conical DEIA is built as

$$
\begin{cases}
\dfrac{\rho \lambda_1^2 L_0^2 (R + 3R_0)}{6 (R + R_0) \left(\lambda_1^2 - 1\right)} \dfrac{d^2 \lambda_1}{dt^2} = \dfrac{\rho \lambda_1 L_0^2 (R + 3R_0)}{6 (R + R_0) \left(\lambda_1^2 - 1\right)^2} \left(\dfrac{d\lambda_1}{dt}\right)^2 + \dfrac{\varepsilon \Phi^2 \lambda_1}{d_0^2} \\
\qquad + \dfrac{P \lambda_1}{\pi d_0 (R + R_0) \sqrt{\lambda_1^2 - 1}} - \mu_0 \dfrac{\lambda_1 - \lambda_1^{-3}}{1 - \dfrac{\lambda_1^2 + \lambda_1^{-2} - 2}{J_0}} \\
\qquad - \displaystyle\sum_{i=1}^{n} \mu_i \dfrac{\lambda_1 \xi_{i1}^{-2} - \lambda_1^{-3} \xi_{i2}^{-2}}{1 - \dfrac{\lambda_1^2 \xi_{i1}^{-2} + \lambda_1^{-2} \xi_{i2}^{-2} + \xi_{i1}^2 \xi_{i2}^2 - 3}{J_i}} \\
\dfrac{d\xi_{i1}}{dt} = -\dfrac{\mu_i}{\eta_i} \dfrac{-\lambda_1^2 \xi_{i1}^{-2} + \xi_{i1}^2 \xi_{i2}^2}{1 - \dfrac{\lambda_1^2 \xi_{i1}^{-2} + \lambda_1^{-2} \xi_{i2}^{-2} + \xi_{i1}^2 \xi_{i2}^2 - 3}{J_i}} \\
\dfrac{d\xi_{i2}}{dt} = -\dfrac{\mu_i}{\eta_i} \dfrac{-\lambda_1^{-2} \xi_{i2}^{-2} + \xi_{i1}^2 \xi_{i2}^2}{1 - \dfrac{\lambda_1^2 \xi_{i1}^{-2} + \lambda_1^{-2} \xi_{i2}^{-2} + \xi_{i1}^2 \xi_{i2}^2 - 3}{J_i}} \quad (i = 1, 2, \ldots, n)
\end{cases} \tag{24}
$$

So far, the dynamic model is built to depict the relationship among λ_1, $\dot{\lambda}_1$ and Φ. This dynamic model can also depict the inherent nonlinearity, complicated electromechanical coupling and time-varying viscoelastic feature of the conical DEIA. In (24), R, R_0, L_0, d_0, and P are easily measured. However, μ_i, J_i $(i = 0, 1, \ldots, n)$ and η_i $(i = 1, 2, \ldots, n)$ are hard to be measured directly. Therefore, we consider these parameters as unknown parameters. In the following developments, we design an experiment to collect the experimental data of the conical DEIA, and then identify the undetermined parameters in (24) by dint of the differential evolution algorithm based on these experimental data.

1.4 Parameter Identification of Dynamic Model

Firstly, we design an excitation voltage for the experiment. Then, the undetermined parameters are identified by dint of the differential evolution algorithm. After balancing the accuracy and computational complexity, the dynamic model (24) with four spring-dashpot units is employed to depict the complex nonlinear characteristics of the DEIA. Moreover, for ease of comparing, the model prediction output λ_1 in (24) is transformed into displacement z_2 according to $z_2 = (R - R_0)\sqrt{\lambda_1^2 - 1} - z_1$.

A. Excitation Voltage

To make the collections of the experimental data easier, the following excitation voltage is designed.

$$\begin{cases} t_m = rem\left(t, \sum_{i=1}^{5} 1/f_i\right) \\ v(t_m) = a_1 \sin(f_1 \pi t_m), & 0 \leqslant t_m \leqslant 1/f_1 \\ v(t_m) = a_2 \sin(f_2 \pi t_m - f_2 \pi / f_1), & 1/f_1 \leqslant t_m \leqslant \sum_{i=1}^{2} 1/f_i \\ v(t_m) = a_3 \sin\left(f_3 \pi t_m - f_3 \pi \sum_{i=1}^{2} 1/f_i\right), & \sum_{i=1}^{2} 1 f_i \leqslant t_m \leqslant \sum_{i=1}^{3} 1/f_i \\ v(t_m) = a_4 \sin\left(f_4 \pi t_m - f_4 \pi \sum_{i=1}^{3} 1/f_i\right), & \sum_{i=1}^{3} 1 f_i \leqslant t_m \leqslant \sum_{i=1}^{4} 1/f_i \\ v(t_m) = a_5 \sin\left(f_5 \pi t_m - f_5 \pi \sum_{i=1}^{4} 1/f_i\right), & \sum_{i=1}^{4} 1 f_i \leqslant t_m \leqslant \sum_{i=1}^{5} 1/f_i \end{cases} \tag{25}$$

where a_i is amplitude; f_i is the frequency; t is the real time; $rem(\alpha, \beta)$ is the remainder of α divided by β. By letting $t_m = rem\left(t, \sum_{i=1}^{5} 1/f_i\right)$, the periodic excitation voltage in $t \in [0, +\infty)$ is generated. Setting different values of a_i and f_i, the excitation voltages with different amplitudes and frequencies are generated within one period.

B. Parameters Identification

For the prestretched status, the displacement of the weight is measured to be $z_1 = 1.26$ cm. The sampling period of the experiment is set to be $T = 0.01$ s. When

***Algorithm:** Model Evolution*

***Input:** Input voltage signal, material and geometrocal parameters*

***Output:** Prediction of time-dependent electromechanical response of the actuator*

begin

Input mechanical parameters of the spring $\boldsymbol{\mu_i}$, $\boldsymbol{J_i}$ (where i=1,...,n) and the relaxation time of the dashpots $\boldsymbol{T_j}$ (where j=1,2,...,n-1)

Input dielectric permitivity ε of the material

Input geometrical parameters $\mathbf{R}$, $\boldsymbol{R_0}$, $\boldsymbol{d_0}$, z_1

*Input voltage signal **(25)***

Initialize the variables λ_1, ξ_{k1}, ξ_{k2} (where k=2,3,4,...,n)

*Call the **ode15s** function for the specified time interval **t***

  Get the voltage value $\boldsymbol{\phi_l}$ at time instant $\boldsymbol{t_l}$

  *Calculate $d\lambda_1/dt$, $d^2\lambda_2/dt^2$, $d\xi_{k1}/dt$, $d\xi_{k2}/dt$ for given $\boldsymbol{\phi_l}$ using system of equation **(24)***

  Integration to find $\lambda_1(t)$, $\xi_{k1}(t)$ and $\xi_{k2}(t)$

end

end

Fig. 7 Algorithm

Table 1 Parameters of DEIA Model

i	$i=0$	$i=1$	$i=2$	$i=3$	$i=4$
μ_i(kPa)	0.1	5277.4	0.1	33.1	571.2
J_i	6.8×10^7	79.9×10^7	80.1×10^7	7.9×10^7	3.7×10^7
$T_i=\eta_i\mu_i$	–	0.01	3945.71	9.82	8484.73

$a_i = 5.5 + 0.5i$ kV ($i = 1, 2, 3, 4, 5$) and $f_i = 0.2i$ Hz, the differential evolution algorithm for the parameters identification is briefly listed in Fig. 7.

In order to conveniently describe the predictive ability of the model, the root-mean-square error e_{rms} and the maximum modelling error e_m are defined.

$$\begin{cases} e_{rms} = \sqrt{\dfrac{1}{n}\sum_{i}^{n}(z_{ei}-z_{mi})^2} \times 100\% \\ e_m = \dfrac{\max(|z_{ei}-z_{mi}|)}{\max(z_{mi})-\min(z_{mi})} \times 100\% \end{cases} \tag{26}$$

where z_{ei} represents the experimental result, z_{mi} represents the model prediction, and n denotes the sampling quantity.

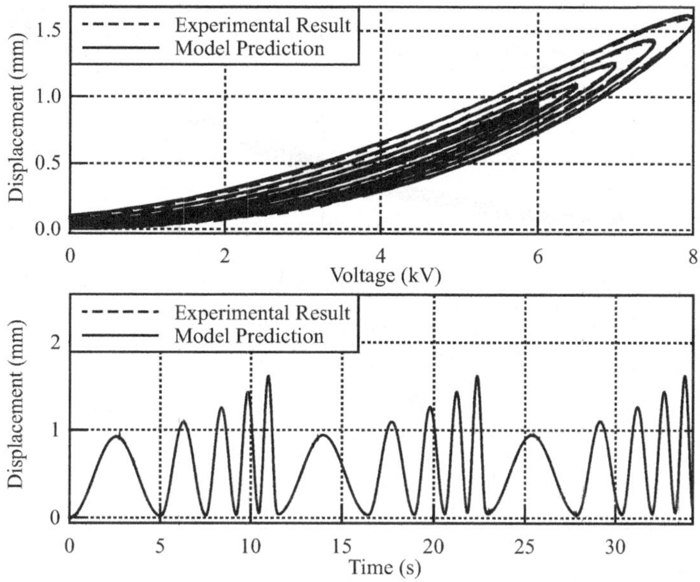

Fig. 8 Contrast of model prediction and experimental result with different driving voltage amplitudes and different frequencies

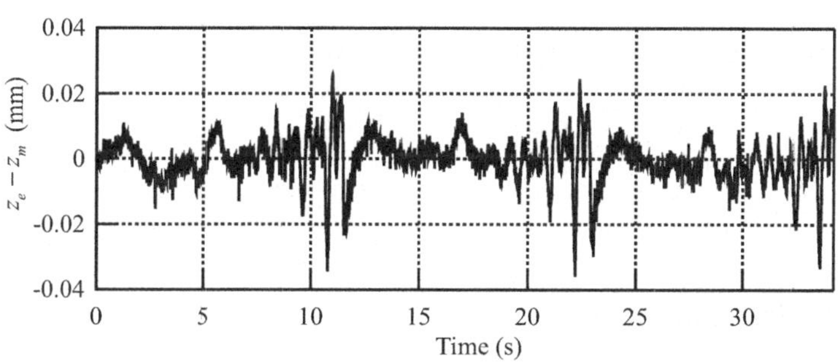

Fig. 9 Error of model prediction and experimental result

Figure 8 shows the contrast of the model prediction and the experimental result. Defining z_e and z_m as the experimental result and the model prediction value of z_2, the modelling error $z_e - z_m$ is given in Fig. 9. Table 1 gives the identified parameters of the dynamic model (24). Moreover, e_{rms} is 0.69% and e_m is 1.60%. Therefore, the dynamic model (24) with the identified parameters is available to depict the complex motion behavior of the DEIA.

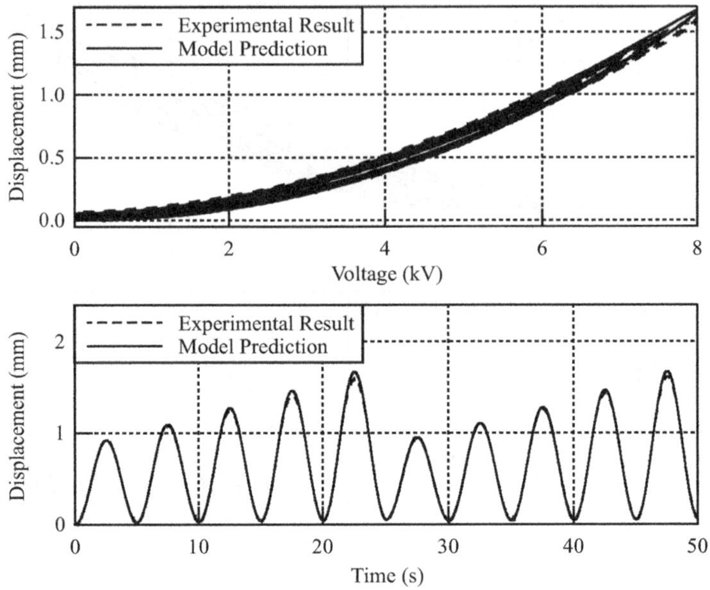

Fig. 10 Contrast of model prediction and experimental result with frequency 0.2 Hz of excitation voltage

Table 2 Modelling error for all experiments w.r.t. different frequencies

	$f = 0.2$	$f = 0.4$	$f = 0.6$	$f = 0.8$	$f = 1.0$
e_{rms}	1.7734	1.6875	1.7013	1.3202	1.2557
e_m	5.4161	4.1992	4.3098	3.6769	2.6517

1.5 Model Validation

Setting different values of a_i and f_i, the generalization capability of the proposed dynamic model of the conical DEIA is validated in this subsection.

A. Model Validation with Different Excitation Voltage Amplitudes

The amplitudes of the excitation voltage are defined as $a_i = 5.5 + 0.5i$ kV $(i = 1, \ldots, 5)$. In addition, the frequencies are defined as $f_i = 0.2,\ 0.4,\ 0.6,\ 0.8,$ 1.0 Hz, respectively. Therefore, the amplitudes of the excitation voltage are various, but its frequency is single in each experiment.

When the frequencies of the excitation voltage respectively are 0.2 Hz, 0.4 Hz, 0.6 Hz, 0.8 Hz and 1.0 Hz, the contrasts of the model prediction and the experimental result are shown in Figs. 10, 11, 12, 13, 14. The modelling errors for all experiments in regard to different frequencies are given in Table 2.

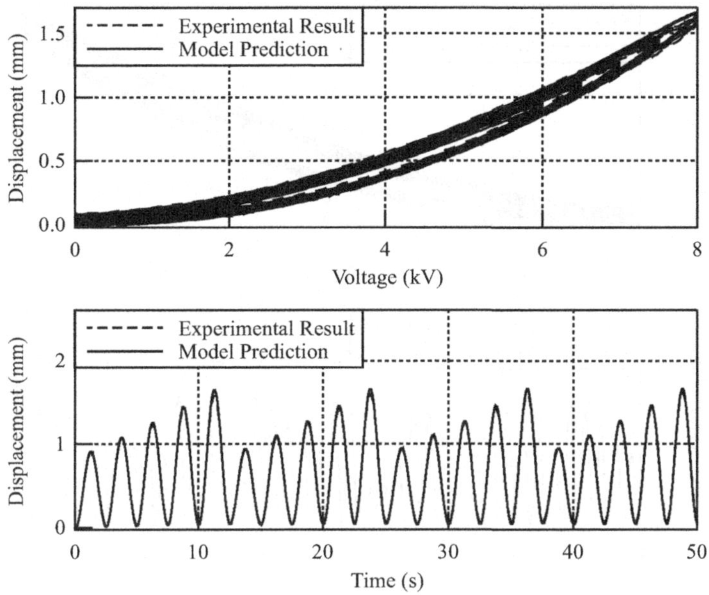

Fig. 11 Contrast of model prediction and experimental result with frequency 0.4 Hz of excitation voltage

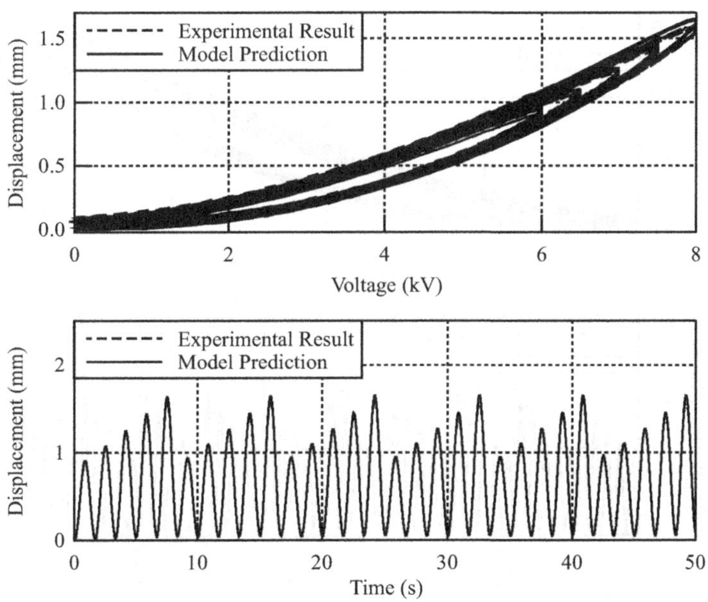

Fig. 12 Contrast of model prediction and experimental result with frequency 0.6 Hz of excitation voltage

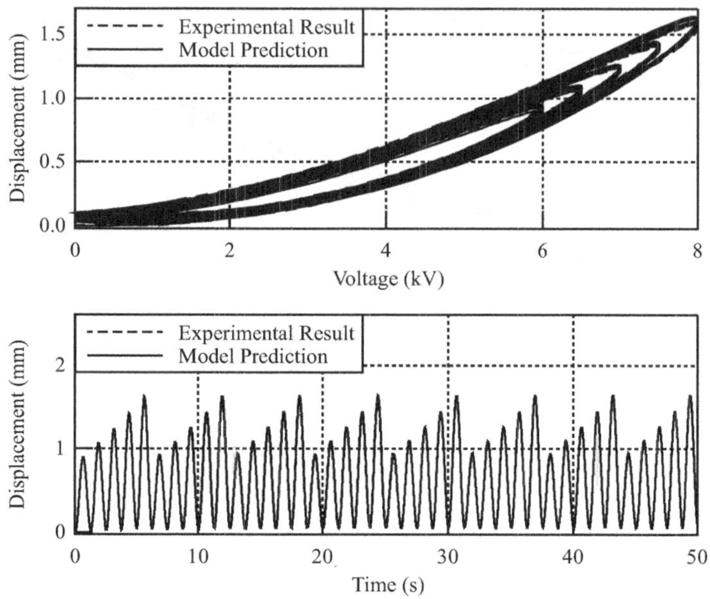

Fig. 13 Contrast of model prediction and experimental result with frequency 0.8 Hz of excitation voltage

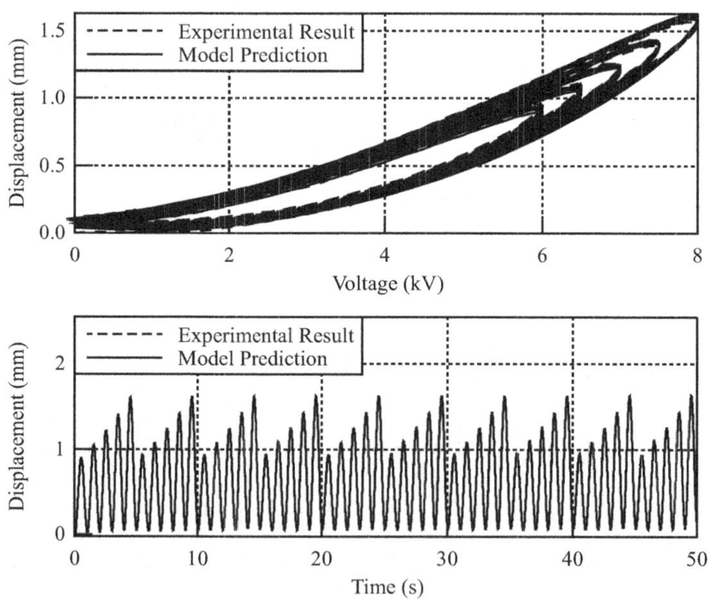

Fig. 14 Contrast of model prediction and experimental result with frequency 1.0 Hz of excitation voltage

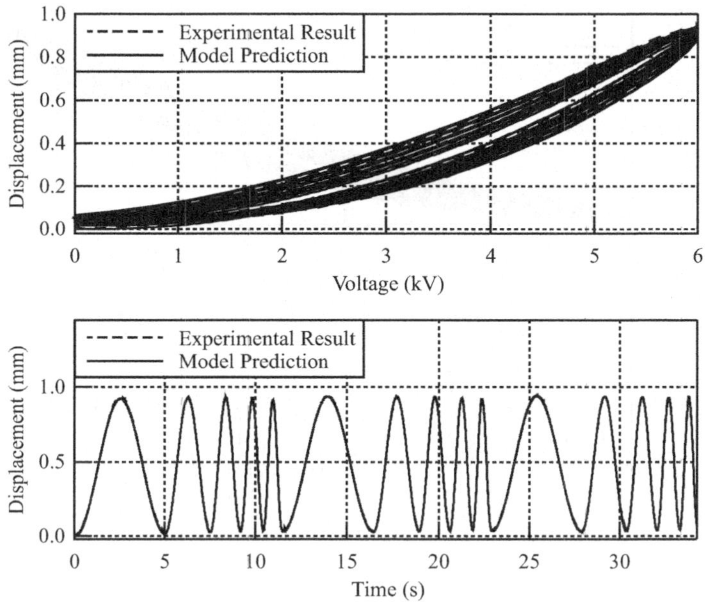

Fig. 15 Contrast of model prediction and experimental result with amplitude 6.0 kV of excitation voltage

Table 3 Modelling error for all experiments w.r.t. different amplitudes

i	$a = 6.0$	$a = 6.5$	$a = 7.0$	$a = 7.5$	$a = 8.0$
e_{rms}	0.6047	0.7842	0.6878	1.2279	1.8889
e_m	1.7454	2.2600	1.5985	1.6069	2.5015

From the above results, e_{rms} of the modelling for any experiment is less than 2%, and e_m for any experiment is less than 6%. Consequently, the generalization capability of the proposed dynamic model of the DEIA is fairly good.

B. Model Validation with Different Excitation Voltage Frequencies

The amplitudes of the driving voltage are defined as $a_i = 6.0, 6.5, 7.0, 7.5, 8.0$ kV, respectively. Meanwhile, the frequencies are defined as $f_i = 0.2i$ Hz. Hence, the frequencies of the excitation voltage are various, but its amplitude is single in each experiment.

When the amplitudes respectively are 6.0 kV, 6.5 kV, 7.0 kV, 7.5 kV and 8.0 kV, the contrasts of the model prediction and the experimental result are shown in Figs. 15, 16, 17, 18, 19. The modelling errors for all experiments in regard to different amplitudes are given in Table 3.

According to the above results, e_{rms} of the modelling for any experiment is less than 2%, and e_m for any experiment is less than 3%. So, the proposed dynamic model has an outstanding performance in the generalization capability.

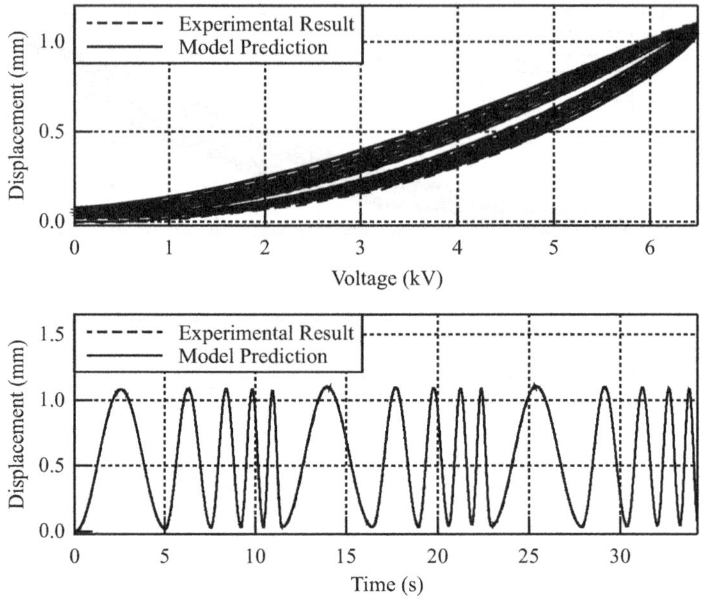

Fig. 16 Contrast of model prediction and experimental result with amplitude 6.5 kV of excitation voltage

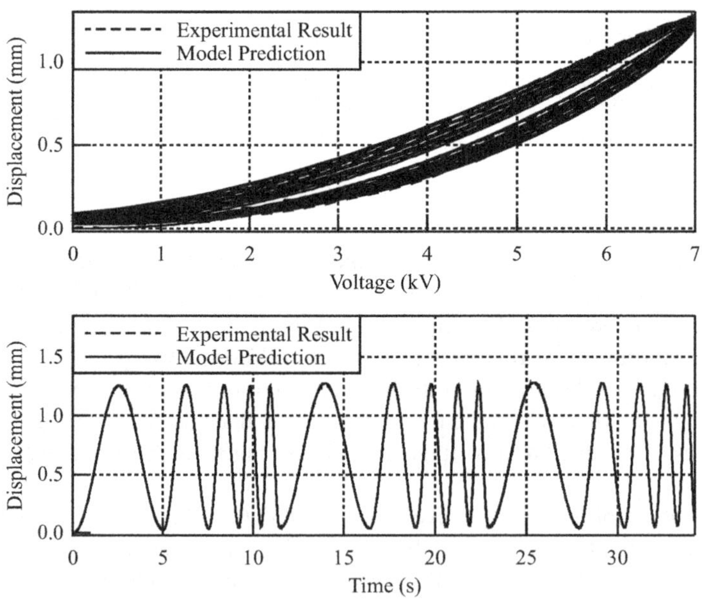

Fig. 17 Contrast of model prediction and experimental result with amplitude 7.0 kV of excitation voltage

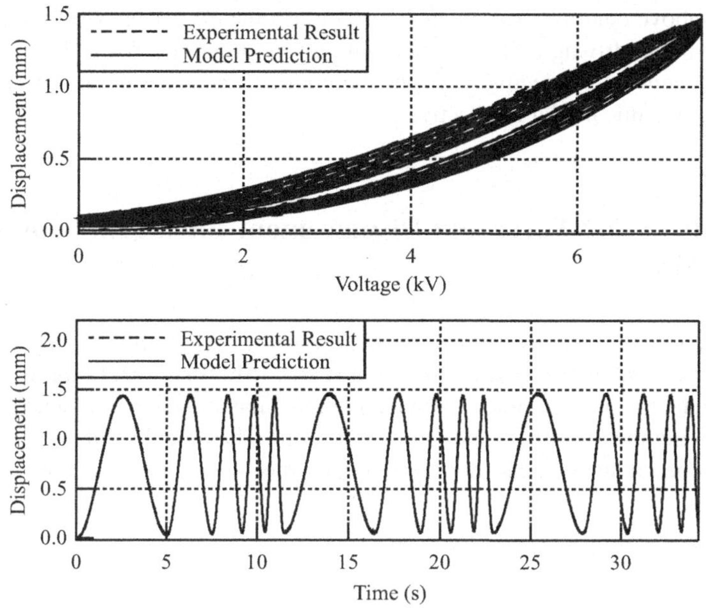

Fig. 18 Contrast of model prediction and experimental result with amplitude 7.5 kV of excitation voltage

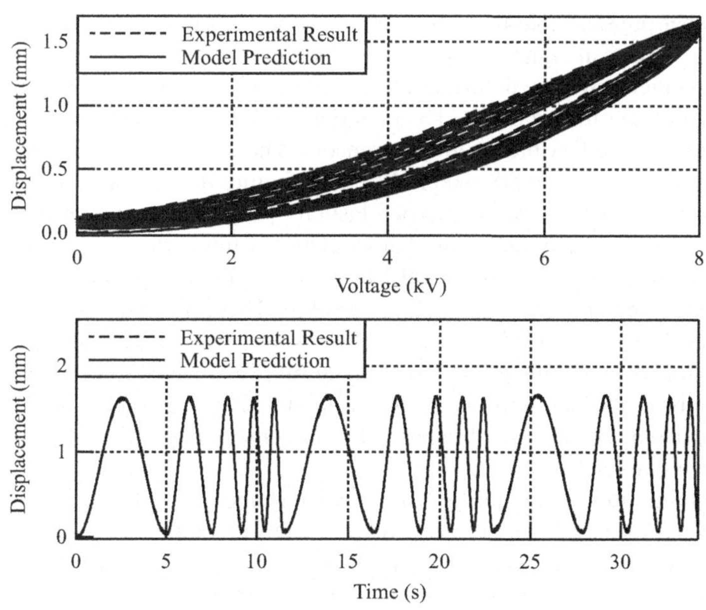

Fig. 19 Contrast of model prediction and experimental result with amplitude 8.0 kV of excitation voltage

In the above contrast experiments, we confirm the effectiveness of the proposed dynamic model driving by the voltage with different amplitudes and also with different frequencies, respectively. From the contrast results, we can know that the proposed dynamic model is effective.

2 Study of Soft Force and Displacement Intelligent Sensor (SFDIS)

In this section, we establish a mathematical model to depict the sensing property of the soft force and displacement intelligent sensor (SFDIS). According to the experimental data, the undetermined parameters in the mathematical model are identified by employing the differential evolution algorithm. The model validation proves that the developed mathematical model can depict the sensing properties of the SFDIS precisely.

2.1 Introduction

The functions of the sensors are to measure the displacement, the temperature, the force or other physical characteristics, and then the sensors convert the measurement results into electrical signals output [22]. The flexible sensors with the abilities of high compliance and large deformation are required in soft robots [23]. As a class of flexible intelligent materials, the DE material exhibits promising potentials for the fabrications of new flexible intelligent sensors. The dielectric elastomer intelligent sensor (DEIS) applies the DE material to manufacture its sensing element [23, 24]. Because the DE material has the hyperelasticity [15], the sensing properties of the DEIS show complicated nonlinear characteristics, which brings great challenges to the mathematical modelling of the DEIS. Therefore, the previous investigations are mostly concentrated at the structural design of the DEIS. For example, a shear force sensor was designed in [25], whose function is to detect the pure shear force in 2-axis. By using different materials and boundaries, [26] devised some stretch sensors, which can be embedded into the clothes or mounted on the bodies. In [27], a DEIS was fabricated by two DE membranes with complementary profiles, which can detect the compressive load.

The configuration of the DEIS is a sandwich shape, two sides of which are compliant electrodes and the middle of which is a DE membrane [24, 28]. The whole DEIS can be seen as a capacitor [25]. Once the dimensions of the DEIS vary, its capacitance also varies correspondingly [29]. Therefore, the force acting on the DEIS or the displacement of the measured object can be calculated reversely according to the variation of the capacitances of the DEIS. By using the charge tracking method, [30] applied the DE strain sensors to measure the motion and behaviour

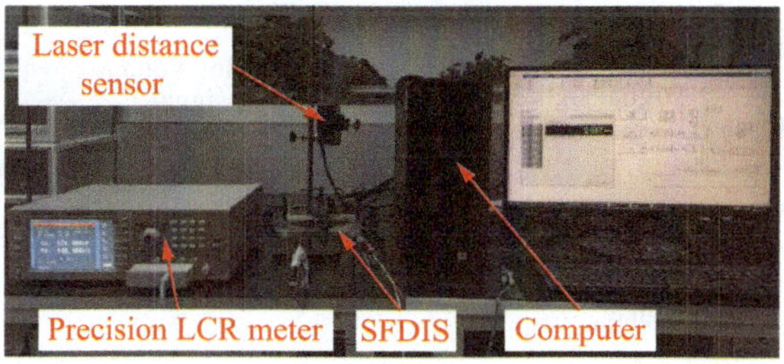

Fig. 20 Picture of experiment platform

of the hand. [31] built a mathematical model to interpret why the sensitivity of the DEIS can be improved by embedding the air chamber. However, these investigations only contribute to characterizing the capacitance of the DEIS, and there are few studies concentrate on the relationships between the capacitance of the DEIS and material characteristics. Furthermore, the mathematical models for describing the force sensing property or the displacement sensing property are insufficient.

The DEIS in this study is manufactured by the PDMS material, and it has the same mechanical structure with the conical DEIA. This DEIS can be deemed as a SFDIS because it can detect the force and displacement simultaneously. In this section, we establish a mathematical model for the SFDIS. The model validation verifies that the proposed model can depict the sensing properties of the SFDIS precisely.

2.2 *Experiment System Description*

The SFDIS has the same mechanical structure with the conical DEIA (see Fig. 2a). Furthermore, the devices in the experiment platform of the SFDIS (see Fig. 20) are basically the same as that of the DEIA (see Fig. 2b), in which the high voltage amplifier is removed and the precision LCR meter (Type: TH2829A; Maker: Tonghui, China) is employed to measure the capacitance of the SFDIS. For that reason, we are not repeating the detailed description of the experiment platform here. The signal transfer diagram of the whole system is shown in Fig. 21.

2.3 *SFDIS Modelling*

A SFDIS (see Fig. 22) is devised and its mathematical model is built in this subsection. To facilitate the understanding, three different statuses of the SFDIS are declared. That is, the relaxing status, the reference status and the detecting status, which are illustrated in Fig. 22a–c, respectively.

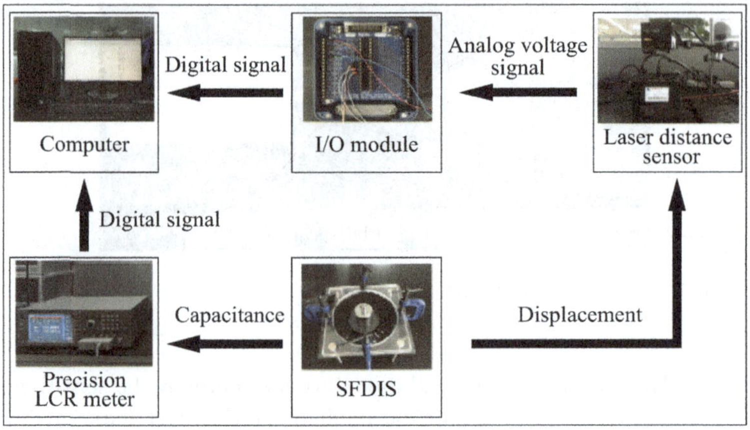

Fig. 21 Signal transfer diagram of experiment platform

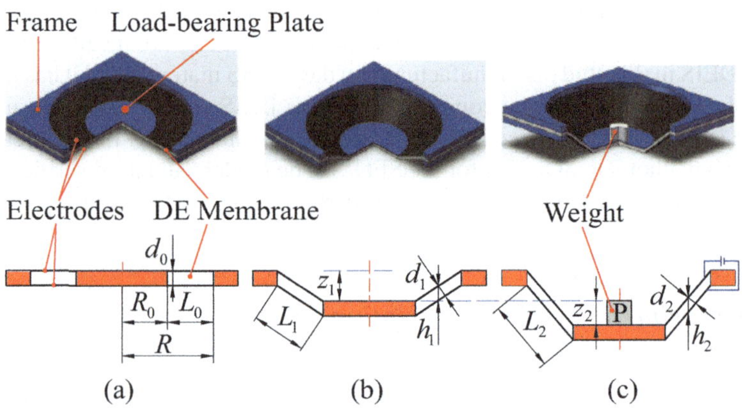

Fig. 22 Statuses of SFDIS. **a** Relaxing status, **b** Reference status, and **c** Detecting status

A. Relaxing Status

The DE membrane is clamped by the frame. The initial thickness of the DE membrane is defined as d_0, and the inner circle radius of the frame is defined as R. A load-bearing plate is stuck to the center of the DE membrane, whose radius is defined as R_0. Two sides of the DE membrane (two annular regions) are coated with the compliant electrodes. Therefore, the radial length of the SFDIS can be defined to be $L_0 = R - R_0$.

B. Reference Status

Subjecting to the gravity of the flexible electrodes, the DE membrane and the load-bearing plate, the load-bearing plate itself will drop a distance that is defined as z_1 to reach the equilibrium position. As thus, the DE membrane is prestretched to be a conical shape. As we defined in Fig. 22b, L_1, d_1 and h_1 are the sizes of the SFDIS in regard to the reference status, in which L_1 represents the generatrix length, d_1 represents the thickness, and h_1 represents the height difference between two sides of the DE membrane.

C. Detecting Status

Once a weight with the gravity P is loaded to the SFDIS, the weight drops down a displacement that is defined as z_2. As we defined in Fig. 22c, L_2, d_2 and h_2 represent the sizes of the SFDIS in regard to the detecting status.

Considering that the mechanical structure of the SFDIS is the same as that of the conical DEIA, (1)–(6) and (8) can be remained in the mathematical modelling of the SFDIS. The capacitance C of the SFDIS can be represented as

$$C = \frac{\varepsilon\varepsilon_0 A}{d_0} = \frac{\varepsilon\varepsilon_0\pi(R^2 - R_0^2)\lambda_1^2}{d_0} \tag{27}$$

where A denotes the surface area of the capacitor, ε_0 is the vacuum dielectric constant, and ε is the relative dielectric constant of the PDMS.

ε changes when the stretch of the PDMS material varies. According to [24] and [32], the relationship between ε and λ_1 can be described as

$$\varepsilon(\lambda_1) = a\lambda_1^2 + c \tag{28}$$

where a and c represent the coefficient of electrostriction and the material constant, respectively.

From (27) and (28), λ_1 is described as

$$\lambda_1 = \sqrt{\frac{-c + \sqrt{c^2 + \dfrac{4aCd_0}{\varepsilon_0\pi(R^2 - R_0^2)}}}{2a}} \tag{29}$$

Based on the geometrical relations among L_2, z_1 and z_2, one can get

$$z_2 = (R - R_0)\sqrt{\lambda_1^2 - 1} - z_1 \tag{30}$$

The SFDIS is regarded as a thermodynamic system [19], and its free energy F can be expressed as

$$F = \pi d_0\left(R^2 - R_0^2\right)W - Pz_2 \tag{31}$$

where W represents the energy density of the PDMS material.

Once the thermodynamic system is in the equilibrium status, its free energy is lowest [19]. Thus, the derivative of F in regard to λ_1 is equal to zero, that is $\mathrm{d}F/\mathrm{d}\lambda_1 = 0$. Then, one can obtain

$$\pi d_0\left(R^2 - R_0^2\right)\frac{\partial W}{\partial\lambda_1} - P\frac{\partial z_2}{\partial\lambda_1} = 0 \tag{32}$$

We select the Gent model [21] to characterize the mathematical relationship between W and λ_1. Therefore, W can be expressed as

$$W = -\frac{\mu J}{2}\ln\left(1 - \frac{\lambda_1^2 + \lambda_2^2 + \lambda_3^2 - 3}{J}\right) \tag{33}$$

where μ denotes the shear modulus, and J denotes the stretch limit.

From (5), (6) and (33), W can also be expressed as

$$W = -\frac{\mu J}{2}\ln\left(1 - \frac{\lambda_1^2 + \lambda_1^{-2} - 2}{J}\right) \tag{34}$$

Submitting (8) and (34) into (32), the gravity of the weight is

$$P = \pi d_0 (R + R_0)\frac{\mu\left(1 - \lambda_1^{-4}\right)\sqrt{\lambda_1^2 - 1}}{1 - \dfrac{\lambda_1^2 + \lambda_1^{-2} - 2}{J}} \tag{35}$$

From (29), (30) and (35), the sensing model of the SFDIS can be represented as

$$\begin{cases} z_2 = (R - R_0)\sqrt{\lambda_1^2 - 1} - z_1 \\ P = \pi d_0 (R + R_0)\dfrac{\mu\left(1 - \lambda_1^{-4}\right)\sqrt{\lambda_1^2 - 1}}{1 - \dfrac{\lambda_1^2 + \lambda_1^{-2} - 2}{J}} \end{cases} \tag{36}$$

So far, the sensing model of the SFDIS is developed to depict its sensing properties. To facilitate the following works, (36) is converted to

$$\begin{cases} z_2 = (R - R_0)\sqrt{\lambda_1^2 - 1} - z_1 \\ m = \dfrac{\pi d_0 (R + R_0)}{g}\dfrac{\mu\left(1 - \lambda_1^{-4}\right)\sqrt{\lambda_1^2 - 1}}{1 - \dfrac{\lambda_1^2 + \lambda_1^{-2} - 2}{J}} \end{cases} \tag{37}$$

where λ_1 is given in (29); m and g are the gravity of the weight and the gravitational acceleration, respectively.

C, d_0, z_1, R and R_0 in (29) and (37) are easy to be detected. However, μ, J, a and c are hard to be measured directly. Thus, these parameters are regarded as unknown parameters. In the following developments, we do an experiment to collect the experimental data of the SFDIS, and then identify the unknown parameters by dint of the differential evolution algorithm based on these experimental data.

Algorithm: *Model Evolution*

Input: *Input the material parameters, geometrocal parameters, capacitance and the mass of the weight*

Output: *Prediction of the sensing property of the SFDS*

begin
*Input the coefficient of electrostriction **a** and material constant **c***
Input the vacuum permittivity $\boldsymbol{\varepsilon_0}$ of the material
*Input mechanical parameters $\boldsymbol{\mu}$, **J***
*Input geometrical parameters **R**, $\boldsymbol{R_0}$, $\boldsymbol{d_0}$, z_1*
*Input the capacitance **C***
*Input the mass **m** of the weight*
*Call the **ode15s** function for the specified time interval **t***
*Calculate the displacement z_2 and the mass **m** of the weight for given capacitance **C** using system of equation **(37)***
end
end

Fig. 23 Algorithm

Table 4 Parameters of SFDIS Model

Parameter	a	c	μ (kPa)	J
Value	−0.956	3.767	475.081	13.649

2.4 *Parameter Identification of Sensing Model*

Based on the experimental data, the unknown parameters in (37) is identified by dint of the differential evolution algorithm. For the reference status, the distance z_1 is detected to be $z_1 = 0.86$ cm. Figure 23 shows the process flow of the differential evolution algorithm.

When the mass of the weight are $m = 50$ g, 70 g, ..., 490 g, the contrast of the experiment result and the model prediction of z_2 is displayed in Fig. 24a; the contrast of the experiment result and the model prediction of m is dispalyed in Fig. 24b. The identified parameters of the SFDIS model are listed in Table 4. Moreover, e_{rms} and e_m for the displacement prediction are 0.013% and 5.626%. Meanwhile, e_{rms} and e_m for the mass prediction are 0.924% and 6.632%.

From the above results, e_{rms} for any experiment is less than 1%, and e_m for any experiment is less than 7%. Hence, the proposed sensing model of the SFDIS is valid.

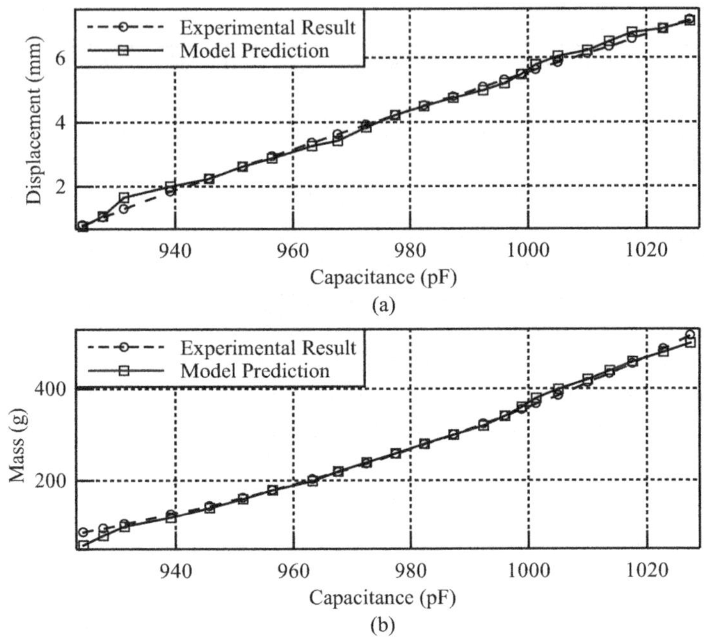

Fig. 24 Contrast of model prediction and experimental result of SFDIS

2.5 Model Validation

The validity of the proposed model is verified. When the mass of the weight are $m = 60$ g, 80 g, ..., 500 g, the contrast of the experiment result and the model prediction of z_2 is displayed in Fig. 25a; the contrast of the experiment result and the model prediction of m is displayed in Fig. 25b. Moreover, e_{rms} and e_m for the displacement prediction are 0.014% and 4.551%. Meanwhile, e_{rms} and e_m for the mass prediction are 0.923% and 5.376%.

Based on the above results, e_{rms} for any experiment is less than 1%, and e_m for any experiment is less than 6%. Therefore, the generalization capability of the developed model of the SFDIS is excellent.

3 Conclusion

Firstly, we propose a dynamic model for the conical DEIA. The contrast of the experiment result and the model prediction output confirms that the dynamic model can depict inherent nonlinearity, complicated electromechanical coupling and time-varing viscoelastic feature of the conical DEIA. Moreover, the dynamic model con-

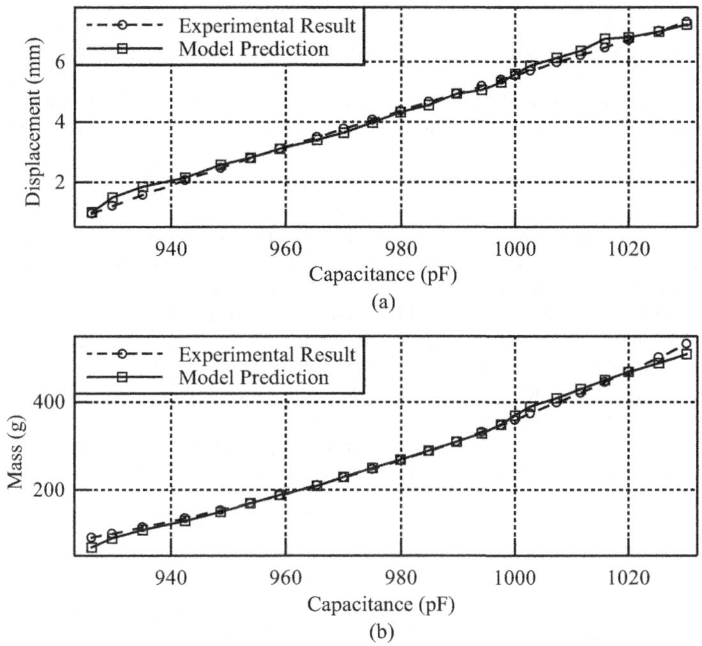

Fig. 25 Contrast of model prediction and experimental result in model validation

tributes to understanding the creep, asymmetric hysteresis, and frequency-dependent hysteresis of the conical DEIA.

Then, a sensing model of the SFDIS fabricated by the PDMS material is developed to depict its sensing properties. This sensor can detect the force and displacement synchronously. The contrast of the experimental result and the model prediction shows that the sensing model of the SFDIS is valid. Furthermore, this work contributes to promoting the applications of the SFDIS in the fields of soft robots or wearable devices.

References

1. Rus, D., Tolley, M.T.: Design, fabrication and control of soft robots. Nature **521**(7553), 467–475 (2015)
2. Gu, G., Zhu, J., Zhu, L., Zhu, X.: A survey on dielectric elastomer actuators for soft robots. Bioinspir. Biomim. **12**(1), 011003 (2017)
3. Zhang, Y., Zhang, N., Hingorani, H., Ding, N., Wang, D., Yuan, C., Zhang, B., Gu, G., Ge, Q.: Fast-response, stiffness-tunable soft actuator by hybrid multimaterial 3d printing. Adv. Funct. Mater. **29**(15), 1806698 (2019)
4. Wehner, M., Truby, R.L., Fitzgerald, D.J., Mosadegh, B., Whitesides, G.M., Lewis, J.A., Wood, R.J.: An integrated design and fabrication strategy for entirely soft, autonomous robots. Nature **536**(7617), 451–455 (2016)

5. Anderson, I.A., Gisby, T.A., McKay, T.G., OBrien, B.M., Calius, E.P.: Multi-functional dielectric elastomer artificial muscles for soft and smart machines. J. Appl. Phys. **112**(4), 041101 (2012)
6. Fukushima, S., Sekiguchi, H., Saito, Y., Iida, W., Nozaki, T., Ohnishi, K.: Artificial replacement of human sensation using haptic transplant technology. IEEE Trans. Ind. Elect. **65**(5), 3985–3994 (2017)
7. Ohnishi, K., Mizoguchi, T.: Real haptics and its applications. IEEE Trans. Elect. Elect. Eng. **12**(6), 803–808 (2017)
8. Pelrine, R.E., Kornbluh, R.D., Joseph, J.P.: Electrostriction of polymer dielectrics with compliant electrodes as a means of actuation. Sens. Actuators A: Phys. **64**(1), 77–85 (1998)
9. Wissler, M., Mazza, E.: Mechanical behavior of an acrylic elastomer used in dielectric elastomer actuators. Sens. Actuators A: Phys. **134**(2), 494–504 (2007)
10. Brochu, P., Pei, Q.: Advances in dielectric elastomers for actuators and artificial muscles. Macromol. Rapid Commun. **31**(1), 10–36 (2010)
11. Gu, G., Zou, J., Zhao, R., Zhao, X., Zhu, X.: Soft wall-climbing robots. Sci. Robot. **3**(25), eaat2874 (2018)
12. Cao, J., Qin, L., Liu, J., Ren, Q., Foo, C.C., Wang, H., Lee, H.P., Zhu, J.: Untethered soft robot capable of stable locomotion using soft electrostatic actuators. Extrem. Mech. Lett. **21**, 9–16 (2018)
13. Kellaris, N., Venkata, V.G., Smith, G.M., Mitchell, S.K., Keplinger, C.: Peano-hasel actuators: muscle-mimetic, electrohydraulic transducers that linearly contract on activation. Sci. Robot. **3**(14), eaar3276 (2018)
14. Shintake, J., Cacucciolo, V., Shea, H., Floreano, D.: Soft biomimetic fish robot made of dielectric elastomer actuators. Soft Robot. **5**(4), 466–474 (2018)
15. Gu, G., Gupta, U., Zhu, J., Zhu, L., Zhu, X.: Modeling of viscoelastic electromechanical behavior in a soft dielectric elastomer actuator. IEEE Trans. Robot. **33**(5), 1263–1271 (2017)
16. Sholl, N., Moss, A., Kier, W.M., Mohseni, K.: A soft end effector inspired by cephalopod suckers and augmented by a dielectric elastomer actuator. Soft Robot. **6**(3), 356–367 (2019)
17. He, T., Cui, L., Chen, C., Suo, Z.: Nonlinear deformation analysis of a dielectric elastomer membrane-spring system. Smart Mater. Struct. **19**(8), 085017 (2010)
18. Rizzello, G., Naso, D., York, A., Seelecke, S.: Modeling, identification, and control of a dielectric electro-active polymer positioning system. IEEE Trans. Control Syst. Technol. **23**(2), 632–643 (2015)
19. Suo, Z.: Theory of dielectric elastomers. Acta Mech. Solida Sin. **23**(6), 549–578 (2010)
20. Roylance, D.: Engineering viscoelasticity. Department of Materials Science and Engineering–Massachusetts Institute of Technology, Cambridge MA, vol. 2139, pp. 14–15 (2001)
21. Gent, A.: A new constitutive relation for rubber. Rubber Chem. Technol. **69**(1), 59–61 (1996)
22. Wilson, J.S.: Sensor technology handbook. Elsevier, New York, USA (2004)
23. Tao, Y., Gu, G., Zhu, L.: Design and performance testing of a dielectric elastomer strain sensor. Int. J. Intell. Robot. Appl. **1**(4), 451–458 (2017)
24. Jeanmistral, C., Sylvestre, A., Basrour, S., Chaillout, J.: Dielectric properties of polyacrylate thick films used in sensors and actuators. Smart Mater. Struct. **19**(7), 075019 (2010)
25. Kim, B., Shin, S.H., Chung, J., Lee, Y., Nam, J., Moon, H., Choi, H.R., Koo, J.: A dual axis shear force film sensor for robotic tactile applications. Proc. SPIE **7976**, 797628 (2011)
26. Obrien, B., Gisby, T.A., Anderson, I.A.: Stretch sensors for human body motion. Proc. SPIE **9056**, 905618 (2014)
27. Bose, H., Fus, E.: Novel dielectric elastomer sensors for compression load detection. Proc. SPIE **9056**, 905614 (2014)
28. Kadooka, K., Imamura, H., Taya, M.: Tactile sensor integrated dielectric elastomer actuator for simultaneous actuation and sensing. Proc. SPIE **9798**, 97982H (2016)
29. Liang, Y., Wan, B., Li, G., Xie, Y., Li, T.: A novel transparent dielectric elastomer sensor for compressive force measurements. Proc. SPIE **9798**, 979830 (2016)
30. Xu, D., Mckay, T., Michel, S., Anderson, I.A.: Enabling large scale capacitive sensing for dielectric elastomers. Proc. SPIE **9056**, 90561A (2014)

31. Zhang, H., Wang, M.Y., Li, J., Zhu, J.: A soft compressive sensor using dielectric elastomers. Smart Mater. Struct. **25**(3), 035045 (2016)
32. Qiang, J., Chen, H., Li, B.: Experimental study on the dielectric properties of polyacrylate dielectric elastomer. Smart Mater. Struct. **21**(2), 025006 (2012)

Design of a 2-DOF Compliant Micropositioning Stage with Large Workspace

Jinqiang Gan, Juncang Zhang, Huafeng Ding, and Andres Kecskemethy

Abstract With the development of mechanical research and manufacturing, micromanipulation has become a hot topic. A micropositioning stage based on a compliant mechanism plays an important role in manipulation at the micro-/nano-meter scale. In this chapter, a 2-DOF compliant micropositioning stage with a large motion range and fine decoupling ability is designed. Its mechanical design is firstly presented and improved. Subsequently, its analytical model is established to analyze the static characteristics. Lastly, the simulation results with finite-element analysis (FEA) demonstrate the established model is effective and the error of amplification ratio is controlled by 5.77%.

Keywords Micromanipulation · Compliant mechanism · Micropositioning stage · Finite-element analysis

J. Gan · J. Zhang · H. Ding (✉)
School of Mechanical Engineering and Electronic Information,
China University of Geosciences, Wuhan 430074, China
e-mail: dinghf@cug.edu.cn

A. Kecskemethy
Faculty of Engineering, University of Duisburg-Essen, 47057 Duisburg, Germany

M. Wu et al. (eds.), *Developments in Advanced Control and Intelligent Automation for Complex Systems*, Studies in Systems, Decision and Control 329,
https://doi.org/10.1007/978-3-030-62147-6_13

Abbreviations

2R PRBM	2R pseudo-rigid-body model
3R PRBM	3R pseudo-rigid-body model
CPFs	Compound parallelogram flexures
DOF	Degree of freedom
FEA	Finite-element analysis
MCPF	Multistage compound parallelogram flexure
MEMS	Micro-electromechanical systems
PRBM	Pseudo-rigid-body model
PZT	Piezoelectric actuator
VCM	Voice coil motor

1 Introduction

A micropositioning technique plays an important role in micromanipulation, which is taken as the basic technique of microgriping, micropuncturing, microinjecting, etc. A micropositioning stage based on the compliant mechanism is widely used in various fields in recent years, such as micro-electromechanical systems (MEMS), micro-assembly, aerospace, optics, etc. [1, 2]. Compliant mechanisms came from nature, and have been developed as a subject since the 1960s. Unlike rigid mechanisms, compliant mechanisms transmit force and motion by using the deformation of the structure within the elastic range of the material [3]. It has the advantages of no friction, no backlash, and no clearance, which contributes to high precision in micropositioning.

Since the flexure hinge is the main part that generates deformation in the compliant mechanisms, researchers have proposed analytical models to study the dynamic characteristics of it. Howell et al. [4] developed a pseudo-rigid-body model (PRBM) for an initially straight and inextensible cantilever beam, which models the mechanism members as rigid with joints at the center of the flexural pivots. Su et al. [5] proposed a 3R pseudo-rigid-body model (3R PRBM) that consists of four rigid links joined by three torsion springs, which is able to approximate exceptionally large deflection. Yu et al. [6] proposed a 2R pseudo-rigid-body model (2R PRBM) that consists of three rigid links joined by two revolute joints and two torsion springs.

There are various kinds of flexure hinges, including rectangular, circular, elliptical, parabolic, hyperbolic, etc. [3]. They are different from one another in shape and characteristics, which can meet the need in different practical applications. In general, the rectangular type has the highest dexterity which is good for the generation of large deformation, but the kinematic accuracy is low. On the contrary, the right-circular type has good kinematic accuracy, but it has limited dexterity and suffers from concentrated stress. Thus, the rectangular type is widely used in translational mechanisms, which mainly comes from a parallelogram mechanism. The right-circular type is chosen to guarantee high positioning accuracy, for example, it is usually adopted in the flexure design of the bridge-type amplification mechanism.

There are two kinds of actuators that are popular in micropositioning stage design, including voice coil motor (VCM) and piezoelectric actuator (PZT). VCM is a kind of electromagnetic actuator, which can convert electrical energy into mechanical energy without friction [7]. VCM converts energy through the interaction between the magnetic field from the permanent magnet and energized coil, by which it can generate strokes at a micrometer and centimeter scale. Thus, VCM is suitable in the micropositioning stage design which needs a large motion range. PZT can convert electrical energy into mechanical energy based on the reverse piezoelectric effect of ionic crystals [8]. Once an external electrical field is applied, the piezoelectric material will elongate to resist the change. PZT has the advantages of large generated force, high response speed, and high resolution. Therefore, PZT has become the most popular actuator in micropositioning research.

Over the past few years, researchers have developed various kinds of micropositioning stages, including 1-DOF stage, multi-DOF stage, serial stage, parallel stage, etc. Since the maximum stroke of the actuator is limited, micropositioning stage design with a large workspace has become a hot topic. There are two main ways to realize large workspace, i.e., decreasing the stiffness in the output direction and using stroke amplifiers. Xu [9] proposed multistage compound parallelogram flexure (MCPF) based on compound parallelogram flexures (CPFs), which decreases the lateral stiffness a lot. Gan et al. [10] used a lever-type amplifier in the design of the stage, which amplifies the stroke from 60 μm to 147.84 μm and 137.96 μm in X-axis and Y-axis respectively. Choi et al. [11] proposed a novel amplifier based on a bridge-type amplifier, which has two mechanical amplification mechanisms arranged in parallel to overcome the reduction of output force. In order to simplify the analysis and control of the structure, the micropositioning stage that can decouple movements generated by different actuators has been widely studied. Tang et al. [12] developed a 3-DOF flexure parallel mechanism with a decoupled kinematic structure, which has small cross-axis error and small parasitic rotation. Xu [9] proposed an XY stage design that contains both lateral and vertical parallelogram mechanisms in each direction, and the stage gains fine decoupling ability.

The major work of this chapter is the design and analytical model of a 2-DOF compliant micropositioning stage. The rest of this chapter is organized as follows. The mechanical design of the 2-DOF positioning stage is outlined in Sect. 2. The quantitative models are developed in Sect. 3 to analyze the stage's static characteristics. The models are verified with FEA in Sect. 4. Finally, Sect. 5 concludes this chapter.

2 Mechanical Design of XY Stage

In this section, mechanisms that usually used in stage design are first introduced. The original scheme is proposed according to the design purpose of a large workspace and

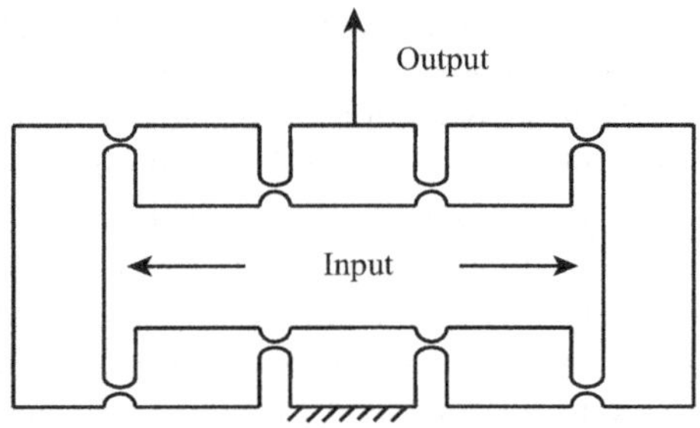

Fig. 1 Input and output of the bridge-type amplifier

decoupled structure. The mechanical design of a novel 2-DOF compliant micropositioning stage is then proposed.

2.1 Introduction of Basic Mechanisms

The bridge-type amplifier as shown in Fig. 1 is one of the most widely used displacement amplifiers in compliant mechanism design. The amplifier is constructed with a right-circular hinge and other elements which can be considered as rigid bodies. With asymmetrical input in lateral direction, the amplifier can generate a vertical output. There are also other kinds of amplifiers, such as a lever-type amplifier, SR mechanism, etc. Compared with other amplifiers, the bridge-type amplification mechanism has the advantage of a large amplification ratio and compact size [13]. Also, the bridge-type amplifier is greatly matched with PZT, which can be installed in the middle of the amplifier.

The parallelogram mechanism as depicted in Fig. 2a is widely used in mechanism design, which comes from rigid mechanism design. The mechanism is designed with slender beams, which guarantee large deformation. With a lateral input, the moving end can generate an output with no torsion. The symmetrical parallelogram mechanism design as shown in Fig. 2b is constructed with a fixed support on both sides. With a lateral input, the moving end will generate a pure translational motion with no parasitic error [3].

The symmetrical parallelogram mechanism can achieve single-DOF motion better compared with the traditional parallelogram mechanism. However, the fixed supports on both sides constraint the motion range of the mechanism. To deal with these problems, and improved parallelogram mechanism named compound parallelogram mechanism as illustrated in Fig. 3a have been proposed and widely used [9, 14]. The compound parallelogram mechanism has two moving ends called the primary stage and secondary stage. With input in the lateral direction, the deformation of the mechanism is shown in Fig. 3b. If the output of the second stage is Δx, the output of the primary stage would be $2\Delta x$. In other words, the vertical stiffness of structure could be reduced into half by adopting a compound parallelogram mechanism. Thus,

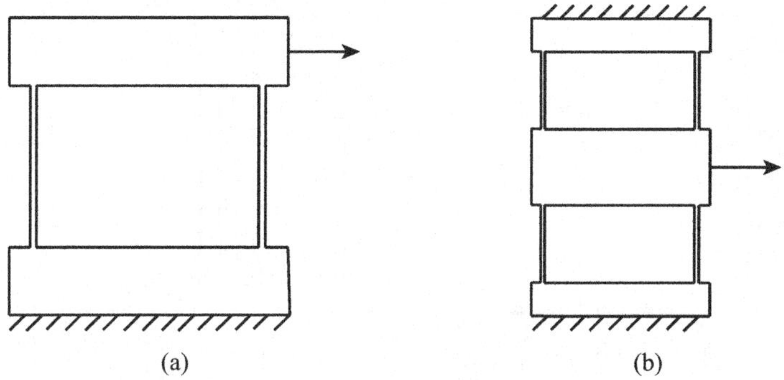

Fig. 2 **a** Parallelogram mechanism. **b** Symmetrical parallelogram mechanism

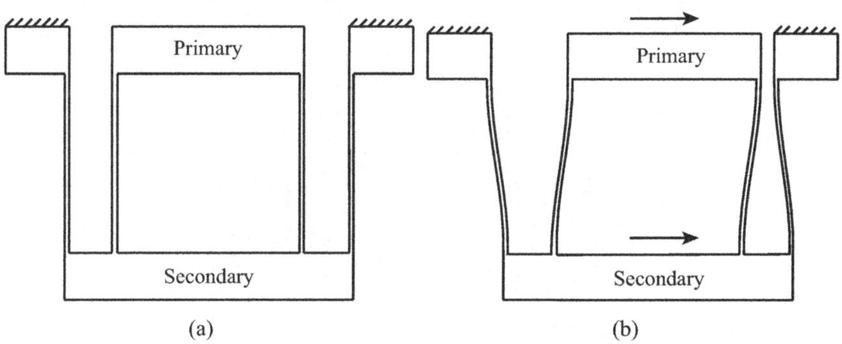

Fig. 3 **a** Compound parallelogram mechanism. **b** Deformed compound parallelogram mechanism

a compound parallelogram mechanism can achieve pure translational motion, as well as a lager motion range.

2.2 Propose of XY Stage

Based on the design purpose of large workspace and decoupled structure, an original scheme of the 2-DOF compliant stage is proposed, as shown in Fig. 4a. The proposed stage is constructed with a bridge-type amplifier, symmetrical parallelogram mechanism, and compound parallelogram mechanism. A bridge-type amplifier is adopted to enlarge the input displacement. The symmetrical parallelogram mechanism beside the output platform can ensure the translational motion in a vertical direction. A compound parallelogram mechanism is adopted to decouple the output motion and enlarge the output range.

The stage in Fig. 4a could decouple the output motion well, but they still exist input coupling which may cause damage to PZT. To deal with this problem, an evolved

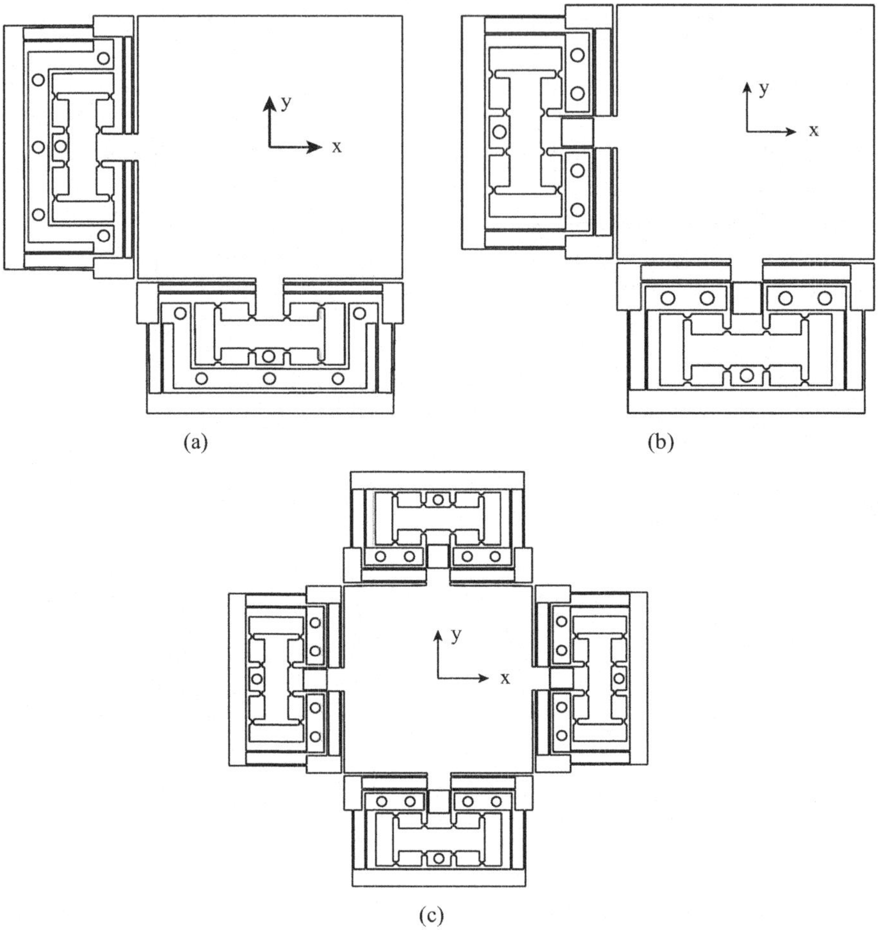

Fig. 4 Mechanical optimization of XY stage

stage is proposed as shown in Fig. 4b. With a parallelogram mechanism between the bridge-type amplifier and output platform, it can ensure a decoupled translation of the XY stage and get rid of undesired bending moment generated at the actuator. What's more, the new structure arranges different parts properly, which brings the profits of compact size and lower costs. Furthermore, to reduce the temperature-gradient effect and get rid of the error caused by torsion, the XY stage is evolved into the symmetric structure as illustrated in Fig. 4c.

3 Static Modeling and Characteristic Analysis

To analyze the output displacement, input compliance, and output compliance of the stage, static models are developed in this part. The modeling is carried out with the following assumptions: (1) the elastic deformation of the micromanipulator only occurs at the flexure hinges and the other parts are considered as rigid bodies; (2) the deformation in the flexure hinges is assumed to be pure bending and the rotation angle is small without expansion and contraction deformation [15].

3.1 *Modeling of Single Flexure Hinges*

In the proposed XY stage, rectangular-type and circular-type flexure hinges are adopted. The rectangular-type has high dexterity and low concentrated stress. It is applied in parallelogram mechanism, so that large deformation could be achieved. The circular-type has lower dexterity and higher concentrated stress, but its kinematic accuracy is good. Therefore, the circular-type is applied in bridge-type amplifier. The compliance modeling of single flexure hinge is the base of stage modeling, and the process is shown as follow.

Both rectangular-type and circular-type flexure hinges could be simplified as flexure beam when talking about its load-deformation relationship. To derive the relationship quantitatively, it's necessary to build approximate differential equation of the flexure beam. The simplified model of flexure beam is depicted in Fig. 5. According to material mechanics, curvature of the neutral layer at any point on flexure beam could be written as

$$\frac{1}{\rho(x)} = \frac{M(x)}{EI} \tag{1}$$

where ρ is curvature radius, $M(x)$ is bending moment, E is elastic modulus, and I is moment of inertia of an area.

Also, curvature radius ρ could be written as follow according to geometry, where $y(x)$ represents the deflection equation.

$$\frac{1}{\rho(x)} = \frac{y''(x)}{[1 + y'^2(x)]^{3/2}} \tag{2}$$

When the deformation is small, $y'(x)$ satisfies: $y'^2(x) \ll 1$. Thus, the following equation is derived according to (1), (2). It's called approximate differential equation of flexure beam.

$$y''(x) = \frac{M(x)}{EI} \tag{3}$$

Integrate the approximate differential equation twice, two more equations could be generated.

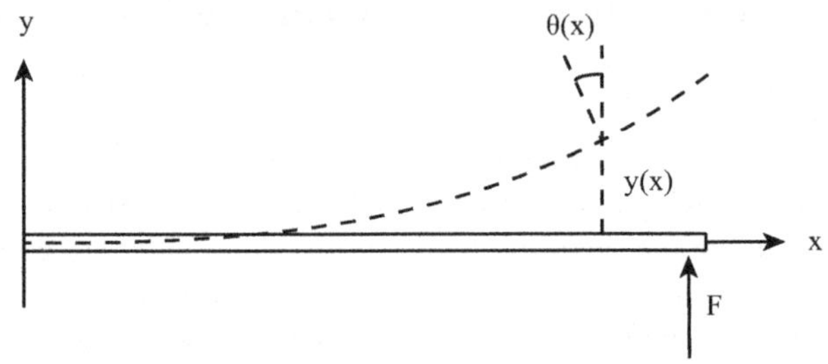

Fig. 5 Simplified model of flexure beam

$$y'(x) = \int \frac{M(x)}{EI} dx + C_1 \tag{4}$$

$$y(x) = \int\int \frac{M(x)}{EI} dx dx + C_1 x + C_2 \tag{5}$$

Considering the boundary conditions (when $x = 0$, $y'(x) = y(x) = 0$), integral constants C_1 and C_2 could be calculated as 0. Then, the following equations could be derived, from which the deformation at any point of a flexure beam could be calculated easily.

$$\theta(x) = y'(x) = \int \frac{M(x)}{EI} dx \tag{6}$$

$$y(x) = \int\int \frac{M(x)}{EI} dx dx \tag{7}$$

According to (6), (7) and material mechanics, rectangular-type flexure hinge's deformation at the end of the hinge can be derived as

$$d_{x1} = \frac{4F_x l^3}{Et^3 w} \tag{8}$$

$$d_{x2} = -\frac{4Ml^3}{Etw^3} \tag{9}$$

$$d_y = \frac{F_y l}{Etw} \tag{10}$$

$$\theta_{z1} = -\frac{4F_x l^3}{Etw^3} \tag{11}$$

$$\theta_{z2} = \frac{12Ml}{Et^3 w} \tag{12}$$

where E, l, t and w represent elastic modulus, length, thickness and height as shown in Fig. 6a. F_x, F_y and M are loads at the center of the hinge's end as shown in Fig. 6b.

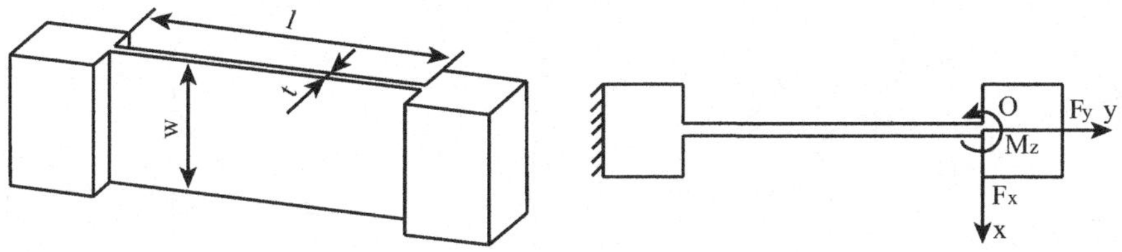

Fig. 6 **a** Dimension of rectangular-type flexure hinge. **b** Rectangular-type flexure hinge with loads at the end

When loads added to the end of the compliant beam, the following two relationships between load F, deformation X, compliance C and stiffness K are effective:

$$X = CF \tag{13}$$

$$F = KX \tag{14}$$

Since the proposed stage is a 2-DOF micropositioning stage, the consideration of stiffness is more than a single direction. There also exists undesired bending during the move, which may influence the consideration of the structure's stiffness. Thus, methods of compliance matrix are adopted in the following compliance modeling. Such a matrix contains several compliance elements, for example, displacement compliance caused by force, displacement compliance caused by moment, rotational compliance caused by force, and rotational compliance caused by moment.

According to (8), (9), (10), (11), (12) and (13), the compliance matrix at the end of rectangular-type flexure hinge could be derived as

$$C_1 = \begin{bmatrix} \frac{4l^3}{Et^3w} & 0 & -\frac{4l^3}{Etw^3} \\ 0 & \frac{l}{Etw} & 0 \\ -\frac{4l^3}{Etw^3} & 0 & \frac{12l}{Et^3w} \end{bmatrix} \tag{15}$$

The dimension and loads of circular-type flexure hinge are illustrated in Fig. 7a, b. E, r, t and w represent elastic modulus, radius, least thickness and height, respectively. The compliance matrix at the end of circular-type flexure hinge could be derived as

$$C_2 = \begin{bmatrix} \frac{9\pi r^{5/2}}{2Et^{5/2}w} + \frac{3\pi r^{3/2}}{2Et^{3/2}w} & 0 & -\frac{9\pi r^{3/2}}{2Et^{5/2}w} \\ 0 & \frac{1}{Ew}[\pi(\frac{r}{t})^{1/2} - \frac{2+\pi}{2}] & 0 \\ -\frac{9\pi r^{3/2}}{2Et^{5/2}w} & 0 & \frac{9\pi r^{1/2}}{2Et^{5/2}w} \end{bmatrix} \tag{16}$$

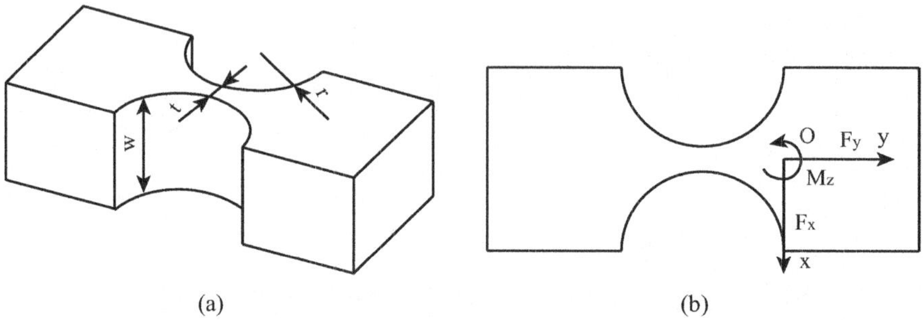

Fig. 7 **a** Dimension of circular-type flexure hinge. **b** Circular-type flexure hinge with loads at the end

3.2 Transform of Compliance Matrix

Since the compliance matrix of a single flexure is generated, the compliance of structure could be derived by integrating matrices together. The integrating progress needs to transform different matrices into one coordinate system. In order to derive the compliance matrix at a random point, a few steps need to be done with the compliant matrix of a single flexure hinge. The transformation from O_i to O_j is shown as Fig. 8, and it can be written as

$$C_{O_j} = T_i^j C_{O_i} (T_i^j)^T \tag{17}$$

where C_{O_j} and C_{O_i} are compliant matrix before and after transformation, and T_i^j is the transformation matrix. T_i^j is derived from the rotation matrix R_i^j and moving matrix P_i^j, and it can be derived as

$$T_i^j = \begin{bmatrix} r_{11} & r_{12} & 0 \\ r_{21} & r_{22} & 0 \\ 0 & 0 & r_{33} \end{bmatrix} \begin{bmatrix} 1 & 0 & p_y \\ 0 & 1 & -p_x \\ 0 & 0 & 1 \end{bmatrix} \tag{18}$$

where r_{ij} represents the element in R_i^j, and p_x, p_y represent the coordinates of moving vector.

3.3 Compliance Matrix of Each Part

In order to analyze the XY stage clearly, the stage is divided into different parts as shown in Fig. 9. Firstly, the stage is divided into a moving platform and four branches I, II, III and IV, which are central-symmetric about point O. The following discussions will be based on branch I. Then, branch I is divided into four parts, including

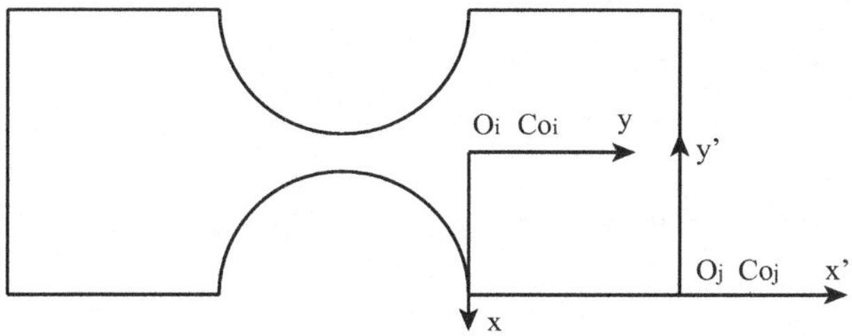

Fig. 8 Transform of compliant matrix

Fig. 9 Different parts of XY stage

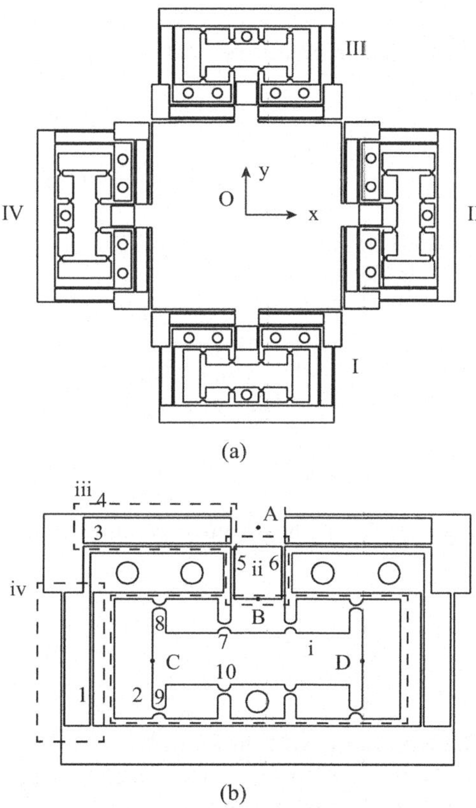

I. bridge-type amplifier, II. parallelogram mechanism, III. symmetrical parallelogram mechanism and IV. compound parallelogram mechanism. The structure of branch I is horizontally symmetric, and the left side of the structure contains ten hinges as shown in Fig. 9b.

A. Compliance of III and iv at Point A

According to (15), (17) and (18), compliance matrix of hinge 1 and hinge 2 at point A could be derived as

$$_1C_A = T^A_{O_1} C_1 (T^A_{O_1})^T \tag{19}$$

$$_2C_A = T^A_{O_2} C_1 (T^A_{O_2})^T \tag{20}$$

Similarly, compliance matrix of hinge 3 and hinge 4 at point A could be derived as

$$_3C_A = T^A_{O_3} C_1 (T^A_{O_3})^T \tag{21}$$

$$_4C_A = T^A_{O_4} C_1 (T^A_{O_4})^T \tag{22}$$

Since hinge 1 and hinge 2 are in series, while hinge 3 and hinge 4 are in parallel, compliance matrix of IV and III at point A could be derived as

$$_{iv}C_A =_1 C_A +_2 C_A \tag{23}$$

$$_{iii}C_A = [(_3C_A)^{-1} + (_4C_A)^{-1}]^{-1} \tag{24}$$

Since III and iv are in series, compliance matrix of the whole of both at point A could be derived as

$$_{iii,iv}C_A =_{iv} C +_{iii} C \tag{25}$$

Considering the symmetrical structure, the compliance matrix of symmetrical parallelogram mechanism and compound parallelogram mechanism at point A is

$$_{side}C_A = [(_{iii,iv}C_A)^{-1} + (R_y(\pi)_{iii,iv}C_A(R_y(\pi))^T)^{-1}]^{-1} \tag{26}$$

where $R_y(\pi)$ is the symmetrical transformation matrix:

$$R_y(\pi) = \begin{bmatrix} -1 & 0 & 0 \\ 0 & 1 & 0 \\ 0 & 0 & -1 \end{bmatrix} \tag{27}$$

B. Compliance of I at Point B

According to (16), (17) and (18), compliance matrix of hinge 7, hinge 8, hinge 9 and hinge 10 at point B could be derived as

$$_7C_B = T^B_{O_7} C_2 (T^B_{O_7})^T \tag{28}$$

$$_8C_B = T^B_{O_8} C_2 (T^B_{O_8})^T \tag{29}$$

$$_9C_B = T^B_{O_9} C_2 (T^B_{O_9})^T \tag{30}$$

$$_{10}C_B = T^B_{O_{10}} C_2 (T^B_{O_{10}})^T \tag{31}$$

Since the four hinges are in series, compliance matrix of the left part of bridge-type amplifier at point B could be derived as

$$_{left}C_B =_7 C_B +_8 C_B +_9 C_B +_{10} C_B \tag{32}$$

Considering the symmetrical structure, compliance matrix of I. bridge-type amplifier at point B is

$$_{bridge}C_B = [(_{left}C_B)^{-1} + (R_y(\pi)_{left}C_B(R_y(\pi))^T)^{-1}]^{-1} \tag{33}$$

C. Compliance of II at Point A

According to (15), (17) and (18), compliance matrix of hinge 5 and hinge 6 at point A could be derived as

$$_5C_A = T^A_{O_5}C_1(T^A_{O_5})^T \tag{34}$$

$$_6C_A = T^A_{O_6}C_1(T^A_{O_6})^T \tag{35}$$

Since the two hinges are in parallel, compliance matrix of II. parallelogram mechanism at point A is

$$_{para}C_A = [(_5C_A)^{-1} + (_6C_A)^{-1}]^{-1} \tag{36}$$

3.4 Output Compliance Matrix of XY Stage

According to (17), (18, (26), (33) and (36), compliance matrix of I, II, III and IV at O could be derived as

$$_{side}C_O = T^O_A\,{}_{side}C_A(T^O_A)^T \tag{37}$$

$$_{bridge}C_O = T^O_B\,{}_{bridge}C_B(T^O_B)^T \tag{38}$$

$$_{para}C_O = T^O_A\,{}_{para}C_A(T^O_A)^T \tag{39}$$

Since I and II are in series, and they are connected with III and IV in parallel, compliant matrix of branch I at O could be derived as

$$_IC_O = [(_{side}C_O)^{-1} + (_{bridge}C_O +_{para} C_O)^{-1}]^{-1} \tag{40}$$

Branch I, II, III and IV are center symmetrical. Thus, compliant matrix of branch II, III and IV could be derived as

$$_{II}C_O = R_z\left(\frac{\pi}{2}\right)_I C_O\left(R_z\left(\frac{\pi}{2}\right)\right)^T \tag{41}$$

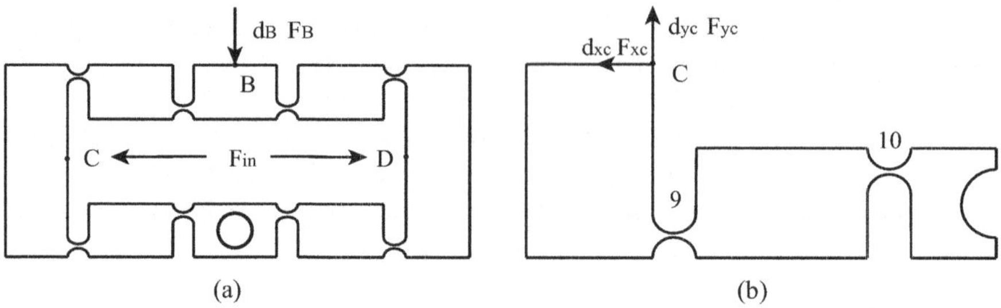

Fig. 10 **a** Loads on bridge-type amplifier. **b** Loads on a quarter of bridge-type amplifier

$$ {}_{III}C_O = R_z(\pi)\,{}_IC_O(R_z(\pi))^T \tag{42} $$

$$ {}_{IV}C_O = R_z\left(-\frac{\pi}{2}\right)_I C_O\left(R_z\left(-\frac{\pi}{2}\right)\right)^T \tag{43} $$

Therefore, the output compliance matrix of XY stage could be derived as

$$ C_{out} = [({}_IC_O)^{-1} + ({}_{II}C_O)^{-1} + ({}_{III}C_O)^{-1} + ({}_{IV}C_O)^{-1}]^{-1} \tag{44} $$

3.5 *Input Compliance of XY Stage*

In order to analyze the compliance of driving point C and D as shown in Fig. 9b, a quarter of bridge-type amplifier is isolated as depicted in Fig. 10b [13]. There are three loads at C, including F_{x_C}, F_{y_C} and M_C. Moment M_C generates no deformation, but can prevent the rigid body from torsion. According to equilibrium conditions of plane force system, F_{x_C} and F_{y_C} could be represented as

$$ F_{x_C} = \frac{F_{in}}{2} \tag{45} $$

$$ F_{y_C} = \frac{F_B}{2} \tag{46} $$

where F_{in} means the input of the XY stage. As shown in Fig. 10a, F_B is the load at B that caused by other parts of the stage.

According to (16), (17) and (18), compliance matrix of hinge 9 and hinge 10 at point C could be derived as

$$ {}_9C_C = T_{O_9}^{C} C_2 (T_{O_9}^{C})^T \tag{47} $$

$$ {}_{10}C_C = T_{O_{10}}^{C} C_2 (T_{O_{10}}^{C})^T \tag{48} $$

Thus, compliance matrix of a quarter of bridge-type amplifier at C could be derived as

$$ {}_{quart}C_C = {}_9C_C + {}_{10}C_C \tag{49} $$

According to (13), the following relationship is efficient:

$$ X_C = \begin{bmatrix} d_{x_C} \\ d_{y_C} \\ 0 \end{bmatrix} = {}_{quart}C_C \begin{bmatrix} F_{x_C} \\ F_{y_C} \\ M_C \end{bmatrix} \tag{50} $$

According to (45) and (46), (50) could be written as

$$ d_{x_C} = 0.5c_{11}F_{in} + 0.5c_{12}F_B + c_{13}M_C \tag{51} $$

$$ 0.5d_{y_B} = 0.5c_{21}F_{in} + 0.5c_{22}F_B + c_{23}M_C \tag{52} $$

$$ 0 = 0.5c_{31}F_{in} + 0.5c_{32}F_B + c_{33}M_C \tag{53} $$

where c_{ij} is the element of ${}_{quart}C_C$.

In order to generate the relationship between $2d_{x_C}$ and F_{in}, one more equation should be added. To other parts of the stage, the following equation is efficient:

$$ d_{y_B} = -b_{22}F_B \tag{54} $$

where b_{22} is an element of compliance matrix of other parts at point B.

Therefore, the relationship of $2d_{x_C}$ and F_{in} could be derived as

$$ 2d_{x_C} = \left(c_{11} - \frac{c_{31}c_{13}}{c_{33}} + \left(c_{12} - \frac{c_{32}c_{13}}{c_{33}} \right) \frac{c_{31}c_{23} - c_{21}c_{33}}{c_{33}b_{22} + c_{33}c_{22} - c_{32}c_{23}} \right) F_{in} \tag{55} $$

Compliance of XY stage at C is

$$ {}_{stage}C_C = \frac{2d_{x_C}}{F_{in}} = c_{11} - \frac{c_{31}c_{13}}{c_{33}} + \left(c_{12} - \frac{c_{32}c_{13}}{c_{33}} \right) \frac{c_{31}c_{23} - c_{21}c_{33}}{c_{33}b_{22} + c_{33}c_{22} - c_{32}c_{23}} \tag{56} $$

Considering the symmetric structure, the input compliance of XY stage could be derived as

$$ C_{in} = 2{}_{stage}C_C = 2\left(c_{11} - \frac{c_{31}c_{13}}{c_{33}} + \left(c_{12} - \frac{c_{32}c_{13}}{c_{33}} \right) \frac{c_{31}c_{23} - c_{21}c_{33}}{c_{33}b_{22} + c_{33}c_{22} - c_{32}c_{23}} \right) \tag{57} $$

3.6 Amplification Ratio of XY Stage

According to (41), (42) and (43), compliance matrix of branch II, III and IV at O could be added up as

$$_{II,III,IV}C_O = [(_{II}C_O)^{-1} + (_{III}C_O)^{-1} + (_{IV}C_O)^{-1}]^{-1} \tag{58}$$

According to (54), (55) and (56), the relationship between the displacement of B and C could be derived as

$$d_{y_B} = \frac{2(c_{31}c_{23} - c_{21}c_{33})b_{22}}{(c_{33}b_{22} + c_{33}c_{22} - c_{32}c_{23})_{stage}C_C} d_{x_C} \tag{59}$$

The load of O is the same with B, and the following relationship could be generated:

$$\frac{d_{y_O}}{o_{22}} = \frac{d_{y_B}}{b_{22}} \tag{60}$$

where o_{22} is an element of $_{II,III,IV}C_O$.

Thus, the following equation could be derived:

$$\frac{d_{y_O}}{2d_{x_C}} = \frac{(c_{31}c_{23} - c_{21}c_{33})o_{22}}{(c_{33}b_{22} + c_{33}c_{22} - c_{32}c_{23})_{stage}C_C} \tag{61}$$

Considering the symmetric structure, the amplification ratio of XY stage could be derived as

$$r = \frac{d_{y_O}}{4d_{x_C}} = \frac{(c_{31}c_{23} - c_{21}c_{33})o_{22}}{2(c_{33}b_{22} + c_{33}c_{22} - c_{32}c_{23})_{stage}C_C} \tag{62}$$

4 Model Verification with FEA

In order to test the performance of the proposed micropositioning stage and to verify the compliance models, the stage's static characteristics are analyzed by FEA with ANSYS Workbench. Al7075-T6 is adopted as the stage's material, the main specifications of the material are described as Table 1. The dimension of the stage and flexure hinges are shown in Fig. 11 and Table 2, where r, t_1, w_1 are the parameters of circular-type flexure hinge, and l_1, l_2, l_3, t_2, w_2 are the parameters of rectangular-type flexure hinge.

The stage is installed through fixing holes as shown in Fig. 9. It's assumed that the stage is driven by PZT. Thus, a pair of symmetrical force that has the same value and opposite directions is added at the input of the bridge-type amplifier. To assess the static performance of the XY stage, the FEA analysis is carried out by applying the input force of 80 N.

Table 1 Main specifications of Al7075-T6

Parameter	Value	Unit
Young's modulus	7.2×10^4	MPa
Density	2.81×10^3	kg/m^3
Poisson's ratio	0.33	
Yield strength	505	MPa
Ultimate strength	570	MPa

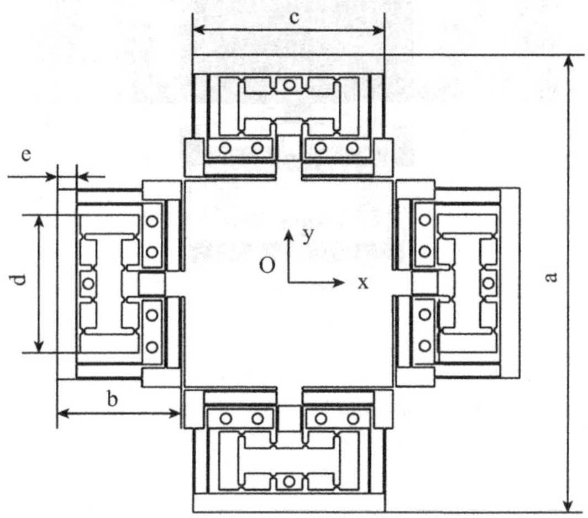

Fig. 11 Parameters of XY stage

Table 2 Main parameters of the stage

Parameter	Value (mm)	Parameter	Value (mm)
a	364	w_1	10
b	97	l_1	21
c	151	l_2	57
d	110	l_3	52
e	15	t_2	0.5
r	2.5	w_2	10
t_1	0.5		

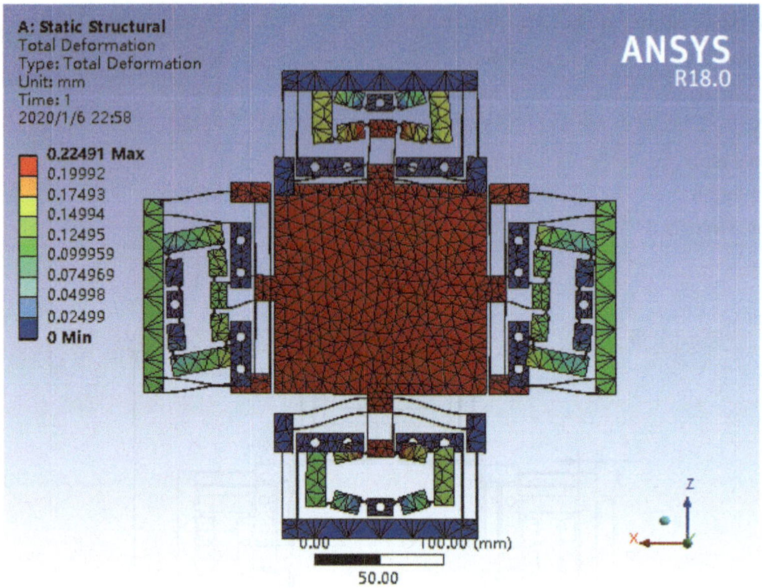

Fig. 12 Static analysis with FEA simulation

The FEA result is shown in Fig. 12. According to FEA simulation, the output displacement is 224.91 μm with the input force 80 N. The amplification ratio of the stage is 3.0, while the ratio generated by the analytic model is 3.1838. Thus, the error of the amplification ratio is 5.77%. It can be observed from Fig. 12 that with a vertical input, there is almost no parasitic error in the lateral direction. The proposed stage can achieve a large translational displacement and has a decoupled structure.

5 Conclusion

This work presents the mechanical design, compliance modeling, and analysis of a novel 2-DOF compliant micropositioning stage with a large workspace. The XY stage is designed with a bridge-type amplifier and multiple parallelogram mechanism. Analytical models are formed in order to analyze the static characteristics of the proposed stage. Compliance modeling of structure and transform of compliance matrix is introduced. FEA analysis is carried out with ANSYS workbench. The simulation result shows that the stage could generate an output of 224.91 μm with an input force of 80 N, and it also has good decoupling ability.

References

1. Cecil, J., Vasquez, D., Powell, D.: A review of gripping and manipulation techniques for micro-assembly applications. Int. J. Prod. Res. **43**(4), 819–828 (2005)
2. Jain, R.K., Majumder, S., Ghosh, B., Saha, S.: Design and manufacturing of mobile micro manipulation system with a compliant piezoelectric actuator based micro gripper. J. Manuf. Syst. **35**, 76–91 (2015)
3. Howell, L.L.: Compliant Mechanisms. Wiley, Hoboken (2001)
4. Howell, L.L., Midha, A.: Parametric deflection approximations for end-loaded, large-deflection beams in compliant mechanisms (1995)
5. Su, H.J.: A pseudorigid-body 3r model for determining large deflection of cantilever beams subject to tip loads. J. Mech. Robot. **1**(2), 021008 (2009)
6. Yu, Y., Feng, Z., Xu, Q.: A pseudo-rigid-body 2r model of flexural beam in compliant mechanisms. Mech. Mach. Theory **55**, 18–33 (2012)
7. Ghosh, A., Corves, B.: Introduction to Micromechanisms and Microactuators. Springer, India (2015)
8. Maluf, N., Williams, K.: Introduction to Microelectromechanical Systems Engineering. Artech House, Boston (2004)
9. Xu, Q.: New flexure parallel-kinematic micropositioning system with large workspace. IEEE Trans. Rob. **28**(2), 478–491 (2011)
10. Gan, J., Zhang, X., Li, H., Wu, H.: Full closed-loop controls of micro/nano positioning system with nonlinear hysteresis using micro-vision system. Sens. Actuators, A **257**, 125–133 (2017)
11. Choi, K., Lee, J.J., Hata, S.: A piezo-driven compliant stage with double mechanical amplification mechanisms arranged in parallel. Sens. Actuators, A **161**(1–2), 173–181 (2010)
12. Tang, X., Chen, I.M.: A large-displacement 3-dof flexure parallel mechanism with decoupled kinematics structure. In: 2006 IEEE/RSJ International Conference on Intelligent Robots and Systems, pp. 1668–1673 (2006)
13. Lobontiu, N., Garcia, E.: Analytical model of displacement amplification and stiffness optimization for a class of flexure-based compliant mechanisms. Comput. Struct. **81**(32), 2797–2810 (2003)
14. Awtar, S., Parmar, G.: Design of a large range xy nanopositioning system. J. Mech. Robot. **5**(2), 021008 (2013)
15. Wang, F., Shi, B., Tian, Y., Huo, Z., Zhao, X., Zhang, D.: Design of a novel dual-axis micromanipulator with an asymmetric compliant structure. IEEE/ASME Trans. Mechatron. **24**(2), 656–665 (2019)

Assistive Robots

Jinhua She, Yasuhiro Ohyama, Edwardo F. Fukushima, and Sho Yokota

Abstract The number of elderly people is increasing rapidly. Accordingly, the health and welfare of the elderly have become a big problem. And many assistive robots have been developed for the elderly to ensure their mental and physical soundness. This chapter explains some robots that we built for this purpose. First, we explain a human-body-motion interface. It allows us to use the motion of the upper body of a user to drive an electric wheelchair. Then, we show an electric cart that helps the elderly to maintain or even improve their physical strength. It features that an optimal pedal load is automatically generated based on a user's physical condition. The normalized longitudinal force is precisely estimated using a simple algorithm based on the equivalent-input-disturbance approach to guarantee driving safety. Finally, we describe the design of a left-right-independent rehabilitation machine for lower limbs. The pedal loads and strokes on the left and right can be adjusted independently. This not only makes it easy to suit different requirements for lower-limb rehabilitation but also mitigates mental distress and excites the volition for rehabilitation.

Keywords Aging · Equivalent input disturbance (EID) · Human body motion · Human machine interface · Motor functions · Pedaling · Rehabilitation

---

J. She (✉)
School of Automation, China University of Geosciences, Wuhan 430074, China
e-mail: she@stf.teu.ac.jp

Hubei Key Laboratory of Advanced Control and Intelligent Automation for Complex Systems, Wuhan 430074, China

J. She · Y. Ohyama · E. F. Fukushima
School of Engineering, Tokyo University of Technology, Hachioji 192-0982, Japan

S. Yokota
Faculty of Science and Engineering, Toyo University, Saitama 350-8585, Japan

M. Wu et al. (eds.), *Developments in Advanced Control and Intelligent Automation for Complex Systems*, Studies in Systems, Decision and Control 329,
https://doi.org/10.1007/978-3-030-62147-6_14

Abbreviations

CoG	center of gravity
DPDC	dynamic parallel distributed compensation
EID	equivalent input disturbance
HBMI	human-body-motion interface
NLF	normalized longitudinal force
PIC	peripheral interface controller
RPE	rating of perceived exertion

1 Introduction

The elderly are referred to as people over 65 years old. According to the statistics of the United Nations in 2019 [1], the percentage of the elderly in the population is 28.4% in Japan, 15.8% in South Korea, and 12.0% in China in 2019. The world's population is growing older, and some countries in Eastern and South-Eastern Asia will experience the fastest increase in the percentage of the elderly between 2020 and 2050, for example, the increase in China and South Korea will be 20% and 23%, respectively.

Along with the continuous increase in the average life span, there is a concomitant marked increase in the number of people who are in fragile health due to aging or being bedridden. It is becoming a very serious social problem. A rapid rise has been seeing in public concern about how to establish an environment for the elderly to ensure their mental and physical soundness. A great number of research and development related to the health and welfare of the elderly has been carrying out in many countries. Many types of electric wheelchairs and electric carts have been developed to improve the ability of old people to get around.

Those electric wheelchairs are mainly controlled by joysticks. However, they require complex wrist movements that might be difficult for the elderly and thus lead to accidents [2]. Other interfaces were studied to solve this problem, for example, a voice [3], a tongue-palate pressure [4], an EMG signal [5], head movement [6], eye movement [7], and face inclinations [8].

While electric wheelchairs are designed for people who completely lost their walking abilities, electric carts focus on expanding people's activity area. Thus, the purposes of the studies on electric wheelchairs and carts should be different. However, almost all of the carts were designed solely as a means of transportation, but no consideration was given to the need for physical exercise [9–12]. As a result, a reliance on a cart for getting around deteriorates a user's walking muscles.

Taking into consideration the difference between electric wheelchairs and carts, we have been carrying out studies on improving operationality for electric wheelchairs [13–16] and on maintaining and/or improving physical strength for electric carts [17–20]. In this chapter, we present a short review of our studies on those topics.

2 Human-Body-Motion-Controlled Electric Wheelchair

Considering that a joystick is not easy to use for the elderly, we explored the possibility of using other means to control an electric wheelchair. In this study, we focused on the use of the motion of the upper body of a user sitting in a wheelchair. There are three possible ways of detecting the motion: a camera image, inclinometers, and pressure distribution on a mat. While a camera image does not impose any loads on a user, an image is influenced largely by light and it is also a problem to find a suitable place in a wheelchair to install it. To measure the inclination of a user accurately, we have to mount sensors on a user's upper body. This inconveniences a user. Pressure distribution on a seat allows us to estimate the body motion. Moreover, it is nonrestrictive and easy to use.

There are two places suitable for pressure measurement: the seat and back of a chair. Preliminary experiments revealed that the pressure distribution on the back showed body movement more clearly than that on the seat. Therefore, we use the center of gravity (CoG) of the pressure distribution on the back of an electric wheelchair to control the movement of the chair, and designed a human-body-motion interface (HBMI) to improve operationality (Fig. 1). The requirement for the HBMI is that the pressure on the back of a wheelchair is measured without giving a user any unpleasant feelings. Thus, we have to carefully adjust the sensitivity of the HBMI to a suitable level. This also eases the estimation of intention.

2.1 *Electric Wheelchair with Human-Body-Motion Interface*

The HBMI has three parts: a BPMS (body pressure measurement system) (Tekscan, Inc.), a motor drive system (JW-1, Yamaha Motor Co., Ltd.), and a notebook personal

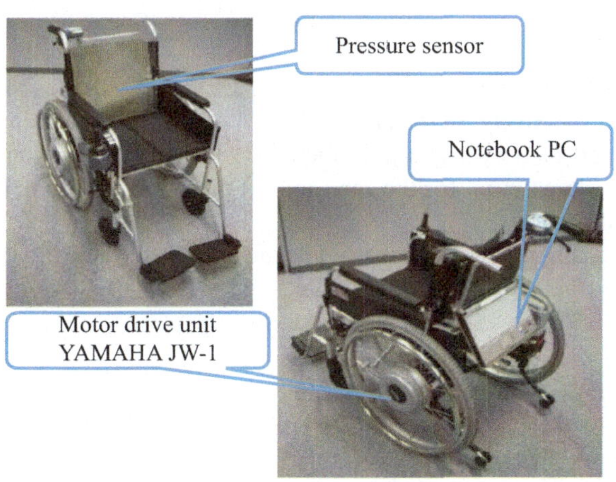

Fig. 1 HBMI for electric wheelchair

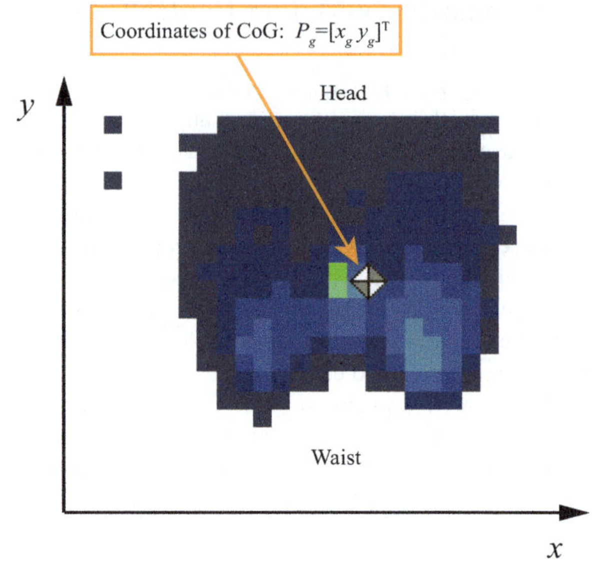

Fig. 2 Coordinates of CoG of pressure distribution on backrest

computer (PC) (Fig. 1). A change in the CoG of the pressure distribution occurs when a user tries to change the state of a wheelchair. The HBMI processes this change to estimate the user's intention.

The processing scheme is shown in Fig. 2. We produce two voltage inputs from the CoG information to drive the wheelchair. The initial position of CoG, P_{g0}, is recorded when the system is turned on.

Then, the voltage inputs are given by

$$V = A(P_{g0} - P_g), \tag{1}$$

$$P_g = \begin{bmatrix} x_g \\ y_g \end{bmatrix}, \; A = \begin{bmatrix} a_{fb} & 0 \\ 0 & a_{lr} \end{bmatrix}, \; V = \begin{bmatrix} v_{fb} \\ v_{lr} \end{bmatrix}, \tag{2}$$

where x_g and y_g in the vector P_g are the coordinates of the x- and y-axes of the CoG, respectively; in the gain matrix, A, a_{fb} (> 0) is a gain factor for fore-and-aft actions and a_{lr} (> 0) is a gain factor for left-right action; and in the vector V, v_{fb} is the voltage input for the fore-and-aft direction and v_{lr} is the voltage input for left-right direction.

Finally, the voltage inputs are sent to a motor drive unit.

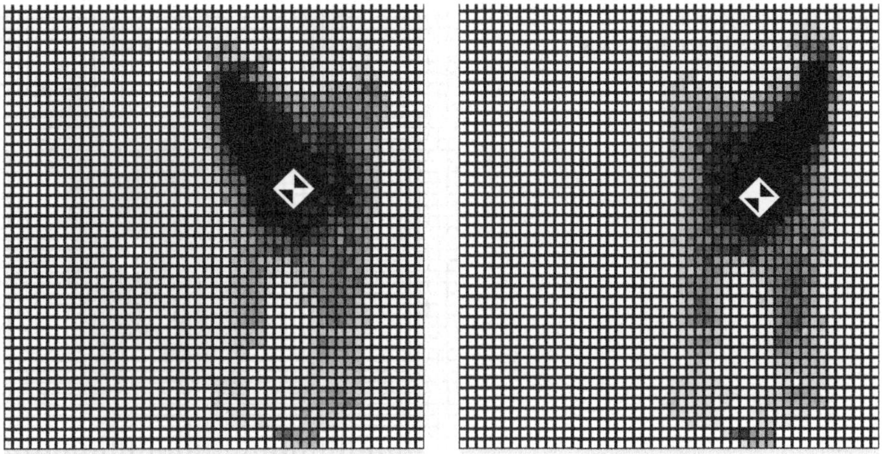

Fig. 3 CoG for taking left and right turns

2.2 *Tuning of Gain A*

Experiments were carried out to verify the designed HBMI using the semantic differential (SD) method for 10 subjects [14].

The tuning of A in (1) is important. Figure 3 shows the CoG that reflects a user's intention of turning left and right, in which the dark and bright places indicate strong and weak pressure, respectively. Small a_{fb} and a_{lr} in A result in a dull response that makes a user feel awkward and hard to drive. On the other hand, large a_{fb} and a_{lr} may make a wheelchair very sensitive to changes in the CoG and difficult to stop the wheelchair. Thus, there is a trade-off between the easy operation and secure stop in tuning A.

We use the following strategies to solve this trade-off problem:

(1) enlarge the changes in CoG of the pressure distribution, and
(2) extract the user's stop intention from the CoG promptly.

The CoG is calculated by

$$x_g = \frac{\sum_{i=0}^{n_x-1}\left(i\sum_{j=0}^{n_y-1}F_{ij}\right)}{\sum_{i=0}^{n_x-1}\sum_{j=0}^{n_y-1}F_{ij}}, \quad y_g = \frac{\sum_{i=0}^{n_y-1}\left(i\sum_{j=0}^{n_x-1}F_{ij}\right)}{\sum_{i=0}^{n_y-1}\sum_{j=0}^{n_x-1}F_{ij}}, \tag{3}$$

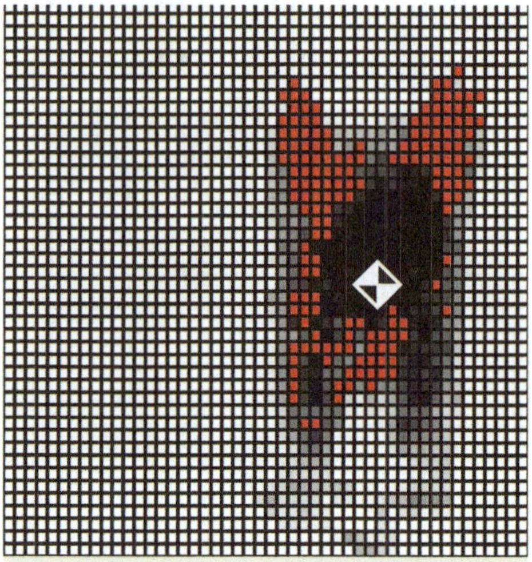

Fig. 4 Extracting cells having pressures larger than thresholds (red)

where n_x and n_y are the numbers of sensor cells in x and y directions on a sensor mat, respectively; and F_{ij} is the pressure at the ij-th cell.

The CoG is mainly determined by the distribution of strong pressures, that is, the dark cells in Fig. 3. Since a body usually contacts with the back of a wheelchair constantly during driving, the changes in the dark area are very small. So, it is important to extract and enlarge the changes. We take the CoG when the system is turned on, P_{g0}, as a threshold, and use $(P_{g0} - P_g)$ to calculate V. This strategy makes the changes in CoG clear and the system precisely understands the user's intention. This is why we do not directly use P_g to produce the control inputs in (1).

A preliminary experiment was carried out for calibration on the pressure distribution. Ten subjects moved their upper body right and left, forward and backward to collect constant contact areas.

P_g was calculated using only cells that exceeded pressure thresholds, which were set to be the difference between the maximum and minimum pressure for each cell. A cell with a pressure less than the threshold was masked, and the pressure was set to be zero. Cells in red in Fig. 4 are those in which the pressures are bigger than the threshold. They were used to calculate P_g. This strategy ensures that P_g reflects even a small body motion.

The motions of an upper-body are classified into seven kinds to control a wheelchair: relaxed sit, lean forward/fore/aft, lean left/right, and twist left/right. The relaxed setting is used to describe the intention of stopping. A self-organizing map (SOM) [21] was used to classify the motions. In the classification, the input to the SOM is the pressures on a 34×44-grid film sensor mat. Those 1496 data form the input vector for the SOM. The output of the SOM is a weighting matrix for all cells. 345 sets of the pressure data of a male subject (height: 170 cm, weight: 60 kg) were used for the learning of the SOM. The learning parameters of the SOM are listed in Table 1. The initial values of the weighting matrix were set randomly.

Table 1 Parameters for SOM learning

Number of learning	Learning rate	Radius	Map size
500000	0.02	6	20 × 20

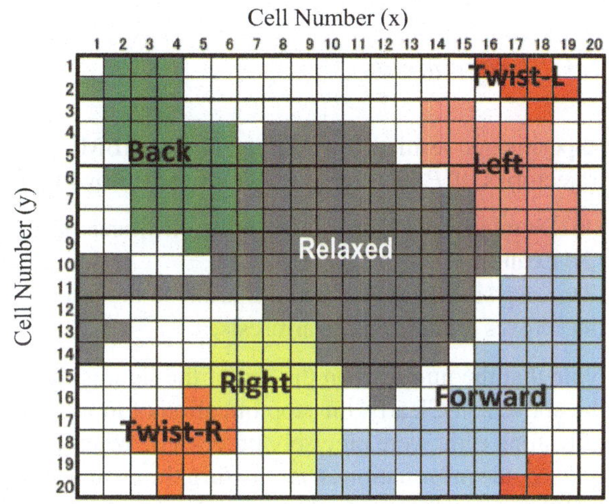

Fig. 5 Identified relationship between cells on sensor map and motion intention

Then, we used the pressure data collected from the experiments to calculate the weighting matrix for all cells and cosine similarity between the learning and experimental results. Finally, an output map corresponding to the seven motions was obtained based on the similarity of each cell on the output map (Fig. 5). This map was used to extract a user's intention.

The above strategies are integrated to process the intention to drive a wheelchair (Fig. 6). All these strategies are combined to ensure that the HBMI is easy to use. The pressure data from the BPMS is sent to the SOM and the process of CoG

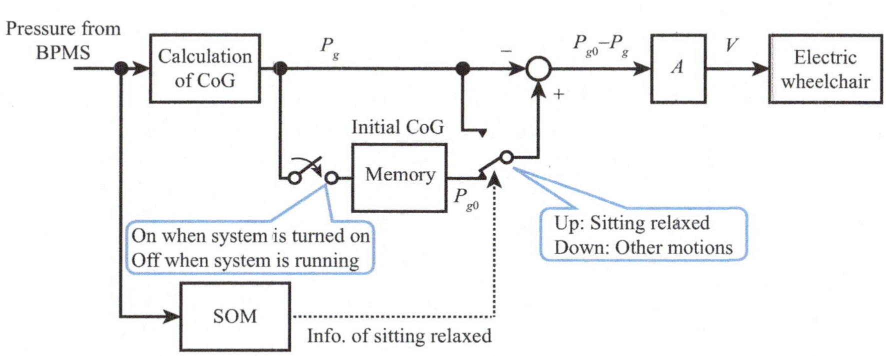

Fig. 6 Intention processing and wheelchair drive

calculation. The SOM detects one of the seven emotions. The CoG calculation first uses the thresholds to delete small pressures, then calculates the CoG. $(P_{g0} - P_g)$ is amplified to produce voltages to drive the wheelchair. If a user is not sitting relaxed, $P_{g0} - P_g \neq 0$ and the wheelchair moves. Otherwise, the wheelchair stops.

2.3 Experiments

Experiments verified the validity of the HBMI. As an example, Fig. 7 shows the trajectories of P_g and $(P_{g0} - P_g)$ for the following motions:

(1) Sit relaxed without putting arms on wheelchair armrests.
(2) Lean right.
(3) Keep sitting relaxed and put arms on the armrests.
(4) Moves arms from the armrests.

Note that the numbers on the top of the figures mean the corresponding motions in the above list. Clearly, the stop intention of motions (1), (3), and (4), and the motion of leaning right were precisely processed. When the SOM detects the state of sitting relaxed, the information is used to turn the switch to the up position in Fig. 6 to ensure that the voltage inputs to the wheelchair are zero so as to stop it.

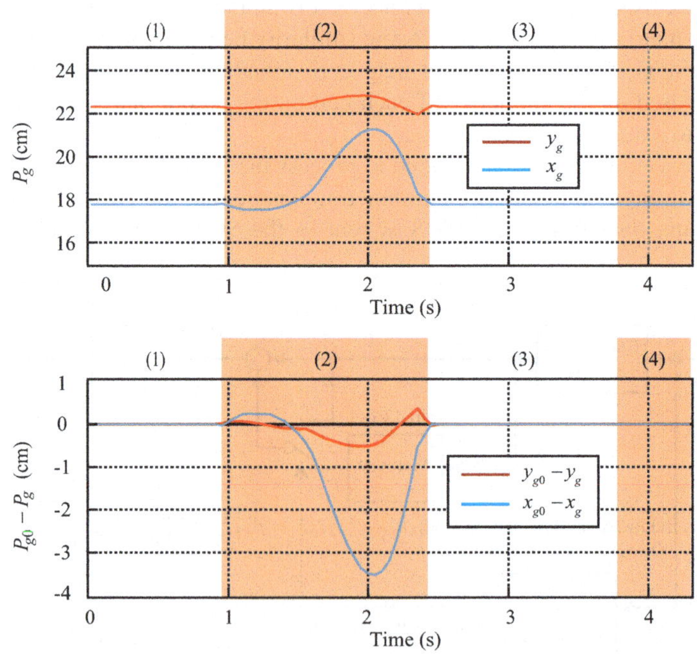

Fig. 7 Trajectories of P_g and $(P_{g0} - P_g)$

2.4 Conclusion

This study focused on building an easily used HBMI to control the motion of an electric wheelchair. The HBMI uses the body motion to generate a command for the motion of the wheelchair. The CoG of the pressure distribution on the back of the wheelchair is detected to catch a user's intention. The gain, A, in the processing of pressure distribution is carefully adjusted to balance the prompt response of the wheelchair to body motion and the precise abstraction of the user's intention. Strategies of thresholds, the SOM, and the amplification of difference $(P_{g0} - P_g)$ improved the estimation precision for the CoG and operating safety for a wheelchair markedly.

The validity of the HBMI has been thoroughly tested through experiments for various conditions.

3 Electric Cart for Maintaining Physical Strength

To help the elderly getting around and also getting physical exercise, we developed a three-wheeled electric cart with a pedal unit. Pedaling was chosen to implement the function of maintaining or improving physical strength. An ergonomically designed pedal unit was mounted on the cart, and an interface board that handles inputs and outputs was built to simplify the design of the system. For the assembled cart system, an impedance model was devised to describe the feeling of pushing the pedals.

A bilateral master-slave H_∞ control system was first built to control the speed of the cart and to provide a driver some physical exercise or assistance [17]. Then, a heart-rate (HR) meter was mounted to estimate a driver's physical condition so as to automatically choose a suitable pedaling load for the driver [18, 20]. On the other hand, it is of great meaning to estimate the normalized longitudinal force (NLF) because the NLF is a key to guaranteeing driving safety under any road and weather conditions. We applied the equivalent-input-disturbance (EID) approach to derive a method of estimating the NLF in a real-time fashion [19, 20].

3.1 Hardware of the Cart System

The Everyday Type-S (Araco Corp., Japan; Cart motor: 24 V, 330 W, DENSO Corp., Japan) was selected as the base of the system. It is a commercially available ready-made three-wheeled electric cart.

Based on the analysis of muscular degeneration due to aging and prevention measures of the degradation of motor functions, we found that pedaling would be an appropriate exercise to work the walking muscles. We mounted two-foot pedals on the electric cart to provide a driver with a way of exercising the walking muscles.

Fig. 8 Photographs of electric cart

The pedals are connected to a geared DC motor (rated voltage: 24 V, rated power: 25 W, Tsukasa Electric Co. Ltd., Japan), which generates a pedaling load. The pedals and motor constitute a pedal unit. Considering that an electrical connection between the pedals and the drive wheels is flexible and provides great potential for further development, we used it rather than a mechanical connection in the system. Based on the statistics of physiological data on the elderly in Japan and the optimal pedaling region given by ergonomics, we finally mounted the pedal unit on an electrical cart as shown in Fig. 8. Note that we also mounted an HR meter to estimate a driver's physical condition.

Two optical encoders measure the rotational angles of the pedal and cart motors (8000 and 2400 pulses per revolution, respectively). An interface board was built to handle all the inputs and outputs of the motors (Fig. 9). The board has a PIC (peripheral interface controller) microcontroller, PIC16C74, and a dual power operational amplifier, TA7272. The PIC was used as an interface between the cart and a controller (a laptop computer, Toshiba Libretto 50) and was programmed to implement the functions of counters (collecting information on the rotational angles from the optical encoders) and D/A converters (sending control inputs for the pedal and cart

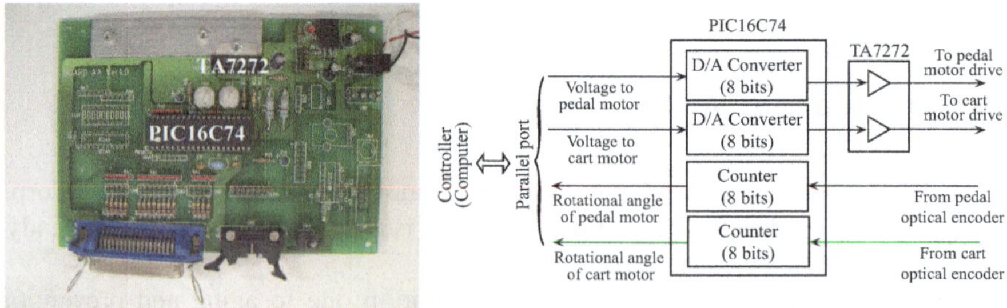

(a) Photo (b) Block diagram

Fig. 9 Interface board. **a** Photo. **b** Block diagram

motors). And TA7272 adjusted the level of the control signals in the range $[-5,\ 5]$ V to that required by the motor drives in the range $[-24,\ 24]$ V.

3.2 *Driver-Adapted Selection of Pedal Load and Control System*

A pedaling load that is responsive to the road conditions provides a driver with an enjoyable and realistic driving experience. So, a time-varying load is added to the pedals. Moreover, since a driver's physical condition changes constantly, it is desirable to select a load that is most suitable to the driver.

The HR of a driver and the Karvonen formula [22, 23] are combined to estimate a driver's rating of perceived exertion (RPE) (r_{PE}). An observation revealed that r_{PE} is linearly related with Borg CR10 (CR: category ratio) scale [24]. Thus, we use r_{PE} to find the most suitable pedal load for the driver. More specifically, we first choose two pedal loads (strenuous: the largest, natural: no load) to give a load range for exercise. An impedance model is used to describe the feeling of pushing the pedals for each load:

$$\frac{dv_{imp}(t)}{dt} = A_p v_{imp}(t) + B_p f(t), \tag{4}$$

where A_p was chosen to be the same of that of the master (pedal unit) ($A_p = -1.49$), and B_p was a constant that determine the feeling:

$$\begin{cases} B_p = 2.00 \text{ (Strenuous load)}, \\ B_p = 3.49 \text{ (Neutral load)}. \end{cases} \tag{5}$$

Then, we select the pedal load based on the relationship shown in Fig. 10.

A driver-adapted bilateral master-slave cart control system is designed by the following steps:

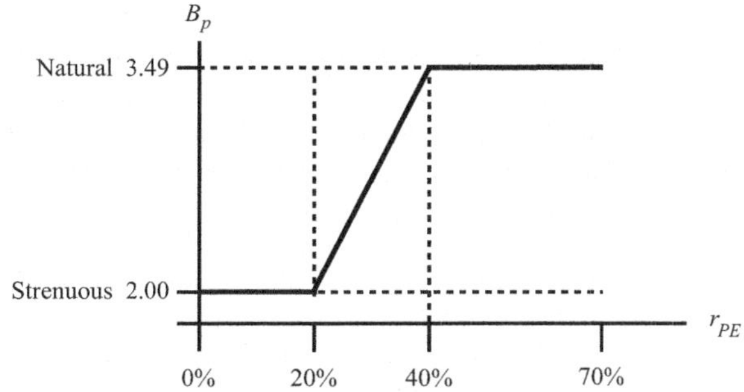

Fig. 10 Relationship between pedal load and r_{PE} just before start of driving

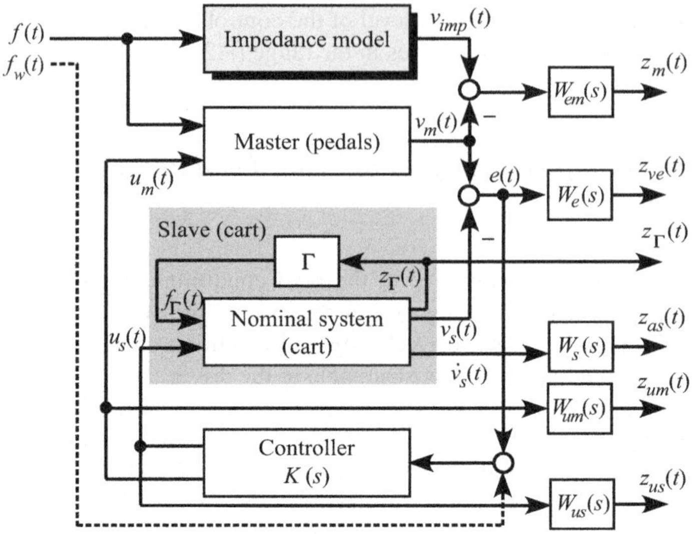

Fig. 11 Block diagram for design of bilateral master-slave cart control system for i $(i = \alpha,\ \beta)$

Step 1: Design a control law, $u_i(t)$ $(= [u_{mi}(t),\ u_{si}(t)]^{\mathrm{T}};\ i = \alpha,\ \beta)$, for each of the strenuous $(i = \alpha)$ and natural $(i = \beta)$ loads, where $u_m(t)$ and $u_s(t)$ are the voltages applied to the pedal and cart motors, respectively.

Step 2: Calculate the control law of the cart control system based on the concept of dynamic parallel distributed compensation (DPDC):

$$u(t) = \lambda u_\alpha(t) + (1 - \lambda)u_\beta(t),\ \lambda = 2 - 5r_{\mathrm{PE}}, \tag{6}$$

where r_{PE} is the RPE of a driver before start of driving.

Each of the control laws, $u_i(t)$ $(i = \alpha,\ \beta)$, is designed based on the block diagram in Fig. 11. In the figure, the master is the pedal unit; the slave is the cart, in which Γ describes the changes of drivers' weight; and $z_\Gamma(t)$ and $f_\Gamma(t)$ are the input and output of Γ, respectively. Since a first-order dynamics is easy for people to operate, speeds ($v_m(t)$ and $v_s(t)$) were selected to be the outputs of both the master and slave. The exogenous input signal is the pushing force on the pedals, $f(t)$. $f_w(t)$ is a disturbance artificially added to the measurement output channel to relax the solvable condition for the design of the controller, $K(s)$, using the H_∞ control method. In the design of $K(s)$, the controlled outputs are

1. $z_m(t)$: the weighted speed error between the impedance model and the pedals, $Z_m(s) = W_{em}(s)[V_{imp}(s) - V_m(s)]$;
2. $z_{ve}(t)$: the weighted speed error between the pedals and the wheels, $Z_{ve}(s) = W_e(s)[V_m(s) - V_s(s)]$;
3. $z_\Gamma(t)$: the input signal of the uncertainty $\Gamma(t)$;
4. $z_{as}(t)$: the weighted acceleration of the cart, $Z_{as}(s) = W_s(s)\dot{V}_s(s)$; and

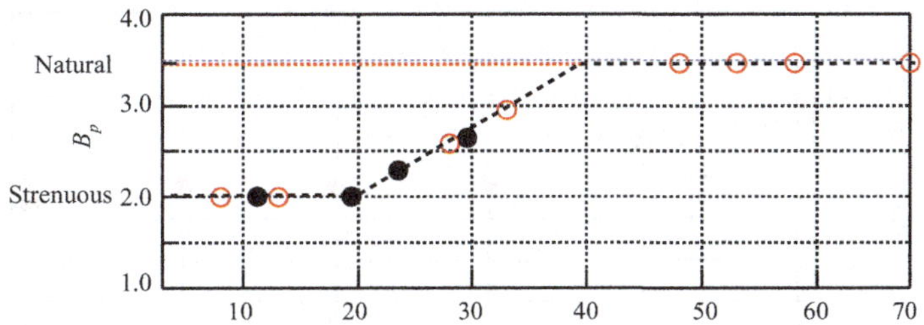

Fig. 12 Verification results of relationship between B_p and r_{PE} (○: Male subject of 21 years old, •: Male subject 83 years old)

5. $z_{um}(t)$ and $z_{us}(t)$: the weighted voltages applied to the pedal and cart motors, $Z_{um}(s) = W_{um}(s)U_m(s)$ and $Z_{us}(s) = W_{us}(s)U_s(s)$;

where $X(s)$ is the Laplace transform of $x(t)$. The evaluation of $z_m(t)$ and $z_{ve}(t)$ suppresses the steady-state error of $[v_{imp}(t) - v_m(t)]$ and $[v_m(t) - v_s(t)]$. Thus, $W_{em}(s)$ and $W_e(s)$ were chosen to have low-pass characteristics. The evaluation of $z_\Gamma(t)$ guarantees the robust stability of the system for the difference in drivers' weight. The evaluation of $z_{as}(t)$ ensures the riding comfort. $W_s(s)$ was selected to be a low-pass filter based on the limits on whole-body vibrations in the fore-and-aft direction [25, 26]. And the evaluation of $z_{um}(t)$ and $z_{us}(t)$ keeps the control inputs to the motors at a low level, and were chosen to be high-pass filters.

The design problem for the controller of load i $(i = \alpha,\ \beta)$ can be stated as follows: *Find a controller* $K(s)$ *such that*

1. the cart control system is internally stable, and
2. the H_∞ norm of the transfer function from $w(t) = [f(t),\ f_w(t)]^{\mathrm{T}}$ to $z(t) = [z_m(t)\ z_{ve}(t)\ z_\Gamma(t)\ z_{as}(t)\ z_{um}(t)\ z_{us}(t)]^{\mathrm{T}}$, $G_{zw}(s)$, is less than 1, that is, $\|G_{zw}\|_\infty < 1$.

A serial experiment confirmed the validity of the driver-adapted bilateral master-slave cart control system. Five students at Tokyo University of Technology (ages: 21–25 yrs., sex: male), and twelve elderly people from volunteer groups in Tokyo (ages: 65–83 yrs., sex: 4 male and 8 female) participated in the experiments. Riding experiments were carried out on a flat road, a 5° uphill road, and a 5° downhill road at the Hachioji campus of Tokyo University of Technology. Experimental results show that a suitable pedal load was selected for each subject in every case. As an example, the relationship between B_p and r_{PE} (Fig. 12) shows that B_p increases as r_{PE} increases. The results agree with the designed rule (Fig. 10).

Figure 13 shows the relationships of the average pushing force on pedals versus r_{PE}, and the average input voltage of the pedal motor versus r_{PE} for a flat road in the steady-state. The figure clearly shows that, as designed, along with the increase in r_{PE}, the pedal load decreases and the applied voltage to the pedal motor increases.

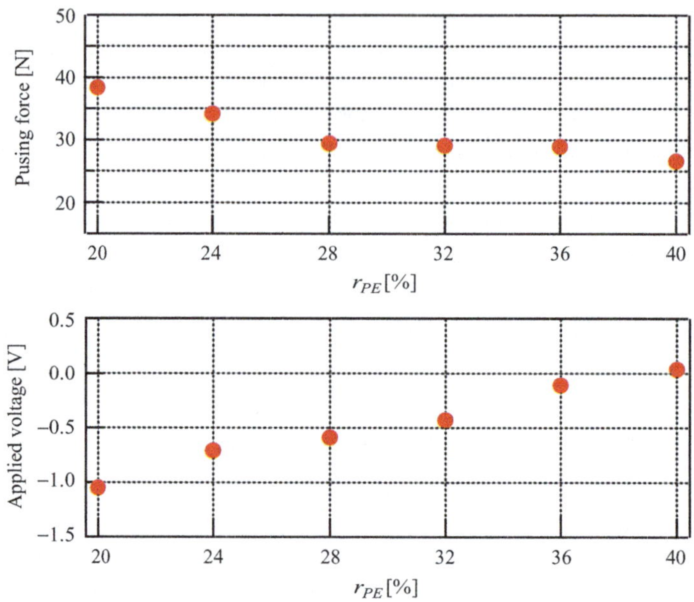

Fig. 13 Average pushing force of pedals and voltage of pedal motor versus r_{PE} for flat road in steady state

A questionnaire was also carried out for the subjects. The results show that more than 80% of the elderly subjects satisfied with the design and the control performance of the cart.

3.3 Estimation of NLF for the Electric Cart Using EID Approach

Electric carts greatly enlarged the area of activity for the elderly. However, along with the increased use of electric carts, the problem of safety has also been taken up as a social problem. Many accidents occur at railroad crossings [27]. One of the reasons seems to be a large change in the NLF between a paved road and steel. The weak back and leg muscles of the elderly electric-cart users result in their slow reaction time for such a change in the NLF. To solve this problem, it is necessary to build a safeguard into an electric-cart control system to guarantee easy and safe driving. One of the key steps to do that is to obtain a precise estimation of the NLF in a real-time fashion.

In this study, we took the NLF to be a state-depend disturbance of the cart dynamics and developed a simple but effective method for the estimation. It is based on the EID approach [28, 29].

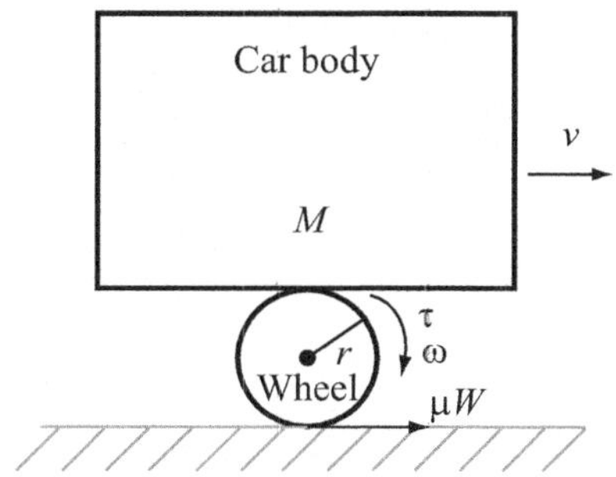

Fig. 14 One-wheel vehicle model of electric cart

A one-wheel vehicle model, which is a half model of the cart, is employed (Fig. 14) in this study. The parameters and variables are

M: mass of cart [kg]
W: weight of cart [N] ($= Mg$)
g: gravitational acceleration [m^2/s]
J: moment of inertia of wheel [$kg \cdot m^2$]
r: effective radius of wheel [m]
B: equivalent bearing and friction coefficient [$Nm \cdot s/rad$]
$\mu(t)$: NLF
$\lambda(t)$: wheel slip rate
$F_d(t)$: longitudinal force [N]
$F_n(t)$: normal (vertical) force [N]

$v(t)$: cart velocity [m/s]
$v_\omega(t)$: circumferential velocity of the wheel [m/s]
$\omega(t)$: angular velocity of the wheel in one-wheel vehicle model [rad/s]
$\tau(t)$: driving torque produced by the motor [Nm]
$\tau_\omega(t)$: friction torque of the wheel [Nm]
$\tau_d(t)$: tractive torque [Nm].

Taking into consideration of the longitudinal motion of the vehicle and ignoring the lateral force give the dynamics of the cart body

$$M\frac{dv(t)}{dt} = F_d(t), \tag{7}$$
$$F_d(t) = \mu(t)F_n(t). \tag{8}$$

For the model in Fig. 14,

$$F_n(t) = W. \tag{9}$$

The maximum $\mu(t)$ is called the tire-road friction coefficient [30].

The dynamics of the wheel are

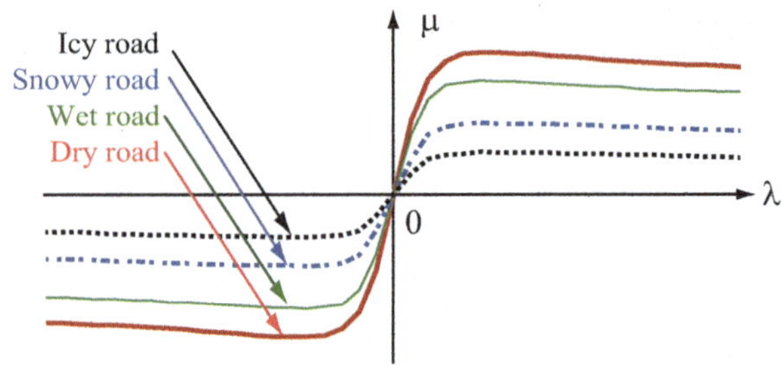

Fig. 15 Typical $\mu \sim \lambda$ relationship for various road conditions [31]

$$J\frac{d\omega(t)}{dt} = \tau(t) - \tau_\omega(t) - \tau_d(t), \tag{10}$$

$$\tau_\omega(t) = B\omega(t), \tag{11}$$

$$\tau_d(t) = rF_d(t). \tag{12}$$

The circumferential velocity of the wheel is

$$v_\omega(t) = r\omega(t). \tag{13}$$

The wheel slip rate is defined to be

$$\lambda(t) = \frac{v_\omega(t) - v(t)}{\max\{v_\omega(t), v(t)\}}. \tag{14}$$

Clearly, $\lambda(t) = 0$ when the tire does not slip at all, $\lambda(t) < 0$ during braking, and $\lambda(t) > 0$ when there is traction. Moreover, $\lambda = -1$ when the wheel is locked, and $\lambda = 1$ when the wheel spins freely. So, $\lambda \in [-1, 1]$.

$\mu(t)$ in (8) depends on the tire and road conditions and is a function of $\lambda(t)$

$$\mu(t) = \mu(v(t), v_\omega(t)) = \mu(\lambda(t)). \tag{15}$$

A typical $\mu \sim \lambda$ relationship (Fig. 15) is nonlinear and has different peak values and slopes for different road conditions. Summarizing the above relationships yields the block diagram Fig. 16. Since the relationship between μ and λ is nonlinear and depends on the road condition, it is difficult to identify the exact relationship in a real-time fashion.

An EID is a signal on the control input channel that produces the same effect on the output of a plant as an actual disturbance does. The EID approach was devised to actively reject disturbances in a servo system [28]. An analysis of the disturbance rejection mechanism reveals that an EID-based control system is a two-degree-of-freedom system, and one degree of freedom is used to produce a precise EID to effectively reject disturbances [29].

In this study, we treat $\tau_d(t)$ in (10) as a state-dependent disturbance

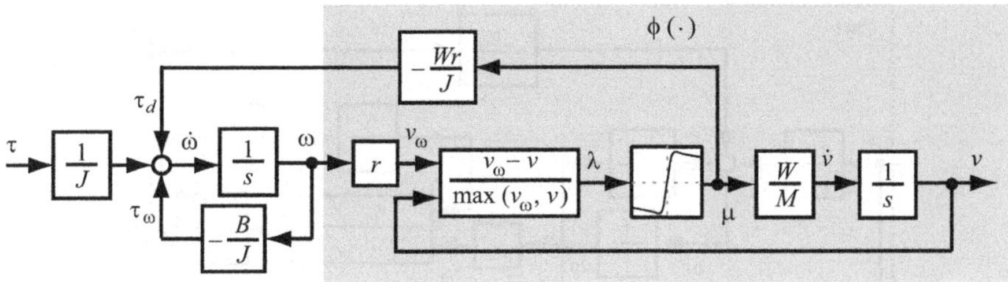

Fig. 16 Block diagram of cart

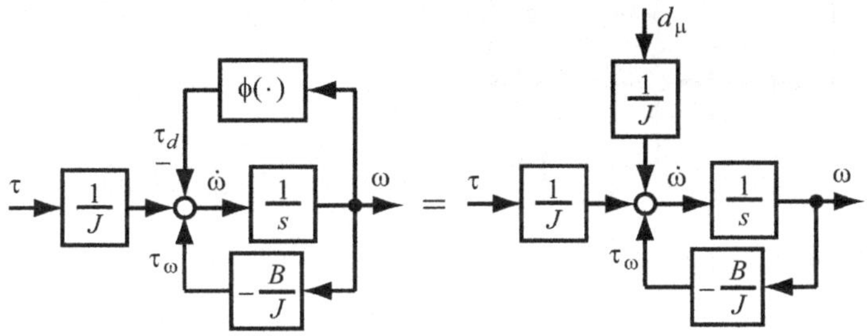

Fig. 17 EID-based formulation of cart dynamics for estimation of μ

$$\tau_d(t) = \phi(\mu(t), M), \tag{16}$$

and define the EID to be

$$d_\mu(t) = -rW\mu(\lambda(t)). \tag{17}$$

As a result, the dynamics of the cart (Fig. 17) becomes

$$\frac{d\omega(t)}{dt} = -\frac{B}{J}\omega(t) + \frac{1}{J}[\tau(t) + d_\mu(t)]. \tag{18}$$

A state observer of the dynamics (18) is

$$\frac{d\hat{\omega}(t)}{dt} = -\frac{B}{J}\hat{\omega}(t) + \frac{1}{J}\tau(t) - L_\mu[\hat{\omega}(t) - \omega(t)], \tag{19}$$

where L_μ is the observer gain, and $\hat{\omega}(t)$ is an estimate of $\omega(t)$. Letting

$$\Delta\omega(t) = \hat{\omega}(t) - \omega(t), \tag{20}$$

then an estimate of $d_\mu(t)$ is derived based on the EID approach [19]

$$\hat{d}_\mu(t) = -JL_\mu(t)\Delta\omega(t). \tag{21}$$

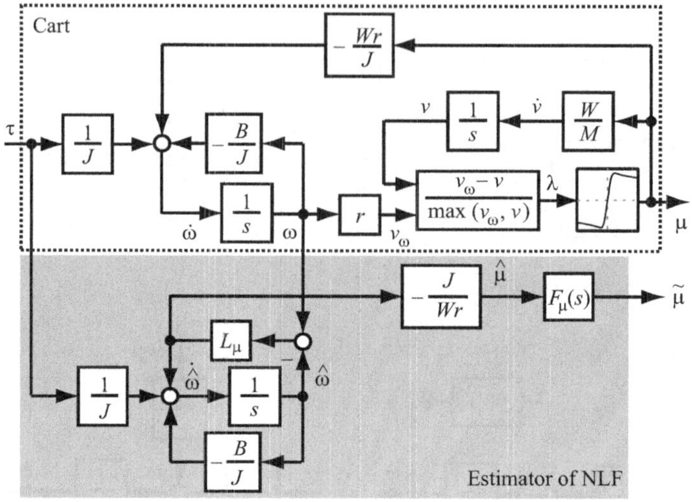

Fig. 18 EID-based estimation system for μ

Thus, an estimate of $\mu(t)$ is given by

$$\hat{\mu}(t) = \frac{J}{rW} L_\mu \Delta\omega(t). \tag{22}$$

A low-pass filter, $F_\mu(s)$, is used to remove noise in the estimate

$$F_\mu(s) = \frac{1}{T_\mu s + 1}. \tag{23}$$

Finally, we obtain the filtered estimate of $\mu(t)$, $\tilde{\mu}(t)$, as the output of $F_\mu(s)$. Collecting the explanation gives an EID-based estimation system for μ in Fig. 18.

An upper bound on the steady-state estimation error in $\mu(t)$ is

$$E_\mu = \frac{\sqrt{rBW}}{B + JL_\mu}. \tag{24}$$

The above equation shows that a larger observer gain L_μ results in a smaller estimation error.

Note that we need the exact value of $1/J$ in the state observer (19) to estimate $\hat{\mu}(t)$. Since J may be different for drivers, we need to estimate it every time when someone starts driving. We use the concept of EID again to estimate $1/J$.

Rewriting (10) yields

$$\frac{d\omega(t)}{dt} = -\frac{B}{J_0}\omega(t) + d_J(t), \tag{25}$$

$$d_J(t) := \frac{1}{J}\tau(t) - \frac{Wr}{J}\mu(\lambda(t)) + \left(\frac{1}{J_0} - \frac{1}{J}\right) B\omega(t). \tag{26}$$

To find an EID estimator for $d_J(t)$, we construct another state observer for (10)

$$\frac{d\omega'(t)}{dt} = -\frac{B}{J_0}\omega'(t) - L_J[\omega'(t) - \omega(t)]. \tag{27}$$

Using the EID approach gives an estimate of $d_J(t)$

$$\hat{d}_J(t) = -L_J[\omega'(t) - \omega(t)]. \tag{28}$$

Assume that $d_J(t)$ is estimated by (28) with a high precision. Estimates at two times, t and $t - \sigma$ ($\sigma \neq 0$), are

$$\hat{d}_J(t) \approx \frac{1}{J}\tau(t) - \frac{Wr}{J}\mu(\lambda(t)) + \frac{J - J_0}{J_0 J} B\omega(t), \tag{29}$$

$$\hat{d}_J(t-\sigma) \approx \frac{1}{J}\tau(t-\sigma) - \frac{Wr}{J}\mu(\lambda(t-\sigma)) + \frac{J - J_0}{J_0 J} B\omega(t-\sigma). \tag{30}$$

Choose a small enough σ and assume that the pavement is the same at t and $t - \sigma$. Then, the continuity of μ and ω ensures

$$\mu(\lambda(t)) \approx \mu(\lambda(t-\sigma)), \tag{31}$$

$$\omega(t) \approx \omega(t-\sigma). \tag{32}$$

An estimate of $1/J$ is derived from (29), (30), and (31):

$$\hat{J}^{-1} = \frac{\hat{d}_J(t-\sigma) - \hat{d}_J(t)}{\tau(t) - \tau(t-\sigma)}. \tag{33}$$

A low-pass filter

$$F_J(s) = \frac{1}{T_J s + 1} \tag{34}$$

is used to smooth the estimate, $\tilde{J}^{-1}$. Note that

$$\tau(t) - \tau(t-\sigma) \neq 0, \tag{35}$$

$$\hat{d}_J(t-\sigma) - \hat{d}_J(t) \neq 0 \tag{36}$$

hold for a cart control system. This guarantees the existence of the estimate of $1/J$ produced by (33).

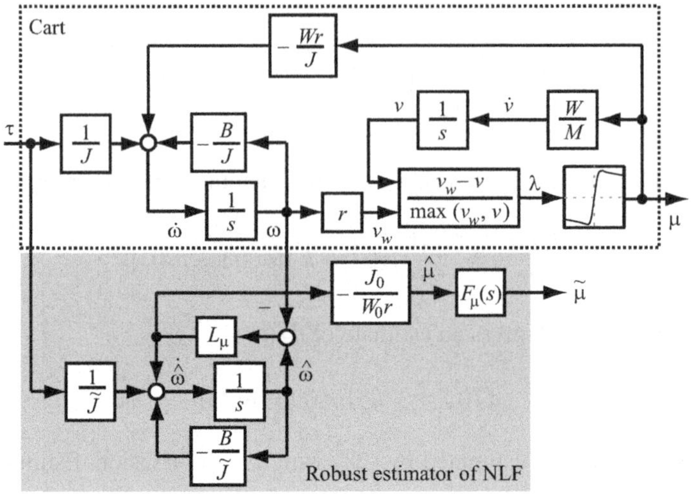

Fig. 19 Robust estimation system for NLF

Finally, combining the estimation of $1/J$ and $d_\mu(t)$ yields a robust estimation system for the NLF in Fig. 19. A two-step procedure is used to estimate μ after a driver turns on a cart: First, estimate $1/J$ using (33) and (34). Then, estimate $\mu(t)$ using (22) and (23).

Serial simulations were performed for verification (see [19]). Results show that the estimated $1/J$ gradually approaches the actual value and the estimated $\mu(t)$ was accurate by using the estimated $1/J$.

Figures 20, 21, and 22 show some typical results. The nominal parameters of the cart are

$$M_0 = 97.0 \text{ kg}, \ J_0 = 5.59 \text{ kg} \cdot \text{m}^2, \ B = 3.21 \text{ Nm} \cdot \text{s/rad}, \ r = 0.243 \text{ m}. \quad (37)$$

The parameters for the estimation of $\mu(t)$ were selected to be

$$L_\mu = 150, \ T_\mu = 0.1 \text{ s}, \quad (38)$$

and the parameters for the estimation of $1/J$ were selected to be

$$L_J = 31.05, \ T_J = 0.5 \text{ s}, \ \sigma = 0.1 \text{ s}. \quad (39)$$

Estimation results are shown in Fig. 20 for the nominal parameters and a dry asphalt road. $\mu > 0$ as the cart accelerates ($t \in [0, 10 \text{ s})$), and $\mu < 0$ as the cart brakes ($t \in [10 \text{ s}, 20 \text{ s}]$). The largest steady-state estimation error is less than 0.003. It indicates that the estimate of μ is precise. The results for a wet asphalt road are as precise as those for a dry asphalt road.

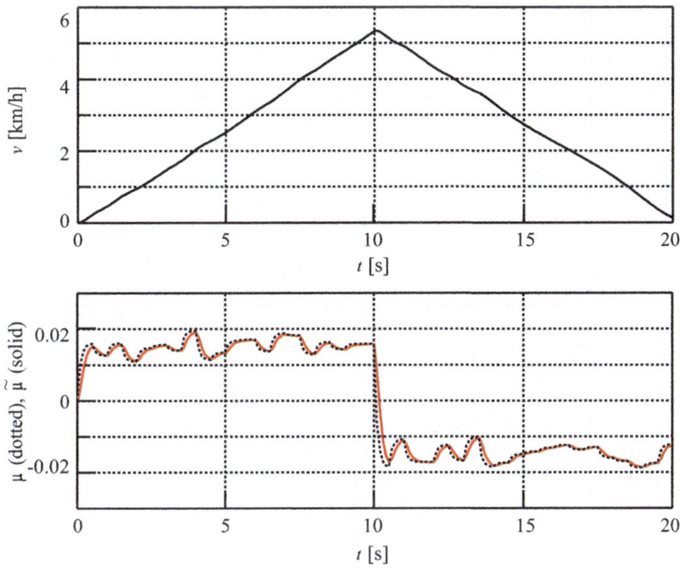

Fig. 20 Cart speed and estimated $\mu(t)$ for nominal case and dry asphalt road

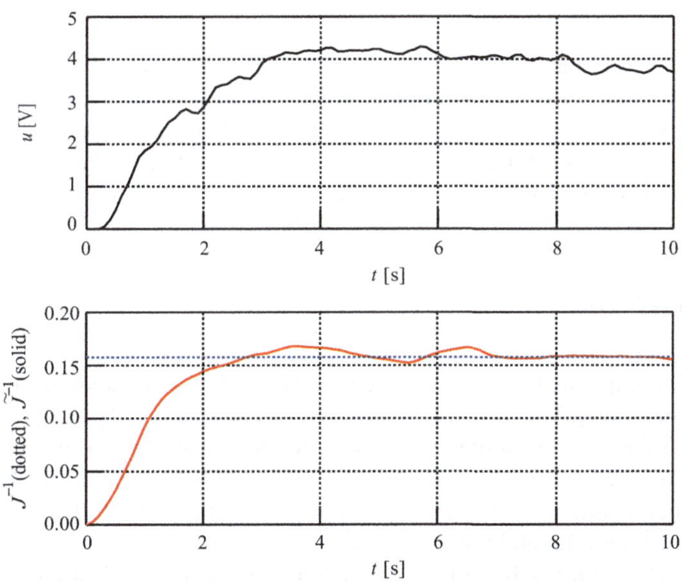

Fig. 21 Estimated $\tilde{J}^{-1}$ obtained with measured cart data for dry asphalt road

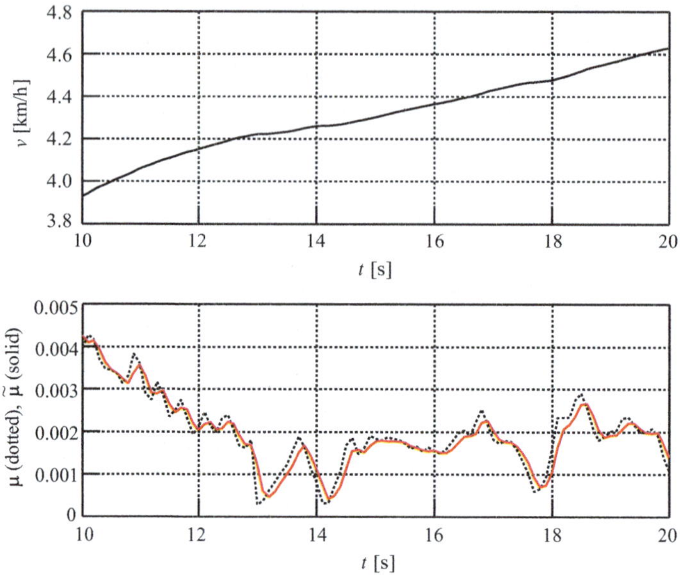

Fig. 22 Estimated μ obtained using $\tilde{J}^{-1}$ for measured cart data for dry asphalt road

Then, we used measured cart data to verify the validity and robustness of our method. In Fig. 21, $u(t)$ is the applied voltage to the motor. It was collected in a driving experiment on a dry asphalt road. It is clear that the estimated $1/J$ gradually approaches the actual value. The estimate at 10 s was used to estimate $\mu(t)$. Figure 22 shows that the estimated $\mu(t)$ is accurate.

3.4 Conclusion

This section explained a three-wheeled electric cart for maintaining the walking ability of the elderly. The cart mounts an ergonomically designed pedal unit to provide exercise for the muscles of the lower limbs. An electrical connection was employed between the pedals and the drive wheels to enhance design flexibility. An interface board, which contains a PIC, was assembled to intelligently deal with the signals between the controller and the cart.

First, the hardware of the cart control system was explained. Then, a pedal load was automatically selected based on a driver's physical condition and a control system was designed using the H_∞ control theory and DPDC to provide sensory stimulation and enjoyable driving experience. The effectiveness of the cart and the system architecture, and the validity of the designed control system were tested under various load and road conditions.

The increased use of electric carts has been accompanied by an increase in the number of traffic accidents. To solve this problem, we developed a method of estimating the NLF based on the EID approach to guarantee the driver's safety. This method has two advantages. First, it does not require any knowledge of the μ-λ relationship. Second, it designs only two linear observers to precisely estimate $\mu(t)$.

4 Design of Left-Right-Independent Lower-Limb Rehabilitation Machine

Commercially available lower-limb rehabilitation machines are basically bisymmetric, and the structures of the machines are usually fixed. Thus, they can hardly meet various requirements for lower-limb rehabilitation. People are required to adapt themselves to the machines to do exercise. To solve this problem, we devised a new kind of rehabilitation machine for the lower limbs. It is a left-right independent one that has a flexible structure. Thus, it allows us to independently adjust each of the pedal loads and strokes to suit different requirements for lower limbs. The basic idea was presented in [32, 33], and a great improvement was made in [34].

4.1 Specification and Mechanism Selection

Rehabilitation is divided into three stages: acute (1–14 days), recovery (up to several months), and functional (several months to years) [35]. Since a patient is constantly given detailed instruction during the first and second stages, this study focuses on providing a means for the patients in the last stage by designing a new human-centered pedaling machine that can easily be adjusted to suit a user's situation.

Figure 23 shows the relationship between a pedal load and the velocity of muscle contraction. The definitions of isometric, eccentric, and concentric exercises [36] are

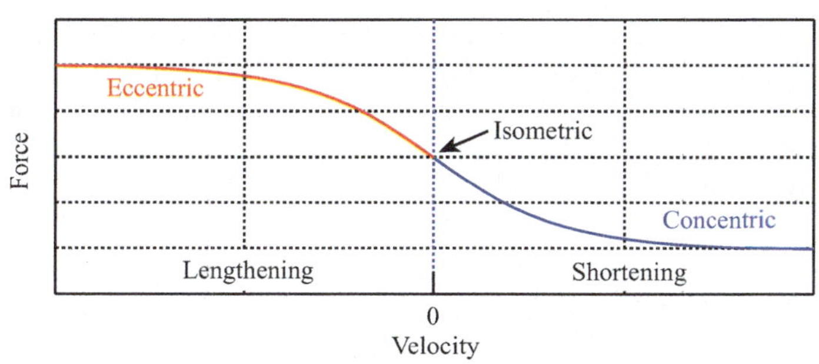

Fig. 23 Relationship between pedal load and velocity of muscle contraction when muscle is shortening or lengthening [36]

also illustrated in the figure. Both static and dynamic strength need to be trained in rehabilitation [37]. As explained in [38], isometric exercise improves static strength. Concentric contraction shortens a muscle when it acts against resistive force, and eccentreic contraction lengthens a muscle when it products force. Both concentric and eccentric exercise improves dynamic strength. There are two parameters to adjust in performing such exercises: load and repetition number. The relationship between the two parameters is given in [39]. The combination of a small load and a large number of repetitions is recommended for safe rehabilitation.

This study considered a linear motion of pedaling rather than a rotational motion because it is easy to design a left-right independent machine to suit different rehabilitation requirements for left and right legs. To design a linear pedal structure, we first need to determine the ranges of pedal load and region. The statics in [40] shows that the maximum average leg-extension force of one leg for a person around 20 years old is about 2900 N. Considering that the value decreases with aging and people who use the machine for rehabilitation have weak lower limbs, we chose the largest pedal load to be

$$P_{\max} = 2000\ \text{N}. \tag{40}$$

Based on a preliminary test, we chose the following stroke of linear motion

$$L = 150\ \text{mm} \tag{41}$$

and the following range for the angle of knee flexion

$$\theta \in [0^\circ,\ 90^\circ]. \tag{42}$$

4.2 System Design

Two completely separated linear pedal units are independently used for left and right legs. Oil dampers were selected to generate pedal loads. Taking the requirements of (40)–(42) into consideration, we finally chose the dampers to be KINECHECK Super K (Meiyu Airmatic Co. Ltd., Japan) (Table 2) [41], which is the one that has the longest stroke among small-size oil dampers. A running block was designed to double the stroke and reduce the force by half. This ensures that the designed machine meets the requirements (40) and (41).

Table 2 Parameters of KINECHECK Super K [41]

Model	Overall length	Weight	Stroke	Force range
5001-31-4	356 mm	658 g	102 mm	23 N ~ 5440 N

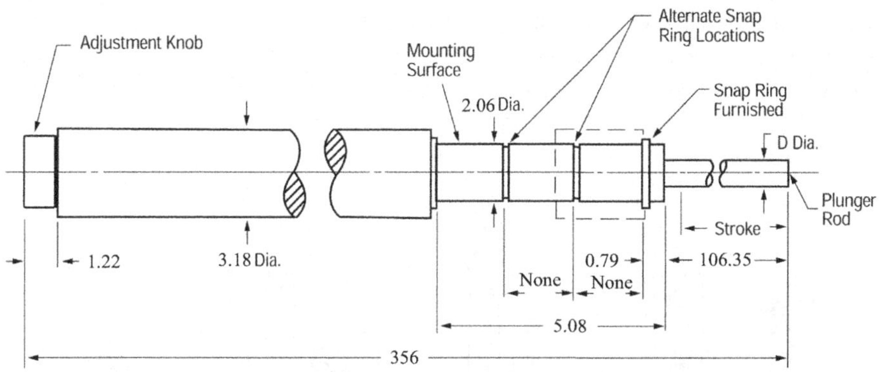

Fig. 24 Dimensions of KINECHECK Super K [41]

Table 3 Parameters of LPR-C-1KNS15 (RO: rated output) [42]

Rated capacity	Nonlinearity	Hysteresis	RO	Weight	Safe overload
1000 N	Within ±0.3%RO	Within ±0.3%RO	Approx. 0.9 mV/V or more	110 g	150%

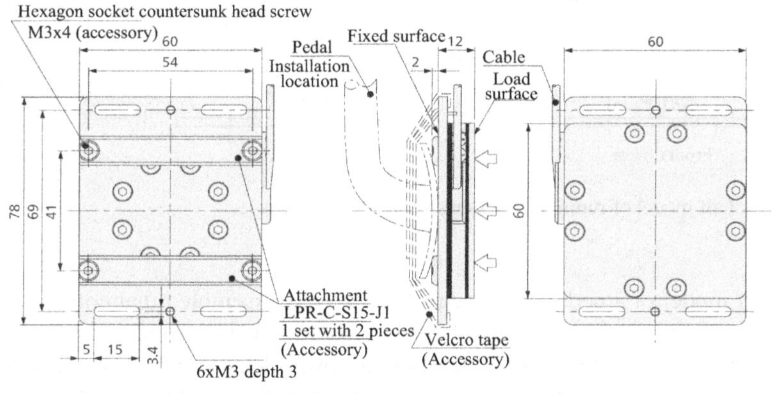

Fig. 25 Dimensions of LPR-C-1KNS15 [42]

The dimensions of the damper are shown in Fig. 24. The pedal load can be regulated by an adjusting knob at the top of a damper. The damper has an automatic recovery capability that makes the stroke slowly return back automatically to the initial position.

To measure the pedal load and displacement for the evaluation of the rehabilitation, an LPR-C-1KNS15 (Kyowa Electronic Instruments Co. Ltd., Japan) (Table 3, Fig. 25) tread force sensor and a DTS-A-100 displacement sensor (Kyowa Electronic Instruments Co. Ltd., Japan) (Table 4) [43] were selected for the machine.

Table 4 Parameters of DTA-A-100 (RO: rated output) [43]

Rated capacity	Nonlinearity	Hysteresis	RO	Weight	Safe overload
100 mm	Within ±0.2%RO	Within ±0.2%RO	2.5 mV/V ± 10%	110 g	1 kΩ ± 20%

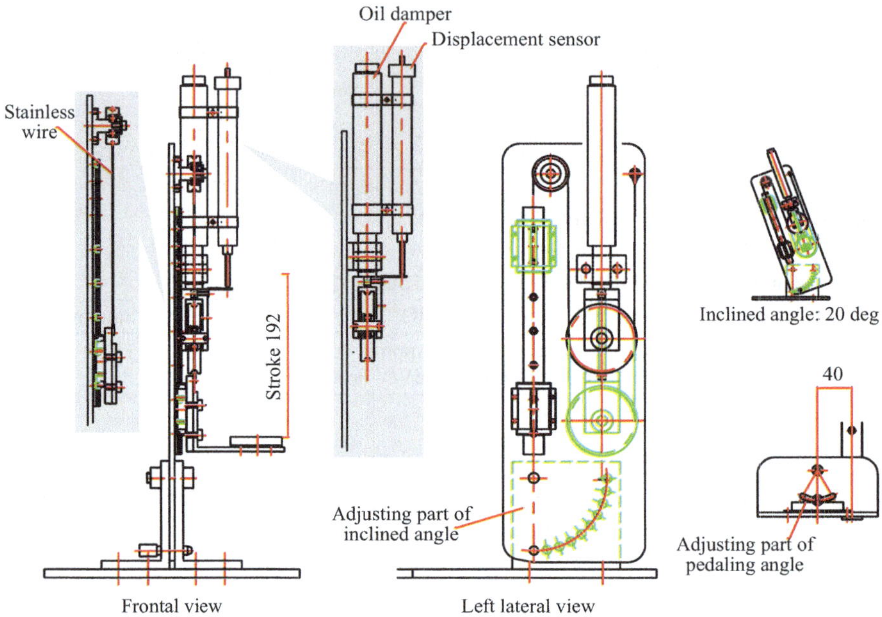

Fig. 26 A half model of running-block-based pedaling system

A half model of a pedaling system (Fig. 26) was assembled that contains the oil damper, a running block, the tread force sensor, and the displacement sensor. A compact recording system, EDX-10A [44] (Kyowa Electronic Instruments Co. Ltd., Japan), was used to record the pedal load and displacement for left and right legs in a real-time fashion.

4.3 *Preliminary Test of Prototype*

A preliminary test was carried out for the prototype (Fig. 27). The running-block set (Fig. 27a) is used to yield the following stoke

$$L_p = 204 \text{ mm}, \tag{43}$$

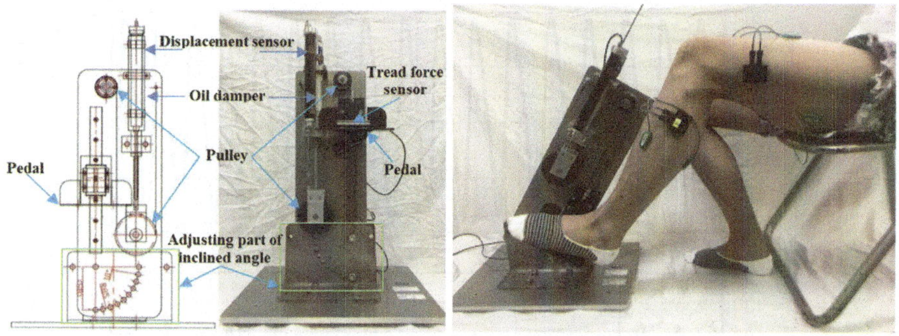

(a) A prototype of a half model (b) Photograph of preliminary test

Fig. 27 Prototype and preliminary test

Table 5 Information on subjects

Age (y/o)	Gender	Weight (kg)	Height (cm)	Health
22	Male	64	167	Good
22	Male	68	170	Good
26	Male	78	168	Good
27	Male	56	169	Good

which satisfies the requirement of (41). The largest pedal load generated by the oil damper was

$$P_p = 2720 \text{ N}, \tag{44}$$

which satisfies (40). The angle of inclination can be adjusted in the range [0°, 90°] with the adjusting interval of 10°.

Preliminary tests were carried out for the prototype. Four subjects participated (Table 5). Each subject sat on a chair in front of the prototype with a suitable distance from the machine. The chair was adjusted to a suitable height for each subject (Fig. 27b). The pedal force was set in the range of 20–300 N, and the angle of inclination was set to 0°-90°. Data were collected by the sensors of the pedal force and the displacement, and stored in a notebook computer (Model: DELL PRECISION M3800; OS: Windows 8.1 Pro 64 bits; CPU: Intel® Core™ i7-4712HQ; RAM: 16.0 GB).

Test results show that the machine worked well and pedaling was carried out smoothly in the ranges for the designed angle and pedal load. Some typical experimental data are shown in Figs. 28, 29 and 30. It is clear that 60° of the angle of inclination is a critical point. Both the pedal load and frequency increased as the angle increased in the range 0°–60° and decreased as the angle increased in the range 60°–90°. This shows that 60° is the best angle of inclination for pedaling.

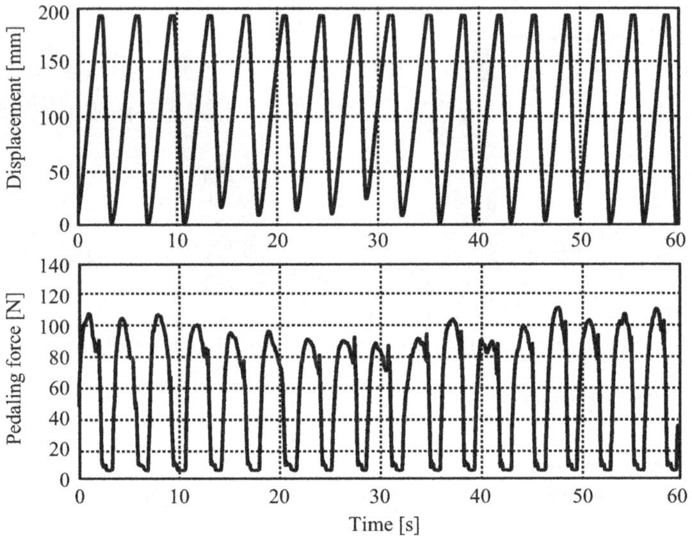

Fig. 28 Typical pedaling results for 40° angle of inclination

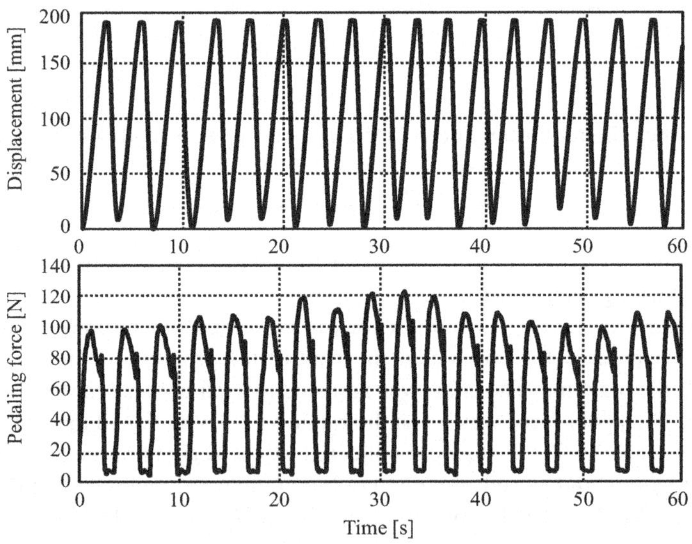

Fig. 29 Typical pedaling results for 60° angle of inclination

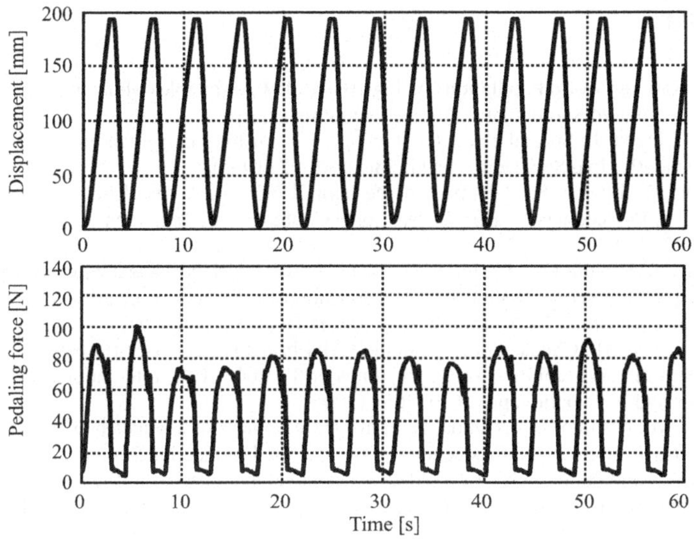

Fig. 30 Typical pedaling results for 80° angle of inclination

The results of the preliminary tests also show that P_{max} in (40) is too large for rehabilitation, choosing $P_{\text{max}} = 500$ N is large enough because a large number of repetitions is required for rehabilitation.

4.4 Conclusion

Commercially available lower-limb rehabilitation machines are basically bisymmetric and have a fixed structure. They can hardly adapt to different needs for lower-limb rehabilitation. To solve this problem, we designed a new type of rehabilitation machine. First, a linear pedaling motion was chosen to make it easy to design a left-right independent machine, and specifications for the design were determined based on statistics, ergonomics, and tests. Then, a pedal-load generator, a force sensor, and a displacement sensor were selected. A pedaling system that meets the specifications was assembled. Finally, a prototype of a half model (for one leg) was built for verification. Preliminary tests for basic functions were carried out and demonstrated the feasibility of the machine.

A full test of the prototype using a heart-rate meter, EMG (electromyography) sensors, and leg-movement sensors in addition to the displacement and force sensors has been carrying out for normal subjects. The results will be reported in the near future.

References

1. United Nations: World Population Ageing 2019 (2019). Available online: https://www.un.org/en/development/desa/population/theme/ageing/index.asp
2. Kono, J., Inada, J.: Special characteristics of tooth brushing movement in elementary school and the elderly. Jpn. Soc. Dent. Health **58**(3), 91–92 (1995)
3. Simpson, R., Levine, S.: Adaptive shared control of a smart wheelchair operated by voice control. In: Proceedings of IEEE International Conference on Intelligent Robots and Systems, pp. 91–92 (1997)
4. Ichinose, Y., Wakumoto, M., Honda, K. et al.: Human interface using a wireless tongue-palate contact pressure sensor system and its application to the control of an electric wheelchair. Trans. Inst. Electron., Inf. Commun. Eng. **J86-D-II**(2), 364–367 (2003)
5. Moon, I., Lee, M., Chu, J. et al.: Wearable EMG-based HCI for electric-powered wheelchair users with motor disabilities. In: Proceedings of the 2005 IEEE International Conference on Robotics and Automation, pp. 2649–2654 (2005)
6. Kamata, M., Nishino, H., Yoshida, H. et al.: Development of a wheelchair of head operation with gyro sensor. In: JSME Proceedings of the Welfare Engineering Symposium: W420 (2001)
7. Eid, M.A., Giakoumidis, N., El Saddik, A.: A novel eye-gaze-controlled wheelchair system for navigating unknown environments: case study with a person with ALS. IEEE Access **4**, 558–573 (2016)
8. Purwanto, D., Mardiyanto, R., Arai, K.: Electric wheelchair control with gaze direction and eye blinking. Artif. Life Robot. **14**(3), 397–400 (2009)
9. Yasunobu, S., Inoue, M.: Intelligent driving system for an electric four-wheeled cart. In: Proceedings of 41st SICE Annual Conference, Osaka, Japan, pp. 1–3 (2002)
10. Sato, M., Tomizawa, T., Kudoh, S., Suehiro, T.: Development of a collision-avoidance assist system for an electric cart. In: Proceedings of 2011 IEEE International Conference on Robotics and Biomimetics, Phuket, Thailand, pp. 337–342 (2011)
11. Tary, J.K., Haideggery, T., Kovácsy, L., Kósiz, K., Botka, B., Rudas, I.J.: Nonlinear order-reduced adaptive controller for a DC motor driven electric cart. In: IEEE 18th International Conference on Intelligent Engineering Systems, Tihany, Hungary, pp. 73–78 (2014)
12. Siradjuddin, I., Amalia, Z., Setiawan, B., Wicaksollo, R.P., Yudaningtyast, E.: Stabilising a cart inverted pendulum system using pole placement control method. In: 15th International Conference on Quality in Research (QiR): International Symposium on Electrical and Computer Engineering, Nusa Dua, Indonesia, pp. 197–203 (2017). https://doi.org/10.1109/QIR.2017.8168481
13. Yokota, S., Ohyama, Y., Hashimoto, H., She, J.: The electric wheelchair controlled by human body motion—Design of the prototype and basic experiment. In: Proceedings of 17th IEEE International Symposium on Robot and Human Interactive Communication, pp. 303–308 (2008)
14. Yokota, S., Hashimoto, H., Ohyama, Y., She, J.: Electric wheelchair controlled by human body motion. IEEJ Trans. Electron. Inf. Syst. **129**(10), 1874–1880 (2009) (in Japanese)
15. Yokota, S., Hashimoto, H., Ohyama, Y., She, J.: Electric wheelchair controlled by human body motion-classification of body motion and improvement of control method. J. Robot. Mechatron. **22**(4), 439–446 (2010)
16. Yokota, S., Hashimoto, H., Ohyama, Y., Chugo, D., She, J., Kobayashi, H.: Improvement of measurement and control scheme on human body motion interface. Am. J. Intell. Syst. **2**(4), 53–59 (2012)
17. She, J., Ohyama, Y., Kobayashi, H.: Master-slave electric cart control system for maintaining/improving physical strength. IEEE Trans. Robot. **22**(3), 481–490 (2006)
18. She, J., Yokota, S., Du, E.Y.: Automatic heart-rate-based selection of pedal load and control system for electric cart. Mechatronics **23**(3), 279–288 (2013)
19. She, J., Makino, K., Ouyang, L., Hashimoto, H., Murakoshi, H., Wu, M.: Estimation of normalized longitudinal force for an electric cart using equivalent-input-disturbance approach. IEEE Trans. Veh. Technol. **63**(8), 3642–3650 (2014)

20. She, J., Ohyama, Y., Wu, M., Hashimoto, H.: Development of electric cart for improving walking ability—application of control theory to assistive technology. Sci. China Inf. Sci. **60**, 123201:1–123201:9 (2017)
21. Kohone, T.: Self-Organizing Maps. Springer, Berlin (1997)
22. Karvonen, M.J., Kentala, E., Mustala, O.: The effects of training on heart rate: a longitudinal study. Ann. Med. Exp. Biol. Fenn. **35**, 307–315 (1957)
23. Hill, D.C., Ethans, K.D., Macleod, D.A., Harrison, D.R.: Exercise stress testing in subacute stroke patients using a combined upper- and lower-limb ergometer. Arch. Phys. Med. Rehabil. **86**, 1860–1866 (2005)
24. Borg, G.: Borg's Perceived Exertion and Pain Scales. Human Kinetics, Champaigne (1998)
25. Griffin, M.J.: Handbook of Human Vibration. Elsevier Ltd., California (1990)
26. Salvendy, G.: Handbook of Human Factors and Ergonomics, 2nd edn. Wiley, New York (1997)
27. Consumer Affairs Agency, Government of Japan: White Paper on Consumer Affairs 2019, 2019 [Online]. Available: https://www.caa.go.jp/en/pu-blication/annual_report/ (Summary in English), https://www.caa.go.jp/polici-es/policy/consumer_research/white_paper/#white_paper_2019 (Full document in Japanese)
28. She, J., Fang, M., Ohyama, Y., Hashimoto, H., Wu, M.: Improving disturbance-rejection performance based on an equivalent-input-disturbance approach. IEEE Trans. Ind. Electron. **55**(1), 380–389 (2008)
29. She, J., Xin, X., Pan, Y.: Equivalent-input-disturbance approach-analysis and application to disturbance rejection in dual-stage feed drive control system. IEEE/ASME Trans. Mechatron. **16**(2), 330–340 (2011)
30. Rajamani, R., Phanomchoeng, G., Piyabongkarn, D., Lew, J.Y.: Algorithms for real-time estimation of individual wheel tire-road friction coefficient. IEEE/ASME Trans. Mechatron. **17**(6), 1183–1195 (2012)
31. Rajamani, R., Piyabongkarn, D., Lew, J.Y., Yi, K., Phanomchoeng, G.: Tire-road friction-coefficient estimation. IEEE Control Syst. **30**(4), 54–69 (2010)
32. She, J., Wu, F., Mita, T., Hashimoto, H., Wu, M.: Design of a new human-centered rehabilitation machine. In: Proceedings of 9th PErvasive Technologies Related to Assistive Environments Conference, pp. 1–4 (2016)
33. She, J., Wu, F., Mita, T., Hashimoto, H., Wu, M., Iliyasu, A.: Design of a new lower-limb rehabilitation machine. J. Adv. Comput. Intell. Intell. Inform. **21**(3), 409–416 (2017)
34. She, J., Wu, F., Hashimoto, H., Mita, T., Wu, M.: Design of a bilaterally asymmetric pedaling machine and its measuring system for medical rehabilitation. In: The 1st International Conference on Human Computer Interaction Theory and Applications, vol. 2, pp. 122–127 (2017)
35. Dugan, S.A.: Exercise in the rehabilitation of the athlete. In: Frontera, W.R., Herring, S. A.L., Micheli, J., Silver, J.K., Young, T.P. (eds.) Clinical Sports Medicine: Medical Management and Rehabilitation. Elsevier, Amsterdam (2007)
36. Thompson, W.R., Bushman, B.A., Desch, J., Kravitz, L.: ACSM's Resources for the Personal Trainer, 3rd edn. Wolters Kluwer (2010)
37. Nicks, D.C., Fleishman, E.A.: What do physical fitness tests measure? A Rev. Factor Anal. Stud., Educ. Psychol. Meas. **22**, 77–94 (1962)
38. Sports fitness advisor: Isometric Exercises & Static Strength Training, Available at: http://www.sport-fitness-advisor.com/isometric-exercises.html Accessed 14 Apr 2020
39. Mitsuru Amano, Sprint Clinic, Knowledge of Strength Training, Available at: http://www10.plala.or.jp/azzurri/sprint/basic_study/knowledge_strength.html Accessed 14 Apr 2020
40. Sato, M.: Ningen Kougaku Kijun Suuchi Suushiki Benran (Handbook of Ergonomic Standards, Numerical Values, and Formulas). Gihodo Shuppan Co., Ltd (1994)
41. Deschner Corporation, Super K and Mini K Kinecheks, Available at: http://deschner.com/products/super-ks-and-mini-ks/ Accessed 14 Apr 2020
42. Kyowa, LPR-C Thin Pedaling Force Transducer, Available at: http://www.kyo-wa-ei.com/eng/product/category/sensors/lpr-c/index.html Accessed 14 Apr 2020

43. Kyowa, Displacement Transducer, Available at: http://www.kyowa-ei.com/eng/product/category/sensors/dtsa/index.html Accessed 14 Apr 2020
44. Kyowa, Compact Recording System EDX-10A, Available at: http://www.kyo-wa-ei.com/eng/product/movie/acquisition/edx-10a_01.html Accessed 14 Apr 2020

Prediction and Control Technology for Renewable Energy

A Short-Term Wind Power Forecasting Method Based on Hybrid-Kernel Least-Squares Support Vector Machine

Min Ding, Min Wu, Ryuichi Yokoyama, Yosuke Nakanishi, and Yicheng Zhou

Abstract Wind power forecasting improves the wind power trade and the wind power dispatch level. Wind speed is closely related to the accuracy of wind energy forecasting. This chapter introduces the process of wind power generation, describes an amplitude-frequency characteristic extraction method for the wind speed, and presents a hybrid-kernel least-squares support vector machine based wind power forecasting method.

Keywords Short-term wind power forecasting · Time series forecasting model · Least-squares support vector machines · Amplitude-frequency characteristic

M. Ding · M. Wu (✉)
School of Automation, China University of Geosciences, Wuhan 430074, China
e-mail: wumin@cug.edu.cn

Hubei Key Laboratory of Advanced Control and Intelligent Automation for Complex Systems, Wuhan 430074, China

R. Yokoyama · Y. Nakanishi · Y. Zhou
Graduate School of Environment and Energy Engineering, Waseda University, Shinjuku-ku, Tokyo 169-8555, Japan

M. Wu et al. (eds.), *Developments in Advanced Control and Intelligent Automation for Complex Systems*, Studies in Systems, Decision and Control 329,
https://doi.org/10.1007/978-3-030-62147-6_15

Abbreviations

HKLSSVM	Hybrid-kernel Least-squares Support Vector Machine
NSGA-II	Multi-objective genetic algorithm for elite strategy
WD	Wavelet decomposition
MWD	Maximum wavelet decomposition
DirRec	Direct-recursive
LSSVM	Least-squares Support Vector Machine
RMSE	Root Mean Squared Error
MAE	Mean absolute error
FCM	Fuzzy-C-Means
ACF	Autocorrelation Function
FFT	Fast Fourier transform

1 Introduction

At present, in the face of the situation that fossil energy in the world is more and more exhausted, and environmental problems are more and more serious, renewable energy technology is constantly developing, and the scale of renewable energy grid connection is constantly expanding [1]. Under large-scale grid connection conditions, due to the intermittent and fluctuating characteristics of solar energy and wind energy, the stability and safety of power grid are challenged [2]. Accurate power prediction provides necessary basis for grid dispatch, unit operation, power plant operation and maintenance.

Due to the outstanding characteristics of wind energy in terms of intermittent and fluctuations, this chapter takes wind power forecasting as an example. At present, there are many clocks in wind power forecasting methods, which are mainly composed of physical methods based on physical principles and statistical methods based on statistical laws. The physical method estimates the wind speed at the hub height of each wind turbine unit based on the topography, and then predicts the wind power based on the theoretical power curve. Statistical methods use one or more models to establish the relationship between historical values of multiple meteorological variables and wind power measurements. As a statistical method, time series method has been widely used in wind power forecasting, but the non-stationary time series data brings challenges to power forecasting.

In the past few decades, various wind power forecasting methods have been proposed. Physical method [3], statistical method [4], intelligent learning method and hybrid method [5] constitute the existing main methods. The physical method is advantageous in long-term forecasting while the statistical method has advantages in short-term forecasting. The intelligent learning method is applicable to short-term forecasting by building a nonlinear relationship between the input data and the power of a wind farm. The hybrid methods incorporate different models and combine them in order to improve the wind power forecasting accuracy.

As a statistical method, time series method has been widely used in wind power forecasting, such as the autoregression [6], the autoregressive moving average [7], autoregressive integrated moving average [8], the random forest model [9] and the Markov chain model [10]. The traditional prediction methods of these time series are not accurate in multi-step time series prediction. In order to improve the forecasting precision, wind power research analyzes the characteristics of time series and uses the hybrid model for short-term prediction [11]. However, the nonstationary power series caused by the uncertainty of wind speed brings a big challenge to the model building.

The decomposition of non-stationary signals effectively improves the prediction accuracy. In [12–14], the empirical mode decomposition (EMD), Hilbert-Huang transform, wavelet decomposition (WD) and mean trend detector were used for time series decomposition and the LSSVM [15] was used for the model building. These results indicated that a hybrid wind power forecasting method based on time series decomposition combined with machine learning method was useful in eliminating the bad influence from the non-stationary time series data.

Among the preceding methods, due to the volatility, the original wind power data have nonlinear and non-stationary characteristics. The WD is an effective decomposition method by decomposing the sequence into components of different frequencies [16]. Components with different frequencies have different amplitudes, which represent detail characteristics and trend characteristics of the original signal. The amplitude-frequency characteristics have been well applied in the classification of decomposed components [17]. For the nonlinear time series forecasting, the direct-recursive (DirRec) method which combines the direct method and the recursive method has good performance [18].

The LSSVM is a kernel-based machine learning method. In LSSVM, the kernel functions map the data into higher dimensional spaces to separate the data more easily. The kernels can be divided into two types: Global kernel and local kernel. Among them, the characteristics of the local kernel are that the data close to each other or nearby has obvious influence on the kernel value. Gaussian kernel is a typical local kernel. On the contrary, the data samples far away from each other will have a significant impact on the kernel value in the case of global kernel. Another typical global kernel is the polynomial kernel. The hybrid kernel contains the characteristics of multiple kernels [19]. Due to the randomness of LSSVM model parameters, the multi-objective functions were used to solve the optimal parameters of the model [20].

Through the review of the literature, decomposition and classification of wind power time series are common means in the wind power prediction. However, the amplitude-frequency characteristics of decomposed signals are seldom analyzed and used as the basis for modeling.

The wind power data with time series characteristics is decomposed into stationary time series component data by maximum wavelet decomposition method. Then, according to the amplitude-frequency characteristics of the decomposed com-

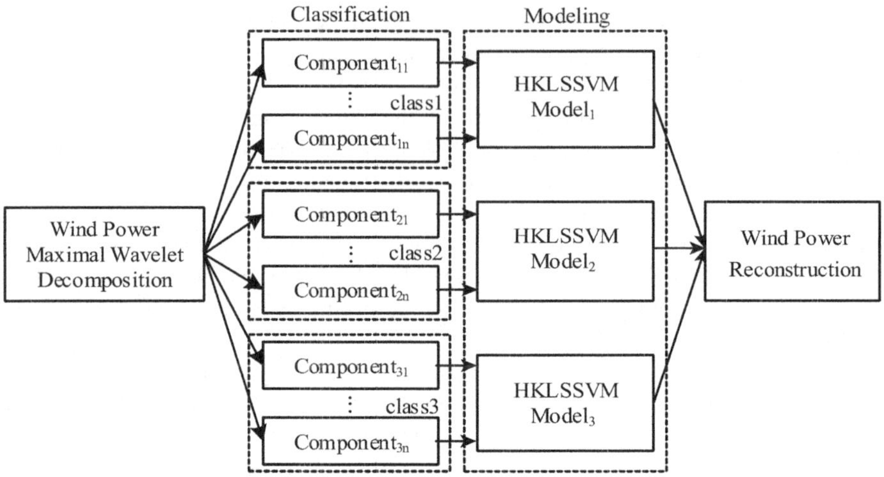

Fig. 1 Framework of the proposed wind power forecasting model

ponents, the stationary time series are divided into three categories by fuzzy C-means (FCM) algorithm and the LSSVM models with three different kernels are established. Finally, the outputs of LSSVM models are reconstructed to get the forecasting wind power. The results show that the proposed model outperforms well. This chapter introduces a time series model of the hybrid least squares support vector machine based on the decomposition, classification and reconstruction process for short-term wind power forecasting.

2 Forecasting Framework

Although the wind is volatile, its statistical characteristics conform to the Weibull distribution, and the wind speed varies within a certain range. Wind energy is converted into electricity by the wind turbine.Then, the prediction model is established by the decomposed time series model to reduce the non-stationary factors in the prediction error:

$$P_m = \frac{1}{2} C_P \rho A V^3 \tag{1}$$

where P_m, C_P, ρ, A and V respectively represent the output power of wind turbine, power factor of turbine, air density, area swept by turbine blade and wind speed, respectively

Wind energy is a kind of non-stationary power supply with strong randomness, due to its intermittency and volatility. The non-stationary of wind power will affect the accuracy of the model. Decomposition is carried out to obtain stationary wind power time series components.

The wind power time series is decomposed into high and low frequency components with different local characteristics, including detail characteristics and trend characteristics. Based on the amplitude-frequency characteristics of wind power time series, a decomposition-classification-reconstruction framework for wind power forecasting is proposed as shown in Fig. 1.

The framework is described as follows:

(1) Firstly, the decomposition is conducted. The maximum wavelet decomposition (MWD) method decomposes the historical time series data of wind power generation. Then these decomposition components are classified into three classes according to their amplitude-frequency characteristic.
(2) Secondly, the forecasting model is built. Different HKLSSVM time series models are built for three classes respectively. And then NSGA-II algorithm optimizes these models for accuracy improving.
(3) Finally, wind power is reconstructed. The final forecasting wind power is obtained by reconstructing output components of all models.

2.1 Wind Power Decomposition Based on Amplitude-Frequency Characteristic

In the WD with J decomposition scale, the original data is decomposed into $D_1, D_2, \ldots, D_J$ and A_J. The appropriate decomposition scale J is the key parameter in the decomposition process. In WD, all wavelets are obtained by scaling and displacement of the mother wavelet, so the chosen of the scale and the wavelet mother function is particularly important, in order to obtain stable wind power series, it is very important to choose wavelet mother and scale.

The MWD method is obtained after WD scale adaptive process to get stable wind power series. The detailed principle process of MWD method: WD method decomposes the time series into multiple decomposition components with a decomposition scale N. If the number of extremal points of components is 1 or 0, the condition is satisfied and the decomposition scale is set as N. If not, the decomposition scale is set as $N+1$.

Different MWD components present different amplitude-frequency characteristics. By the range frequency, the frequency characteristic of an MWD components are distinguished. And then, the amplitude characteristics of different MWD components with a certain frequency range are calculated.

The MWD component amplitude decreases with the increase of frequency and vice versa. Fuzzy C-means (FCM) algorithm classifies these MWD components into three classes, which are high-frequency low-amplitude component, high-frequency medium-amplitude component and low-frequency high-amplitude component. Among them, the high-frequency low-amplitude component contains the detailed feature with noise, while the low-frequency high-amplitude component contains the trend feature. Next, different models are built for different classes.

3 Time Series Forecasting Models for Different Classes

In this section, the forecasting models are built for different classes with different amplitude-frequency characteristics. First, the direct-recursive (DirRec) time series model is introduced. Then, a hybrid kernel least-squares support vector machine (HKLSSVM) is proposed to build DirRec time series model. Finally, the NSGA-II is used to optimize the parameters of HKLSSVM.

3.1 DirRec Time Series Forecasting Model

A time series forecasting uses the previous value to estimate the next value. In mathematical form DirRec model is written as:

$$\begin{cases} \hat{y}_{t+1} = f_1\left(y_1, \ldots, y_t\right) \\ \hat{y}_{t+2} = f_2\left(y_1, \ldots, y_t, \ldots, \hat{y}_{t+1}\right) \\ \vdots \\ \hat{y}_{t+n} = f_n\left(y_1, \ldots, y_t, \ldots, \hat{y}_{t+n-1}\right) \end{cases} \tag{2}$$

where $\hat{y}_{t+i}$ is the time series forecasting result, r is regressor size, $f_i(\cdot)$ is the autoregression model.

In the case of 5-step forecasting, the forecasting output of the first 5 steps should be obtained respectively. Similarly, in the case of n-step forecasting, the forecasting output of the first n steps should be obtained respectively. The 5-step, 10-step, 15-step, and n-step are shown in Fig. 2.

In order to forecast $\hat{y}_{t+1}$, the model is trained by using the input matrix $\boldsymbol{X}_{\text{train}}$ and the training target matrix $\boldsymbol{Y}_{\text{train}}$ and then the test input matrix $\boldsymbol{X}_{\text{test}}$ is input into the model to obtain $\hat{y}_{t+1}$.

$$\boldsymbol{X}_{\text{train}} = \left| \begin{matrix} y_1 & y_2 & \cdots & y_p \\ y_2 & y_3 & \cdots & y_{p+1} \\ \vdots & \vdots & \ddots & \vdots \\ y_{t-p} & y_{t-p+1} & \cdots & y_{t-1} \end{matrix} \right| \begin{matrix} \rightarrow \\ \rightarrow \\ \vdots \\ \rightarrow \end{matrix} \left| \begin{matrix} y_{p+1} \\ y_{p+2} \\ \vdots \\ y_t \end{matrix} \right| = \boldsymbol{Y}_{\text{train}} \tag{3}$$

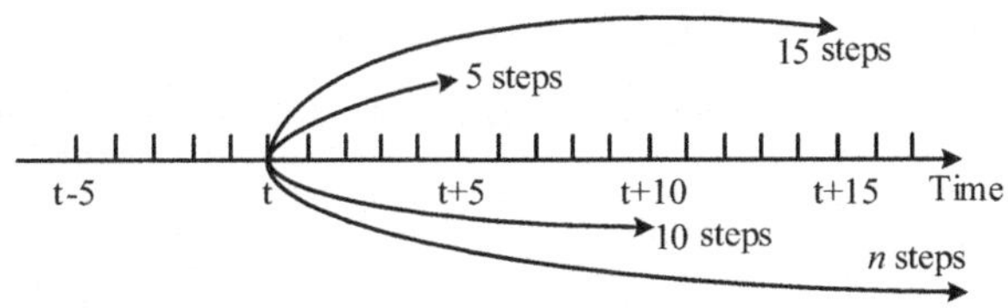

Fig. 2 Diagrammatic sketch of multi-step ahead forecasting

$$\boldsymbol{X}_{\text{test}} = \begin{vmatrix} y_1 & y_2 & \cdots & y_p \\ y_2 & y_3 & \cdots & y_{p+1} \\ \vdots & \vdots & \ddots & \vdots \\ y_{t-p+1} & y_{t-p+2} & \cdots & y_t \end{vmatrix} \tag{4}$$

where p represents the dimension of the input matrix and $p \leq t - 2$.

Similarly, in order to obtain $\hat{y}_{t+2}, \ldots, \hat{y}_{t+n}$, the training time series should be set as all time series before $t + 1, \ldots, t + n$, and corresponding adjustments should be made to the input matrix $\boldsymbol{X}_{\text{train}}$, the target matrix $\boldsymbol{Y}_{\text{train}}$, and the test input matrix $\boldsymbol{X}_{\text{test}}$.

The forecasting models for $\hat{y}_{t+1}$ is represented as:

$$\mathbf{Y} = f_{\text{TS}}\left(\mathbf{X}\right) \tag{5}$$

where $f_{\text{TS}}(\cdot)$ is a nonlinear mapping, which comes from the spatial relationship between training data and target data in the future.

3.2 HKLSSVM Time Series Forecasting Model

Given a set of samples $\{x_i, y_i\}_{i=1}^{N}$,where the input vector is x_i and its class label is y_i, the single-kernel LSSVM model f_{SK} can then be defined as:

$$f_{\text{SK}}\left(x\right) = \sum_{i=1}^{N} \alpha_i k\left(x_i, x\right) + b \tag{6}$$

where $(\alpha_1, \ldots, \alpha_N)$ is the weight vector, b is the bias and $k(\cdot, \cdot)$ is the kernel function.

LSSVM has different kernel functions according to different attributes, generally divided into global kernel and local kernel. The high frequency timing components need good local expression ability, and the low frequency timing components need good global expression ability. Therefore, the corresponding kernel functions are selected according to the characteristics of different components to improve the forecasting accuracy of the model.

There are several kernel functions, such as the linear kernel k_{LIN}, the polynomial kernel k_{POL}, and the Gaussian kernel k_{GAU}:

$$k_{\text{LIN}}\left(x_i, x_j\right) = \left\langle x_i, x_j \right\rangle \tag{7}$$

$$k_{\text{POL}}\left(x_i, x_j\right) = \left(\left\langle x_i, x_j \right\rangle + 1\right)^q, \; q \text{ is natural number} \tag{8}$$

$$k_{\text{GAU}}\left(x_i, x_j\right) = \exp\left(\frac{-\left\|x_i - x_j\right\|_2^2}{2s^2}\right), \; s > 0. \tag{9}$$

These three different kernel functions have different expression abilities, respectively. Generally speaking, the time delay of time series decreases with the increase of time series frequency. Therefore, the model with strong local expression ability has good effect on the time series with low time delay, while the model with strong global expression ability has good effect on the time series with high time delay.

According to the expression ability of kernel functions and the characteristics of time series components, the Gaussian kernel, the polynomial kernel and the linear kernel are used for the first, second and third classes, respectively. The HKLSSVM model f_{HK} can be expressed as follows:

$$f_{\text{HK}} = \begin{cases} \sum_{i=1}^{N} \alpha_i k_{\text{GAU}}\left(x_i, x\right) + b_1, & \text{Class 1} \\ \sum_{i=1}^{N} \alpha_i k_{\text{POL}}\left(x_i, x\right) + b_2, & \text{Class 2} \\ \sum_{i=1}^{N} \alpha_i k_{\text{LIN}}\left(x_i, x\right) + b_3, & \text{Class 3.} \end{cases} \tag{10}$$

The forecasting results of different kernel functions for MWD components [21] (i.e., $C_1, C_2, \cdots, C_{10}$) are shown in Fig. 3. The first 20 samples and the last 20 samples are the actual and forecasting components, respectively. In C_1 to C_3, Gaussian kernel LSSVM is better than linear and polynomial kernel LSSVM. In C_4 to C_9, polynomial kernel LSSVM works the best. The results show that the proposed LSSVM with proper kernel is effective for time series components with different frequencies.

3.3 *Optimized HKLSSVM Time Series Model*

The forecasting diagram of n-step leading time series is shown in Fig. 4. The optimal parameters of the HKLSSVM forecasting model are obtained by cross-validation of the time series in the forecasting intervals [t-3n+1, t-2n], [t-2n+1, t-n].

The NSGA-II optimization flowchart of HKLSSVM time series model is shown in Fig. 5. NSGA-II optimizes the parameters of the HKLSSVM model so as to minimize the cross-validation cost function constructed by cross-validation errors.

The main optimization processes of the HKLSSVM time series model is described as the following steps [21]:

Step 1: Data preparation

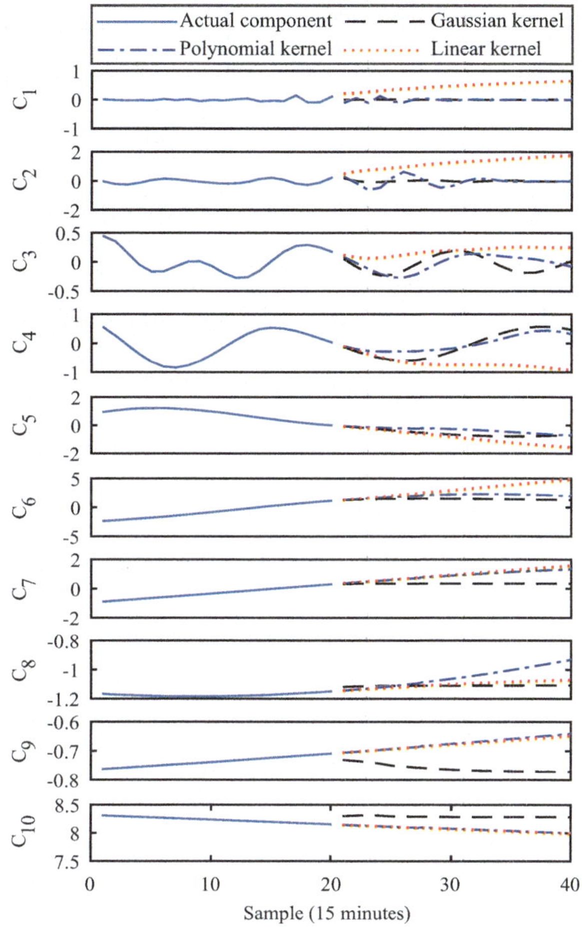

Fig. 3 Forecasting results for each component by different kernel functions

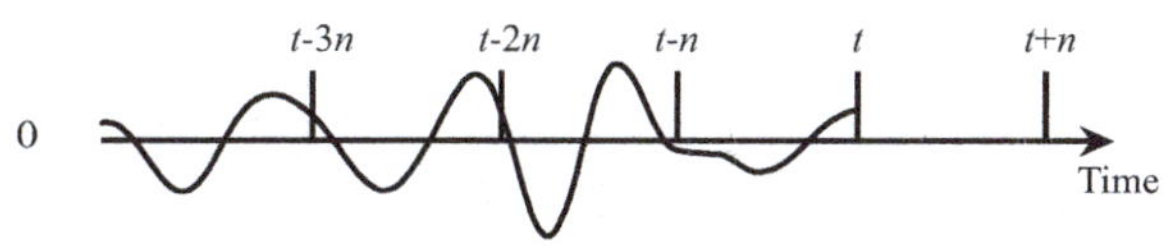

Fig. 4 Diagrammatic sketch of n-step ahead time series forecasting

In n-step ahead time series forecasting, the time series $\boldsymbol{Y}_t\{y_1, y_2, \ldots, y_t\}$ is divided into $\boldsymbol{Y}_{t-n}\{y_1, y_2, \ldots, y_{t-n}\}$, $Y_{t-2n}\{y_1, y_2, \ldots, y_{t-2n}\}$, and $\boldsymbol{Y}_{t-3n}\{y_1, y_2, \ldots, y_{t-3n}\}$. They are used as historical data to forecast the n-step ahead time series respectively.

Step 2: Initial NSGA-II parameters

Set the range of the optimized parameters, including LSSVM regularization parameter, the range of time-series length p, and kernel function parameter. In this study, the initial solution population is 100. A random initial population of binary code is created.

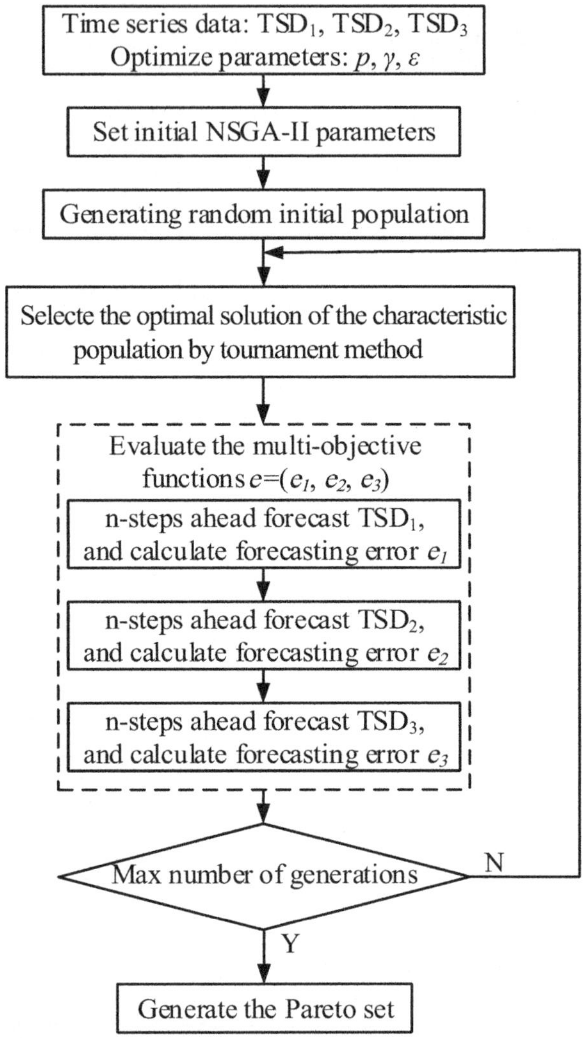

Fig. 5 NSGA-II flowchart of the LSSVM time series model

Step 3: NSGA-II operators

Then, the optimal solution of the characteristic population is selected by tournament method. Finally, a mating pool is created based on the principle of generating offspring solution by non dominated sorting.

In order to ensure the diversity of different generation solutions, crossover and mutation operators are used to generate offspring and parent populations. The best individuals are selected from the combined offspring and parents according to the non dominated order as the new population. The crossover operator and mutation operator use the probabilities of 0.9 and 0.1 respectively.

Step 4: Multi-objective cost function

$\hat{Y}_1\{\hat{y}_{t-3n+1}, \ldots, \hat{y}_{t-2n}\}$, $\hat{Y}_2\{\hat{y}_{t-2n+1}, \ldots, \hat{y}_{t-n}\}$, and $\hat{Y}_3\{\hat{y}_{t-n+1}, \ldots, \hat{y}_t\}$ are forecasted time series data. $\hat{Y}_1\{y_{t-3n+1}, \ldots, y_{t-2n}\}$, $\hat{Y}_2\{y_{t-2n+1}, \ldots, y_{t-n}\}$, and $\hat{Y}_3\{y_{t-n+1}, \ldots, y_t\}$ are actual time series data. The root-mean-square error (RMSE) of the cross-validation e_1, e_2, e_3 between the actual time series and forecasting time series are calculated as follow:

$$e_1 = \sqrt{\frac{\|\boldsymbol{Y}_1 - \hat{\boldsymbol{Y}}_1\|_2^2}{n}} \tag{11}$$

$$e_2 = \sqrt{\frac{\|\boldsymbol{Y}_2 - \hat{\boldsymbol{Y}}_2\|_2^2}{n}} \tag{12}$$

$$e_3 = \sqrt{\frac{\|\boldsymbol{Y}_3 - \hat{\boldsymbol{Y}}_3\|_2^2}{n}} \tag{13}$$

The objective function of NAGS-II is

$$\min\{e_1, e_2, e_3\}. \tag{14}$$

Step 5: Stop criterion

Step 3 is repeated until the stop criterion (200 generations for the current study) is satisfied.

4 Experiment Design and Experiment Results

In this section, the proposed method is verified by experiments designed. In the experiment, two benchmark models are used as evaluation criteria. The performance of the proposed model is compared with two benchmark models.

4.1 Experiment Design

In order to analyze the performance of proposed the model, a comparative test between the proposed model and the other two models was carried out on Windows 7 PC and MATLAB 8.5 version. The proposed method was compared with EMDLSSVM and WD-LSSVM benchmark models for short-term wind power forecasting.

Through the actual production data and the historical generation data to carry on the comparison test, among them set 1 is the actual production data, which is obtained from the statistics of wind farms located in Shanxi Province, China in May 2016, 15

Table 1 Statistic of Set 1 and Set 2

Statistical parameter	Set 1	Set 1
Minimum	0	0
Maximum	48.99	17.53
Mean	23.53	4.26
Median	24.58	2.88
Standard deviation	17.21	4.17

min is the sampling time. Set 2 is the historical wind power generation data from a wind farm in Sotavento in October 2017, 10 min is the sampling time. The statistical property for the two collections are shown below in Table 1.

4.2 Evaluation Criteria

In order to make a comparison, the mean absolute error (MAE) and RMSE are used as criteria to evaluate the prediction ability of the proposed model. RMSE and Mae have a negative correlation with the performance of the model, that is, the smaller is the better. Where RMSE and Mae are in the range of [0, 1]. The equations for RMSE and Mae are as follows:

$$RMSE = \sqrt{\frac{1}{l}\sum_{i=1}^{l}\left(\frac{P_i - \hat{P}_i}{Cap_i}\right)^2} \tag{15}$$

$$MAE = \frac{1}{l}\sum_{i=1}^{l}\left(\frac{\left|P_i - \hat{P}_i\right|}{Cap_i}\right) \tag{16}$$

where l is the sample number, P_i is ith actual power, $\hat{P}_i$ is ith forecasting power, Cap_i is ith installed capacity of wind farm.

4.3 Experimental Results

The capability of the proposed method is evaluated by simulation experiments. First, wind power time series is decomposed by MWD to get the MWD components. FCM method was used to classify the decomposition components and obtain three amplitude frequency components. Then, the two sets were tested by multi-steps ahead forecasting, and the results were analyzed. Finally, the results of the proposed method is compared with the results of benchmark methods.

Table 2 Amplitude-frequency characteristics of MWD components

Characteristic	C1	C2	C3	C4	C5	C6	C7	C8	C9	C10
Square sum function of amplitude	0.457	205	434	1	3	4	1	524	761	44538
Upper Frequency (1/900 Hz)	512	282	127	72	36	20	11	7	5	4
Lower Frequency (1/900 Hz)	171	98	50	28	14	6	2	2	2	1

Table 3 Forecasting performance of different models for Set 1 and Set 2 (%)

Model	Data set	5-step ahead		10-step ahead		15-step ahead	
		RMSE	MAE	RMSE	MAE	RMSE	MAE
Proposed model	Set 1	8.78	6.59	14.57	11.54	17.20	13.37
	Set 2	8.09	5.96	12.52	9.40	14.88	11.63

4.3.1 MWD results and analysis

Figure 6 shows the MWD components, i.e., C_1, C_2 ... C_{10}. C_{10} with the lowest frequency satisfies the MWD stop criterion. In order to obtain the bandwidth of these components, fast Fourier transform (FFT) is used in the experiment. Because the sampling time is 15 min, the frequency is 1/900 Hz. By caculating the amplitude and frequency range of each components, the amplitude-frequency characteristics of these components are shown in Table 2. From C_1 to C_{10},the square sum function of amplitude is increasing, while the upper and lower frequencies are decreasing, respectively. The MWD components have obvious amplitude-frequency characteristics.

According to the amplitude-frequency characteristics, proper kernel functions are choosen for different HKLSSVM time series models. NSGA-II HKLSSVM Model $_{11}$, Model $_{12}$ and Model $_{13}$ are constructed by gaussian kernel for class 1. NSGA-II HKLSSVM models $_{21}$ through $_{26}$ are constructed by polynomial kernel for class 2. NSGA-II HKLSSVM model $_{31}$ is constructed by linear kernel function for class 3.

Three kinds of wind power time series forecasting are tested. They are 5-step ahead, 10-step ahead, and 15-step ahead forecasting scenarios. Figures 7, and 8 represent the forecasting results for two sets, respectively. Table 3 shows the RMSE and MAE of the forecasting results. The forecasting wind power and the actual wind power are quite consistent. But, because of the time lag,bthe forecasting errors increase when the step increases. The proposed method uses the advantages of LSSVM moedels with different kernel functions and gets good forecasting results.

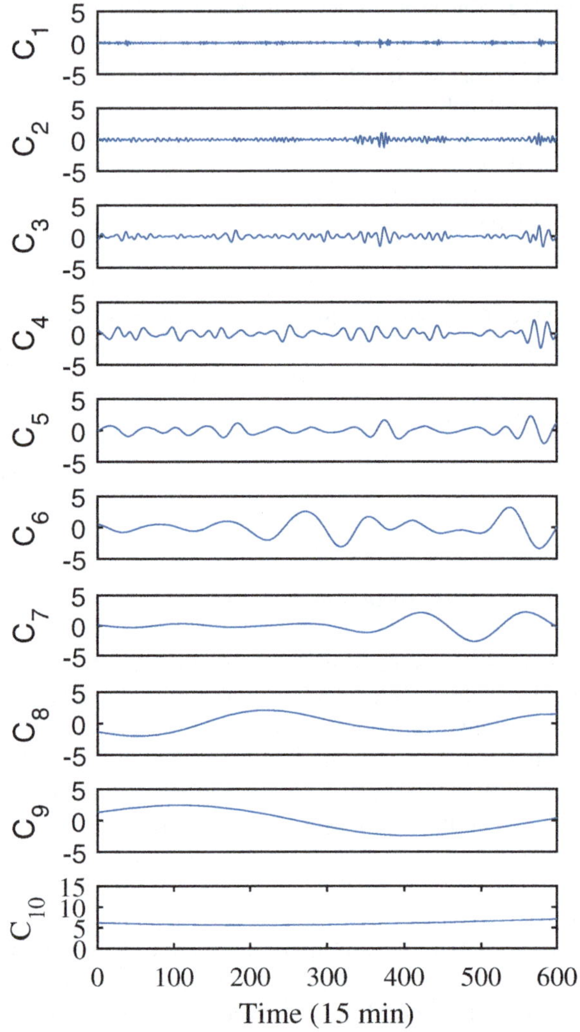

Fig. 6 Decomposition results of MWD

5 Conclusion

In this paper, a multi-step time series forecasting method based on decomposition, classification and reconstruction is proposed for short-term wind power forecasting. First, the wind power time series is decomposed by MWD and the decomposed components are classified by FCM based on the analysis of amplitude-frequency characteristics. Then, the forecasting models are established to predict different classes. Finally, NSGA-II optimizes the forecasting models. From the experimental results, the proposed method has a high level of short-term wind power prediction ability.

Fig. 7 5th step, 10th step, 15th step forecasting results for Set 1

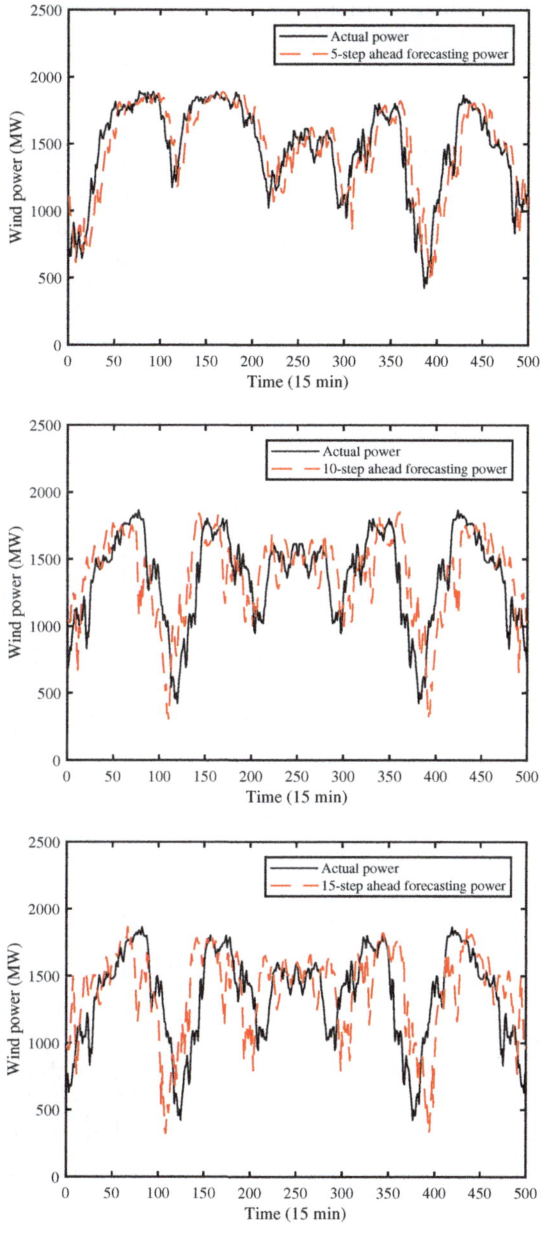

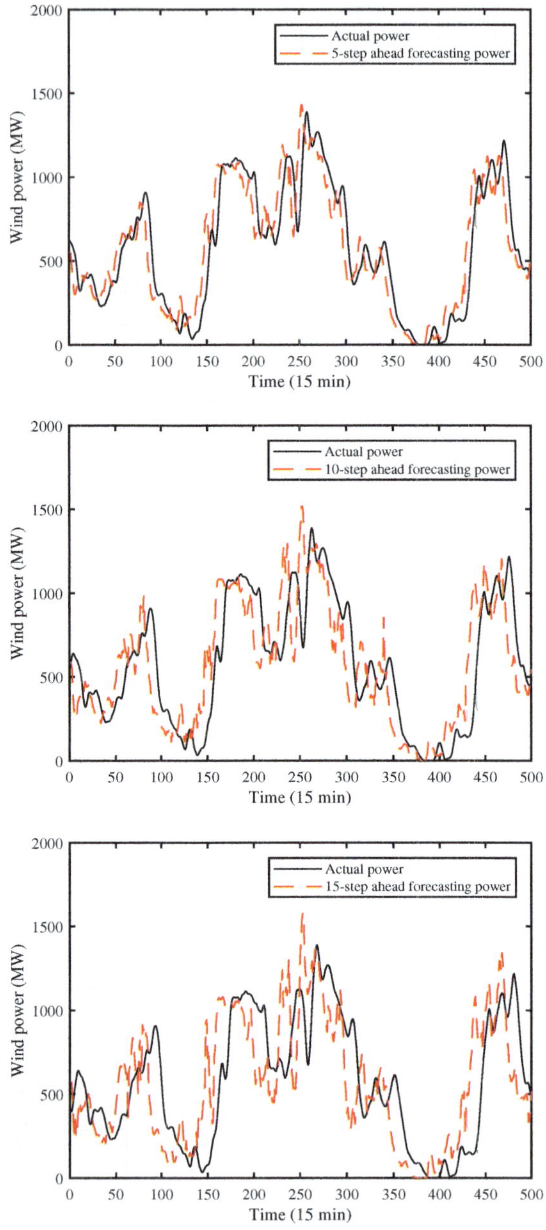

Fig. 8 5th step, 10th step, 15th step forecasting results for Set 2

References

1. Council, Global Wind, Energy.: Global wind statistics 2018, 2017
2. Jiang, X., Chen, H., Xiang, T.: Assessing the effect of wind power peaking characteristics on the maximum penetration level of wind power. IET Gener., Transm. Distrib. **9**(16), 2466–2473 (2015)
3. Lange, M., Focken, U.: Physical Approach to Short-Term Wind Power Prediction. Springer, Berlin (2006)
4. Sideratos, G., Hatziargyriou, N.D.: An advanced statistical method for wind power forecasting. IEEE Trans. Power Syst. **22**(1), 258–265 (2007)
5. Okumus, I., Dinler, A.: Current status of wind energy forecasting and a hybrid method for hourly predictions. Energy Convers. Manag. **123**, 362–371 (2016)
6. Karaku, O., Kuruolu, E.E., Altnkaya, M.A.: One-day ahead wind speed/power prediction based on polynomial autoregressive model. IET Renew. Power Gener. **11**(11), 1430–1439 (2017)
7. De Oliveira, R.T.A., Assis, T.F.O.D., Firmino, P.R.A.: Copulas-based time series combined forecasters. Inf. Sci. **376**, 110–124 (2017)
8. Xue, Y., Yu, C., Li, K.: Adaptive ultra-short-term wind power prediction based on risk assessment. CSEE J. Power Energy Syst. **2**(3), 59–64 (2016)
9. Lahouar, A., Slama, J.B.H.: Hour-ahead wind power forecast based on random forests. Renew. Energy **109**, 529–541 (2017)
10. Li, D., Yan, W., Li, W.: A two-tier wind power time series model considering day-to-day weather transition and intraday wind power fluctuations. IEEE Trans. Power Syst. **31**(6), 4330–4339 (2016)
11. Wang, C., Zhang, H., Fan, W.: A new wind power prediction method based on chaotic theory and Bernstein Neural Network. Energy **117**, 259–271 (2016)
12. Zhang, W., Dang, H., Simoes, R.: A new solar power output prediction based on hybrid forecast engine and decomposition model. ISA Trans. **81**, 105–120 (2018)
13. Wu, J.L., Ji, T.Y., Li, M.S.: Multistep wind power forecast using mean trend detector and mathematical morphology-based local predictor. IEEE Trans. Sustain. Energy **6**(4), 1216–1223 (2015)
14. Wu, Q., Peng, C.: Wind power generation forecasting using least squares support vector machine combined with ensemble empirical mode decomposition, principal component analysis and a bat algorithm. Energies **9**(4), 261 (2016)
15. Suykens, J.A.K.: Least Squares Support Vector Machines. World Scientific, Singapore (2002)
16. Wang, H., Li, G., Wang, G.: Deep learning based ensemble approach for probabilistic wind power forecasting. Appl. Energy **188**, 56–70 (2017)
17. Xie, H., Ding, M., Chen, L.: Short-term wind power prediction by using empirical mode decomposition based GA-SVR. In: Proceedings of the 36th Chinese Control Conference, 2017, pp. 9175–9180
18. Sorjamaa, A., Lendasse, A.: Time series prediction using DirRec strategy. Eur. Symp. Esann. DBLP **6**, 143–148 (2006)
19. Gnen, M., Alpaydn, E.: Multiple kernel learning algorithms. J. Mach. Learn. Res. **12**(Jul), 2211–2268 (2011)
20. Wu, Q., Law, R.: Complex system fault diagnosis based on a fuzzy robust wavelet support vector classifier and an adaptive Gaussian particle swarm optimization. Inf. Sci. **180**(23), 4514–4528 (2010)
21. Ding, M., Zhou, H., Xie, H., et al.: A time series model based on hybrid-kernel least-squares support vector machine for short-term wind power forecasting, ISA Transactions (2020). https://doi.org/10.1016/j.isatra.2020.09.002